Cambridge O Level

Mathematics
Second edition

Michael Handbury, Jean Matthews
Series editor: Brian Seager
Contributing author: Elaine Dorsett

Endorsement indicates that a resource has passed Cambridge International's rigorous quality-assurance process and is suitable to support the delivery of a Cambridge International syllabus. However, endorsed resources are not the only suitable materials available to support teaching and learning, and are not essential to be used to achieve the qualification. Resource lists found on the Cambridge International website will include this resource and other endorsed resources. Any example answers to questions taken from past question papers, practice questions, accompanying marks and mark schemes included in this resource have been written by the authors and are for guidance only. They do not replicate examination papers. In examinations the way marks are awarded may be different. Any references to assessment and/or assessment preparation are the publisher's interpretation of the syllabus requirements. Examiners will not use endorsed resources as a source of material for any assessment set by Cambridge International. While the publishers have made every attempt to ensure that advice on the qualification and its assessment is accurate, the official syllabus, specimen assessment materials and any associated assessment guidance materials produced by the awarding body are the only authoritative source of information and should always be referred to for definitive guidance. Cambridge International recommends that teachers consider using a range of teaching and learning resources based on their own professional judgement of their students' needs. Cambridge International has not paid for the production of this resource, nor does Cambridge International receive any royalties from its sale. For more information about the endorsement process, please visit www.cambridgeinternational.org/endorsed-resources.

Cambridge International copyright material in this publication is reproduced under licence and remains the intellectual property of Cambridge Assessment International Education.

The Publishers and authors of this second edition would like to thank John Jeskins and Heather West for permission to reuse content they wrote for the first edition.

The Publishers would like to thank the following for permission to reproduce copyright materials:

Photo credits: p.79 © jamenpercy – Fotolia

Every effort has been made to trace all copyright holders, but if any have been inadvertently overlooked, the Publishers will be pleased to make the necessary arrangements at the first opportunity.

Although every effort has been made to ensure that website addresses are correct at time of going to press, Hodder Education cannot be held responsible for the content of any website mentioned in this book. It is sometimes possible to find a relocated web page by typing the address of the home page for a website in the URL window of your browser.

Hachette UK's policy is to use papers that are natural, renewable and recyclable products and made from wood grown in well-managed forests and other controlled sources. The logging and manufacturing processes are expected to conform to the environmental regulations of the country of origin.

Orders: please contact Hachette UK Distribution, Hely Hutchinson Centre, Milton Road, Didcot, Oxfordshire, OX11 7HH. Telephone: +44 (0)1235 827827. Email education@hachette.co.uk Lines are open from 9 a.m. to 5 p.m., Monday to Friday. You can also order through our website: www.hoddereducation.com

ISBN: 9781398373877

© Brian Seager, Michael Handbury, Jean Matthews, John Jeskins, Heather West and Hodder & Stoughton Limited 2023

First published in 2016
This edition published in 2023 by
Hodder Education,
An Hachette UK Company
Carmelite House
50 Victoria Embankment
London EC4Y 0DZ

www.hoddereducation.com

Impression number   10 9 8 7 6 5 4 3 2 1

Year   2026 2025 2024 2023

All rights reserved. Apart from any use permitted under UK copyright law, no part of this publication may be reproduced or transmitted in any form or by any means, electronic or mechanical, including photocopying and recording, or held within any information storage and retrieval system, without permission in writing from the publisher or under licence from the Copyright Licensing Agency Limited. Further details of such licences (for reprographic reproduction) may be obtained from the Copyright Licensing Agency Limited, www.cla.co.uk

Cover photo © Krathin – stock.adobe.com

Illustrations by Integra Software Services Ltd.

Typeset by Integra Software Services Ltd.

Printed in Italy

A catalogue record for this title is available from the British Library.

CONTENTS

INTRODUCTION xi

1 NUMBER 1
Types of number 1
Prime factors 2
Common factors and common multiples 4

2 SETS 6
The definition of a set 6
The universal set 6
Venn diagrams 7
The relationship between sets 8
The complement of a set 9
Subsets 11
Problem solving with Venn diagrams 12
Alternative ways to define a set 12
Three-set problems 13

3 POWERS AND ROOTS 16
Squares and square roots 16
Cubes and cube roots 17

4 FRACTIONS, DECIMALS AND PERCENTAGES 20
Fractions 20
Fraction of a quantity 21
Equivalent fractions 22
Fractions and decimals 23
Terminating and recurring decimals 23
Fractions, decimals and percentages 25

5 ORDERING 27
Ordering integers 27
Inequalities 27
Ordering decimals 28
Ordering fractions 29
Ordering fractions, decimals and percentages 30

6 THE FOUR OPERATIONS 31
Numbers below zero 31
Adding and subtracting with negative numbers 34
Multiplying and dividing with negative numbers 35
Combining operations 36
Order of operations 37
Multiplying integers 39

	Multiplying decimals	40
	Dividing integers	41
	Dividing decimals	41
	Adding and subtracting fractions	43
	Multiplying fractions	45
	Dividing fractions	47
	Review exercise 1	50
7	**INDICES 1**	52
	Simplifying numbers with indices	52
	Multiplying numbers in index form	53
	Dividing numbers in index form	54
	Negative indices	55
	Fractional indices	56
8	**STANDARD FORM**	58
	Standard form	58
	Calculating with numbers in standard form	59
9	**ESTIMATION**	61
	Estimating lengths	61
	Rounding to a given number of decimal places	62
	Rounding to a given number of significant figures	63
	Estimating answers to problems	65
	Working to a sensible degree of accuracy	68
10	**LIMITS OF ACCURACY**	70
	Bounds of measurement	70
	Sums and differences of measurements	71
	Multiplying and dividing measurements	73
11	**RATIO AND PROPORTION**	76
	Ratio	76
	Using ratio to find an unknown quantity	78
	Dividing a quantity in a given ratio	80
	Proportion	81
	Review exercise 2	84
12	**RATES**	86
	Common measures of rate	86
	Speed	87
	Applying other measures of rate	88
13	**PERCENTAGES**	91
	Fractions, decimals and percentages	91
	Percentage of a quantity	91
	Expressing one quantity as a percentage of another	93

Percentage change 95
Percentage increase and decrease 96
Percentage increase and decrease using a multiplier 97
Finding the original quantity 98
Wages and salaries 100
Discount, profit and loss 100
Repeated percentage change 102
Compound and simple interest 102

14 USING A CALCULATOR 106
Order of operations 106
Standard form on your calculator 107
Checking accuracy 109
Interpreting the calculator display 111
Calculating with time 112
Fractions on a calculator 114

15 TIME 116
The 24-hour clock 116
Calculating with time 117
Converting between hours and minutes and hours written
 as a decimal 119
Time zones 120

16 MONEY 122
Value for money 122
Currency conversion 124

17 EXPONENTIAL GROWTH AND DECAY 126
Exponential growth 126
Exponential decay 126

18 SURDS 129
Simplifying surds 129
Manipulation of expressions of the form $a + b\sqrt{c}$ 130
Rationalising denominators 132

Review exercise 3 135

19 INTRODUCTION TO ALGEBRA 137
Letters for unknowns 137
Substituting numbers into algebraic expressions 138
Using harder numbers when substituting 139

20 ALGEBRAIC MANIPULATION 141
Simplifying algebraic expressions 141
Simplifying more complex algebraic expressions 143
Expanding a single bracket 144
Factorising algebraic expressions 144

Factorising expressions of the form $ax + bx + kay + kby$ 146
Expanding a pair of brackets 147
Expanding more complex brackets 149
Factorising expressions of the form $a^2x^2 - b^2y^2$ 150
Factorising expressions of the form $x^2 + bx + c$ 151
Factorising expressions of the form $a^2 + 2ab + b^2$ 153
Factorising quadratic expressions of the form $ax^2 + bx + c$ 154
Factorising expressions of the form $ax^3 + bx^2 + cx$ 157
Completing the square 157

21 ALGEBRAIC FRACTIONS 159
Simplifying algebraic fractions 159
Adding and subtracting algebraic fractions 160

22 INDICES 2 162
Simplifying algebraic expressions using indices 162
Using the laws of indices with numerical and algebraic expressions .. 164

Review exercise 4 166

23 EQUATIONS 168
Writing formulas 168
Writing equations 169
Solving simple linear equations 170
Solving equations with a bracket 171
Solving equations with the unknown on both sides 172
Solving equations involving fractions 173
Solving harder equations involving fractions 174
Solving linear simultaneous equations 175
Solving harder simultaneous equations 177
Solving quadratic equations of the form $x^2 + bx + c = 0$
  by factorisation 179
Solving quadratic equations of the form $ax^2 + bx + c = 0$
  by factorisation 180
Solving quadratic equations by completing the square 181
Solving quadratic equations using the quadratic formula 183
Changing the subject of formulas 185
Changing the subject of harder formulas 187
Solving equations involving algebraic fractions 189

24 INEQUALITIES 191
Solving inequalities 191
Showing regions on graphs 193
Representing regions satisfying more than one inequality 195

25 SEQUENCES 200
Number patterns 200
Linear sequences 202
The *n*th term 203

Finding the formula for the nth term 203
The nth term of quadratic, cubic and exponential sequences 205

26 PROPORTION 208
Proportion 208
Proportion as a formula 209
Other types of proportion 210

Review exercise 5 215

27 GRAPHS IN PRACTICAL SITUATIONS 217
Conversion graphs 217
Travel graphs 220
Rate of change on a distance–time graph 223
Rate of change on a speed–time graph 226
Area under a speed–time graph 229

28 GRAPHS OF FUNCTIONS 233
Quadratic graphs 233
Using graphs to solve equations 235
Cubic graphs 237
Reciprocal graphs 239
Other graphs 240
Exponential graphs 244
Growth and decay 244
Estimating the gradient to a curve 246

29 SKETCHING CURVES 248
Families of graphs 248
Linear graphs 248
Quadratic graphs 249
Cubic graphs 251
Reciprocal graphs 252
Exponential graphs 254

30 FUNCTIONS 256
Function notation 256
Domain and range 256
Inverse functions 258
Composite functions 261

31 COORDINATE GEOMETRY 263
The gradient of a straight-line graph 263
Line segments 264
The length of a line segment 264
The general form of the equation of a straight line 265
Parallel and perpendicular lines 267

Review exercise 6 270

32 GEOMETRICAL TERMS ... 272
Dimensions ... 272
Angles ... 272
Lines ... 273
Bearings ... 274
Triangles ... 275
Quadrilaterals ... 276
Polygons ... 277
Solids ... 278
Nets of 3-D shapes ... 280
Congruence and similarity ... 282
Circles ... 285

33 GEOMETRICAL CONSTRUCTIONS ... 287
Measuring angles ... 287
Constructing a geometrical figure using compasses ... 290
Nets of 3-D shapes ... 291

34 SCALE DRAWINGS ... 295
Scale drawings and maps ... 295
Bearings ... 298

35 SIMILARITY ... 301
Similar shapes ... 301
The areas and volumes of similar shapes ... 304

36 SYMMETRY ... 307
Line symmetry ... 307
Rotational symmetry ... 308
Symmetry properties of shapes and solids ... 310

37 ANGLES ... 312
Angles formed by straight lines ... 312
Angles formed within parallel lines ... 314
The angles in a quadrilateral ... 318
The angles in a polygon ... 320

38 CIRCLE THEOREMS ... 323
Symmetry properties of circles ... 323
Angles in a circle ... 325
The alternate segment theorem ... 332

39 UNITS OF MEASURE ... 335
Basic units of length, mass and capacity ... 335
Area and volume measures ... 336

40 MENSURATION 338
The perimeter of a 2-D shape 338
The area of a rectangle 340
The area of a triangle 341
The area of a parallelogram 343
The area of a trapezium 345
The area of shapes made from rectangles and triangles 348
The circumference of a circle 350
The area of a circle 352
Arc length and sector area 353
The volume of a prism 356
The surface area of a prism 359
The volume of a pyramid, a cone and a sphere 361
The surface area of a pyramid, a cone and a sphere 364
The area and volume of compound shapes 367

Review exercise 7 372

41 PYTHAGORAS' THEOREM AND TRIGONOMETRY ... 374
Pythagoras' theorem 374
Trigonometry 378
The sine and cosine functions for obtuse angles 387
Non-right-angled triangles 389
Finding lengths and angles in three dimensions 398
The angle between a line and a plane 404

42 TRANSFORMATIONS 408
The language of transformations 408
Reflection 408
Rotation 411
Translation 413
Enlargement 415
Recognising and describing transformations 420
Combining transformations 425

43 VECTORS 428
Vectors and translations 428
Vector notation 430
Multiplying a vector by a scalar 435
Addition and subtraction of column vectors 435
Position vectors 437
The magnitude of a vector 438
Vector geometry 439

Review exercise 8 444

Photocopying is prohibited

44 PROBABILITY ... 446
The probability of a single event ... 446
The probability of an event not occurring ... 448
Estimating from a population ... 450
Relative frequency and probability ... 450
The probability of combined events ... 453
Independent events ... 456
Tree diagrams for combined events ... 457
The probability of dependent events ... 459

45 CATEGORICAL, NUMERICAL AND GROUPED DATA ... 462
Collecting and grouping data ... 462
Surveys ... 463
Designing a questionnaire ... 465
Two-way tables ... 465
Averages and range ... 468
Mean and range ... 469
Which average to use when comparing data ... 471
Working with larger data sets ... 474
Working with grouped and continuous data ... 477

46 STATISTICAL DIAGRAMS ... 483
Bar charts and pictograms ... 483
Pie charts ... 487
Scatter diagrams ... 490
Cumulative frequency diagrams ... 494
Median, quartiles and percentiles ... 497
Histograms ... 502

Review exercise 9 ... 507

GLOSSARY ... 509

INDEX ... 514

Answers are available at www.hoddereducation.com/cambridgeextras.

INTRODUCTION

For the teacher

This book is intended to be used by students preparing for Cambridge O Level Mathematics (Syllabus D) for examination from 2025. It is the second edition of this book. It has been updated both to cover the revised syllabus, and to incorporate several helpful suggestions from students and teachers using the first edition.

The structure of the book closely follows the contents of the syllabus, as shown by the mapping table on the next page. It is possible to work through the chapters in order, but more variety can be achieved by moving between different areas of the syllabus, and the following features will help in planning this.

Each chapter starts with a **By the end of this chapter you will be able to:** box, which repeats the relevant statements from the syllabus.

Immediately after this is a **Check you can:** box, which lists the knowledge and skills required of students before beginning the chapter. This will help you decide whether topics from earlier in the book need to be covered or revised first. Sometimes, work from a later chapter may be required to complete the topic. If so, that part of the chapter can be delayed until later, or the extra knowledge or technique needed could be introduced early. It will then be consolidated in working through the later chapter.

Throughout each chapter there are numerous **Examples** with worked solutions, illustrating each aspect of the topic and helping students understand how to tackle the formative **Exercises** that follow. The questions in the exercises give considerable opportunity to reinforce learning.

Each chapter ends with a **Key points** box, which lists a summary of the points that students should have learned and understood.

Summative **Review exercises** appear at various stages throughout the book. These contain questions from past examination papers, which include a reference to the past paper they are taken from. Questions without a past paper reference in the Review exercises are practice questions written by the author. Each question has been assigned a mark tariff by the author, and is mapped to the chapters that students should have studied before completing the question. This means that students can complete the questions at the appropriate stage of their learning journey, even if they are not following the chapter order. Alternatively, they can be used at the end of the course for revision.

The syllabus requires that some of the work presented in this book be done without a calculator. This is clearly indicated in the book and non-calculator questions are included.

There are also regular tips and guidance contained in the **Note** boxes, helping students avoid common pitfalls and clarifying what is in the text.

Brian Seager

INTRODUCTION

Answers and mark schemes

You can download numerical answers to all questions from www.hoddereducation.com/Cambridgeextras.

Mark schemes with worked solutions for the Review exercises are available in the Boost eBook: Teacher edition.

Cambridge Assessment International Education bears no responsibility for the example answers and mark schemes for questions taken from its past question papers which are contained in this publication. All answers and mark schemes have been written by the authors.

Syllabus mapping

Chapter		Syllabus section	Chapter		Syllabus section
1	Number	1.1	24	Inequalities	2.6
2	Sets	1.2	25	Sequences	2.7
3	Powers and roots	1.3	26	Proportion	2.8
4	Fractions, decimals and percentages	1.4	27	Graphs in practical situations	2.9
5	Ordering	1.5	28	Graphs of functions	2.10
6	The four operations	1.6	29	Sketching curves	2.11
7	Indices 1	1.7	30	Functions	2.12
8	Standard form	1.8	31	Coordinate geometry	3.1–3.7
9	Estimation	1.9	32	Geometrical terms	4.1
10	Limits of accuracy	1.10	33	Geometrical constructions	4.2
11	Ratio and proportion	1.11	34	Scale drawings	4.3
12	Rates	1.12	35	Similarity	4.4
13	Percentages	1.13	36	Symmetry	4.5
14	Using a calculator	1.14	37	Angles	4.6
15	Time	1.15	38	Circle theorems	4.7–4.8
16	Money	1.16	39	Units of measure	5.1
17	Exponential growth and decay	1.17	40	Mensuration	5.2–5.5
18	Surds	1.18	41	Pythagoras' theorem and trigonometry	6.1–6.4
19	Introduction to algebra	2.1	42	Transformations	7.1
20	Algebraic manipulation	2.2	43	Vectors	7.2–7.4
21	Algebraic fractions	2.3	44	Probability	8.1–8.3
22	Indices 2	2.4	45	Categorical, numerical and grouped data	9.1–9.3
23	Equations	2.5	46	Statistical diagrams	9.4–9.7

For the student

How to use this book

The chapters in this book follow the syllabus. Each chapter begins with the green box shown in the margin.

Following the 'By the end of this chapter' statements there will be a blue 'Check you can' box.

> **BY THE END OF THIS CHAPTER YOU WILL BE ABLE TO:**
> - which lists what you will learn, as detailed in the syllabus.

> **CHECK YOU CAN:**
> - which lists what you need to know before you start the chapter. If you are not certain, it is a good idea to revise these points first.

Throughout the book, key terms are written in blue. These are defined in the glossary at the back of the book.

> Key points and mathematical rules are highlighted, like this sentence.

In the chapters, you will also find:

> **Examples**
> Which show you how to apply what you are learning. In each case, a question is followed by a worked solution. These are models to help you answer the questions in the exercises that follow the Example boxes.

> **Notes**
> These contain tips and advice about what you are learning.

 This icon shows questions that you should do without a calculator, so you can practise non-calculator questions.

Where you can use a calculator, you should make sure that you know how to use your own calculator. The buttons may look different from those shown in this book.

Each chapter ends with:

> **Key points**
> - which lists what you should have learned and understood before you move on. You can also use this as a revision checklist when you prepare for your examinations.

In various places throughout the book, you will find **Review exercises** that cover the previous chapters. You can work through these exercises during the course to summarise your learning, or at the end of the course as revision. Next to each question are the chapters that you should have studied before attempting that question. Marks for each question are shown in square brackets.

INTRODUCTION

Command words

The table below shows the command words that may appear in examinations for this syllabus. You will see these words in the questions throughout this book.

Command word	What it means
Calculate	work out from given facts, figures or information
Construct	make an accurate drawing
Describe	state the points of a topic / give characteristics and main features
Determine	establish with certainty
Explain	set out purposes or reasons / make the relationships between things clear / say why and/or how and support with relevant evidence
Give	produce an answer from a given source or recall/memory
Plot	mark point(s) on a graph
Show (that)	provide structured evidence that leads to a given result
Sketch	make a simple freehand drawing showing the key features
State	express in clear terms
Work out	calculate from given facts, figures or information with or without the use of a calculator
Write	give an answer in a specific form
Write down	give an answer without significant working

The information in this section is taken from the Cambridge International syllabus. You should always refer to the appropriate syllabus document for the year of examination to confirm the details and for more information. The syllabus document is available on the Cambridge International website at **www.cambridgeinternational.org**.

Explore the book cover: how are ferns mathematical?

A fractal is a complex mathematical shape. Very basically, it is a shape within which the same shape is repeated many times. Examples in nature include the leaves of ferns. Research 'fractals in nature' to discover more.

1 NUMBER

Types of number

Integers

Integers are positive and negative whole numbers: …, –2, –1, 0, 1, 2, …

Natural numbers

Natural numbers are integers that can be used for counting: 1, 2, 3, 4, 5, …

Rational numbers

A rational number is a number that can be written as a fraction $\frac{a}{b}$ where a and b are integers and $b \neq 0$.

Rational numbers include:

- all integers
- all terminating decimals, for example, 5.81 can be written as $\frac{581}{100}$
- all recurring decimals, for example, $0.\dot{6} = 0.666\,666\ldots$ can be written as $\frac{2}{3}$.

Irrational numbers

An irrational number is a number that cannot be written as a fraction, such as $\sqrt{2}$ or π.

They give decimals that do not terminate or recur.

> **BY THE END OF THIS CHAPTER YOU WILL BE ABLE TO:**
> - identify and use: natural numbers, integers (positive, zero and negative), prime numbers, square numbers, cube numbers, common factors, common multiples, rational and irrational numbers, and reciprocals.

> **CHECK YOU CAN:**
> - recognise factors and multiples of a number
> - write a product using index notation, for example $5 \times 5 \times 5 = 5^3$.

> **Note**
> Recurring decimals can be shown by placing dots above the digits, e.g.
> $0.\dot{3} = 0.333\,333\ldots$
> $1.\dot{3}0\dot{7} = 1.307\,307\,307\ldots$

> **Note**
> An irrational number multiplied by a non-zero rational number is still irrational.

Example 1.1

Question

Sort the numbers in the list below into rational and irrational numbers. Show how you decide.

$0.8652 \quad \sqrt{12} \quad 4\pi \quad \frac{67}{19} \quad \sqrt{64}$

Solution

Rational numbers can be written as fractions, so 0.8652, $\frac{67}{19}$ and $\sqrt{64}$ are rational.

$0.8652 = \frac{8652}{10\,000}$ $\qquad \sqrt{64} = 8$

Irrational numbers cannot be written as fractions, so $\sqrt{12}$ and 4π are irrational.

$\sqrt{12} = 3.464\,101\ldots \qquad 4\pi = 12.566\,370\ldots$

1 NUMBER

Reciprocals

> **Note**
> The number $\frac{3}{1}$ is the same as 3.

The reciprocal of a number is $\frac{1}{\text{the number}}$. So the reciprocal of 3 is $\frac{1}{3}$, of $\frac{1}{3}$ is 3 and of $\frac{2}{3}$ is $\frac{3}{2}$.

Exercise 1.1

1 State which of these numbers are
   a integers
   b natural numbers.
   $-7 \quad 0.6 \quad 27 \quad \sqrt{8} \quad 1534 \quad 0 \quad \frac{4}{5} \quad -12$

2 State which of these numbers are rational, showing how you decide.
   a $\frac{17}{20}$ b 0.46 c $\sqrt{\frac{2}{25}}$
   d 5π e 3.14159 f $-0.2\dot{3}\dot{4}$
   g $\sqrt{\frac{4}{25}}$ h $\sqrt{225}$ i $2\sqrt{3} + \sqrt{3}$

3 State which of these numbers are rational, showing how you decide.
   a $\sqrt{169}$ b 0.49 c $5 + \sqrt{3}$
   d -2.718 e $5\pi + 2$ f $\frac{4\pi}{3\pi}$
   g $\sqrt{27}$ h $\sqrt{1\frac{7}{9}}$ i $-6\sqrt{2}$

4 Write down an irrational number between each pair of numbers.
   a 3 and 4 b 10 and 11 c 19 and 20

5 Write down the reciprocals of these numbers.
   a 5 b $\frac{1}{4}$
   c $\frac{2}{3}$ d $\frac{7}{5}$

Prime factors

The **factors** of a number are all of the numbers that divide exactly into that number.

A **prime number** is a number with only two factors.

The factors of 7 are 1 and 7, so 7 is a prime number.

The factors of 12 are 1, 2, 3, 4, 6 and 12, so 12 is not a prime number.

The only factor of 1 is 1, so 1 is not a prime number.

Any number that is not prime can be written as the product of its prime factors.

The prime factors of a number can be found either by using a factor tree or by dividing repeatedly by prime numbers.

Prime factors

Example 1.2

Question

Write 60 as the product of its prime factors.

Solution

Factor tree method

It doesn't matter how you start the factor tree, the ends of the branches will be the same.

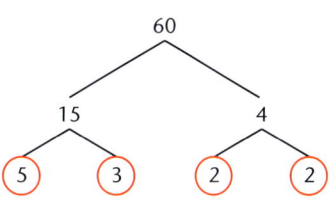

Division method

```
2 | 60
2 | 30
3 | 15
5 |  5
  |  1
```

Divide by the prime numbers, starting with 2, until the result is 1.

The prime factors of 60 are 2, 3 and 5.
Written as a product of its prime factors, $60 = 2 \times 2 \times 3 \times 5$.
Using index notation, $60 = 2^2 \times 3 \times 5$.

Exercise 1.2

1. Write down all of the factors of each of these numbers.
   a 8 b 15 c 27 d 54
2. Write down the first ten prime numbers.
3. Write down the prime factors of each of these numbers.
   a 12 b 20 c 55 d 84
4. Express each of these numbers as the product of its prime factors.
   a 48 b 72 c 210 d 350
   e 75 f 275 g 120 h 198
5. Express each of these numbers as the product of its prime factors.
   a 495 b 260 c 2700 d 1078
   e 420 f 1125 g 112 h 1960
6. a Write each of these square numbers as the product of its prime factors.
      i 25 ii 36 iii 100 iv 144
   b Comment on what you notice about each of the products in **a**.
7. a Write 96 as the product of its prime factors.
   b Find the smallest positive integer k such that $96k$ is a square number.
8. a Write 392 as the product of its prime factors.
   b Find the smallest positive integer k such that $392k$ is a cube number.

Note

If calculators are permitted, you can use the FACT button to find prime factors.

1 NUMBER

Common factors and common multiples

A **common factor** of two numbers is a number that is a factor of *both* of them.

2 is a common factor of 8 and 12 because $8 \div 2 = 4$ and $12 \div 2 = 6$.

The **highest common factor (HCF)** of two numbers is the highest number that is a factor of both numbers.

The highest common factor of 8 and 12 is 4.

A **multiple** of a number is the product of the number and any integer.

A **common multiple** of two numbers is a number that is a multiple of *both* of them.

20 is a common multiple of 2 and 5 because $2 \times 10 = 20$ and $5 \times 4 = 20$.

The **lowest common multiple (LCM)** of two numbers is the lowest number that is a multiple of both numbers.

The lowest common multiple of 2 and 5 is 10.

Example 1.3

Question

a Find the highest common factor (HCF) of 84 and 180.
b Find the lowest common multiple (LCM) of 84 and 180.

Solution

First write each number as the product of its prime factors.

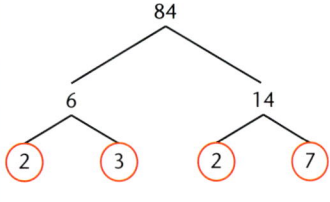

$84 = 2 \times 2 \times 3 \times 7 = 2^2 \times 3 \times 7$ $180 = 2 \times 2 \times 3 \times 3 \times 5 = 2^2 \times 3^2 \times 5$

a To find the highest common factor, find all numbers that appear in *both* lists.
$84 = \underline{2 \times 2 \times 3} \times 7$
$180 = \underline{2 \times 2 \times 3} \times 3 \times 5$

The highest common factor will be no higher than the smaller of the two numbers.

The highest common factor of 84 and 180 is $2 \times 2 \times 3 = 12$.

This means that 12 is the highest number that is a factor of both 84 and 180.

b To find the lowest common multiple, find all of the prime numbers that appear in each list and use the higher power of each.

$84 = 2^2 \times 3 \times 7$ The lowest common multiple will be no lower than the larger of the two numbers.
$180 = 2^2 \times 3^2 \times 5$

The lowest common multiple of 84 and 180 is $2^2 \times 3^2 \times 5 \times 7 = 1260$.

This means that 1260 is the lowest number that is a multiple of both 84 and 180.

It is the lowest number that has both 84 and 180 as factors.

Exercise 1.3

1 For each pair of numbers:
   a 18 and 24   b 64 and 100   c 50 and 350   d 72 and 126
   i   express each number as the product of its prime factors
   ii find the highest common factor (HCF)
   iii find the lowest common multiple (LCM).

2 Find the highest common factor (HCF) and lowest common multiple (LCM) of each pair of numbers.
   a 27 and 63   b 20 and 50   c 48 and 84   d 50 and 64   e 42 and 49

3 For each pair of numbers:
   a 260 and 300   b 340 and 425   c 756 and 2100   d 1980 and 2376
   i   express each number as the product of its prime factors
   ii find the highest common factor (HCF)
   iii find the lowest common multiple (LCM).

4 Find the HCF and LCM of each pair of numbers.
   a 5544 and 2268   b 2016 and 10 584

5 a Find the highest common factor of 45, 60 and 75.
   b Find the lowest common multiple of 45, 60 and 75.

6 A rectangle measures 240 mm by 204 mm.
   It is split up into identical squares.
   Find the largest possible side length for the squares.

7 There are two lighthouses near a port.
   The light in the first lighthouse flashes every 22 seconds.
   The light in the second lighthouse flashes every 16 seconds.
   At 10 p.m. one evening both lights are switched on.
   What is the next time that the lights flash at the same time?

8 Buses to Shenley leave the bus station every 40 minutes.
   Buses to Winley leave every 15 minutes.
   At 8.15 a.m. buses to both Shenley and Winley leave the bus station.
   When is the next time that buses to both places leave at the same time?

Key points

- Numbers can be classified as natural numbers, integers, rational and irrational numbers.
- The reciprocal of a number is $\frac{1}{\text{the number}}$.
- Prime numbers have only 1 and themselves as factors.
- The highest common factor (HCF) and the lowest common multiple (LCM) can be found when the numbers are split into prime factors.

Photocopying is prohibited

2 SETS

BY THE END OF THIS CHAPTER YOU WILL BE ABLE TO:

- understand and use set language, notation and Venn diagrams to describe sets and represent relationships between sets.

CHECK YOU CAN:

- recall the meaning of the terms *integer*, *prime number*, *multiple* and *factor*.

The definition of a set

A set is a collection of numbers, shapes, letters, points or other objects. They form a set because they fulfil certain conditions.

The notation for a set is a pair of curly brackets: {…}.

The individual members of a set are called **elements**.

A set can be defined by giving a rule which satisfies all the elements, or by giving a list of the elements.

For example,

A = {integers from 1 to 10}   or A = {1, 2, 3, 4, 5, 6, 7, 8, 9, 10}

Exercise 2.1

List the elements of the following sets.

1. {integers from 11 to 18}
2. {the first five prime numbers}
3. {the factors of 12}
4. {the multiples of 8 less than 50}
5. {vowels}

Note

If you are giving a rule for a set then it must be precise. In A above, do not just use 'integers'.

The universal set

All the sets in Exercise 2.1 are **finite sets**. This means that they have a fixed number of elements.

Sets can also have an infinite number of elements.

For example,

{multiples of 8}   or   {prime numbers}

Most of the sets you will be dealing with will be finite sets.

Sometimes we make sure we are dealing with a finite set by defining a **universal set**.

A universal set is a set from which – for a particular situation – all other sets will be taken.

For example, if you define the universal set as positive integers less than 50, then the set in Exercise 2.1 question **4** could simply be defined as {multiples of 8} since you are only considering integers less than 50.

Similarly, if you define the universal set as positive integers less than 12, then the set in question **2** could have been defined as {prime numbers}.

Notation

The symbol for a universal set is $\mathscr{E}$

The symbol $\in$ means 'is an element of'.

The symbol $\notin$ means 'is not an element of'.

So $\quad 3 \in \{\text{prime numbers}\}$

and $\quad 4 \notin \{\text{prime numbers}\}$.

Venn diagrams

Venn diagrams are a way of showing sets and the relationships between sets. Venn diagrams were introduced in 1880 by John Venn.

In a Venn diagram, the universal set is shown by a rectangle. Other sets are drawn as circles or ovals within the rectangle.

The diagrams here show three typical Venn diagrams.

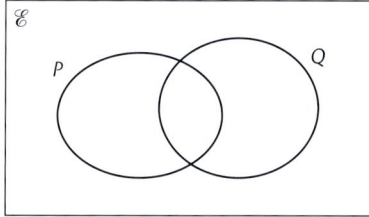

This Venn diagram shows two sets, P and Q, where there are some elements that are in both sets.

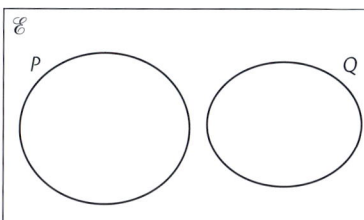

This Venn diagram shows two sets, P and Q, where there are no elements that are in both sets.

Sets P and Q are **disjoint**.

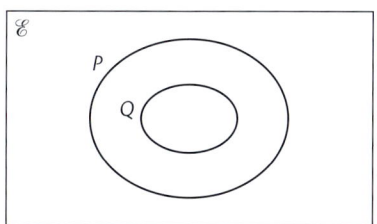

This Venn diagram shows two sets, P and Q, where all the elements of set Q are also in set P.

Set Q is a **subset** of set P.

Venn diagrams are not restricted to two sets; they can have three or more sets.

Sometimes Venn diagrams need to be drawn and the elements of the sets filled in.

2 SETS

Example 2.1

$\mathcal{E}$ = {integers from 1 to 20} P = {factors of 12} Q = {prime numbers}

Question

a Draw a Venn diagram to show the elements of the universal set and its subsets P and Q.
b List the elements that are in both set P and set Q.

Solution

a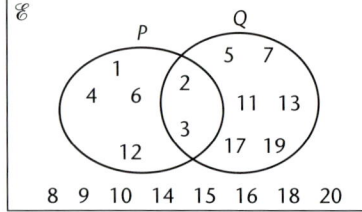

Note
Do not forget to fill in the elements of $\mathcal{E}$ that are in neither P nor Q.

b 2, 3

We do not put an element in a Venn diagram more than once. So, for example, if

A = {letters in the word *label*}

the set is {l, a, b, e} and you put those four letters in the Venn diagram. You would not put the letter *l* in twice.

Exercise 2.2

Draw a Venn diagram to show the elements of the universal set and its subsets P and Q.

1 $\mathcal{E}$ = {integers from 1 to 20}
   P = {factors of 18}
   Q = {odd numbers}

2 $\mathcal{E}$ = {natural numbers less than 25}
   P = {multiples of 4}
   Q = {prime numbers}

3 $\mathcal{E}$ = {first 15 letters of the alphabet}
   P = {letters in the word *golf*}
   Q = {letters in the word *beam*}

4 $\mathcal{E}$ = {positive integers less than 21}
   P = {even numbers}
   Q = {multiples of 4}

5 $\mathcal{E}$ = {days of the week}
   P = {days with six letters}
   Q = {days beginning with S}

6 $\mathcal{E}$ = {multiples of 3 less than 50}
   P = {factors of 36}
   Q = {odd numbers}

The relationship between sets

The **intersection** of two sets, P and Q, is all the elements that are in both set P and set Q.

You can write this as $P \cap Q$.

In Example 2.1, $P \cap Q$ = {2, 3}.

The **union** of two sets, P and Q, is all the elements that are in set P or set Q or both.

You can write this as $P \cup Q$.

In Example 2.1, $P \cup Q = \{1, 2, 3, 4, 5, 6, 7, 11, 12, 13, 17, 19\}$

If a set has no elements, it is called the **empty set**.

The symbol for the empty set is Ø.

It may seem a trivial idea, but it is quite important when dealing with some sets.

For example, if

$\mathscr{E}$ = {polygons}   P = {triangles}   Q = {quadrilaterals}

then the Venn diagram looks like this.

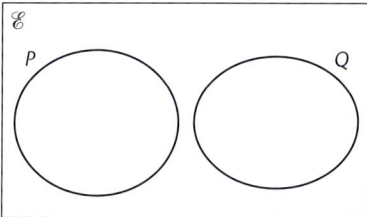

The sets are disjoint and so the intersection, $P \cap Q = $ Ø.

Exercise 2.3

For each of the questions in Exercise 2.2, find

a $P \cap Q$ b $P \cup Q$.

The complement of a set

A' means those elements of the universal set which are *not* in set A.

It is called the **complement** of A.

For example, if

$\mathscr{E}$ = {positive integers less than 13} and A = {factors of 12} = {1, 2, 3, 4, 6, 12}

then

$A' = \{5, 7, 8, 9, 10, 11\}$

2 SETS

Example 2.2

Question

On separate copies of the diagram, shade these sets.
a P'
b $(P \cap Q)'$
c $(P \cup Q)'$

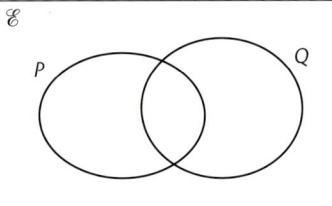

Solution

a 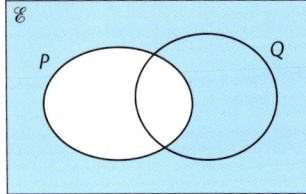 b 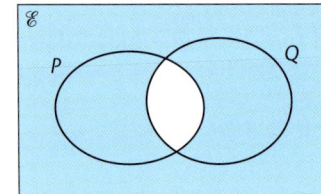 c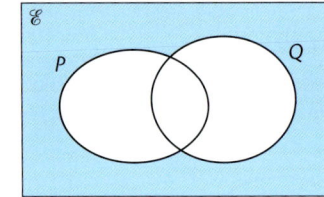

Exercise 2.4

For each of the questions in Exercise 2.2, list the elements of these sets.

a Q'
b $(P \cup Q)'$

Exercise 2.5

1 On separate copies of the diagram, shade these sets.
   a P'
   b $(P \cap Q)'$
   c $(P \cup Q)'$

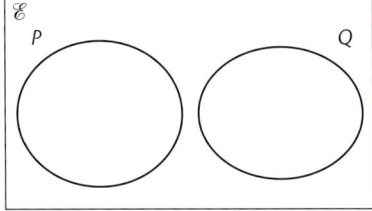

2 On separate copies of the diagram, shade these sets.
   a $P \cap Q$
   b $P \cup Q$
   c P'

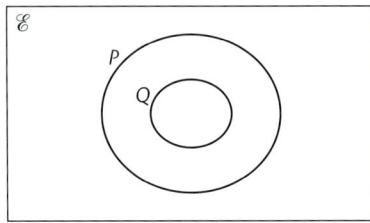

3 $\mathscr{E}$ = {polygons}
   A = {quadrilaterals}
   B = {regular polygons}
   a Draw a Venn diagram to represent these sets.
   b Describe, in words, the set $A \cap B$.

4 a On separate copies of the diagram, shade these sets.
      i $(P \cap Q)'$ ii $P' \cup Q'$
   b What do you notice?

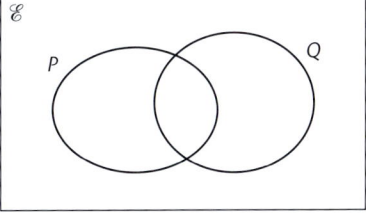

5 a On separate copies of the diagram, shade these sets.
      i $(P \cup Q)'$ ii $P' \cap Q'$
   c What do you notice?

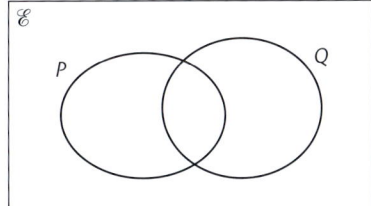

Subsets

Example 2.3

Question
Find all the subsets of $P = \{a, b, c, d\}$.

Solution
Subsets with one element: $\{a\}, \{b\}, \{c\}, \{d\}$
Subsets with two elements: $\{a, b\}, \{a, c\}, \{a, d\}, \{b, c\}, \{b, d\}, \{c, d\}$
Subsets with three elements: $\{a, b, c\}, \{a, b, d\}, \{a, c, d\}, \{b, c, d\}$
In addition to these 14 subsets,
the set P itself is regarded as a subset: $\{a, b, c, d\}$
the empty set also is regarded as a subset: ∅
This makes 16 subsets in total.

A set with 4 elements has 16 or 2^4 subsets.

> In general, a set with n elements has 2^n subsets.

Notation

The symbol $\subseteq$ means 'is a subset of'.

So $A \subseteq B$ means set A is a subset of set B. It includes the possibility that set A could be the set B itself or the empty set.

The symbol $\not\subseteq$ means 'is not a subset of'. So $A \not\subseteq B$ means set A is not a subset of set B.

2 SETS

Problem solving with Venn diagrams

If A is a set, then $n(A)$ means the number of elements in set A.

For example, if $A = \{a, b, c, d\}$, then $n(A) = 4$.

When you are using a Venn diagram to solve a problem, you can write the number of elements in the subsets rather than filling in all the elements.

Example 2.4

Question

There are 32 students in a class.

They can choose to study history (H), or geography (G), or both or neither:
- 18 study history
- 20 study geography
- 8 study both history and geography.

a Draw a Venn diagram to show this information.
b Find the number of students who study neither history nor geography.

Solution

a Since 8 students study both,

10 students study history but not geography

12 students study geography but not history.

b The number of students who study history, or geography or both is

$10 + 8 + 12 = 30$.

So the number of students who study neither history nor geography is

$32 - 30 = 2$.

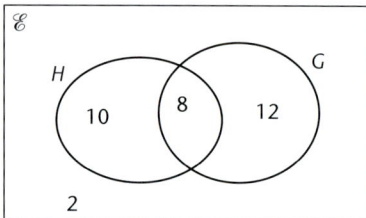

Alternative ways to define a set

> **Note**
>
> In definitions like this the colon (:) is read as 'where'.
>
> So $A = \{1, 2, 3, 4, 5, \ldots\}$ and is an infinite set.
>
> $B = \{(x, y) : y = 2x + 1\}$ is read as 'set B consists of the points (x, y), where $y = 2x + 1$'.

If a set is finite, you can define it by listing its elements.

For example, $P = \{1, 4, 9, 16, 25\}$.

However, if a set is infinite, you cannot list all the elements, so you define it by giving the rule used to form it.

There are various ways to do this. For example,

$A = \{x : x \text{ is a natural number}\}$

$B = \{(x, y) : y = 2x + 1\}$

$C = \{x : 2 \leqslant x \leqslant 5\}$

Exercise 2.6

1. Find all the subsets of {p, q, r}.
2. How many subsets does the set {5, 6, 7, 8, 9, 10} have?
3. Muna has 35 books in her electronic book reader,
   20 are crime stories (C)
   12 are books of short stories (S)
   7 are books of short crime stories.
   a Copy and complete this Venn diagram to show the number of books of each type in Muna's e-reader.
   b Find
     i n(C ∩ S') ii n(C ∪ S)'.

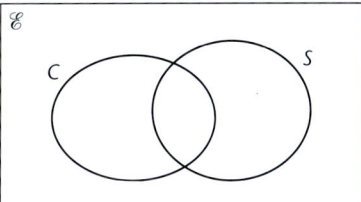

4. In a class of 30 students,
   22 study physics (P)
   19 study chemistry (C)
   6 study neither physics nor chemistry.
   a Find the number of students who study both physics and chemistry.
   b Show the information in a Venn diagram.
5. In a sports club with 130 members,
   50 play tennis but not soccer
   40 play soccer but not tennis
   15 play neither soccer nor tennis.
   a Draw a Venn diagram to show this information.
   b Find the number of members who play both tennis and soccer.
6. For two sets A and B, n(A) = 20, n(B) = 30 and n($A \cup B$) = 38.
   By drawing a Venn diagram or otherwise, find n($A \cap B$).
7. For two sets A and B, n(A) = 20, n(B) = 30 and n($A \cap B$) = 11.
   By drawing a Venn diagram or otherwise, find n($A \cup B$).
8. Two sets P and Q are such that n($P \cap Q$) = 0.
   Show the sets P and Q on a Venn diagram.
9. Two sets P and Q are such that n($P \cap Q$) = n(P).
   Show the sets P and Q on a Venn diagram.
10. Two sets P and Q are such that n($P \cup Q$) = n(P).
    Show the sets P and Q on a Venn diagram.
11. $\mathscr{E}$ = {$x : x$ is an integer} and P = ($x : -2 \leqslant x < 4$).
    List the elements of P.

Three-set problems

Example 2.5

Question

In a sixth form of 200 students, three of the subjects students can study are mathematics (M), technology (T) and psychology (P).

110 students study mathematics, 85 study technology and 70 study psychology
45 study mathematics and technology   19 study technology and psychology
35 study mathematics and psychology   9 study all three subjects.

2 SETS

a Copy and complete this Venn diagram.
b Find
   i   the number of students who study none of the three subjects
   ii n(M ∩ T ∩ P′)
   iii n[M ∩ (T ∪ P)].

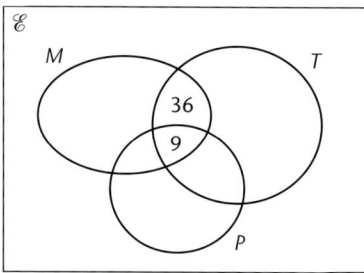

Solution

a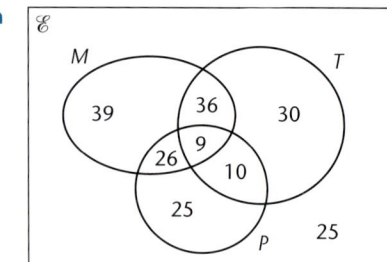

b i   200 − (39 + 36 + 9 + 26 + 30 + 10 + 25) = 200 − 175 = 25
   ii 36
   iii 36 + 9 + 26 = 71

Exercise 2.7

1 ℰ = {integers from 2 to 15 inclusive}
   A = {prime numbers}
   B = {multiples of 4}
   C = {multiples of 2}
   a Draw a Venn diagram to represent the sets ℰ, A, B and C.
   b Find
      i   n($B \cup C$)′   ii n($A \cup B$) ∩ C′.

2 Copy the diagram.
   Insert a, b, c and d in the correct subsets in the Venn diagram, given the following information.
   i $a \in P \cap Q \cap R$
   ii $b \in (P \cup Q \cup R)'$
   iii $c \in (P \cup Q)' \cap R$
   iv $d \in P \cap Q \cap R'$

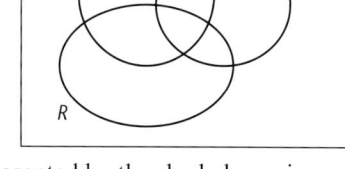

3 Use set notation to describe the sets represented by the shaded area in these Venn diagrams.
   a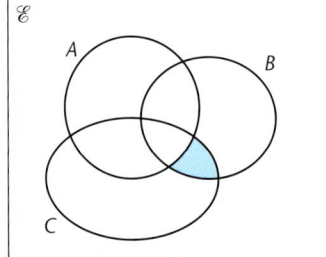
   b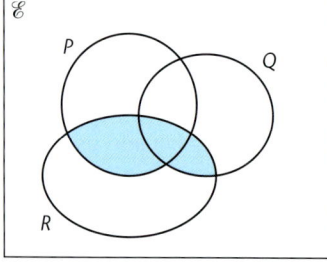

4 ℰ = {integers}
   E = {multiples of 2}
   T = {multiples of 3}
   F = {multiples of 4}
   a Use set notation to express M = {multiples of 12}, as simply as possible, in terms of E, T and F.
   b Simplify E ∪ F.

5 The Venn diagram shows the number of elements in each subset.
   a Find n(B ∩ C).
   b You are given that n(A ∪ B) = n(C). Find x.

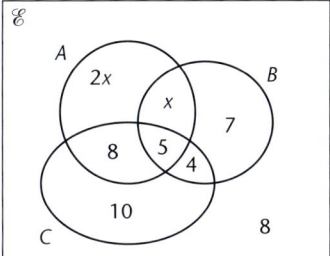

6 The sets P, Q and R are subsets of the universal set ℰ.
   Q ⊆ P
   Q ∩ R = ∅
   P ∩ R ≠ ∅
   Draw a Venn diagram to show the sets P, Q, R and ℰ.

7 Copy and complete these statements about sets P, Q and R.
   a P ∩ … = ∅
   b R … Q
   c R ∪ … = Q
   d n(R ∩ Q) = n(…)
   e n(P) + n(…) = n(P ∪ R)

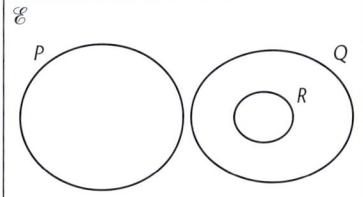

Key points
- A set can be defined by giving a rule or description, or by listing the elements; for example, {1, 2, 3, 4}, {(x, y): y = mx + c}.
- Venn diagrams can be used to represent sets and their elements.
- Set notation:

Universal set	ℰ
Empty set	∅
Number of elements in set A	n(A)
Is an element of	∈
Is not an element of	∉
Complement of set A	A′
A is a subset of B	A ⊆ B
A is not a subset of B	A ⊄ B
Union of A and B	A ∪ B
Intersection of A and B	A ∩ B

3 POWERS AND ROOTS

BY THE END OF THIS CHAPTER YOU WILL BE ABLE TO:
- calculate with the following: squares, square roots, cubes, cube roots and other powers and roots of numbers.

CHECK YOU CAN:
- multiply numbers, with and without a calculator
- find the area of a square.

You should already have met the topics in this chapter. This chapter is a quick reminder.

Squares and square roots

As you can see in the diagram below, the square with side 3 has an area of $3 \times 3 = 9$ squares and the square with side 4 has an area of $4 \times 4 = 16$ squares.

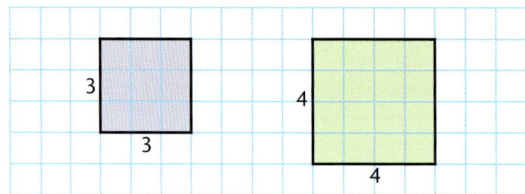

Note
You can say that the square of 3 is 9, or that 3 squared is 9 and write it as $3^2 = 9$.

The integers 1, 4, 9, 16, 25, … are the squares of the integers 1, 2, 3, 4, 5, … .

Because $16 = 4^2$, the positive square root of 16 is 4. It is written as $\sqrt{16} = 4$.

Similarly $\sqrt{36} = 6$ and $\sqrt{81} = 9$.

You can use your calculator to find squares and square roots.

Example 3.1

Question
Work out these using a calculator.
a 47^2 b $\sqrt{729}$

Note
Some calculators operate slightly differently. Learn which keys you need to use to do these operations on your calculator.

Solution
a $47^2 = 2209$ On your calculator, you need to press [4][7][$x^\bullet$][2][=]

b $\sqrt{729} = 27$ On your calculator, you need to press [√][7][2][9][=]

You also need to be able to find squares and square roots without a calculator.

Here is a list of the squares of the integers 1 to 15. You should learn these square numbers. You will also need them for finding square roots without a calculator.

Integer	1	2	3	4	5	6	7	8	9	10	11	12	13	14	15
Square number	1	4	9	16	25	36	49	64	81	100	121	144	169	196	225

Example 3.2

Question
Work out $8^2 - 4^2$.

Solution
Work out the squares first.
$8^2 = 8 \times 8 = 64$
$4^2 = 4 \times 4 = 16$
So $8^2 - 4^2 = 64 - 16$
$= 48$

Exercise 3.1

1. Write down the square of these numbers.
   a 7   b 12   c 5   d 10   e 9
   f 8   g 11   h 3   i 6   j 4

2. Write down the positive square root of these numbers.
   a 49   b 121   c 81   d 36   e 25
   f 169   g 144   h 225   i 100   j 196

3. Work out these.
   a 13^2 b 11^2 c 14^2 d 6^2 e 9^2

4. Work out these.
   a $\sqrt{81}$ b $\sqrt{144}$ c $\sqrt{16}$ d $\sqrt{100}$ e $\sqrt{64}$

5. Work out these.
   a $\sqrt{529}$ b $\sqrt{256}$ c $\sqrt{324}$ d $\sqrt{841}$ e $\sqrt{784}$

6. Work out these.
   a 20^2 b 25^2 c 13^2 d 24^2 e 33^2

7. Work out these.
   a $6^2 - 5^2$ b $2^2 + 3^2$ c $7^2 - 4^2$ d $3^2 - 2^2$
   e $4^2 + 5^2$ f $6^2 - 3^2$ g $5^2 - 4^2 - 3^2$ h $13^2 - 5^2 - 9^2$

8. Work out these.
   a $21^2 - 9^2$
   b $24^2 + 7^2 - 10^2$
   c $17^2 - 15^2 + 11^2$
   d $20^2 + 21^2 + 22^2$

Cubes and cube roots

The cube in the diagram has a volume of $2 \times 2 \times 2 = 8$.

The cube of a number is the number multiplied by itself, and then by itself again.

The integers 1, 8, 27, 64, 125, 216, … are the cubes of the integers 1, 2, 3, 4, 5, 6, … .

Because $8 = 2^3$ the cube root of 8 is 2. It is written as $\sqrt[3]{8} = 2$.

Similarly $\sqrt[3]{27} = 3$ and $\sqrt[3]{64} = 4$.

You need to be able to find cubes and cube roots of some numbers without a calculator.

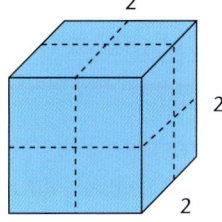

Note
You can say that the cube of 2 is 8, or that 2 cubed is 8 and write it as $2^3 = 2 \times 2 \times 2 = 8$.

3 POWERS AND ROOTS

Here is a list of the cubes of the integers 1 to 5, and 10. You should learn these cube numbers.

Integer	1	2	3	4	5	10
Cube number	1	8	27	64	125	1000

Example 3.3

Question
Work out $\sqrt[3]{125}$ without a calculator.

Solution
$\sqrt[3]{125} = 5$ You know that $5^3 = 125$, so you also know that the cube root of 125 is 5.

You can also use your calculator to find cubes and cube roots.

Example 3.4

Question
Work out these using a calculator.
a 15^3 b $\sqrt[3]{4913}$

Note
Some calculators operate slightly differently.
Learn which keys you need to use to do these operations on your calculator.

Solution
a $15^3 = 3375$ On your calculator press $\boxed{1}\boxed{5}\boxed{x^3}\boxed{=}$

b $\sqrt[3]{4913} = 17$ On your calculator press $\boxed{\sqrt[3]{}}\boxed{4}\boxed{9}\boxed{1}\boxed{3}\boxed{=}$

Note
You also need to be able to use your calculator to find other powers and roots.
On many calculators these buttons are labelled $x^\blacksquare$ and $\sqrt[\blacksquare]{x}$.
Find them and learn how to use them to work out results such as
$2.5^4 = 39.0625$ and $\sqrt[5]{693.43957} = 3.7$.

Exercise 3.2

1 Write down the cube of these numbers.
   a 4   b 5   c 3   d 10   e 2
2 Write down the cube root of these numbers.
   a 1   b 64   c 1000
3 Work out these.
   a 7^3 b 9^3 c 20^3 d 25^3 e 1.5^3
   f 2.7^3 g 5.4^3

Cubes and cube roots

4 Work out these.
  a $\sqrt[3]{343}$ **b** $\sqrt[3]{729}$ **c** $\sqrt[3]{1331}$ **d** $\sqrt[3]{1000000}$
  e $\sqrt[3]{216}$ **f** $\sqrt[3]{1728}$ **g** $\sqrt[3]{512}$

5 Work these out. Give your answers to 2 decimal places.
  a $\sqrt[3]{56}$ **b** $\sqrt[3]{230}$ **c** $\sqrt[3]{529}$
  d $\sqrt[3]{1100}$ **e** $\sqrt[3]{7526}$

6 Find the length of a cube whose volume is 45 cm³.
  Give your answer to 2 decimal places.

7 Work out these.
  a 3.4^6 **b** 7.5^4 **c** 0.7^5

8 Work out these.
  a $\sqrt[4]{282.5761}$ **b** $\sqrt[5]{4181.95493}$ **c** $\sqrt[6]{0.015625}$

9 Work out these.
  a $4^2 \times \sqrt[4]{390625}$ **b** $\sqrt[5]{40.84101} \times 2.4^4$ **c** $\dfrac{1.2^4}{\sqrt[3]{32.768}}$

Key points

- A square number is the result of multiplying a number by itself.
- 4^2 means 4×4. So, $4^2 = 16$.
- The square root of a number is the positive number that multiplies by itself to give that number.
- $\sqrt{}$ means 'the square root of'. So, $\sqrt{16} = 4$.
- A cube number is the result of multiplying a number by itself, and then by itself again. $4^3 = 4 \times 4 \times 4 = 64$.
- The fact that $4^3 = 64$ means also that the cube root of 64 is 4. This is written as $\sqrt[3]{64} = 4$.
- You should know the square numbers for 1^2 to 15^2. For example, know that $15^2 = 225$ and that $\sqrt{225} = 15$.
- You should know how to find and use the square root and cube root buttons on your calculator.
- You should know how to find and use the buttons on your calculator for other powers and roots.

4 FRACTIONS, DECIMALS AND PERCENTAGES

BY THE END OF THIS CHAPTER YOU WILL BE ABLE TO:
- use the language and notation of the following in appropriate contexts: proper fractions, improper fractions, mixed numbers, decimals and percentages
- recognise equivalence and convert between these forms.

CHECK YOU CAN:
- identify a fraction of a shape
- understand and use decimal notation
- find the highest common factor of two numbers.

Note

The denominator of the mixed number is the same as the denominator of the improper fraction.

Fractions

A fraction is a number written in the form $\frac{a}{b}$, where a and b are integers.

The value on the top of the fraction is known as the **numerator**.

The value on the bottom of the fraction is known as the **denominator**.

The fraction $\frac{4}{7}$ is a **proper fraction** because the numerator is smaller than the denominator.

The fraction $\frac{7}{4}$ is an **improper fraction** because the numerator is larger than the denominator.

The fraction $1\frac{3}{4}$ is a **mixed number** because it is formed from an integer and a proper fraction.

An improper fraction can be written as a mixed number.

$$\frac{7}{4} = \frac{4}{4} + \frac{3}{4} = 1\frac{3}{4}$$

Example 4.1

Question

a Write $\frac{17}{6}$ as a mixed number.
b Write $3\frac{4}{5}$ as an improper fraction.

Solution

a Divide the numerator by the denominator and write the remainder as a fraction over the denominator.

$17 \div 6 = 2$ remainder 5

so $\frac{17}{6} = 2\frac{5}{6}$

b Multiply the integer by the denominator of the fraction and add the numerator.

$3 \times 5 + 4 = 19$

so $3\frac{4}{5} = \frac{19}{5}$

Exercise 4.1

1 State whether each of these is a proper fraction, an improper fraction or a mixed number.

  a $\frac{19}{18}$ b $1\frac{2}{3}$ c $\frac{6}{8}$ d $8\frac{5}{6}$ e $\frac{15}{16}$

2 Write each of these improper fractions as a mixed number.

  a $\frac{11}{8}$ b $\frac{11}{5}$ c $\frac{9}{4}$ d $\frac{7}{2}$ e $\frac{15}{7}$

  f $\frac{10}{3}$ g $\frac{19}{8}$ h $\frac{23}{4}$ i $\frac{33}{10}$ j $\frac{37}{9}$

3 Write each of these mixed numbers as an improper fraction.
  a $1\frac{1}{8}$ b $2\frac{5}{8}$ c $3\frac{3}{4}$ d $5\frac{1}{2}$ e $3\frac{2}{9}$
  f $2\frac{2}{5}$ g $3\frac{2}{3}$ h $2\frac{1}{10}$ i $2\frac{3}{8}$ j $4\frac{6}{7}$

Fraction of a quantity

A fraction can be used to describe a share of a quantity.

The denominator shows how many parts the quantity is divided into.

The numerator shows how many of those parts are required.

Example 4.2

Question
a Work out $\frac{1}{8}$ of 56.   b Work out $\frac{5}{8}$ of 56.

Solution
a Finding $\frac{1}{8}$ of 56 is the same as dividing 56 into 8 parts.
  So $\frac{1}{8}$ of 56 = 56 ÷ 8 = 7
b $\frac{5}{8}$ means $5 \times \frac{1}{8}$.
  So $\frac{5}{8}$ of 56 = 5 × 7 = 35

To find a fraction of a quantity, divide by the denominator and multiply by the numerator.

Exercise 4.2

1 Work out these.
  a $\frac{3}{4}$ of 64   b $\frac{2}{3}$ of 96   c $\frac{5}{7}$ of 35   d $\frac{9}{10}$ of 160   e $\frac{11}{12}$ of 180

2 In a school of 858 students, $\frac{6}{11}$ are boys.
  How many boys are there?

3 In an election, the winning party got $\frac{5}{9}$ of the votes.
  There were 28 134 votes in total.
  How many votes did the winning party get?

4 A shop offers $\frac{1}{4}$ off everything as a special offer.
  A mobile phone normally costs $168.
  How much does it cost with the special offer?

5 Which is larger, a $\frac{7}{10}$ share of $120 or a $\frac{7}{8}$ share of $104?
  Show how you decide.

6 Which is larger, a $\frac{3}{8}$ share of $192 or a $\frac{2}{5}$ share of $180?
  Show how you decide.

Photocopying is prohibited

4 FRACTIONS, DECIMALS AND PERCENTAGES

Equivalent fractions

These squares can be divided into equal parts in different ways.

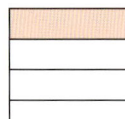 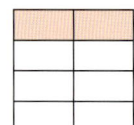 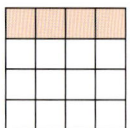

The fraction represented by the shaded parts is $\frac{1}{4}$ or $\frac{2}{8}$ or $\frac{4}{16}$.

These three fractions are equal in value and are **equivalent fractions**.

$\frac{1}{4} = \frac{2}{8} = \frac{4}{16}$ but $\frac{1}{4}$ is in its **simplest form**.

Example 4.3

Question

Write each fraction in its simplest form.

a $\frac{6}{10}$ b $\frac{18}{24}$

Solution

a The highest common factor of 6 and 10 is 2, so divide the numerator and denominator by 2.
$\frac{6}{10} = \frac{3}{5}$ in its simplest form.

b The highest common factor of 18 and 24 is 6, so divide the numerator and denominator by 6.
$\frac{18}{24} = \frac{3}{4}$ in its simplest form.

Note

18 and 24 also have common factors of both 2 and 3, so $\frac{9}{12}$ and $\frac{6}{8}$ are also equivalent to $\frac{18}{24}$, but $\frac{3}{4}$ is the simplest form of this fraction.

Exercise 4.3

1 Fill in the missing numbers in each set of equivalent fractions.

a $\frac{1}{4} = \frac{\square}{8} = \frac{\square}{12} = \frac{5}{\square}$

b $\frac{1}{5} = \frac{\square}{10} = \frac{\square}{20} = \frac{7}{\square}$

c $\frac{2}{5} = \frac{4}{\square} = \frac{\square}{25} = \frac{12}{\square}$

d $\frac{2}{9} = \frac{4}{\square} = \frac{\square}{36} = \frac{6}{\square}$

e $\frac{1}{7} = \frac{2}{\square} = \frac{\square}{35}$

f $\frac{4}{9} = \frac{16}{\square} = \frac{\square}{72}$

2 Express each fraction in its simplest form.

a $\frac{8}{10}$ b $\frac{2}{12}$ c $\frac{15}{21}$ d $\frac{12}{16}$

e $\frac{14}{21}$ f $\frac{25}{30}$ g $\frac{20}{40}$ h $\frac{18}{30}$

i $\frac{16}{24}$ j $\frac{150}{300}$ k $\frac{20}{120}$ l $\frac{500}{1000}$

m $\frac{56}{70}$ n $\frac{64}{72}$ o $\frac{60}{84}$ p $\frac{120}{180}$

3 A bag contains 96 balls.
36 of the balls are red.
What fraction of the balls are red? Give your answer in its simplest form.

4 Akbar drove 64 km of a 120 km journey on motorways.
What fraction of the journey did he drive on motorways?
Give your answer in its simplest form.

5 The table gives information about the members of a club.

	Male	Female
Adult	90	60
Child	55	45

Answer these questions, giving each answer as a fraction in its simplest form.
a What fraction of the members are adult males?
b What fraction of the members are female?
c What fraction of the members are children?

Fractions and decimals

We can use place value in a decimal to convert the decimal to a fraction.

Example 4.4

Question
Convert the decimal 0.245 to a fraction in its simplest form.

Solution

Units	.	Tenths	Hundredths	Thousandths
0	.	2	4	5

$0.245 = \frac{2}{10} + \frac{4}{100} + \frac{5}{1000}$

$= \frac{200}{1000} + \frac{40}{1000} + \frac{5}{1000}$ Convert each fraction to its equivalent with a denominator of 1000.

$= \frac{245}{1000}$ The HCF of 245 and 1000 is 5, so divide the numerator and the denominator by 5 to simplify the fraction.

$= \frac{49}{200}$

0.245 is equivalent to $\frac{49}{200}$

Note
You can use the place value of the final digit to write the decimal directly as a single fraction. The final digit here, 5, represents thousandths so 0.245 is equivalent to $\frac{245}{1000}$.

 Exercise 4.4

Convert each of these decimals to a fraction in its simplest form.

1 0.7
2 0.29
3 0.85
4 0.07
5 0.312
6 0.056
7 0.008
8 0.02
9 0.545
10 0.1345

Terminating and recurring decimals

In Chapter 1, you learnt that both terminating and recurring decimals were rational numbers.

So both terminating and recurring decimals can be written as fractions.

Also, all fractions can be written as a terminating or a recurring decimal.

You can convert the fraction $\frac{5}{8}$ to a decimal using division.

$\frac{5}{8} = 5 \div 8 = 0.625$

4 FRACTIONS, DECIMALS AND PERCENTAGES

This is a **terminating decimal** because it finishes at the digit 5.

You can convert the fraction $\frac{1}{6}$ to a decimal using division.

$\frac{1}{6} = 1 \div 6 = 0.166666...$

This is a **recurring decimal** because the digit 6 repeats indefinitely.

Dot notation for recurring decimals

Dot notation can be used when writing recurring decimals.

Dots are placed over the digits that recur.

For example,

$\frac{1}{3} = 0.333333...$ is written using dot notation as $0.\dot{3}$.

$\frac{1}{6} = 0.166666...$ is written using dot notation as $0.1\dot{6}$.

$\frac{124}{999} = 0.124124124...$ is written using dot notation as $0.\dot{1}2\dot{4}$ or $0.\overline{124}$.

Example 4.5

Question

a Convert $\frac{13}{25}$ to a decimal.

b Convert $\frac{7}{11}$ to a decimal.

c Convert $0.\dot{6}$ to a fraction using $\frac{1}{3} = 0.\dot{3}$.

d Convert $0.\dot{3}\dot{6}$ to a fraction.

Solution

a $\frac{13}{25} = 13 \div 25 = 0.52$

b $\frac{7}{11} = 7 \div 11 = 0.636363...$ In this case the digits 6 and 3 recur so you can write the answer using dot notation.

$\frac{7}{11} = 0.\dot{6}\dot{3}$

c $0.\dot{6} = 2 \times 0.\dot{3} = 2 \times \frac{1}{3} = \frac{2}{3}$

d Let $N = 0.363636...$

Multiply each side by 100

$100N = 36.363636...$

Subtract

$99N = 36$, $N = \frac{36}{99} = \frac{4}{11}$

Exercise 4.5

1 Convert each of these fractions to a decimal.

  a $\frac{3}{8}$ b $\frac{5}{16}$ c $\frac{11}{40}$ d $\frac{79}{250}$ e $\frac{3}{80}$

2 Convert each of these fractions to a recurring decimal. Write your answers using dot notation.

  a $\frac{2}{3}$ b $\frac{5}{6}$ c $\frac{1}{12}$ d $\frac{4}{15}$ e $\frac{16}{33}$

3 Given that $0.\dot{1} = \frac{1}{9}$, write each of these as a fraction.
   a $0.\dot{2}$ b $0.\dot{3}$ c $0.\dot{5}$

4 Given that $\frac{1}{27} = 0.\dot{0}3\dot{7}$ and $\frac{1}{11} = 0.\dot{0}\dot{9}$, find the decimal equivalent of each of these.
   a $\frac{2}{27}$ b $\frac{5}{27}$ c $\frac{10}{27}$ d $\frac{2}{11}$ e $\frac{6}{11}$

5 Convert each of these recurring decimals to fractions in their simplest form.
   a $0.\dot{4}$ b $0.\dot{4}\dot{5}$ c $0.\dot{7}\dot{3}$

Fractions, decimals and percentages

The term **per cent** means 'out of 100'.

For example 75% means 75 out of every 100 or $\frac{75}{100}$.

$\frac{75}{100}$ can be written in decimal form as 0.75.

So 75% is equivalent to $\frac{75}{100}$ and 0.75.

You can find fraction and decimal equivalents of all percentages.

There are some fraction, decimal and percentage equivalents that are useful to remember.

Fraction	Decimal	Percentage
$\frac{1}{2}$	0.5	50%
$\frac{1}{4}$	0.25	25%
$\frac{3}{4}$	0.75	75%
$\frac{1}{10}$	0.1	10%
$\frac{1}{5}$	0.2	20%

Example 4.6

Question

a Convert $\frac{3}{8}$ to a percentage.

b Convert 65% to a fraction in its simplest form.

Solution

a $\frac{3}{8} = 3 \div 8 = 0.375$ Convert to a decimal by dividing.
   $0.375 \times 100 = 37.5$ Multiply by 100 for percentage.
   So $\frac{3}{8} = 37.5\%$

b $65\% = \frac{65}{100}$ The HCF of 65 and 100 is 5, so divide the numerator and the denominator by 5 to simplify the fraction.
   $65\% = \frac{13}{20}$

Note

You can convert directly from a fraction to a percentage using multiplication.

4 FRACTIONS, DECIMALS AND PERCENTAGES

Exercise 4.6

1. Convert each of these percentages to a fraction. Write your answers in their simplest form.
   a 35%   b 65%   c 8%   d 120%

2. Convert each of these percentages to a decimal.
   a 16%   b 27%   c 83%   d 7%
   e 31%   f 4%   g 17%   h 2%
   i 150% j 250% k 9%   l 12.5%

3. Convert each of these fractions to a decimal.
   a $\frac{1}{100}$ b $\frac{17}{100}$ c $\frac{2}{50}$ d $\frac{8}{5}$ e $\frac{1}{8}$
   f $\frac{5}{8}$ g $\frac{3}{20}$ h $\frac{17}{40}$ i $\frac{5}{16}$

4. Convert each of the decimals you found in question 3 to a percentage.

5. Convert each of these fractions to a percentage. Give your answers correct to 1 decimal place.
   a $\frac{1}{6}$ b $\frac{5}{6}$ c $\frac{1}{12}$ d $\frac{5}{12}$ e $\frac{3}{70}$

6. Write three fractions that are equivalent to 40%.

7. Write three fractions that are equivalent to 0.125.

8. a Convert 160% to a decimal.
   b Convert 160% to a mixed number in its simplest form.

9. The winning party in an election gained $\frac{7}{12}$ of the votes.
   What percentage is this? Give your answer correct to the nearest 1%.

10. Imran saves $\frac{2}{9}$ of his earnings.
    What percentage is this? Give your answer correct to the nearest 1%.

11. In a survey about the colour of cars, 22% of the people said they preferred red cars, $\frac{3}{20}$ of the people said they preferred silver cars and $\frac{6}{25}$ said they preferred black cars.
    Which colour car was most popular? Show how you decide.

12. In class P, $\frac{3}{7}$ of the students are boys.
    In class Q, 45% of the students are boys.
    Which class has the higher proportion of boys? Show how you decide.

> **Key points**
> - A proper fraction has a numerator smaller than the denominator.
> - An improper fraction has a numerator larger than the denominator. It can also be written as a mixed number formed from an integer and a proper fraction.
> - Fractions of equal value are equivalent.
> - The fraction with no common factors in its numerator and denominator is in its simplest form.
> - Fractions are equivalent to terminating or recurring decimals.
> - Percentages are fractions written out of 100.
> - Equivalent fractions, decimals and percentages can be converted from one to another.

5 ORDERING

Ordering integers

BY THE END OF THIS CHAPTER YOU WILL BE ABLE TO:
- order quantities by magnitude and demonstrate familiarity with the symbols =, ≠, >, <, ≥ and ≤.

Example 5.1

Question
Put these masses in order, smallest first.
1.2 kg   1500 g   175 g   2 kg   0.8 kg

Solution
Write each value with the same units first. (It is usually easier to use the smallest unit.)
1200 g   1500 g   175 g   2000 g   800 g
Then order them, writing your answer using the original units.
So order is
175 g   0.8 kg   1.2 kg   1500 g   2 kg.

CHECK YOU CAN:
- use the symbols = and ≠ correctly
- order whole numbers
- work with negative numbers
- convert between metric units:
  - m, cm and mm
  - kg and g
  - litres (l), cl and ml
- convert between fractions, decimals and percentages.

Exercise 5.1

1 Write each set of temperatures in order, lowest first.
   a −2 °C   7 °C   0 °C   −5 °C   3 °C
   b −2 °C   5 °C   1 °C   2 °C   −1 °C
   c 7 °C   −7 °C   4 °C   −9 °C   −3 °C
   d 9 °C   4 °C   −2 °C   7 °C   −8 °C
   e −4 °C   5 °C   −2 °C   3 °C   −7 °C

2 Write each set of lengths in order of size, smallest first.
   a 2.42 m   1600 mm   284 cm   9 m   31 cm
   b 423 cm   6100 mm   804 cm   3.2 m   105 mm

3 Write each set of masses in order of size, smallest first.
   a 4000 g   52 000 g   9.4 kg   874 g   1.7 kg
   b 4123 g   2104 g   3.4 kg   0.174 kg   2.79 kg

4 Write each set of capacities in order of size, smallest first.
   a 2.4 litres   1600 ml   80 cl   9 litres   51 cl
   b 3.1 litres   1500 ml   180 cl   1 litre   51.5 ml

Inequalities

$a < b$ means 'a is less than b'.

$a \leq b$ means 'a is less than or equal to b'.

$a > b$ means 'a is greater than b'.

$a \geq b$ means 'a is greater than or equal to b'.

Photocopying is prohibited

5 ORDERING

Exercise 5.2

1 Rewrite these: insert > or < in each part as appropriate.
- a 7 °C is ... than 2 °C.
- b −10 °C is ... than 5 °C.
- c −3 °C is ... than 0 °C.
- d −7 °C is ... than −12 °C.
- e 10 °C is ... than 0 °C.
- f −2 °C is ... than −5 °C.
- g 4 °C is ... than −1 °C.
- h −2 °C is ... than 2 °C.

Ordering decimals

Example 5.2

Question

Put these decimals in order of size, smallest first.

0.412   0.0059   0.325   0.046   0.012

Solution

Add zeros to make the decimals all the same length.

0.4120   0.0059   0.3250   0.0460   0.0120

Now, remove the decimal points and any zeros in front of the digits.

4120   59   3250   460   120

The order in size of these values is the order in size of the decimals.

The order is

0.0059   0.012   0.046   0.325   0.412

Note
As you get used to this, you can omit the second step.

Exercise 5.3

1 Put these numbers in order of size, smallest first.
- a 462, 321, 197, 358, 426, 411
- b 89 125, 39 171, 4621, 59 042, 6317, 9981
- c 124, 1792, 75, 631, 12, 415
- d 9425, 4257, 7034, 5218, 6641, 1611
- e 1 050 403, 1 030 504, 1 020 504, 1 040 501, 1 060 504, 1 010 701

2 Put these decimals in order of size, smallest first.
- a 0.123, 0.456, 0.231, 0.201, 0.102
- b 0.01, 0.003, 0.1, 0.056, 0.066
- c 0.0404, 0.404, 0.004 04, 0.044, 0.0044
- d 0.71, 0.51, 0.112, 0.149, 0.2
- e 0.913, 0.0946, 0.009 16, 0.090 11, 0.091

3 Put these numbers in order of size, smallest first.
- a 3.12, 3.21, 3.001, 3.102, 3.201
- b 1.21, 2.12, 12.1, 121, 0.12
- c 7.023, 7.69, 7.015, 7.105, 7.41
- d 5.321, 5.001, 5.0102, 5.0201, 5.02
- e 0.01, 12.02, 0.0121, 1.201, 0.0012
- f 8.097, 8.79, 8.01, 8.1, 8.04

Ordering fractions

To put fractions in order, convert them to equivalent fractions all with the same denominator, and order them by the numerator.

Example 5.3

Question
Which is the bigger fraction, $\frac{3}{4}$ or $\frac{5}{6}$?

Solution
First, find a common denominator. 24 is an obvious one, as $4 \times 6 = 24$, but a smaller one is 12. $\frac{3}{4} = \frac{9}{12}$ $\frac{5}{6} = \frac{10}{12}$
$\frac{10}{12}$ is bigger than $\frac{9}{12}$, so $\frac{5}{6}$ is bigger than $\frac{3}{4}$.

Note
Multiplying the two denominators together will always work to find a common denominator, but the lowest common multiple of the denominators is sometimes smaller.

Alternatively, you can convert each fraction to a decimal and compare the decimals as before.

Example 5.4

Question
Put these fractions in order, smallest first.
$\frac{3}{10}$ $\frac{1}{4}$ $\frac{9}{20}$ $\frac{2}{5}$ $\frac{1}{2}$

Solution
Use division to convert the fractions to decimals ($3 \div 10$, $1 \div 4$, etc.).
0.3   0.25   0.45   0.4   0.5
Now make the number of decimal places the same.
0.30   0.25   0.45   0.40   0.50
So order is $\frac{1}{4}$ $\frac{3}{10}$ $\frac{2}{5}$ $\frac{9}{20}$ $\frac{1}{2}$.

Exercise 5.4

1 Write each pair of fractions, inserting > or < as appropriate.

a $\frac{2}{3}\ldots\frac{7}{9}$ b $\frac{5}{6}\ldots\frac{7}{8}$ c $\frac{3}{8}\ldots\frac{7}{20}$ d $\frac{3}{4}\ldots\frac{5}{8}$ e $\frac{7}{9}\ldots\frac{5}{6}$ f $\frac{3}{10}\ldots\frac{4}{15}$

2 Write each of these sets of fractions in order, smallest first.

a $\frac{7}{10}$ $\frac{3}{4}$ $\frac{11}{20}$ $\frac{3}{5}$ b $\frac{7}{12}$ $\frac{3}{4}$ $\frac{7}{8}$ $\frac{5}{6}$ c $\frac{13}{15}$ $\frac{2}{3}$ $\frac{3}{10}$ $\frac{2}{5}$ $\frac{1}{2}$

d $\frac{13}{16}$ $\frac{5}{8}$ $\frac{3}{4}$ $\frac{7}{16}$ $\frac{1}{2}$ e $\frac{2}{5}$ $\frac{1}{2}$ $\frac{9}{20}$ $\frac{17}{40}$ $\frac{3}{8}$ f $\frac{11}{16}$ $\frac{7}{8}$ $\frac{3}{4}$ $\frac{17}{32}$

Note
Try both methods shown in the examples. The first is useful when calculators are not allowed.

Ordering fractions, decimals and percentages

Example 5.5

Question

Put these numbers in order, smallest first.

$\frac{1}{4}$ 3%   0.41 $\frac{11}{40}$ 0.35

Solution

Convert them all to decimals.

0.25,   0.03,   0.41,   0.275,   0.35

So the order is

0.03   0.25   0.275   0.35   0.41.

Write the numbers in their original form.

3% $\frac{1}{4}$ $\frac{11}{40}$ 0.35   0.41

Note

A common error is to write 3% as 0.3, rather than 0.03.

Exercise 5.5

1 Put these numbers in order, smallest first.

$\frac{4}{5}$, 88%, 0.83, $\frac{17}{20}$, $\frac{7}{10}$

2 Put these numbers in order, smallest first.

$\frac{3}{8}$, 35%, 0.45, $\frac{2}{5}$, $\frac{5}{12}$

3 Put these numbers in order, smallest first.

$\frac{3}{5}$, 30%, 0.7, $\frac{3}{4}$, $\frac{2}{3}$

4 Soccer teams United, City and Rovers have all played the same number of games.
United have won $\frac{3}{8}$ of the games they have played, City have won 35% and Rovers have won 0.4.
Put the teams in order of the number of matches they have won.

5 A group of boys were asked to name their favourite sport.
$\frac{2}{7}$ chose soccer, 0.27 chose rugby and 28% chose gymnastics.
List the sports in the order of their popularity, the most popular first.

Key points

- Know what the symbols =, ≠, >, <, ⩾, ⩽ mean and how to use them.
- When ordering values with different metric units, change each value so they all have the same unit.
- When comparing decimals, make them all have the same number of digits after the decimal point by adding zeros to the ends.
- When comparing fractions, change each so they all have the same denominator. You can also compare fractions by changing each to a decimal and then comparing the decimals.
- The easiest way to compare a mixture of fractions, decimals and percentages is to change each to a decimal and then compare the decimals.

6 THE FOUR OPERATIONS

Numbers below zero

Some numbers are less than zero. These are called **negative numbers**. They are written as ordinary numbers with a negative sign in front.

Negative numbers are used in many situations.

Thermometer measuring temperature

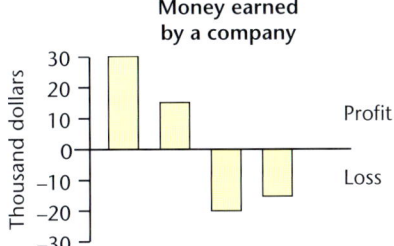
Money earned by a company

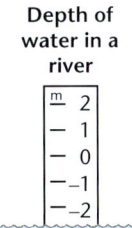
Depth of water in a river

Buttons in a lift

3 Car park
2
1
0 Ground floor
−1 Basement

> **BY THE END OF THIS CHAPTER YOU WILL BE ABLE TO:**
> - use the four operations for calculations with integers, fractions and decimals, including correct ordering of operations and use of brackets.

> **CHECK YOU CAN:**
> - add, subtract, multiply and divide integers without a calculator
> - add and subtract decimals without a calculator
> - find equivalent fractions
> - read whole numbers from a scale.

> **Note**
> Remember that zero is neither positive nor negative.

Example 6.1

Question
Suravi measured the daytime and night-time temperatures in her garden for 2 days. Here are her results.

Day	Monday daytime	Monday night-time	Tuesday daytime	Tuesday night-time
Temperature (°C)	7	−2	3	−5

a How much did the temperature change between each reading?
b The daytime temperature on Wednesday was 4 °C warmer than the Tuesday night-time temperature.
  What was the Wednesday daytime temperature?
Use the temperature scale to help you.

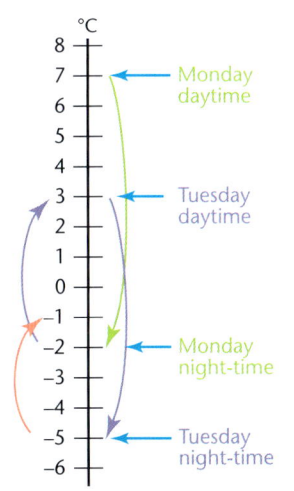

Solution
a From 7 °C to −2 °C you go down 9 °C.
  From −2 °C to 3 °C you go up 5 °C.
  From 3 °C to −5 °C you go down 8 °C.

b Find −5 °C on the scale and move 4 °C up.
  You get to −1 °C.

Photocopying is prohibited

6 THE FOUR OPERATIONS

Example 6.2

Question
Abal's bank account is overdrawn by $75.
How much must he put in this account for it to be $160 in credit?

Solution
Abal's bank account is overdrawn by $75.
This means Abal owes the bank $75. It can be shown as −$75.
$160 in credit means that there is $160 in the account. It can be shown as +$160.
From −75 to +160 is 235, so he must put $235 into the account.

Exercise 6.1

1 Copy and complete the table.

	Start temperature/°C	Move/°C	End temperature/°C
a	4	Up 3	
b	−2	Down 4	
c	10	Down 14	
d	−5	Down 3	
e	−10		−2
f	10		−9
g	−4		2
h		Up 7	10
i		Down 6	−9
j		Up 2	−8

2 The floors in a shopping centre are numbered −3, −2, −1, 0, 1, 2, 3, 4, 5.
   a Ali parks his car on floor −2. He takes the lift and goes up 6 floors. Which floor is he now on?
   b While shopping, Ubah goes down 3 floors, ending on floor −1. Which floor did she start from?

3 What is the difference in temperature between each of these?
   a 17°C and −1°C
   b −8°C and 12°C
   c −19°C and −5°C
   d 30°C and −18°C
   e 13°C and −5°C
   f −10°C and 15°C
   g −20°C and −2°C
   h 25°C and −25°C

4 To the nearest degree, the hottest temperature ever recorded on Earth was 58°C in 1922 and the coldest ever recorded was −89°C in 1983. What is the difference between these temperatures?

5 Geta's bank account is $221 in credit.
   She puts $155 into the account.
   Later she spends $97 on clothes and $445 on some electronic goods.
   What will Geta's bank account show now?

6 The table shows the heights above sea level, in metres, of seven places.

Place	Height/m
Mount Everest	8 863
Bottom of Lake Baikal	−1 484
Bottom of Dead Sea	−792
Ben Nevis	1 344
Mariana Trench	−11 022
Mont Blanc	4 807
World's deepest cave	−1 602

What is the difference in height between the highest and lowest places?

7 The highest temperature ever recorded in England was 38.5 °C in 2006 and the lowest ever recorded was −26.1 °C in 1982. What is the difference between these temperatures?

8 The table shows the coldest temperatures ever recorded on each continent.

Continent	Lowest temperature
Africa	−23.9 °C
Antarctica	−89.2 °C
Asia	−71.2 °C
Australia	−23.0 °C
North America	−60.0 °C
South America	−32.8 °C
Europe	−58.1 °C

What is the difference between the highest and lowest of these temperatures?

9 This is an extract from a tide table for Dungeness, Washington State, USA.

Day	High/Low	Tide time	Height/feet above normal
1	High	2.03 a.m.	7.2
	Low	9.28 a.m.	−1.8
	High	2.27 p.m.	7.5
	Low	9.53 p.m.	5.6
2	High	2.48 a.m.	7.2
	Low	10.09 a.m.	−2.2
	High	3.00 p.m.	7.7
	Low	10.42 p.m.	5.4

What is the difference in height between the highest high tide and the lowest low tide?

Adding and subtracting with negative numbers

A negative number is a number less than zero.

A number line is very useful when adding or subtracting with negative numbers.

This number line shows $-2 + 4 = 2$

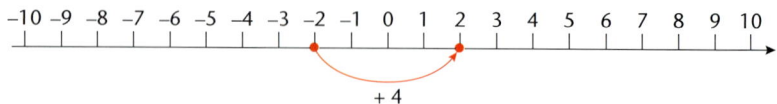

This number line shows $5 - 7 = -2$

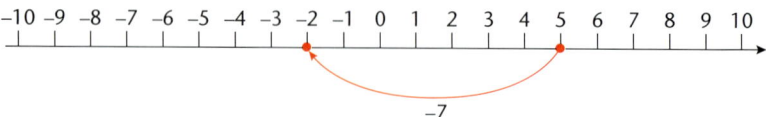

Patterns can also help.

$-2 + 4 = 2$	$-2 + 3 = 1$	$-2 + 2 = 0$	$-2 + 1 = -1$
$-2 + 0 = -2$	$-2 - 1 = -3$	$-2 - 2 = -4$	
$1 - 4 = -3$	$1 - 3 = -2$	$1 - 2 = -1$	$1 - 1 = 0$
$1 - 0 = 1$	$1 - -1 = 2$	$1 - -2 = 3$	$1 - -3 = 4$

This helps to show that

- adding a negative number is the same as subtracting a **positive number**
- subtracting a negative number is the same as adding a positive number.

Example 6.3

Question

Work out these.

a $-2 + -6$ b $-2 - -5$

Solution

a $-2 + -6 = -8$ Adding a negative number is like subtracting a positive number.

So $-2 + -6$ is the same as $-2 - 6$.

b $-2 - -5 = 3$ Subtracting a negative number is like adding a positive number.

So $-2 - -5$ is the same as $-2 + 5$.

Multiplying and dividing with negative numbers

 ## Exercise 6.2

Work out these.

1. $4 + -5$
2. $6 + -2$
3. $-5 - -6$
4. $6 - -3$
5. $2 + -1$
6. $-1 - -4$
7. $4 + 3 - -2 - 5$
8. $9 + -2 - 3 - -2$
9. $-4 - -2 - 2 - -5$
10. $2 + -3 - -4 - 7$

When adding and subtracting many numbers, it is best to add up the positive numbers and the negative numbers separately, then find the difference between the totals and give the answer the sign of the larger total.

Multiplying and dividing with negative numbers

Look at these patterns.

$5 \times 4 = 20 \qquad 5 \times 3 = 15 \qquad 5 \times 2 = 10 \qquad 5 \times 1 = 5 \qquad 5 \times 0 = 0$
$5 \times -1 = -5 \qquad 5 \times -2 = -10 \qquad 5 \times -3 = -15 \qquad 5 \times -4 = -20$

$-3 \times 4 = -12 \qquad -3 \times 3 = -9 \qquad -3 \times 2 = -6 \qquad -3 \times 1 = -3 \qquad -3 \times 0 = 0$
$-3 \times -1 = 3 \qquad -3 \times -2 = 6 \qquad -3 \times -3 = 9 \qquad -3 \times -4 = 12$

This suggests these rules:

$+ \times + = +$	and	$- \times - = +$
$+ \times - = -$	and	$- \times + = -$

Example 6.4

Questions

Work out these.

a 6×-4 \qquad b -7×-3 \qquad c -5×8

Solution

a $\quad 6 \times 4 = 24 \qquad (+ \times - = -)$
$\quad$ So $6 \times -4 = -24$

b $\quad 7 \times 3 = 21 \qquad (- \times - = +)$
$\quad$ So $-7 \times -3 = 21$

c $\quad 5 \times 8 = 40 \qquad (- \times + = -)$
$\quad$ So $-5 \times 8 = -40$

6 THE FOUR OPERATIONS

From the previous page, you can see that $-3 \times 4 = -12$.

It follows that $-12 \div 4 = -3$ and $-12 \div -3 = 4$.

This suggests these rules:

$$+ \div + = + \qquad \text{and} \qquad - \div - = +$$
$$+ \div - = - \qquad \text{and} \qquad - \div + = -$$

You now have a complete set of rules for multiplying and dividing positive and negative numbers.

$$+ \times + = + \qquad - \times - = + \qquad + \div + = + \qquad - \div - = +$$
$$+ \times - = - \qquad - \times + = - \qquad + \div - = - \qquad - \div + = -$$

> **Note**
> These rules only apply to multiplying and dividing, not to adding and subtracting.

Here is another way of thinking of these rules:

Signs different = answer is negative.

Signs the same = answer is positive.

You can extend the rules to calculations with more than two numbers.

If there is an even number of negative signs, the answer is positive.

If there is an odd number of negative signs, the answer is negative.

Combining operations

As with positive numbers, multiplying and dividing is done before addition and subtraction, unless there are brackets.

Example 6.5

Question

Work out these.

a $(-3 \times -4) + (-2 \times 3)$

b $6 - 2 \times 3$

c $\dfrac{5 \times -4 + 3 \times -2}{-6 + 4}$

Solution

a $(-3 \times -4) + (-2 \times 3) = +12 + -6$ Brackets are not really needed here, as multiplication should be done first.

$\qquad = 12 - 6$

$\qquad = 6$

b $6 - 2 \times 3 = 6 - 6 = 0$ Do the multiplication first.

c $\dfrac{5 \times -4 + 3 \times -2}{-6 + 4} = \dfrac{-20 + -6}{-2}$ Work out the numerator and the denominator separately.

$\qquad = \dfrac{-26}{-2}$ Then divide.

$\qquad = 13$

Exercise 6.3

Work out these.

1. $(-4 \times -3) - (-2 \times 1)$
2. $(-7 \times -2) + (4 \times -2)$
3. $(-15 \div 2) - (4 \times -6)$
4. $-4 + 3 + 2 + 3 + 4 - 5 - 6 - 9 + 1$
5. $\dfrac{-2 + 12}{-5}$
6. $\dfrac{-4 \times -3}{-4 + 3}$
7. $\dfrac{-4 \times 5}{-6 + 4}$
8. $(16 \div 2) - (-2 \times 4)$
9. $-2 \times 3 + -3 \times 4$
10. $-1 \times -4 + -7 \times -8$
11. $(24 \div -3) - (-5 \times -4)$
12. $-6 - 2 - 3 + 5 - 7 + 4 - 2 + 8$
13. $-12 \div -4 - 24 \div -6$
14. $6 + (5 \times -2 \times -5)$
15. $\dfrac{-3 \times 7}{-2}$
16. $\dfrac{-7 \times -12}{-8 + 4}$
17. $\dfrac{8 \times -6}{-4 + -8}$
18. $\dfrac{9 \times 4}{-3 \times -6}$
19. $4 \times 2 - 5 \times -4$
20. $4 + 5 \times -2 - 6$

Order of operations

When working out the answer to a calculation involving more than one operation, it is important that you use the correct order of operations.

For example, if you want to work out $2 + 3 \times 4$, should you carry out the addition or the multiplication first?

> This is the correct order of operations when carrying out any calculation:
> - first, work out anything in brackets
> - then, work out any powers (such as squares or square roots)
> - then do any multiplication or division
> - finally, do any addition or subtraction.

So to work out $2 + 3 \times 4$, you should first do the multiplication, then the addition.

$2 + 3 \times 4 = 2 + 12 = 14$

If you want the addition to be done first, then brackets are needed in the calculation.

$(2 + 3) \times 4 = 5 \times 4 = 20$

Some calculations are written like fractions, for example $\dfrac{6 \times 4}{5 + 3}$.

In this case, the fraction line works in the same way as brackets.

6 THE FOUR OPERATIONS

First, evaluate the numerator, then the denominator, and then do the division.

$$\frac{6 \times 4}{5 + 3} = \frac{24}{8} = 3$$

This calculation could also be written as $6 \times 4 \div (5 + 3)$.

Example 6.6

Question
Work out these.

a $7 + 4 \div 2 - 5$

b $(6 - 2)^2 + 5 \times -3$

Solution

a $7 + 4 \div 2 - 5$

$= 7 + 2 - 5$ First, work out the division.

$= 4$ The addition and subtraction can be done in a single step.

b $(6 - 2)^2 + 5 \times -3$

$= 4^2 + 5 \times -3$ First, work out the brackets.

$= 16 + 5 \times -3$ Next, work out the power.

$= 16 + -15$ Then the multiplication.

$= 1$ And finally the addition.

Exercise 6.4

Work out these.

1 $2 \times 5 + 4$

2 $2 \times (5 + 4)$

3 $2 + 5 \times 4$

4 $2 + 5^2$

5 5×4^2

6 $13 - 2 \times 5$

7 $(13 - 2) \times 5$

8 2×3^3

9 $(2 \times 5)^2$

10 $(5 - 3) \times (8 - 3)$

11 $6 + \frac{8}{2}$

12 $\frac{6 + 8}{2}$

13 $\frac{12}{4 \times 3}$

14 $\frac{12}{4} \times 3$

15 $\frac{6 + 4}{5} - 2$

16 $\frac{20}{5 + 3}$

17 $\frac{20}{5} + 3$

18 $3 \times 5 - \frac{6 \times 4}{8}$

19 $(20 - 2 \times 6)^2$

20 $12 - 6 \times 3$

21 $4 \times 3 - 5 \times 4$

22 3×2^3

23 $\frac{-7 \times -12}{-8 + 4}$

24 $\frac{8 \times -6}{-4 + -8}$

25 $\frac{9 \times 4}{-3 + -6}$

26 $4 \times 2 - 5 \times -4$

27 $4 + 5 \times -2 - 6$

28 Hassan says that $12 \div 2 + 4 = 2$.
   a Explain what he has done wrong.
   b Work out the correct answer.

29 Aisha says that $3 \times 2^2 = 36$.
   a Explain what she has done wrong.
   b Work out the correct answer.

30 Write out each of the following calculations, with brackets if necessary, to give the answers stated.
   a 3 + 6 × 5 − 1 to give
      i 44   ii 32   iii 27
   b 6 + 4² − 16 ÷ 2 to give
      i 6   ii 3   iii 92
   c 12 − 8 ÷ 4 + 4 to give
      i 11   ii 5   iii 14
   d 18 + 12 ÷ 6 − 3 to give
      i 2   ii 17   iii 22   iv 10

Multiplying integers

There are a number of methods that can be used for multiplying three-digit by two-digit integers. These methods can be adapted for use with integers of any size.

Example 6.7

Question
Work out 352 × 47.

Solution
Method 1: Long multiplication

```
          3  5  2
    ×        4  7
    1  4  0  8  0    (352 × 40)
       2  4  6  4    (352 × 7)
    1  6  5  4  4
```

Method 2: Grid method

×	300	50	2
40	12000	2000	80
7	2100	350	14

= 14080
= 2464
  16544

Exercise 6.5

Work out these.

1 138 × 13   2 581 × 23   3 614 × 14   4 705 × 32   5 146 × 79
6 615 × 46   7 254 × 82   8 422 × 65   9 428 × 64   10 624 × 75

11 A theatre has 42 rows of seats. There are 28 seats in each row.
    How many seats are there altogether?
12 A packet contains 36 pens. A container holds 175 of these packets.
    How many pens are in the container?

6 THE FOUR OPERATIONS

Multiplying decimals

Use this method for multiplying two decimal numbers:
- count the number of decimal places in the numbers you are multiplying
- ignore the decimal points and do the multiplication
- put a decimal point in your result so that there are the same number of decimal places in your answer as in the original calculation.

Example 6.8

Question

Work out these.

a 0.4×0.05
b 4.36×0.52

Solution

a $0.\text{④} \times 0.\text{⑤}$ — There are three decimal places in the calculation, so there will be three in the answer.

$4 \times 5 = 20$ — Multiply the figures, ignoring the decimal points.

$0.4 \times 0.05 = 0.0\text{②}0$ — Insert a decimal point so that there are three decimal places in the answer. In this case we need to add an extra zero between the decimal point and the 2.

$0.4 \times 0.05 = 0.02$

b $4.\text{③⑥} \times 0.\text{⑤②}$ — There are four decimal places in the calculation, so there will be four in the answer.

$436 \times 52 = 22\,672$ — Multiply the figures, ignoring the decimal points. Use whichever method you prefer.

$4.36 \times 0.52 = 2.2672$ — Insert a decimal point so that there are four decimal places in the answer.

Exercise 6.6

1. Given that $63 \times 231 = 14\,553$, write down the answers to these.
   a 6.3×2.31
   b 63×23.1
   c 0.63×23.1
   d 63×0.231
   e $6.3 \times 23\,100$

2. Given that $12.4 \times 8.5 = 105.4$, write down the answers to these.
   a 124×8.5
   b 12.4×0.85
   c 0.124×8.5
   d 1.24×8.5
   e 0.124×850

3. Work out these.
   a 0.3×5
   b 0.6×0.8
   c 0.2×0.4
   d 0.02×0.7
   e 0.006×5
   f 0.07×0.09

4. Work out these.
   a 42×1.5
   b 5.9×6.1
   c 10.9×2.4
   d 2.34×0.8
   e 5.46×0.7
   f 6.23×1.6

5. Work out these.
   a $0.5 + 0.2 \times 0.3$
   b $0.1 \times 0.8 - 0.03$
   c $0.2 + 0.4^2$
   d $0.9 \times 0.8 - 0.2 \times 0.7$

Dividing integers

There are a number of methods that can be used for dividing three-digit by two-digit integers. These methods can be adapted for use with integers of any size.

> **Example 6.9**
>
> **Question**
> Work out $816 \div 34$.
>
> **Solution**
> **Method 1: Long division**
> $816 \div 34 = 24$
>
> ```
>        24
>    34) 816
>      - 680    (34 × 20)
>        ---
>        136
>      - 136    (34 × 4)
>        ---
>          0
> ```
>
> The answer is 24.
>
> **Method 2: Chunking**
> $816 \div 34 = 24$
>
> ```
>     816
>   - 340    (34 × 10)
>     ---
>     476
>   - 340    (34 × 10)
>     ---
>     136
>   - 136    (34 × 4)
>     ---
>       0
> ```
>
> The answer is $10 + 10 + 4 = 24$.

 Exercise 6.7

Work out these.

1. $987 \div 21$
2. $684 \div 18$
3. $864 \div 16$
4. $924 \div 28$
5. $544 \div 32$
6. $352 \div 22$
7. $855 \div 45$
8. $992 \div 31$
9. $918 \div 27$
10. $576 \div 18$
11. Eggs are packed in boxes of 12.
    How many boxes are needed for 828 eggs?
12. A group of 640 people are going on a bus trip.
    Each bus can carry 54 people.
    How many buses are needed?

Dividing decimals

> Use this method for dividing by a decimal:
>
> - first, write the division as a fraction
> - make the denominator an integer by multiplying the numerator and the denominator by the same power of 10
> - cancel the fraction to its simplest form if possible; this gives you easier numbers to divide
> - divide to find the answer.

6 THE FOUR OPERATIONS

Example 6.10

Question

Work out these.

a $0.9 \div 1.5$

b $25.7 \div 0.08$

Solution

a $0.9 \div 1.5 = \dfrac{0.9}{1.5}$ Write the division as a fraction.

$= \dfrac{9}{15}$ Make the denominator an integer by multiplying both the numerator and the denominator by 10.

$= \dfrac{3}{5}$ Simplify.

$\dfrac{3}{5} = 5\overline{)3.0}^{\,0.6}$ Divide, adding an extra zero after the decimal point to complete the division.

$0.9 \div 1.5 = 0.6$

b $25.7 \div 0.08 = \dfrac{25.7}{0.08}$ Write the division as a fraction.

$= \dfrac{2570}{8}$ Make the denominator an integer by multiplying both the numerator and the denominator by 100.

$= \dfrac{1285}{4}$ Simplify.

Use whichever method you prefer for the division.

$\dfrac{1285}{4} = 4\overline{)1285.00}^{\,321.25}$ Divide, adding extra zeros after the decimal point to complete the division.

$25.7 \div 0.08 = 321.25$

Exercise 6.8

1 Given that $852 \div 16 = 53.25$, write down the answers to these.
   a $852 \div 1.6$
   b $8.52 \div 16$
   c $85.2 \div 1.6$
   d $85.2 \div 0.16$
   e $852 \div 0.016$

2 Given that $482 \div 25 = 19.28$, write down the answers to these.
   a $4.82 \div 2.5$
   b $0.482 \div 2.5$
   c $48.2 \div 0.25$
   d $4.82 \div 250$
   e $4.82 \div 0.025$

3 Work out these.
   a $8 \div 0.2$
   b $1.2 \div 0.3$
   c $5.6 \div 0.7$
   d $9 \div 0.3$
   e $15 \div 0.03$
   f $6.5 \div 1.3$

4 Work out these.
   a $1.55 \div 0.05$
   b $85.8 \div 0.11$
   c $5.55 \div 1.5$
   d $0.68 \div 1.6$
   e $87.6 \div 0.24$
   f $1.35 \div 1.8$

5 Work out these.
   a $0.6 + 0.2 \div 0.1$
   b $\dfrac{0.5 + 0.7}{0.3 \times 0.2}$
   c $\dfrac{1.8}{0.3^2}$
   d $\dfrac{2.6 \times 0.5}{0.1^2}$

Adding and subtracting fractions

In the diagram, each rectangle is divided into 20 small squares.

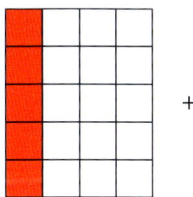 + 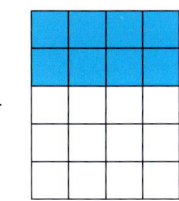 =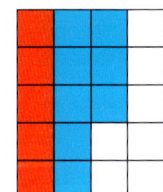

The diagram shows $\frac{1}{4} + \frac{2}{5}$.

The result of the addition has 13 squares shaded, or $\frac{13}{20}$.

> Use this method to add two fractions:
> - change the fractions to equivalent fractions with the same denominator
> - add the numerators.

$\frac{1}{4} = \frac{5}{20}$ and $\frac{2}{5} = \frac{8}{20}$

$\frac{1}{4} + \frac{2}{5} = \frac{5}{20} + \frac{8}{20} = \frac{13}{20}$

The method is the same for subtracting two fractions. Change them to equivalent fractions with the same denominator, then subtract the numerators.

If the fractions are mixed numbers, change them to improper fractions first.

Example 6.11

Question

Work out these.

a $\frac{2}{3} + \frac{5}{6}$ b $\frac{3}{4} - \frac{1}{3}$ c $2\frac{3}{5} + 1\frac{1}{3}$ d $3\frac{1}{4} - 1\frac{5}{6}$

Solution

a $\frac{2}{3} + \frac{5}{6} = \frac{4}{6} + \frac{5}{6}$ Write the fractions with a common denominator.
The lowest common multiple of 3 and 6 is 6, so use 6 as the common denominator.

$= \frac{9}{6}$ The result is an improper fraction and not in its simplest form.

$= \frac{3}{2} = 1\frac{1}{2}$ Simplify the fraction and write it as a mixed number.

b $\frac{3}{4} - \frac{1}{3} = \frac{9}{12} - \frac{4}{12}$ The lowest common multiple of 4 and 3 is 12, so use 12 as the common denominator.

$= \frac{5}{12}$

6 THE FOUR OPERATIONS

c $2\frac{3}{5} + 1\frac{1}{3} = \frac{13}{5} + \frac{4}{3}$ First, change the mixed numbers to improper fractions.

$\phantom{2\frac{3}{5} + 1\frac{1}{3}} = \frac{39}{15} + \frac{20}{15}$ The lowest common multiple of 5 and 3 is 15.

$\phantom{2\frac{3}{5} + 1\frac{1}{3}} = \frac{59}{15} = 3\frac{14}{15}$ The result is an improper fraction, so write it as a mixed number.

d $3\frac{1}{4} - 1\frac{5}{6} = \frac{13}{4} - \frac{11}{6}$ First, change the mixed numbers to improper fractions.

$\phantom{3\frac{1}{4} - 1\frac{5}{6}} = \frac{39}{12} - \frac{22}{12}$ The lowest common multiple of 4 and 6 is 12.

$\phantom{3\frac{1}{4} - 1\frac{5}{6}} = \frac{17}{12} = 1\frac{5}{12}$ The result is an improper fraction, so write it as a mixed number.

> **Note**
> You should always give your final answer as a fraction in its simplest form.

Exercise 6.9

1 Add these fractions.
 a $\frac{2}{7} + \frac{4}{7}$ b $\frac{1}{3} + \frac{1}{6}$ c $\frac{2}{3} + \frac{1}{4}$
 d $\frac{1}{5} + \frac{3}{4}$ e $\frac{3}{8} + \frac{1}{5}$ f $\frac{3}{4} + \frac{2}{5}$

2 Subtract these fractions.
 a $\frac{2}{7} - \frac{1}{7}$ b $\frac{5}{6} - \frac{1}{3}$ c $\frac{2}{3} - \frac{1}{4}$
 d $\frac{11}{12} - \frac{2}{3}$ e $\frac{5}{8} - \frac{1}{3}$ f $\frac{7}{9} - \frac{5}{12}$

3 Add these fractions. Write your answers as simply as possible.
 a $\frac{3}{10} + 1\frac{2}{5}$ b $1\frac{1}{10} + \frac{3}{5}$ c $2\frac{1}{5} + 1\frac{1}{10}$ d $4\frac{1}{2} + 2\frac{3}{10}$ e $1\frac{3}{10} + \frac{2}{5}$

4 Subtract these fractions. Write your answers as simply as possible.
 a $5\frac{1}{10} - \frac{7}{10}$ b $1\frac{5}{6} - \frac{2}{3}$ c $1\frac{1}{5} - \frac{7}{10}$ d $1\frac{1}{2} - \frac{3}{4}$ e $2\frac{7}{10} - 1\frac{4}{5}$

5 Work out these.
 a $3\frac{1}{2} + 2\frac{1}{5}$ b $4\frac{7}{8} - 1\frac{3}{4}$ c $6\frac{5}{12} - 3\frac{1}{3}$ d $4\frac{3}{4} + 2\frac{5}{8}$ e $5\frac{5}{6} - 1\frac{1}{4}$

6 Work out these.
 a $4\frac{7}{9} + 2\frac{5}{6}$ b $4\frac{7}{13} - 4\frac{1}{2}$ c $7\frac{2}{5} - 1\frac{3}{4}$ d $5\frac{2}{7} - 3\frac{1}{2}$ e $4\frac{1}{12} - 3\frac{1}{4}$

Multiplying fractions

Multiplying a fraction by a whole number

In this diagram, $\frac{1}{8}$ is red. 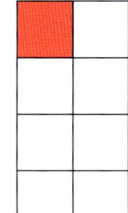 In this diagram, five times as much is red, so $\frac{1}{8} \times 5 = \frac{5}{8}$.

This shows that to multiply a fraction by an integer, you multiply the numerator by the integer. Then simplify by cancelling and changing to a mixed number if possible.

Multiplying a fraction by a fraction

In this diagram, $\frac{1}{4}$ is red.

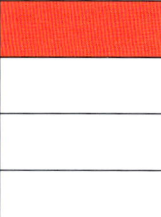

In this diagram, $\frac{1}{3}$ of the red is dotted.

This is $\frac{1}{12}$ of the original rectangle.

So $\frac{1}{3} \times \frac{1}{4} = \frac{1}{12}$.

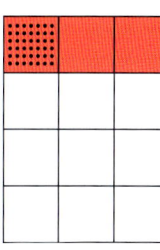

> To multiply fractions, multiply the numerators and multiply the denominators, then simplify if possible.

6 THE FOUR OPERATIONS

Example 6.12

Question
Work out these.

a $\frac{7}{12} \times 2$ 	 b $\frac{3}{4} \times \frac{6}{7}$ 	 c $2\frac{2}{3} \times 1\frac{5}{6}$

Solution

a $\frac{7}{12} \times 2 = \frac{14}{12}$ Multiply the numerator by 2.

   $= \frac{7}{6} = 1\frac{1}{6}$ Simplify the fraction and write it as a mixed number.

b $\frac{3}{\cancel{4}_2} \times \frac{\cancel{6}^3}{7} = \frac{3}{2} \times \frac{3}{7}$ Cancel the common factor of 2 in the numerator and the denominator.

   $= \frac{9}{14}$ The result is a proper fraction in its simplest form.

c $2\frac{2}{3} \times 1\frac{5}{6} = \frac{8}{3} \times \frac{11}{6}$ First change the mixed numbers to improper fractions.

   $= \frac{\cancel{8}^4}{3} \times \frac{11}{\cancel{6}_3}$ Cancel the common factor of 2 in the numerator and the denominator.

   $= \frac{4}{3} \times \frac{11}{3}$ Multiply the numerators and multiply the denominators.

   $= \frac{44}{9} = 4\frac{8}{9}$ The result is an improper fraction, so write it as a mixed number.

> **Note**
> Cancelling common factors before multiplying makes the arithmetic simpler. If you multiply first, you may have to cancel to give a fraction in its simplest form.

Exercise 6.10

Work out these.

Write each answer as an integer, a proper fraction or a mixed number in its simplest form.

1 $\frac{1}{2} \times 4$ 2 $7 \times \frac{1}{2}$ 3 $9 \times \frac{1}{3}$ 4 $\frac{3}{4} \times 12$ 5 $\frac{2}{5} \times 5$

6 $24 \times \frac{5}{12}$ 7 $\frac{2}{3} \times 4$ 8 $\frac{4}{9} \times 2$ 9 $\frac{4}{5} \times 3$ 10 $\frac{1}{5} \times 3$

11 $\frac{1}{4} \times \frac{2}{3}$ 12 $\frac{2}{3} \times \frac{3}{5}$ 13 $\frac{4}{9} \times \frac{1}{2}$ 14 $\frac{1}{3} \times \frac{2}{3}$ 15 $\frac{5}{6} \times \frac{3}{5}$

16 $\frac{3}{7} \times \frac{7}{9}$ 17 $\frac{1}{2} \times \frac{5}{6}$ 18 $\frac{3}{10} \times \frac{5}{11}$ 19 $\frac{2}{3} \times \frac{5}{8}$ 20 $\frac{3}{5} \times \frac{5}{12}$

21 $4\frac{1}{2} \times 2\frac{1}{6}$ 22 $1\frac{1}{2} \times 3\frac{2}{3}$ 23 $4\frac{1}{5} \times 1\frac{2}{3}$ 24 $3\frac{1}{3} \times 2\frac{2}{5}$ 25 $3\frac{1}{5} \times 1\frac{2}{3}$

Dividing fractions

When you work out 6 ÷ 3, you are finding how many 3s there are in 6.

When you work out $6 \div \frac{1}{3}$, you are finding out how many $\frac{1}{3}$s there are in 6.

In this diagram, $\frac{1}{3}$ of the rectangle is red.

This diagram shows 6 of these rectangles.

You can see that 18 of the red squares will fit into this diagram.

So $6 \div \frac{1}{3} = 6 \times 3 = 18$

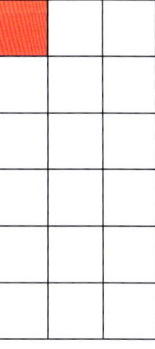

You can see that dividing by $\frac{1}{3}$ is the same as multiplying by 3.

$\frac{1}{3}$ is known as the **reciprocal** of 3.

This can be extended to division by a non-unit fraction.

For example, when you work out $4 \div \frac{2}{3}$, you are finding out how many $\frac{2}{3}$s there are in 4.

In this diagram, $\frac{2}{3}$ of the rectangle is red.

This diagram shows 4 of these rectangles.

You can see that 6 of these 2-square shapes will fit into this diagram.

You can think of the calculation in two steps.

$4 \div \frac{1}{3} = 4 \times 3 = 12$ (there are 12 small squares in the 4 rectangles).

So, $\quad 4 \div \frac{2}{3} = 12 \div 2 = 6$

Or, $\quad 4 \div \frac{2}{3} = 4 \times \frac{3}{2} = \frac{12}{2} = 6$.

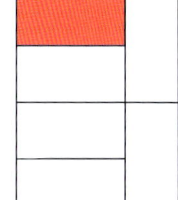

You can see that dividing by $\frac{2}{3}$ is the same as multiplying by $\frac{3}{2}$.

$\frac{2}{3}$ is known as the reciprocal of $\frac{3}{2}$.

To find the reciprocal of any fraction, turn it upside down.

The reciprocal of $\frac{a}{b}$ is $\frac{b}{a}$.

> Use this method for dividing by a fraction:
> - find the reciprocal of the fraction you are dividing by
> - multiply by the reciprocal.

6 THE FOUR OPERATIONS

Example 6.13

Question
Work out these.

a $\quad \frac{8}{9} \div 2$
b $\quad \frac{9}{10} \div \frac{3}{4}$
c $\quad 2\frac{3}{4} \div 1\frac{5}{8}$

Solution

a $\quad \frac{8}{9} \div 2 = \frac{8}{9} \div \frac{2}{1}$ — 2 is the same as $\frac{2}{1}$

$\qquad = \frac{8}{9} \times \frac{1}{2}$ — The reciprocal of $\frac{2}{1}$ is $\frac{1}{2}$, so multiply by this reciprocal.

$\qquad = \frac{\cancel{8}^4}{9} \times \frac{1}{\cancel{2}_1}$ — Cancel the common factor of 2 in the numerator and the denominator.

$\qquad = \frac{4}{9} \times 1$

$\qquad = \frac{4}{9}$

b $\quad \frac{9}{10} \div \frac{3}{4} = \frac{9}{10} \times \frac{4}{3}$ — The reciprocal of $\frac{3}{4}$ is $\frac{4}{3}$, so multiply by this reciprocal.

$\qquad = \frac{\cancel{9}^3}{\cancel{10}_5} \times \frac{\cancel{4}^2}{\cancel{3}_1}$ — Cancel the common factors of 2 and 3 in the numerator and the denominator.

$\qquad = \frac{3}{5} \times \frac{2}{1}$

$\qquad = \frac{6}{5} = 1\frac{1}{5}$ — The result is an improper fraction, so write it as a mixed number.

c $\quad 2\frac{3}{4} \div 1\frac{5}{8} = \frac{11}{4} \div \frac{13}{8}$ — First, change the mixed numbers to improper fractions.

$\qquad = \frac{11}{4} \times \frac{8}{13}$ — The reciprocal of $\frac{13}{8}$ is $\frac{8}{13}$, so multiply by this reciprocal.

$\qquad = \frac{11}{\cancel{4}_1} \times \frac{\cancel{8}^2}{13}$ — Cancel the common factor of 4 in the numerator and the denominator.

$\qquad = \frac{22}{13} = 1\frac{9}{13}$ — The result is an improper fraction, so write it as a mixed number.

Note
Remember that when you have a fraction calculation involving mixed numbers, you should first change them to improper fractions.
Always check that your final answer is given as a mixed number in its simplest form.

Exercise 6.11

Work out these.

Write each answer as an integer, a proper fraction or a mixed number in its simplest form.

1 $9 \div \frac{1}{3}$
2 $\frac{3}{4} \div \frac{1}{2}$
3 $\frac{2}{3} \div 3$
4 $\frac{1}{3} \div \frac{3}{4}$
5 $\frac{4}{9} \div 2$
6 $\frac{4}{5} \div 4$
7 $\frac{2}{3} \div \frac{1}{3}$
8 $\frac{5}{6} \div 10$
9 $\frac{7}{9} \div \frac{1}{9}$
10 $\frac{4}{5} \div \frac{3}{10}$
11 $\frac{1}{4} \div \frac{3}{8}$
12 $\frac{2}{5} \div \frac{1}{5}$
13 $\frac{2}{5} \div \frac{7}{10}$
14 $2\frac{1}{3} \div 1\frac{1}{3}$
15 $2\frac{2}{5} \div 1\frac{1}{2}$
16 $\frac{3}{18} \div 1\frac{1}{4}$
17 $2\frac{1}{4} \div 3\frac{1}{2}$
18 $\left(3\frac{1}{2} + 2\frac{4}{5}\right) \times 2\frac{1}{2}$
19 $5\frac{1}{3} \div 3\frac{3}{5} + 2\frac{1}{3}$
20 $\left(2\frac{4}{5} + 3\frac{1}{4}\right) \div \left(3\frac{2}{3} - 2\frac{3}{4}\right)$

Dividing fractions

Key points

- Negative numbers are less than zero.
- When multiplying and dividing numbers, an even number of negative numbers gives a positive answer. An odd number gives a negative answer.
- The correct order of operations is: brackets; powers; multiplication and division; addition and subtraction.
- When multiplying decimals, the number of decimal places in the answer is equal to the total number of decimal places in the numbers being multiplied.
- To divide decimals, write each as a fraction. Multiply both the numerator and denominator by a power of 10 to make the denominator an integer. Now divide.
- To add or subtract fractions, replace the fractions with equivalents with the same denominator. Then, add or subtract the numerators.
- To multiply fractions, multiply the numerators and multiply the denominators.
- To divide fractions, multiply the first fraction by the reciprocal of the fraction you are dividing by.

REVIEW EXERCISE 1

1 **a** Express 36 as the product of its prime factors. [1]
   b Find the highest common factor (HCF) of 36 and 120. [2]
   c Find the lowest common multiple (LCM) of 36 and 120. [1]

2 **a** Express 99 as the product of prime factors. [1]
   b Expressed as the product of prime factors,
   $$p = 2^{n+2} \times 3^n \times 5 \quad \text{and} \quad q = 2^n \times 3^{n+1} \times 5^2$$
   where n is a positive integer.
   i The lowest common multiple (LCM) of p and q is $2^n \times 3^n \times R$.
   Express R as the product of prime factors. [2]
   ii Express $p + q$ as the product of prime factors. [2]

Cambridge O Level Mathematics Syllabus D (4024) Paper 11 Q24, November 2020

3 In a group of 35 people,
22 are wearing spectacles,
10 are wearing a hat,
6 are wearing spectacles and a hat.
By drawing a Venn diagram, or otherwise, find the number of people who are wearing neither spectacles nor a hat. [2]

Cambridge O Level Mathematics Syllabus D (4024) Paper 12 Q7, November 2018

4 **a** $\mathscr{E} = \{x : x \text{ is an integer } 1 \leq x \leq 16\}$
   $A = \{x : x \text{ is an even number}\}$
   $B = \{x : x \text{ is a square number}\}$
   $C = \{x : x \text{ is a factor of } 100\}$
   i Complete the Venn diagram. [3]

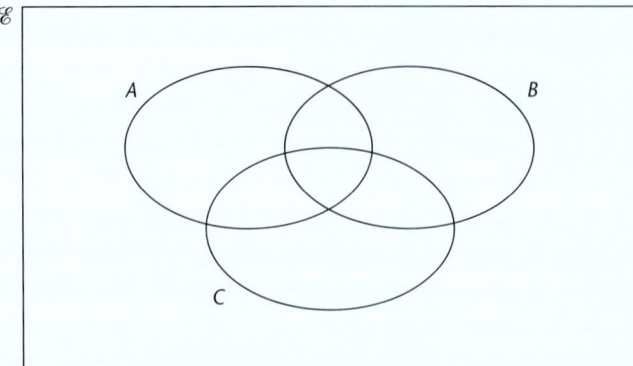

   ii Find $n(A' \cup B)$. [1]
   iii $p \in A \cap C$
   Write down all the possible values of p. [1]

Cambridge O Level Mathematics Syllabus D (4024) Paper 22 Q5, November 2019

Review exercise 1

5 √7   17   25   27   √36

Write down from this list
a   a prime number [1]
b   a cube number [1]
c   a square number [1]

6   Write $0.3\dot{6}$ as a fraction in its simplest form. [3]

7   Write $\frac{2}{11}$ as a decimal. [1]

8   Write these numbers in order of size, starting with the smallest.

$\frac{3}{4}$ 0.83 $\frac{17}{20}$ 82%   0.8 [2]

Cambridge O Level Mathematics Syllabus D (4024) Paper 11 Q2, November 2021

9   The table shows some information about the temperatures in a city.

Date	Maximum temperature	Minimum temperature
1 February	−10 °C	T °C
1 March	4 °C	−5 °C

a   Find the difference between the maximum and minimum temperatures on 1 March. [1]
b   The minimum temperature, T °C, on 1 February was 13 degrees lower than the minimum temperature on 1 March.
Find T. [1]

Cambridge O Level Mathematics Syllabus D (4024) Paper 12 Q5, June 2016

10 Work out these.
a $-7 + 3 \times 4$ [1]
b $8 \div (7 - 9)$ [1]

11 a   Work out $\frac{3}{7} + \frac{2}{5}$. [1]
    b   Find $\frac{2}{3}$ of $\frac{6}{11}$, giving your answer as a fraction in its simplest form. [1]

Cambridge O Level Mathematics Syllabus D (4024) Paper 11 Q3, June 2021

12 Work out these.
a 0.04×0.05 [1]
b $4.2 \div 0.02$ [1]

7 INDICES 1

BY THE END OF THIS CHAPTER YOU WILL BE ABLE TO:
- understand and use indices (positive, zero, negative and fractional)
- understand and use the rules of indices.

CHECK YOU CAN:
- work with squares, cubes, square roots and cube roots
- write a fraction in its simplest form.

Simplifying numbers with indices

Indices (or powers) are a form of mathematical shorthand.

You know that 4^2 means 4×4.

The notation 4^2 is known as **index notation**.

2 is the **index** and 4 is the **base**.

Note
The plural of 'index' is 'indices'.

Index notation can be used as shorthand for any power of a number.

$3 \times 3 \times 3 \times 3$ can be written as 3^4 using index notation.

$2 \times 2 \times 2 \times 2 \times 2 \times 2 \times 2 \times 2$ can be written as 2^8 using index notation.

Example 7.1

Question

Write each of these using index notation.
a $5 \times 5 \times 5 \times 5$
b $6 \times 6 \times 6 \times 4 \times 6 \times 4 \times 6 \times 4$

Solution

a $5 \times 5 \times 5 \times 5 = 5^4$
   5 is multiplied 4 times, so the index is 4 and the base is 5.
b $6 \times 6 \times 6 \times 4 \times 6 \times 4 \times 6 \times 4 = 6^5 \times 4^3$
   6 is multiplied 5 times so the index is 5 and the base is 6.
   4 is multiplied 3 times so the index is 3 and the base is 4.
   The two terms with different bases cannot be combined.

Exercise 7.1

1 Write these using index notation.
   a $6 \times 6 \times 6 \times 6$
   b $7 \times 7 \times 7$
   c $8 \times 8 \times 8 \times 8 \times 8$
   d $4 \times 4 \times 4 \times 4$
   e $2 \times 2 \times 2 \times 2 \times 2 \times 2$
   f 10×10
2 Write these using index notation.
   a $5 \times 4 \times 4 \times 4 \times 5$
   b $3 \times 3 \times 5 \times 5 \times 5$
   c $2 \times 2 \times 2 \times 3 \times 3 \times 4 \times 4 \times 4 \times 4$
   d $7 \times 7 \times 7 \times 8 \times 8 \times 9 \times 9 \times 9$
3 Work out the value of these.
   a 4^3
   b 2^6
   c 3^4
   d 10^5

Multiplying numbers in index form

When calculations involve more than one term with the same base, they can be simplified.

$3^4 \times 3^8 = (3 \times 3 \times 3 \times 3) \times (3 \times 3 \times 3 \times 3 \times 3 \times 3 \times 3 \times 3)$

$ = 3 \times 3 \times 3 \times 3 \times 3 \times 3 \times 3 \times 3 \times 3 \times 3 \times 3 \times 3$

$ = 3^{12}$

The indices are added.

$3^4 \times 3^8 = 3^{4+8}$

$ = 3^{12}$

This gives a rule for multiplying numbers in index form.

$n^a \times n^b = n^{a+b}$

$(3^4)^3 = 3^4 \times 3^4 \times 3^4$

$ = 3^{4+4+4}$

$ = 3^{3 \times 4}$

$ = 3^{12}$

This gives the rule

$(n^a)^b = n^{a \times b}$

Example 7.2

Question

Write these in a simpler form using indices.

a $4^3 \times 4^8$ b $9^2 \times 9 \times 9^4$ c $2^5 \times 3^2 \times 2^4$ d $(2^3)^2$

Solution

a $4^3 \times 4^8 = 4^{3+8} = 4^{11}$ — The indices are added.

b $9^2 \times 9 \times 9^4 = 9^{2+1+4}$ — Note that 9 is the same as 9^1.

$ = 9^7$ — The bases are all the same so the three indices are added.

c $2^5 \times 3^2 \times 2^4 = 2^{5+4} \times 3^2$ — Only terms with the same base can be combined.

$ = 2^9 \times 3^2$ — There are two terms in the final answer.

d $(2^3)^2 = 2^{3 \times 2}$

$ = 2^6$

Photocopying is prohibited

7 INDICES 1

Exercise 7.2

Write these in a simpler form using indices.

1. $5^2 \times 5^3$
2. $6^2 \times 6^7$
3. $10^3 \times 10^4$
4. $3^6 \times 3^5$
5. $8^3 \times 8^2$
6. $4^2 \times 4^3$
7. $9^2 \times 9^7$
8. $6^2 \times 6^6$
9. $4^2 \times 4^5 \times 4$
10. $5^2 \times 6^2 \times 5^5$
11. $2^5 \times 3 \times 2^4 \times 3^7$
12. $7 \times 8^2 \times 7^3 \times 8^3 \times 7^4$
13. $(5^2)^4$
14. $(6^3)^3$
15. $(8^4)^2$

Dividing numbers in index form

Divisions can be simplified by cancelling common factors.

$$2^6 \div 2^4 = \frac{\cancel{2} \times \cancel{2} \times \cancel{2} \times \cancel{2} \times 2 \times 2}{\cancel{2} \times \cancel{2} \times \cancel{2} \times \cancel{2}}$$

$$= 2 \times 2$$

$$= 2^2$$

The indices are subtracted.

$$2^6 \div 2^4 = 2^{6-4}$$

$$= 2^2$$

This gives a rule for dividing numbers in index form.

$$n^a \div n^b = n^{a-b}$$

Example 7.3

Question

Write these in a simpler form using indices.

a $8^6 \div 8^2$ b $\dfrac{7^9}{7 \times 7^3}$

Solution

a $8^6 \div 8^2 = 8^{6-2} = 8^4$ The indices are subtracted.

b $\dfrac{7^9}{7 \times 7^3} = \dfrac{7^9}{7^4}$ First simplify the denominator by adding the indices.

$\quad\quad\quad = 7^{9-4}$ Then subtract the indices.

$\quad\quad\quad = 7^5$

Exercise 7.3

1. Write these in a simpler form using indices.
   a $10^5 \div 10^2$
   b $3^5 \div 3^2$
   c $8^4 \div 8^2$
   d $7^5 \div 7^3$
   e $6^3 \div 6^2$

2. Work out these, giving your answers in index form.
   a $\dfrac{3^9}{3^4 \times 3^2}$
   b $\dfrac{2^4 \times 2^3}{2^5}$
   c $\dfrac{5^4 \times 5^5}{5^2 \times 5^3}$
   d $\dfrac{4^{12}}{4^5 \times 4^4}$
   e $\dfrac{2^5 \times 2^6}{2^4}$
   f $\dfrac{6^5 \times 6^4}{6^2 \times 6}$

Negative indices

The index laws can also be used for numbers with negative indices.

Using the laws of indices

$5^0 \div 5^3 = 5^{0-3} = 5^{-3}$.

You also know that $5^0 = 1$, so

$5^0 \div 5^3 = 1 \div 5^3 = \frac{1}{5^3}$.

It follows that $5^{-3} = \frac{1}{5^3}$.

This gives a general law:

$$a^{-n} = \frac{1}{a^n}$$

Example 7.4

Question

Write these as fractions.

a 6^{-2} b 16^{-1} c n^{-3} d $\left(\frac{1}{4}\right)^{-2}$

Solution

a $6^{-2} = \frac{1}{6^2}$ b $16^{-1} = \frac{1}{16^1}$ c $n^{-3} = \frac{1}{n^3}$ d $\left(\frac{1}{4}\right)^{-2} = 4^2 = 16$
  $= \frac{1}{36}$ $= \frac{1}{16}$

Exercise 7.4

1 Write each of these as an integer or a fraction.
   a 6^{-1} b 3^{-2} c 1^{-5}
   d 5^{-2} e 10^{-3}

2 Write each of these as an integer or a fraction.
   a $\left(\frac{1}{2}\right)^{-1}$ b $\left(\frac{1}{3}\right)^{-2}$ c $\left(\frac{2}{3}\right)^{-1}$
   d $\left(\frac{3}{4}\right)^{-2}$ e $\left(\frac{2}{5}\right)^{-3}$

3 Evaluate these.
   a $3^4 \times 3^{-3}$ b $5^2 \times 5^{-1}$ c $6^8 \times 6^{-6}$
   d $2^{-3} \times 2^0$

7 INDICES 1

Fractional indices

Is there an index for a square or cube root?

Suppose that $a^b = \sqrt[3]{a}$.

Then $(a^b)^3 = \left(\sqrt[3]{a}\right)^3$ Cube both sides.

The cube of the cube root of a is equal to a, so the right-hand side of the equation is equivalent to a.

$a^{3b} = a = a^1$ Use the index law for powers on the left-hand side of the equation.

So $3b = 1$ Equate the powers.

$b = \frac{1}{3}$

So $a^{\frac{1}{3}} = \sqrt[3]{a}$.

This can be extended to give a general law:

$$a^{\frac{1}{n}} = \sqrt[n]{a}$$

The index laws can be combined to work with more complex fractional powers.

For example, $a^{\frac{3}{2}} = \left(a^{\frac{1}{2}}\right)^3 = \left(\sqrt{a}\right)^3$

Also $a^{\frac{3}{2}} = (a^3)^{\frac{1}{2}} = \sqrt{a^3}$

A scientific calculator has a power key. It may look like this $\boxed{x^\bullet}$.

Find out how to use this key on your calculator to evaluate fractional and negative indices.

Example 7.5

Question

a Write $\sqrt[5]{n}$ using index notation.

b Evaluate $16^{\frac{1}{4}}$.

c Evaluate $25^{-\frac{1}{2}}$.

d Evaluate $125^{\frac{2}{3}}$.

Solution

a $\sqrt[5]{n} = n^{\frac{1}{5}}$

b $16^{\frac{1}{4}} = 2$ because $2^4 = 16$ so $16^{\frac{1}{4}} = (2^4)^{\frac{1}{4}} = 2^1 = 2$ Check this result on your calculator.

c $25^{-\frac{1}{2}} = \left(25^{\frac{1}{2}}\right)^{-1} = 5^{-1} = \frac{1}{5}$ Check this result on your calculator.

d $125^{\frac{2}{3}} = \left(\sqrt[3]{125}\right)^2 = 5^2 = 25$ Check this result on your calculator.

Note

Either the cube root or the square can be calculated first, the final result will be the same. It is often easier to work out the root first.

Fractional indices

Exercise 7.5

1. Write each of these in index form.
   a. the cube root of n
   b. $\sqrt[6]{x}$
   c. $\sqrt[3]{m^5}$
   d. $\sqrt[5]{p^4}$

Work out these. Give your answers as whole numbers or fractions.

2. a. 4^{-1} b. $4^{\frac{1}{2}}$ c. 4^0 d. 4^{-2} e. $4^{\frac{3}{2}}$

3. a. $8^{\frac{1}{3}}$ b. 8^{-1} c. $8^{\frac{4}{3}}$ d. $\left(\frac{1}{8}\right)^{-2}$ e. 8^1

4. a. 9^{-1} b. $9^{\frac{1}{2}}$ c. 9^0 d. 9^{-2} e. $9^{\frac{3}{2}}$

5. a. $27^{\frac{1}{3}}$ b. $27^{\frac{4}{3}}$ c. 27^{-1} d. $\left(\frac{1}{27}\right)^{\frac{1}{3}}$ e. 27^0

6. a. $64^{\frac{1}{2}}$ b. $64^{-\frac{1}{3}}$ c. 64^0 d. $64^{-\frac{2}{3}}$ e. $64^{\frac{5}{6}}$

7. a. $16^{\frac{1}{2}}$ b. $16^{-\frac{1}{4}}$ c. 16^0 d. $16^{\frac{3}{2}}$ e. $16^{-\frac{7}{4}}$

8. a. $3^0 + 3^3$ b. $4 + 4^{\frac{1}{2}}$ c. $2^{-1} - 4^{-1}$ d. $36^{\frac{1}{2}} - 36^0$

9. a. $2^2 \times 9^{\frac{1}{2}}$ b. $2^5 \times 8^{\frac{1}{3}}$ c. $81^{\frac{1}{4}} \times 3^{-2}$ d. $9^{\frac{1}{2}} \times 6^2 \times 4^{-1}$

10. a. $2^2 + 3^0 + 16^{\frac{1}{2}}$ b. $\left(\frac{3}{4}\right)^{-2} \times 27^{\frac{2}{3}}$ c. $4^2 \div 9^{\frac{1}{2}}$ d. $4^2 - 8^{\frac{1}{3}} + 9^0$

Use your calculator to work out these.
Give the answers as whole numbers or fractions.

11. a. $25^{\frac{3}{2}}$ b. $36^{\frac{1}{2}}$ c. $125^{\frac{2}{3}} \times 8^{\frac{2}{3}}$ d. $49^{\frac{3}{2}} \times 81^{-\frac{1}{4}}$

12. a. $5^{-2} \times 10^5 \times 16^{-\frac{1}{4}}$ b. $5^3 - 25^{\frac{1}{2}} - \left(\frac{2}{5}\right)^{-2}$
    c. $\left(\frac{4}{5}\right)^2 \times 128^{-\frac{3}{7}}$ d. $125^{-\frac{1}{3}} - 121^{\frac{1}{2}} + 216^{\frac{1}{3}}$

Key points
- Index notation is a mathematical shorthand with a base number and index (power).
- To multiply numbers with the same base, add the indices.
- To divide numbers with the same base, subtract the indices.
- The reciprocal of a number has a negative index.
- Any number with index zero is equal to 1.
- Fractional indices represent roots.

8 STANDARD FORM

BY THE END OF THIS CHAPTER YOU WILL BE ABLE TO:
- use the standard form $A \times 10^n$ where n is a positive or negative integer and $1 \leq A < 10$
- convert numbers into and out of standard form
- calculate with values in standard form.

CHECK YOU CAN:
- understand and use index notation.

Note
You can write down the answer without any intermediate steps.
Move the decimal point until the number is between 1 and 10.
Count the number of places the point has moved: that is the power of 10.

Standard form

Standard form is a way of making very large numbers and very small numbers easy to deal with.

In standard form, numbers are written as a number between 1 and 10 multiplied by a power of 10, for example 6.3×10^5.

Large numbers

Example 8.1

Write these numbers in standard form.

Question	Solution
a 500 000	a $500\,000 = 5 \times 100\,000 = 5 \times 10^5$
b 6 300 000	b $6\,300\,000 = 6.3 \times 1\,000\,000 = 6.3 \times 10^6$
c 45 600	c $45\,600 = 4.56 \times 10\,000 = 4.56 \times 10^4$

Small numbers

Example 8.2

Question

Write these numbers in standard form.

a 0.000 003   b 0.000 056   c 0.000 726

Note
You can write down the answer without any intermediate steps.
Move the decimal point until the number is between 1 and 10.
Count the number of places the point has moved, put a minus sign in front and that is the power of 10.

Solution

a $0.000\,003 = \frac{3}{1\,000\,000} = 3 \times \frac{1}{1\,000\,000} = 3 \times \frac{1}{10^6} = 3 \times 10^{-6}$

b $0.000\,056 = \frac{5.6}{100\,000} = 5.6 \times \frac{1}{100\,000} = 5.6 \times \frac{1}{10^5} = 5.6 \times 10^{-5}$

c $0.000\,726 = \frac{7.26}{10\,000} = 7.26 \times \frac{1}{10\,000} = 7.26 \times \frac{1}{10^4} = 7.26 \times 10^{-4}$

Exercise 8.1

1. Write these numbers in standard form.
   - a 7000
   - b 84 000
   - c 563
   - d 6 500 000
   - e 723 000
   - f 27
   - g 53 400
   - h 693
   - i 4390
   - j 412 300 000
   - k 8 million
   - l 39.2 million

2. Write these numbers in standard form.
   - a 0.003
   - b 0.056
   - c 0.0008
   - d 0.000 006 3
   - e 0.000 082
   - f 0.0060
   - g 0.000 000 38
   - h 0.78
   - i 0.003 69
   - j 0.000 658
   - k 0.000 000 000 56
   - l 0.000 007 23

3. These numbers are in standard form. Write them as ordinary numbers.
   - a 5×10^4
   - b 3.7×10^5
   - c 7×10^{-4}
   - d 6.9×10^6
   - e 6.1×10^{-3}
   - f 4.73×10^4
   - g 2.79×10^7
   - h 4.83×10^{-5}
   - i 1.03×10^{-2}
   - j 9.89×10^8
   - k 2.61×10^{-6}
   - l 3.7×10^2
   - m 3.69×10^3
   - n 6.07×10^{-4}
   - o 5.48×10^{-7}
   - p 1.98×10^9

4. A billion is a thousand million.
   In 2015, the population of the world was approximately 7.2 billion.
   Write the population of the world in standard form.

Calculating with numbers in standard form

When you need to multiply or divide numbers in standard form, you can use your knowledge of the laws of indices.

Example 8.3

Question

Work out these. Give your answers in standard form.

a $(7 \times 10^3) \times (4 \times 10^4)$

b $(3 \times 10^8) \div (5 \times 10^3)$

Solution

a $(7 \times 10^3) \times (4 \times 10^4) = 7 \times 4 \times 10^3 \times 10^4$
$= 28 \times 10^7$
$= 2.8 \times 10^8$

b $(3 \times 10^8) \div (5 \times 10^3) = \dfrac{3 \times 10^8}{5 \times 10^3}$
$= 0.6 \times 10^5$
$= 6 \times 10^4$

Note

$10^3 \times 10^4 = 1000 \times 10000$
$= 10 000 000$
$= 10^7$

Note

$\dfrac{3}{5} = 0.6$ and

$\dfrac{10^8}{10^3} = \dfrac{100\,000\,000}{1\,000}$
$= 100 000$
$= 10^5$

When you need to add or subtract numbers in standard form it is much safer to change to ordinary numbers first.

8 STANDARD FORM

Example 8.4

Question

Work out these. Give your answers in standard form.

a $(7 \times 10^3) + (1.4 \times 10^4)$
b $(7.2 \times 10^5) + (2.5 \times 10^4)$
c $(5.3 \times 10^{-3}) - (4.9 \times 10^{-4})$

Solution

a
$7\,000$
$+14\,000$
$\overline{21\,000} = 2.1 \times 10^4$

b
$720\,000$
$+25\,000$
$\overline{745\,000} = 7.45 \times 10^5$

c
$0.005\,30$
$-0.000\,49$
$\overline{0.004\,81} = 4.81 \times 10^{-3}$

Exercise 8.2

 1 Work out these. Give your answers in standard form.

a $(4 \times 10^3) \times (2 \times 10^4)$
b $(6 \times 10^7) \times (2 \times 10^3)$
c $(7 \times 10^3) \times (8 \times 10^2)$
d $(9 \times 10^7) \div (3 \times 10^4)$
e $(4 \times 10^3) \times (1.3 \times 10^4)$
f $(4.8 \times 10^3) \div (1.2 \times 10^{-2})$
g $(8 \times 10^6) \times (9 \times 10^{-2})$
h $(4 \times 10^8) \div (8 \times 10^2)$
i $(7 \times 10^{-4}) \times (8 \times 10^{-3})$
j $(5 \times 10^{-5}) \div (2 \times 10^4)$
k $(4 \times 10^3) + (6 \times 10^4)$
l $(7 \times 10^6) - (3 \times 10^3)$
m $(6.2 \times 10^5) - (3.7 \times 10^4)$
n $(4.2 \times 10^9) + (3.6 \times 10^8)$
o $(7.2 \times 10^6) - (4.2 \times 10^4)$
p $(7.8 \times 10^{-5}) + (6.1 \times 10^{-4})$

2 Work out these. Give your answers in standard form.

a $(6.2 \times 10^5) \times (3.8 \times 10^7)$
b $(6.3 \times 10^7) \div (4.2 \times 10^2)$
c $(6.67 \times 10^8) \div (4.6 \times 10^{-3})$
d $(3.7 \times 10^{-4}) \times (2.9 \times 10^{-3})$
e $(1.69 \times 10^8) \div (5.2 \times 10^3)$
f $(5.8 \times 10^5) \times (3.5 \times 10^3)$
g $(5.2 \times 10^6)^2$
h $(3.1 \times 10^{-4})^2$
i $(3.72 \times 10^6) - (2.8 \times 10^4)$
j $(7.63 \times 10^5) + (3.89 \times 10^4)$
k $(5.63 \times 10^{-3}) - (4.28 \times 10^{-4})$
l $(6.72 \times 10^{-3}) + (2.84 \times 10^{-5})$
m $(4.32 \times 10^{-5}) - (4.28 \times 10^{-3})$
n $(7.28 \times 10^8) + (3.64 \times 10^6)$

Key points

- A number written in standard form looks like $A \times 10^n$, where $1 \leq A < 10$ and n is a positive or negative integer.
- Without a calculator, know how to change numbers of any size into standard form and how to change standard form numbers into ordinary numbers.
- When adding or subtracting numbers in standard form without a calculator, change each to an ordinary number, perform the calculation and change the answer back to standard form.
- When multiplying or dividing numbers in standard form without a calculator, combine the numbers and the powers separately. Then, if necessary, change the answer back to standard form.
- Know how to input and read standard form numbers on a calculator.

9 ESTIMATION

Estimating lengths

> **BY THE END OF THIS CHAPTER YOU WILL BE ABLE TO:**
> - round values to a specified degree of accuracy
> - make estimates for calculations involving numbers, quantities and measurements
> - round answers to a reasonable degree of accuracy in the context of a given problem.

To estimate lengths, compare with measures you know.

Example 9.1

Question

Estimate the height of the lamp post.

Solution

7.5 m

Any answer from 6.5 to 8.5 m would be acceptable.

The lamp post is about four times the height of the man.

> **Note**
> The height to an adult person's waist is about 1 m (100 cm) and their full height is about 1.7 m (170 cm).

> **CHECK YOU CAN:**
> - use a protractor and ruler
> - round to the nearest whole number, 10, 100, etc.
> - use a calculator efficiently
> - find the area of a rectangle
> - find the volume of a cuboid.

Exercise 9.1

1 Estimate
   a the height of the fence
   b the length of the fence.

2 Estimate the height of the tree.

> **Note**
> Make sure that you include units with your answer and that they are sensible. For example, the length of a motorway should be in kilometres, not metres.

> **Note**
> When estimating, do not try to be too accurate. What is wanted is a rough estimate and anything about right is acceptable. Decide if it is bigger or smaller than a measure you know and work from there.

3 The car is 3.8 m long. Estimate the length of the truck.

Photocopying is prohibited

9 ESTIMATION

Rounding to a given number of decimal places

Find the digit in the **decimal place** to be rounded. Look at the next digit.

- If that is less than 5, leave the decimal as it is.
- If it is 5 or more, round the decimal up by 1.

You ignore any digits further to the right.

Example 9.2

Question

Write each of these numbers correct to 2 decimal places (2 d.p.).
- **a** 9.368
- **b** 0.0438
- **c** 84.655

Solution

a 9.37 — The digits in the first two decimal places are 36.
The digit in the third decimal place is 8, which is more than 5, so you add 1 to the 6.

b 0.04 — The digits in the first two decimal places are 04.
The digit in the third decimal place is 3, which is less than 5, so you leave the 4 as it is.

c 84.66 — The digit in the third decimal place is 5, so you add 1 to the digit in the second decimal place.

Exercise 9.2

1. Write each of these numbers correct to 1 decimal place.
   - **a** 4.62
   - **b** 5.47
   - **c** 4.58
   - **d** 8.41
   - **e** 0.478
   - **f** 0.1453
   - **g** 82.16
   - **h** 2.97
   - **i** 6.15
   - **j** 0.4512
   - **k** 5.237
   - **l** 48.024
   - **m** 0.8945
   - **n** 7.6666
   - **o** 9.9876

2. Write each of these numbers correct to 2 decimal places.
   - **a** 5.481
   - **b** 12.0782
   - **c** 0.214
   - **d** 0.5666
   - **e** 9.017
   - **f** 78.044
   - **g** 7.0064
   - **h** 0.0734
   - **i** 1.5236
   - **j** 2.1256
   - **k** 9.424
   - **l** 0.8413
   - **m** 0.283
   - **n** 0.851
   - **o** 7.093
   - **p** 18.6306
   - **q** 7.1111
   - **r** 8.081
   - **s** 4.656
   - **t** 3.725

3. Work out these. Give your answers to 2 decimal places.
   - **a** 1.38×6.77
   - **b** 4.7×3.65
   - **c** $9.125 \div 3.1$
   - **d** 16.4×3.29

4. Work out these. Give your answers to 1 decimal place.
   - **a** $1 \div 8$
   - **b** $3 \div 7$
   - **c** $4 \div 11$
   - **d** $5 \div 13$

5. Find the square root of 55. Write this value correct to 1 decimal place.

6. Work out these. Give your answers to 3 decimal places.
   - **a** $1 \div 3$
   - **b** $2 \div 7$
   - **c** $3 \div 11$
   - **d** $4 \div 13$
   - **e** $28 \div 6$

7. Find the mean of these numbers.
   Give your answer correct to 1 decimal place.
   4, 6, 8, 9, 11, 3, 2, 15

Note

The mean of a set of numbers is the sum of the numbers divided by how many numbers there are.

8 Find the mean of these numbers.
   Give the answer to the nearest 10.
   351, 521, 791, 831, 941, 1171, 1351
9 A rectangle is 3.4 cm by 5.2 cm. Work out the area correct to 1 decimal place.
10 A square has sides of 6.35 cm. Work out the area correct to 2 decimal places.
11 A cube has edges of 4.82 cm. Find its volume correct to 1 decimal place.
12 The river Nile is 4169 miles long. 1 mile is approximately 1.6 kilometres. How long is the Nile in kilometres correct to the nearest kilometre?

Rounding to a given number of significant figures

Significant figures are counted from left to right, starting from the first non-zero digit.

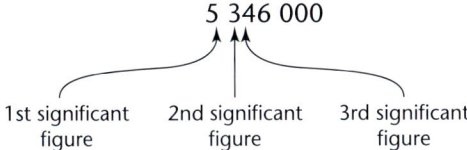

You use a similar technique to the one you use for rounding to a given number of decimal places.

You look at the first digit that is not required. If this is less than 5, you round down. If it is 5 or more, you round up.

Rounding 5 346 000 to

1 significant figure   gives 5 000 000
2 significant figures   gives 5 300 000
3 significant figures   gives 5 350 000

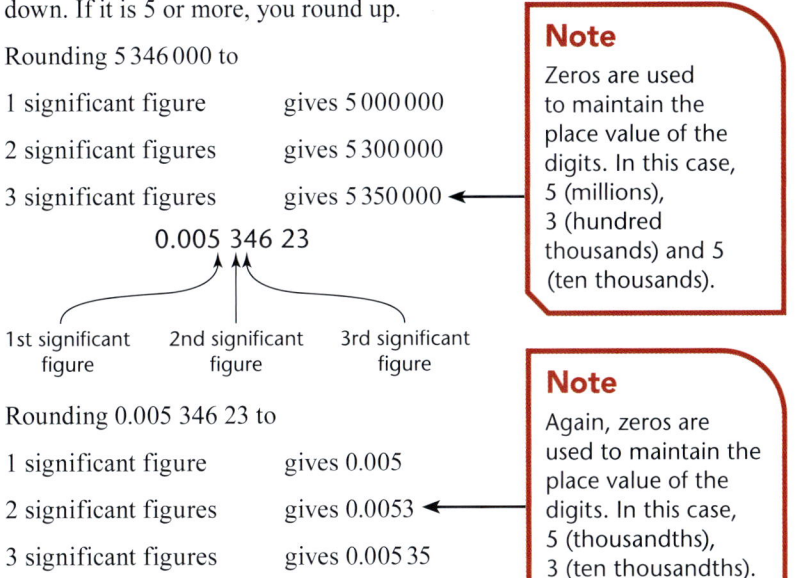

Rounding 0.005 346 23 to

1 significant figure   gives 0.005
2 significant figures   gives 0.0053
3 significant figures   gives 0.005 35

Note
Zeros are used to maintain the place value of the digits. In this case, 5 (millions), 3 (hundred thousands) and 5 (ten thousands).

Note
Again, zeros are used to maintain the place value of the digits. In this case, 5 (thousandths), 3 (ten thousandths).

Note
Zeros to the right of the first non-zero digit *are* significant. For example, 0.070 56 = 0.0706 to 3 significant figures.

9 ESTIMATION

Example 9.3

Question
a Round 52 617 to 2 significant figures.
b Round 0.072 618 to 3 significant figures.
c Round 17 082 to 3 significant figures.

Solution

a 52 617 = 53 000 to 2 s.f.

To round to 2 significant figures, look at the third significant figure.
It is 6, so the second significant figure changes from 2 to 3.
Remember to add zeros for placeholders.

b 0.072 618 = 0.0726 to 3 s.f.

The first significant figure is 7.
To round to 3 significant figures, look at the fourth significant figure.
It is 1, so the third significant figure does not change.

c 17 082 = 17 100 to 3 s.f.

The 0 in the middle here is a significant figure.
To round to 3 significant figures, look at the fourth figure.
It is 8, so the third figure changes from 0 to 1.
Remember to add zeros for placeholders.

> **Note**
> Always state the accuracy of your answers when you have rounded them.

Example 9.4

Question
Write each of these numbers to 1 significant figure.
a 5126   b 5.817   c 0.047 15   d 146 500   e 1968

Solution
a 5000   The second significant figure is 1, so the 5 stays as it is.
            You use three zeros to show the size.
b 6   The second significant figure is 8, so you change the 5 to 6.
            No zeros are needed.
c 0.05   The second significant figure is 7, so the 4 becomes 5.
            You keep the zero before the 5 to show the size.
d 100 000 The second significant figure is 4 so the 1 stays as it is.
            You use zeros to show the size.
e 2000   The second figure is 9, so the 1 changes to 2.

> **Note**
> Zeros are used to show size. A common error is to use the wrong number of zeros.

Exercise 9.3

1. Write each of these to 1 significant figure.
   - a 4.2
   - b 6.45
   - c 7.9
   - d 4.25
   - e 5.6
   - f 62.1
   - g 45.9
   - h 26.5
   - i 314
   - j 4579

2. Write each of these to 1 significant figure.
   - a 4271
   - b 26.45
   - c 910.9
   - d 64.25
   - e 0.002 156
   - f 6.21
   - g 4.59
   - h 0.002 65
   - i 0.003 14
   - j 0.0459

3. Round each of these numbers to 2 significant figures.
   - a 17.6
   - b 184.2
   - c 5672
   - d 97 520
   - e 50.43
   - f 0.172
   - g 0.0387
   - h 0.006 12
   - i 0.0307
   - j 0.994

4. Round each of these numbers to 3 significant figures.
   - a 8.261
   - b 69.77
   - c 16 285
   - d 207.51
   - e 12 524
   - f 7.103
   - g 50.87
   - h 0.4162
   - i 0.038 62
   - j 3.141 59

5. Over 3 days a shop had 4700 customers.
   On average, how many customers were there each day, correct to 2 significant figures?

6. At a soccer match, the attendance was 15 870 men, 10 740 women and 8475 juniors.
   A newspaper reported that there were 35 000 people at the match.
   Explain this.

Estimating answers to problems

To estimate answers to problems, you can round each number to 1 significant figure and then carry out the calculation.

Example 9.5

Question

Jim buys 24 golf balls at $3.75 each.
What is the approximate cost?

Solution

20 × 4 = $80   Round to 1 significant figure.

9 ESTIMATION

Example 9.6

Question

Estimate the result of each of these calculations by rounding to 1 significant figure.

a 3.7×2.1 b 47×63 c 4.32×8.76 d 2.95×0.321 e $0.815 \div 3.41$

Solution

a $4 \times 2 = 8$ b $50 \times 60 = 3000$ c $4 \times 9 = 36$ d $3 \times 0.3 = 0.9$
e $0.8 \div 3 = 0.266 = 0.27$ (2 s.f.) or 0.3 (1 s.f.)

Note

Do not try to be too accurate with your answer. Give the answer to 1 or 2 significant figures.

Example 9.7

Question

Estimate the result of each of these calculations.
In each case, say whether the estimated result is bigger or smaller than the actual result and explain your answer.

a 4.7×3.9 b 516×2.34 c $79.6 \div 23.7$

Solution

a $5 \times 4 = 20$

The estimated result is bigger than the actual result, because both the rounded values used in the calculation are bigger than the actual values.

b $500 \times 2 = 1000$

The estimated result is smaller than the actual result, because both the rounded values used in the calculation are smaller than the actual values.

c $80 \div 20 = 4$

The estimated result is bigger than the actual result, because the rounded value of the number being divided is bigger than the actual value, and the rounded value of the number it is divided by is smaller than the actual value.

Exercise 9.4

In this exercise show the approximations you use to get your estimate.

1 At the school fair, Taha sold 245 ice-creams at 85c each.
   Estimate his takings.
2 A rectangle measures 5.8 cm by 9.4 cm.
   Estimate its area.
3 Salma bought 18 packets of crisps at 32c each.
   Estimate how much she spent.
4 A cuboid has edges of 3.65 cm, 2.44 cm and 2.2 cm.
   Estimate its volume.

5 A rectangle has an area of 63.53 cm² and its length is 9.61 cm.
   Estimate the width of the rectangle.
6 An aeroplane flew 3123 km in 8.4 hours.
   Estimate how far it flew on average in each hour.
7 Pasha went to see her local soccer team 23 times during the season.
   Tickets cost $18.50 for each game.
   Estimate how much she spent.
8 At Duha's ice-cream stall, he sold 223 ice-creams at 67c each.
   Estimate how much money he took.
9 To find the area of a circle of radius 2.68 metres, Rihanna worked out
   $3.14 \times 2.68 \times 2.68$.
   Estimate the result of this calculation.
10 Estimate the result of each of these calculations.
   a 5.89×1.86 b $19.25 \div 3.8$ c 36.87×22.87
   d $9.7 \div 3.5$ e 2.14×0.82 f 3.14×7.92
   g 113.5×2.99 h 4.93×0.025
11 Estimate the result of each of these calculations.
   a $3.6 \times 14.9 \times 21.5$ b 0.89×5.2 c 61.33×11.79
   d $198.5 \times 63.1 \times 2.8$ e 9.87×0.0657 f 0.246×0.789
   g 46×82 h 7.05^2 i $43.7 \times 18.9 \times 29.3$
   j 917×38 k 3.1×14.9 l $47 \times (21.7 + 39.2)$
12 Estimate the result of each of these calculations.
   In each case, say whether the estimated result is bigger or smaller than
   the actual result and explain your answer.
   a 4.21×81.6 b $189 \div 11.4$ c 16.3×897
13 a Estimate the area of a rectangle with sides of length 4.7 cm and 6.8 cm.
   b Is the estimated area greater or smaller than the actual area?
      Explain your answer.
14 The area of a rectangle is 17.3 cm² and the length is 6.3 cm.
   a Estimate the width.
   b Is the estimated width greater or less than the actual width?
      Explain your answer.

Estimating answers to problems involving division and square roots

It is sometimes more convenient to use numbers other than the given values rounded to 1 significant figure.

When division is involved, you can round the numbers so that there are common factors.

When rounding numbers to estimate a square root, round to the nearest perfect square.

9 ESTIMATION

> ### Example 9.8
>
> **Question**
> Find approximate answers to these.
> a $83.7 \div 2.87$ b $0.73 \times \sqrt{18.2}$
>
> **Solution**
> a Round to $90 \div 3 = 30$
>   or $81 \div 3 = 27 = 30$ to 1 s.f.
>   (Rounding to $80 \div 3$ is more difficult, but would still give 30 to 1 s.f.)
> b Round to $0.7 \times \sqrt{16} = 0.7 \times 4 = 2.8 = 3$ to 1 s.f.

Exercise 9.5

Estimate the result of each of these calculations.

Show the approximations you use to get your estimate.

1 $31.3 \div 4.85$

2 $289 \div 86$

3 $44.669 \div 8.77$

4 $45 \div 6.8$

5 $(1.8 \times 2.9) \div 3.2$

6 $\dfrac{14.56 \times 22.4}{59.78}$

7 $\sqrt{4.9 \times 5.2}$

8 $\dfrac{\sqrt{8.1 \times 1.9}}{1.9}$

9 $(0.35 \times 86.3) \div 7.9$

10 $0.95 \div 4.8$

11 $32 \times \sqrt{124}$

12 $\dfrac{62 \times 9.7}{10.12 \times 5.1}$

13 $44.555 \div 0.086$

14 $\sqrt{84}$

15 $\dfrac{1083}{8.2}$

16 $\dfrac{2.46}{18.5}$

17 $\dfrac{29}{41.6}$

18 $\dfrac{283 \times 97}{724}$

19 $\dfrac{614 \times 0.83}{3.7 \times 2.18}$

20 $\dfrac{6.72}{0.051 \times 39.7}$

21 $\sqrt{39 \times 80}$

Working to a sensible degree of accuracy

Measurements and calculations should always be expressed to a suitable degree of accuracy.

For example, it would be silly to say that a train journey took 4 hours 46 minutes and 13 seconds, but sensible to say that it took $4\frac{3}{4}$ hours or about five hours.

In the same way, saying that the distance the train travelled was 93 kilometres 484 metres and 78 centimetres would be giving the measurement to an unnecessary degree of accuracy.

It would more sensibly be stated as 93 km.

> As a general rule, the answer you give after a calculation should not be given to a greater degree of accuracy than any of the values used in the calculation.

Example 9.9

Question

Bilal measured the length and width of a table as 1.84 m and 1.32 m.
He calculated the area as $1.84 \times 1.32 = 2.4288 \, m^2$.
How should he have given the answer?

Solution

Bilal's answer has four places of decimals (4 d.p.) so it is more accurate than the measurements he took.

His answer should be $2.43 \, m^2$.

Exercise 9.6

1. Write down sensible values for each of these measurements.
   a. 3 minutes 24.8 seconds to boil an egg
   b. 2 weeks, 5 days, 3 hours and 13 minutes to paint a house
   c. A book weighing 2.853 kg
   d. The height of a door is 2 metres 12 centimetres and 54 millimetres

2. Work out these and give the answers to a reasonable degree of accuracy.
   a. Find the length of the side of a square field with area $33 \, m^2$.
   b. A book has 228 pages and is 18 mm thick.
      How thick is Chapter 1 which has 35 pages?
   c. It takes 12 hours to fly between two cities on an aeroplane travelling at 554 km/h.
      How far apart are the cities?
   d. The total weight of 13 people in a lift is 879 kg.
      What is their average weight?
   e. The length of a strip of card is 2.36 cm, correct to 2 decimal places.
      The width is 0.041 cm, correct to 3 decimal places.
      Calculate the area of the card.

Key points

- Know how to round a number to the nearest 100, 1000, and so on.
- When rounding to a given number of decimal places, count the number of places from the decimal point. This is where the rounded answer will end. If the next digit along is less than five, leave the rounded answer as it is. If the next digit along is greater than or equal to five, add one to the final digit of the rounded answer.
- When rounding to a given number of significant figures, count the number of figures from the first non-zero digit starting at the left of the number. This is where the digits will end in the answer. As with decimal places, check the next digit along to see if the final digit needs altering.
- After rounding to a number of significant figures, make sure the answer has the same order of magnitude as the original number.
- A zero within a series of digits should be counted as one of the decimal places or one of the significant figures.
- To find an estimate to the answer to a problem, round each of the numbers in the problem to one significant figure and then perform the calculation.
- Know that an answer should not be given to a greater degree of accuracy than any of the values used in the calculation.

10 LIMITS OF ACCURACY

BY THE END OF THIS CHAPTER YOU WILL BE ABLE TO:
- give upper and lower bounds for data rounded to a specified accuracy
- find upper and lower bounds of the results of calculations which have used data rounded to a specified accuracy.

CHECK YOU CAN:
- round to a given number of decimal places
- round to a given number of significant figures
- convert one metric unit to another.

Note
Many people are confused about the upper bound. The convention is that the lower bound is contained in the interval and the upper bound is in the next, higher interval.

Bounds of measurement

Ali measures the length of a sheet of paper. She says the length is 26 cm to the nearest centimetre. What does this mean?

It means the length is nearer to 26 cm than the closest measurements on each side, 25 cm and 27 cm.

Any measurement that is nearer to 26 cm than to 25 cm or 27 cm will be counted as 26 cm. This is the marked interval on the number line.

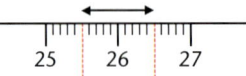

The boundaries of this interval are 25.5 cm and 26.5 cm. These values are exactly halfway between one measurement and the next. Usually when rounding to a given number of decimal places or significant figures, you would round 25.5 up to 26 and 26.5 up to 27.

- The interval for 26 cm to the nearest centimetre is m cm where $25.5 \leq m < 26.5$.
- 25.5 cm is called the **lower bound** of the interval.
- 26.5 cm is called the **upper bound** of the interval but it is not actually included in the interval.

Example 10.1

Question

Por won the 200 m race in a time of 24.2 seconds to the nearest tenth of a second. Complete the sentence below.

Por's time was between … seconds and … seconds.

Solution

As the measurement is stated to the nearest tenth of a second, the next possible times are 24.1 seconds and 24.3 seconds.

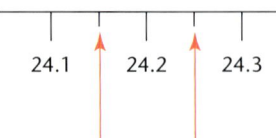

Halfway between: 24.15   24.25

Por's time was between 24.15 seconds and 24.25 seconds.

Exercise 10.1

1. Give the upper and lower bounds of each of these measurements.
   a. Given to the nearest centimetre
      i. 27 cm ii. 30 cm iii. 128 cm
   b. Given to the nearest 10 cm
      i. 10 cm ii. 30 cm iii. 150 cm
   c. Given to the nearest millimetre
      i. 5.6 cm ii. 0.8 cm iii. 12.0 cm
   d. Given to the nearest centimetre
      i. 1.23 m ii. 0.45 m iii. 9.08 m
   e. Given to the nearest hundredth of a second
      i. 10.62 seconds ii. 9.81 seconds iii. 48.10 seconds

2. Complete each of these sentences.
   a. A mass given as 57 kg to the nearest kilogram is between … kg and … kg.
   b. A height given as 4.7 m to 1 decimal place is between … m and … m.
   c. A volume given as 468 ml to the nearest millilitre is between … ml and … ml.
   d. A winning time given as 34.91 seconds to the nearest hundredth of a second is between … seconds and … seconds.
   e. A mass given as 0.634 kg to the nearest gram is between … kg and … kg.

3. A drain cleaner uses a pole made up of ten identical flexible pieces. Each piece is 1 metre long, measured to the nearest centimetre. What length of drain can you be sure that the drain cleaner could reach?

4. A rectangle measures 12 cm by 5 cm, with both measurements correct to the nearest centimetre.
   a. Work out the greatest possible perimeter of the rectangle.
   b. Work out the smallest possible area of the rectangle.

Sums and differences of measurements

Two kitchen cupboards have widths of 300 mm and 500 mm correct to the nearest millimetre.

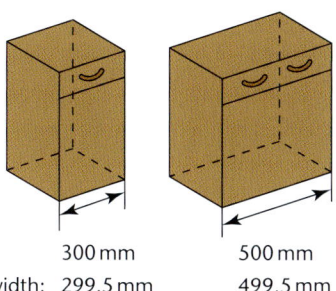

300 mm   500 mm
Lower bound of width: 299.5 mm   499.5 mm
Upper bound of width: 300.5 mm   500.5 mm

If the cupboards are put next to each other then

- the lower bound of w, their joint width = 299.5 + 499.5 = 799 mm
- the upper bound of w, their joint width = 300.5 + 500.5 = 801 mm
- so 799 mm $\leq w \leq$ 801 mm.

10 LIMITS OF ACCURACY

> To find the lower bound of a sum, add the lower bounds.
>
> To find the upper bound of a sum, add the upper bounds.

However, the difference between the widths of the kitchen cupboards is largest when the smallest possible width of the smaller cupboard is subtracted from the largest possible width of the larger cupboard.

The difference is least when the largest possible width of the smaller cupboard is subtracted from the smallest possible width of the larger cupboard. So

- the upper bound of the difference in their widths = 500.5 − 299.5 = 201 mm
- the lower bound of the difference in their widths = 499.5 − 300.5 = 199 mm.

> To find the upper bound of a difference, subtract the lower bound of the smaller value from the upper bound of the larger value.
>
> To find the lower bound of a difference, subtract the upper bound of the smaller value from the lower bound of the larger value.

Example 10.2

Question

A piece of red string is 35.2 cm long to the nearest millimetre.
A piece of blue string is 12.6 cm long to the nearest millimetre.
a What is the minimum length of the two pieces laid end to end?
b What is the lower bound of the difference in the lengths?

Solution

For the red string: LB = 35.15 cm   UB = 35.25 cm
For the blue string: LB = 12.55 cm   UB = 12.65 cm

a Minimum total length = sum of lower bounds
= 35.15 + 12.55 = 47.7 cm

b LB of difference in lengths = LB of longer piece − UB of shorter piece
= 35.15 − 12.65 = 22.5 cm

Exercise 10.2

1. Find the upper bound of the sum of each of these pairs of measurements.
   a. 29.7 seconds and 31.4 seconds (both to 3 significant figures)
   b. 11.04 s and 13.46 s (both to the nearest hundredth of a second)
   c. 6.42 m and 5.97 m (both to the nearest centimetre)
   d. 1.248 kg and 0.498 kg (both to the nearest gram)
   e. 86 mm and 98 mm (both to the nearest millimetre)

2. Find the lower bound of the sum of measurements in question **1**.

3. Find the upper bound of the difference between each of these pairs of measurements.
   a. 947 g and 1650 g (both to the nearest g)
   b. 16.4 cm and 9.8 cm (both to the nearest mm)
   c. 24.1 s and 19.8 s (both to the nearest 0.1 s)

d 14.86 s and 15.01 s (both to the nearest $\frac{1}{100}$ second)
   e 12 700 m and 3800 m (both to the nearest 100 m)
4 Find the lower bound of the difference between measurements in question **3**.
5 A piece of paper 21.0 cm long is taped on to the end of another piece 29.7 cm long.
   Both measurements are given to the nearest millimetre.
   What is the upper bound of the total length?
6 Two stages of a relay race are run in times of 14.07 seconds and 15.12 seconds, both to the nearest 0.01 second.
   a Calculate the upper bound of the total time for these two stages.
   b Calculate the upper bound of the difference between the times for these two stages.
7 A triangle has sides of length 7 cm, 8 cm and 10 cm.
   All measurements are correct to the nearest centimetre.
   Work out the upper and lower bounds of the perimeter of the triangle.
8 Given that $p = 5.1$ and $q = 8.6$, both correct to 1 decimal place, work out the largest and smallest possible values of
   a $p + q$
   b $q - p$.
9 Fah is fitting a new kitchen.
   She has an oven which is 595 mm wide, to the nearest millimetre.
   Will it definitely fit in a gap which is 60 cm wide, to the nearest centimetre?

Multiplying and dividing measurements

The measurements of a piece of A4 paper are given as 21.0 cm and 29.7 cm to the nearest millimetre. What are the upper and lower bounds of the area of the piece of paper?

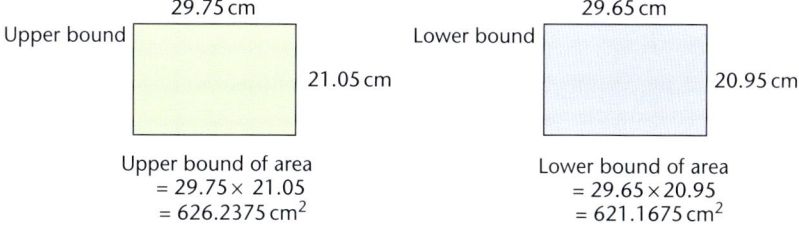

Upper bound of area
= 29.75 × 21.05
= 626.2375 cm²

Lower bound of area
= 29.65 × 20.95
= 621.1675 cm²

When multiplying
- to find the upper bound, multiply the upper bounds
- to find the lower bound, multiply the lower bounds.

When dividing, however, the situation is different.

Dividing by a larger number makes the answer smaller.

10 LIMITS OF ACCURACY

> **When dividing**
> - to find the upper bound, divide the upper bound by the lower bound
> - to find the lower bound, divide the lower bound by the upper bound.

Example 10.3

Question

Gan cycles 14.2 km (to 3 significant figures) in a time of 46 minutes (to the nearest minute). What are the upper and lower bounds of his average speed in kilometres per hour?
Give your answer to 3 significant figures.

Solution

Gan cycles 14.2 km.
This distance lies between 14.15 km and 14.25 km.
Gan takes 46 minutes.
This time lies between 45.5 minutes and 46.5 minutes.

$$\text{Upper bound of speed} = \frac{\text{upper bound of distance}}{\text{lower bound of time}}$$
$$= \frac{14.25}{45.5} \text{ km/minute}$$
$$= \frac{14.25}{45.5 \div 60} \text{ km/h}$$
$$= 18.8 \text{ km/h to 3 significant figures}$$

$$\text{Lower bound of speed} = \frac{\text{lower bound of distance}}{\text{upper bound of time}}$$
$$= \frac{14.15}{46.5 \div 60} \text{ km/h}$$
$$= 18.3 \text{ km/h to 3 significant figures}$$

> **Note**
> To find the upper bound of any combined measurement, work out which of the upper and lower bounds of the given measurements you need to use to give you the greatest result.
> If you aren't sure, experiment!

Exercise 10.3

1. Pencils have a width of 8 mm, to the nearest millimetre.
   What is the smallest total width of 10 pencils?
2. Find the upper and lower bounds of the floor area of a rectangular room with these measurements.
   a 5.26 m by 3.89 m to the nearest centimetre
   b 7.6 m by 5.2 m to the nearest 10 cm
3. Calculate the upper and lower bounds of the distance travelled for each of these pairs of speed and time. Give your answers to 3 significant figures.
   a 92.4 cm/second for 12.3 seconds (both values to 3 s.f.)
   b 1.54 m/second for 8.20 seconds (both values to 3 s.f.)
   c 57 km/h for 2.5 hours (both values to 2 s.f.)
   d 5.61 m/second for 2.08 seconds (both values to 3 s.f.)
4. Find the upper and lower bounds of the average speed for these measured times and distances. Give your answers to 3 significant figures.
   a 6.4 cm in 1.2 seconds (both values to 2 s.f.)
   b 106 m in 10.0 seconds (both values to 3 s.f.)

Multiplying and dividing measurements

5 Calculate, to 3 significant figures, the minimum width of a rectangle with these dimensions.
   a Area = 210 cm² to 3 significant figures,
      length = 17.8 cm to the nearest millimetre
   b Area = 210 cm² to 2 significant figures,
      length = 19.2 cm to the nearest millimetre

6 The population of a town is 108 000 to the nearest 1000.
   Its area is given as 129 km² to the nearest km².
   Calculate the upper and lower bounds of its population density, giving your answers to 3 significant figures.

   $$\text{Population density} = \frac{\text{population}}{\text{area}}$$

7 Work out the largest and smallest possible areas of a rectangle measuring 27 cm by 19 cm, where both lengths are correct to the nearest centimetre.

8 A bar of gold is a cuboid and measures 8.6 cm by 4.1 cm by 2.4 cm.
   All the measurements are correct to the nearest millimetre.
   a Work out the largest and smallest possible volumes of the bar of gold.
   b The density of gold is 19.3 g/cm³, correct to the nearest 0.1 g/cm³.
      Work out the greatest and least possible masses of the gold bar.

   $$\text{Density} = \frac{\text{mass}}{\text{volume}}$$

9 Use the formula $P = \frac{V^2}{R}$ to work out the upper and lower bounds of P when $V = 6$ and $R = 1$ and both values are correct to the nearest whole number.

10 In May 1976, Don Quarrie ran 100 m in 9.9 s (correct to 1 d.p.) and in July 1983, Calvin Smith ran it in 9.93 s (correct to 2 d.p.).
   Using these figures, explain which runner may have run the faster time.

11 A water tank measures 80 cm by 75 cm by 90 cm, correct to the nearest centimetre.
   Could it hold exactly 550 litres?

Note
You will learn more about density and population density in Chapter 12.

Key points
- A value to a given unit of accuracy has an upper bound that is the value plus half the given unit and a lower bound that is the value minus half the given unit.
- When calculating with two values, A and B, each given to a unit of accuracy, the upper bound (U) and lower bound (L) of the answer can be found using the following.

   $U(A + B) = U(A) + U(B)$ $L(A + B) = L(A) + L(B)$
   $U(A - B) = U(A) - L(B)$ $L(A - B) = L(A) - U(B)$
   $U(A \times B) = U(A) \times U(B)$ $L(A \times B) = L(A) \times L(B)$
   $U\left(\dfrac{A}{B}\right) = \dfrac{U(A)}{L(B)}$ $L\left(\dfrac{A}{B}\right) = \dfrac{L(A)}{U(B)}$

11 RATIO AND PROPORTION

BY THE END OF THIS CHAPTER YOU WILL BE ABLE TO:

understand and use ratio and proportion to:
- give ratios in their simplest form
- divide a quantity in a given ratio
- use proportional reasoning and ratios in context.

CHECK YOU CAN:
- find common factors
- simplify fractions
- understand enlargement
- change between metric units.

Ratio

Simplifying ratios

When something is divided into a number of parts, ratios are used to compare the sizes of those parts.

For example, if ten sweets are divided between two people so that one person gets six sweets and the other person gets four sweets, the sweets have been divided in the ratio 6 : 4.

As with fractions, ratios can be simplified.

For example, the fraction $\frac{6}{4} = \frac{3}{2}$; similarly, the ratio 6 : 4 = 3 : 2.

When simplifying ratios you need to look for common factors.

When the ratio you are using involves measurements, the units must be the same for both measurements when simplifying.

Example 11.1

Question

Write each of these ratios in its simplest form.

a 1 millilitre : 1 litre
b 1 kilogram : 200 grams

Note

'In its simplest form' means simplify as much as possible.

When the units are the same, you do not need to include them in the ratio.

Solution

a 1 millilitre : 1 litre = 1 millilitre : 1000 millilitres   Same units.
                       = 1 : 1000
b 1 kilogram : 200 grams
                = 1000 grams : 200 grams   Same units.
                = 5 : 1   Divide each by 200.

Exercise 11.1

1 Write each of these ratios in its simplest form.
   a 6 : 3   b 25 : 75   c 30 : 6   d 10 : 15   e 7 : 35
   f 15 : 12   g 24 : 8   h 4 : 48   i 3 : 27   j 9 : 81

2 Write each of these ratios in its simplest form.
   a 20 minutes : 1 hour   b 50 g : 1 kg   c 300 ml : 1 litre
   d 2 kg : 600 g   e 2 minutes : 30 seconds

3 Write each of these ratios in its simplest form.
   a 50 g : 1000 g
   b 30 cents : $2
   c 4 m : 75 cm
   d 300 ml : 2 litres
4 Write each of these ratios in its simplest form.
   a 5 : 15 : 25
   b 6 : 12 : 8
   c 8 : 32 : 40
5 Write each of these ratios in its simplest form.
   a 50 cents : $2.50 : $5
   b 20 cm : 80 cm : 1.20 m
   c 600 g : 750 g : 1 kg
6 Mali, Polly and Dave invest $500, $800 and $1000 respectively in a business.
   Write the ratio of their investments in its simplest form.

Writing a ratio in the form 1 : n

It is sometimes useful to have a ratio with 1 on the left.

A common scale for a scale model is 1 : 24.

The scale of a map or enlargement is often given as 1 : n.

To change a ratio to this form, divide both numbers by the number on the left.

Example 11.2

Question

Write these ratios in the form 1 : n.
a 2 : 5
b 8 mm : 3 cm
c 25 mm : 1.25 km

Solution

a 2 : 5 = 1 : 2.5   Divide each side by 2.
b 8 mm : 3 cm = 8 mm : 30 mm   Write each side in the same units.
              = 1 : 3.75   Divide each side by 8.
c 25 mm : 1.25 km = 25 : 1 250 000   Write each side in the same units.
                  = 1 : 50 000   Divide each side by 25.

Note

1 : 50 000 is a common map scale. It means that 1 cm on the map represents 50 000 cm, or 500 m, on the ground.

Exercise 11.2

1 Write each of these ratios in the form 1 : n.
   a 2 : 6
   b 3 : 15
   c 6 : 15
   d 4 : 7
   e 20 cents : $1.50
   f 4 cm : 5 m
   g 10 : 2
   h 2 mm : 1 km
2 Write each of these ratios in the form 1 : n.
   a 2 : 8
   b 5 : 12
   c 2 mm : 10 cm
   d 2 cm : 5 km
   e 100 : 40
3 On a map a distance of 8 mm represents a distance of 2 km.
   What is the scale of the map in the form 1 : n?
4 A passport photo is 35 mm long.
   An enlargement of the photo is 21 cm long.
   What is the ratio of the passport photo to the enlargement in the form 1 : n?

11 RATIO AND PROPORTION

Using ratio to find an unknown quantity

Sometimes you know one of the quantities in the ratio, but not the other.

If the ratio is in the form $1:n$, you can work out the second quantity by multiplying the first quantity by n. You can work out the first quantity by dividing the second quantity by n.

Example 11.3

Question

A map is drawn to a scale of 1 cm : 2 km.

a On the map, the distance between Amhope and Didburn is 5.4 cm. What is the real distance in kilometres?

b The length of a straight railway track between two stations is 7.8 km. How long is this track, in centimetres, on the map?

Solution

a The real distance, in kilometres, is twice as large as the map distance, in centimetres.

So multiply by 2.

$2 \times 5.4 = 10.8$

Real distance = 10.8 km

b The map distance, in centimetres, is half of the real distance, in kilometres.

So divide by 2.

$7.8 \div 2 = 3.9$

Map distance = 3.9 cm

Sometimes you have to work out quantities using a ratio that is not in the form $1:n$.

To work out an unknown quantity, convert to an equivalent ratio involving the known quantity.

Using ratio to find an unknown quantity

Example 11.4

Question

Two photos are in the ratio 2 : 5.
a What is the height of the larger photo?
b What is the width of the smaller photo?

← 15 cm →

8 cm

Solution

a The multiplier for the ratio of the heights is 4, since for the smaller photo 2 × 4 = 8. Use the same multiplier for the larger photo.

Height of the larger photo = 20 cm.

b The multiplier for the ratio of the widths is 3, since for the larger photo 5 × 3 = 15. Use the same multiplier for the smaller photo.

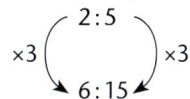

Width of the smaller photo = 6 cm.

Exercise 11.3

1 The ratio of helpers to babies in a crèche must be 1 : 4.
   a There are six helpers on Tuesday.
     How many babies can there be?
   b There are 36 babies on Thursday.
     How many helpers must there be?
2 Sanjay is mixing pink paint. To get the shade he wants, he mixes red and white paint in the ratio 1 : 3.
   a How much white paint should he mix with 2 litres of red paint?
   b How much red paint should he mix with 12 litres of white paint?
3 For a recipe, Chetan mixes water and lemon curd in the ratio 2 : 3.
   a How much lemon curd should he mix with 20 ml of water?
   b How much water should he mix with 15 teaspoons of lemon curd?
4 To make a solution of a chemical, a scientist mixes 3 parts of the chemical with 20 parts of water.
   a How much water should she mix with 15 ml of the chemical?
   b How much of the chemical should she mix with 240 ml of water?
5 An alloy is made by mixing 2 parts of silver with 5 parts of nickel.
   a How much nickel must be mixed with 60 g of silver?
   b How much silver must be mixed with 120 g of nickel?
6 Sachin and Rehan share a flat.
   They agree to share the rent in the same ratio as their wages.
   Sachin earns $300 a month and Rehan earns $400 a month.
   Sachin pays $90. How much does Rehan pay?

11 RATIO AND PROPORTION

Dividing a quantity in a given ratio

Use this method to divide a quantity in a given ratio:
- find the total number of shares by adding the parts of the ratio
- divide the quantity by the total number of shares to find the multiplier
- multiply each part of the ratio by the multiplier.

Example 11.5

Question

To make fruit punch, orange juice and grapefruit juice are mixed in the ratio 5 : 3.
Jo wants to make 1 litre of punch.
a   How much orange juice does she need, in millilitres?
b   How much grapefruit juice does she need, in millilitres?

Solution

5 + 3 = 8   First, work out the total number of shares.
1000 ÷ 8 = 125   Convert 1 litre to millilitres and divide by 8 to find the multiplier.

A table is often helpful for this sort of question.

	Orange	Grapefruit
Ratio	5	3
Amount	5 × 125 = 625 ml	3 × 125 = 375 ml

Note

To check your answers, add the parts together.
Together they should equal the total quantity.
For example, 625 ml + 375 ml = 1000 ml ✔

a   She needs 625 ml of orange juice.
b   She needs 375 ml of grapefruit juice.

Exercise 11.4

 1 Paint is mixed in the ratio 3 parts red to 5 parts white to make 40 litres of pink paint.
   a   How much red paint is used?
   b   How much white paint is used?

 2 To make a solution of a chemical, a scientist mixes 1 part of the chemical with 5 parts of water. She makes 300 ml of the solution.
   a   How much of the chemical does she use?
   b   How much water does she use?

 3 Amit, Bree and Chris share $1600 between them in the ratio 2 : 5 : 3. How much does each receive?

 4 A 600 g bar of brass is made using the metals copper and zinc in the ratio 2 : 1.
   How much of each metal is used?

 5 In a class of 36 students the ratio of boys to girls is 5 : 4.
   How many boys are there?

 6 To get home from work Mehr runs and walks.
   The distances he runs and walks are in the ratio 2 : 3.
   He works 2 km from home.
   How far does he run?

Proportion

 7 Orange squash needs to be mixed in the ratio 1 part concentrate to 6 parts water.
How much concentrate is needed to make 3.5 litres of orange squash?

8 There are 572 senators in a national assembly.
The numbers of senators in the Blue, Orange and Green parties are in the ratio $6:3:2$.
How many senators are there in each of the parties?

9 Sally makes breakfast cereal by mixing bran, currants and wheat germ in the ratio $8:3:1$ by mass.
   a How much bran does she use to make 600 g of the cereal?
   b One day, she has only 20 g of currants.
   How much cereal can she make? She has plenty of bran and wheat germ.

Proportion

Direct proportion

In direct proportion, both quantities increase at the same rate.

For example, if you need 200 g of flour to make 10 cupcakes, you need 400 g of flour to make 20 cupcakes.

The quantity of flour is multiplied by 2.

The number of cupcakes is multiplied by 2.

Both quantities are multiplied by the same number, called the **multiplier**.

To find the multiplier, you divide either pair of quantities.
For example, $\frac{400}{200} = 2$ or $\frac{20}{10} = 2$.

So a quick way to solve this type of problem is to find the multiplier and use it to find the unknown quantity.

> ### Example 11.6
>
> **Question**
> A car uses 20 litres of fuel when making a journey of 160 kilometres.
> How many litres of fuel would it use to make a similar journey of 360 kilometres?
>
> **Solution**
> You could solve this problem using ratios, but here is a method just using proportion.
> Write the distance travelled in the two journeys as a fraction.
> You want to find the fuel needed for a 360 kilometre journey, so make 360 the numerator and 160 the denominator.
> Cancel the fraction so that it is in its simplest form.
>
> $\frac{360}{160} = \frac{9}{4}$
>
> $20 \times \frac{9}{4} = 45$ Multiply the amount of fuel needed for the first journey by the multiplier.
>
> 45 litres of fuel would be used to make a journey of 360 kilometres.

11 RATIO AND PROPORTION

Exercise 11.5

For each of these questions
- a write down the multiplier
- b calculate the required quantity.

1. An express train travels 165 metres in 3 seconds.
   How far would it travel in 8 seconds?
2. An aeroplane travels 216 kilometres in 27 minutes.
   How far did it travel in 12 minutes?
3. £50 is worth $90.
   How much is £175 worth?
4. A ladder that is 7 metres long has 28 rungs.
   How many rungs would there be in a ladder that is 5 metres long?
5. A piece of string 27 metres long has a mass of 351 grams.
   What is the mass of 15 metres of the string?
6. A piece of wood with a volume of $2.5\,m^3$ has a mass of 495 kilograms.
   What is the mass of a piece of this wood with a volume of $0.9\,m^3$?

Inverse proportion

In inverse proportion, as one quantity increases, the other decreases.

In such cases, you need to divide by the multiplier, rather than multiply.

Example 11.7

Question
Three excavators can dig a hole in 8 hours.
How long would it take four excavators to dig the hole?

Solution
Clearly, as more excavators are to be used, the digging will take less time.
The multiplier is $\frac{4}{3}$.

$8 \div \frac{4}{3} = 6$ Divide the known time by the multiplier to find the unknown time.

It will take four excavators 6 hours to dig the hole.

Exercise 11.6

For each of these questions
- a write down the multiplier
- b calculate the required quantity.

1. A journey takes 18 minutes at a constant speed of 32 kilometres per hour.
   How long would the journey take at a constant speed of 48 kilometres per hour?
2. It takes a team of 8 workers six weeks to paint a bridge.
   How long would take 12 workers to paint the same bridge?
3. A pool is normally filled using four inlet valves in a period of 18 hours.
   One of the inlet valves is broken.
   How long will it take to fill the pool using just three inlet valves?

4 A journey can be completed in 44 minutes at an average speed of 50 kilometres per hour.
   How long would the same journey take at an average speed of 40 kilometres per hour?
5 A quantity of hay is enough to feed 12 horses for 15 days.
   How long would the same quantity of hay feed 20 horses?
6 It takes three harvesters 6 hours to harvest a crop of wheat.
   How long would it take to harvest the wheat with only two harvesters?
7 On the outward leg of a journey a cyclist travels at an average speed of 12 kilometres per hour for a period of 4 hours.
   The return journey took 3 hours.
   What was the cyclist's average speed for the return journey?
8 It takes a team of 18 people 21 weeks to dig a canal.
   How long would it take 14 people to dig the canal?
9 A tank can be emptied using six pumps in a period of 18 hours.
   How long will it take to empty the tank using eight pumps?
10 A gang of 9 bricklayers can build a wall in 20 days.
   How long would a gang of 15 bricklayers take to build the wall?

Key points
- Simplify ratios by dividing by common factors whenever you can.
- To simplify ratios involving measures, both quantities must be in the same unit.
- If a quantity is in the form 1 : n, you can work out the second quantity by multiplying the first quantity by n. You can work out the first quantity by dividing the second quantity by n.
- To find an unknown quantity, each part of the ratio must be multiplied by the same number, called the multiplier.
- To divide a quantity in a given ratio, first find the total number of shares, then divide the total quantity by the total number of shares to find the multiplier. You can then multiply each part of the ratio by the multiplier.
- To solve proportion problems, use a pair of known values to find the multiplier. Then, use the multiplier to find the unknown quantity.

REVIEW EXERCISE 2

1 Evaluate these expressions.
  a 5^3 [1]
  b 2^4 [1]
  c $3^2 \times 5^2$ [1]

2 Write these expressions using index notation.
  a $3 \times 3 \times 3 \times 3$ [1]
  b $2 \times 3 \times 5 \times 3 \times 2 \times 2$ [1]

3 a Express 0.043×100^2 in standard form. [1]
  b Evaluate $\dfrac{1.2 \times 10^7}{2 \times 10^{-3}}$, giving your answer in standard form. [2]

Cambridge O Level Mathematics Syllabus D (4024) Paper 11 Q9, November 2020

4 Write these numbers in standard form.
  a 326 000 [1]
  b 0.0219 [1]
  c 64 [1]

5 a Write these numbers in order of size, starting with the smallest.
  2.1×10^{-3} 4.2×10^{-4} 1.7×10^{-5} 3.5×10^{-4} [1]
  b $P = 6 \times 10^{10}$ $Q = 5 \times 10^9$
  Evaluate the following.
  Give each answer in standard form.
  i $P - Q$ [1]
  ii PQ [1]

Cambridge O Level Mathematics Syllabus D (4024) Paper 12 Q12, June 2019

6 By writing each number correct to 1 significant figure, estimate the value of
$$\frac{6013 \times 0.0405}{\sqrt{8.986}}.$$ [2]

Cambridge O Level Mathematics Syllabus D (4024) Paper 11 Q8, November 2020

7 A car has a mass of 2400 kg, correct to the nearest hundred kilograms.
A caravan has a mass of 1460 kg, correct to the nearest ten kilograms.
Calculate the lower bound for the total mass of the car and caravan. [2]

Cambridge O Level Mathematics Syllabus D (4024) Paper 11 Q22, June 2021

8 The table shows the populations, correct to 2 significant figures, of some African countries in 2014.

Country	Population
Nigeria	
Sudan	3.6×10^7
Chad	1.1×10^7
Namibia	2.2×10^6

a In 2014, the population of Nigeria was 177 156 000.
Complete the table with the population of Nigeria using standard form, correct to 2 significant figures. [2]

b Complete the following.
The population of Chad was ………. times the population of Namibia. [1]

Cambridge O Level Mathematics Syllabus D (4024) Paper 12 Q22a & b, June 2016

9

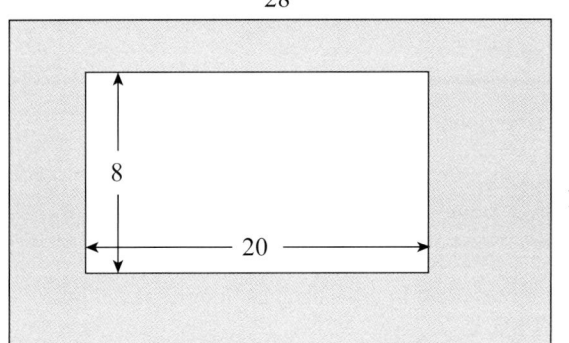

NOT TO SCALE

A rectangle 20 cm by 8 cm is cut from a rectangle 28 cm by 15 cm.
Each measurement is given correct to the nearest centimetre.
Calculate the upper bound for the area of the shaded region. [3]

Cambridge O Level Mathematics Syllabus D (4024) Paper 22 Q2b, June 2020

10 Some sweets are shared between Georgia and Hassan in the ratio 5 : 2.
If Georgia gets 30 sweets, how many sweets are shared? [1]

11 a 200 grams of a spice cost 85 cents.
Find the cost, in dollars, of 1 kilogram of this spice. [1]

b You are given that $60 : x = 3 : 2$.
Find x. [1]

Cambridge O Level Mathematics Syllabus D (4024) Paper 12 Q2, November 2018

12 RATES

BY THE END OF THIS CHAPTER YOU WILL BE ABLE TO:
- use common measures of rate
- apply other measures of rate
- solve problems involving average speed.

CHECK YOU CAN:
- change between metric units.

Common measures of rate

In this section, you will look at rate of pay and rate of flow.

Example 12.1

Question
Nila has a part-time job. Her rate of pay is $18.70 per hour.
How much does she earn in a week where she works for 15 hours 30 minutes?

Solution
Number of hours worked = 15.5
Earnings = $18.70 × 15.5
         = $289.85

Rate of flow is a compound measure linking volume or mass with time.

$$\text{Average rate of flow} = \frac{\text{volume or mass}}{\text{total time taken}}$$

Rates of flow are measured in units such as kilograms per hour (kg/h) or litres per hour (l/h).

Example 12.2

Question
It takes half an hour to fill a 1000 litre tank.
What is the average rate of flow?

Solution
$$\text{Rate of flow} = \frac{1000 \text{ l}}{0.5 \text{ h}} = 2000 \text{ l/h}$$

Another example of a rate is the rate of exchange between currencies. Currency conversion is a section in Chapter 16, which is all about money.

Exercise 12.1

1. Water from a tap flows at 25 litres/minute.
   How long will it take for this tap to fill a bath that holds 500 litres?
2. Flour is loaded into a lorry which holds 20 tonnes.
   It takes 40 minutes to load the lorry.
   What is the rate of flow, in kilograms per minute?

3 An oil tank contains 1500 litres.
   Oil is used at a rate of 2 litres/hour.
   How much is left in the tank after 7 days?
4 Hanif is paid at a rate of $5.70 per hour.
   One week he works 42 hours.
   How much does he earn in this week?
5 Jo receives $216 one week when she works 25 hours.
   a What is her rate of pay per hour?
   b At this rate, how much would she earn for a week in which she works 16 hours?

Speed

Speed is a compound measure linking distance and time.

$$\text{Average speed} = \frac{\text{total distance travelled}}{\text{total time taken}}$$

The units of your answer will depend on the units you begin with. Speed has units of 'distance per time', for example, km/h.

The formula for speed can be rearranged to find the distance travelled or the time taken for a journey.

$$\text{Distance} = \text{speed} \times \text{time} \qquad \text{Time} = \frac{\text{distance}}{\text{speed}}$$

Note
You may find the d.s.t. triangle helpful. Cover up the quantity you are trying to find. What is left shows you how to multiply or divide.

Example 12.3

Question
How many minutes does it take to walk 2 km at a speed of 5 km/h?

Solution
$\text{Time} = \frac{\text{distance}}{\text{speed}} = \frac{2}{5}$ hour

$\frac{2}{5}$ hour $= \frac{2}{5} \times 60$ minutes $= 24$ minutes

Exercise 12.2

1 Find the average speed of a car which travels 75 kilometres in $1\frac{1}{2}$ hours.
2 Find the average speed of a runner who covers 180 m in 40 seconds.
3 A cyclist rides 0.6 kilometres in 3 minutes.
   Calculate her average speed, in kilometres per hour.
4 A bus travels at 5 m/s on average.
   How many kilometres per hour is this?
5 A runner's average speed in an 80 m race is 7 m/s.
   Find the time she takes for the race, to the nearest 0.1 second.

12 RATES

Applying other measures of rate

There are several other situations in which you need to use a rate. The same basic principles can be applied. The formulas for these will be given in the question.

Here are some examples.

Pressure

Pressure is a compound measure linking force and area.

$$\text{Pressure} = \frac{\text{force}}{\text{area}}$$

Usually, force is measured in newtons (N). When the area is given in square metres (m²), the pressure is measured in N/m².

> **Example 12.4**
>
> **Question**
> A heavy box is placed on a table.
> The base of the box is a rectangle measuring 50 cm by 40 cm.
> The box exerts a force of 120 N on the table.
> Calculate the pressure exerted on the table, in N/m².
>
> **Solution**
> Area = $0.5 \times 0.4 = 0.2 \, m^2$
> Force = 120 N
> Pressure = $\frac{\text{force}}{\text{area}}$
> $= \frac{120}{0.2}$
> $= 600 \, N/m^2$
>
> **Note**
> Remember to change the units from cm to m.

Density

Another example of a compound measure is density, which links mass and volume.

$$\text{Density of a substance} = \frac{\text{mass}}{\text{volume}}$$

It is measured in units such as grams per cubic centimetre (g/cm³).

Example 12.5

Question

The density of gold is 19.3 g/cm³.

Calculate the mass of a gold bar with a volume of 30 cm³.

Solution

Density = $\frac{\text{mass}}{\text{volume}}$

so mass = density × volume.

The mass of the gold bar is density × volume = 19.3 × 30 = 579 g.

Note

As with speed, there is a triangle that can be helpful with these questions.

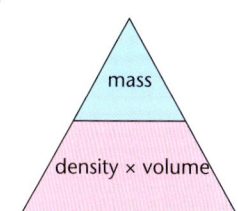

Population density

Population density is another example of a compound measure. It gives an idea of how heavily populated an area is. It is measured as the number of people per square kilometre.

Population density = $\frac{\text{number of people}}{\text{area}}$

Example 12.6

Question

In a small town, 300 people live in an area of 2.4 km².

Find the population density of the town.

Solution

Population density = $\frac{300}{2.4}$ = 125 people/km²

Note

Once again, a triangle can be helpful with these questions.

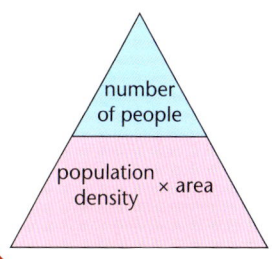

Exercise 12.3

In this exercise, use the formulas for pressure, density and population density that you have been given in this section.

1. Calculate the pressure that a force of 200 N exerts on an area of 2.5 m².
2. Calculate the density of a stone of mass 350 g and volume 40 cm³.
3. Waring has a population of 60 000 in an area of 8 square kilometres. Calculate its population density.
4. Trenton has a population of 65 000 in an area of 5.8 square kilometres. Calculate its population density, correct to the nearest thousand.
5. A foam plastic ball with volume 20 cm³ has density 0.3 g/cm³. What is its mass?
6. A town has a population of 200 000. Its population density is 10 000 people per square mile. What is the area of the town?

12 RATES

7 A rubber ball with volume 28.3 cm³ has density 0.7 g/cm³.
What is its mass?

8 A town has a population of 276 300.
Its population density is 9800 people per square mile.
What is the area of the town?

9 Find the force needed to exert a pressure of 15 N/m² on an area of 0.8 m².

> **Key points**
>
> - Average rate of flow = $\dfrac{\text{volume or mass}}{\text{total time taken}}$
>
>   This is an example of one of the basic rates that you should know.
>
> - Average speed = $\dfrac{\text{total distance travelled}}{\text{total time taken}}$
>
>   You are expected to know this relationship and to use it to solve problems.
>
> - Other examples of compound measures include:
>   - pressure (measured in units such as N/m²)
>   - density (measured in units such as g/cm³)
>   - population density (measured in units such as people/km²).
>
>   You do not need to memorise the formula for these compound measures but you should be able to use them correctly.

13 PERCENTAGES

Fractions, decimals and percentages

Since 'per cent' means 'out of 100', the fraction equivalent to a percentage is found by writing the percentage as a fraction with a denominator of 100.

So $13\% = \frac{13}{100}$, $5\% = \frac{5}{100}$, $140\% = \frac{140}{100}$, $2.3\% = \frac{2.3}{100}$, etc.

Sometimes you need to write equivalent fractions to give a fraction in its simplest form.

So $13\% = \frac{13}{100}$, but $5\% = \frac{1}{20}$, $140\% = 1\frac{2}{5}$, $2.3\% = \frac{23}{1000}$, etc.

The decimal equivalents can be found by carrying out the division shown in the fraction.

So $13\% = \frac{13}{100} = 0.13$, $5\% = \frac{5}{100} = 0.05$, $140\% = \frac{140}{100} = 1.4$,
$2.3\% = \frac{2.3}{100} = 0.023$

Exercise 13.1

1. Convert each of these percentages to a fraction in its simplest form.
   a 75%   b 12%   c 6%   d 80%

2. Convert each of these percentages to a decimal.
   a 27%   b 86%   c 3%   d 8%
   e 150%   f 140%   g 250%   h 174%
   i 2.7%   j 3.6%   k 12.8% l 17.3%

Percentage of a quantity

Using fractions

This is often the best method to use when calculating percentages without a calculator.

When you are calculating simple percentages such as 10%, 20% or 5%, first find 10% and then multiply or divide the result to find the percentage you want.

> **BY THE END OF THIS CHAPTER YOU WILL BE ABLE TO:**
> - calculate a given percentage of a quantity
> - express one quantity as a percentage of another
> - calculate percentage increase or decrease
> - calculate with simple and compound interest
> - calculate using reverse percentages.

> **CHECK YOU CAN:**
> - understand what is meant by a percentage
> - convert between fractions, decimals and percentages.

13 PERCENTAGES

Example 13.1

Question

a Find 15% of $240. b Find 31% of $6.80.

Solution

a 10% of $240 = 240 ÷ 10
     = $24
   5% of $240 = half of $24
     = $12
   So 15% of $240 = $24 + $12
     = $36

b 10% of 680c = 680 ÷ 10
     = 68c
   30% of 680 = 68 × 3
     = 204
   So 30% of $6.80 = 204c
   1% of 680c = 680 ÷ 100 = 6.8c
   So 31% of $6.80 = 204c + 6.8c
     = 210.8c
     = $2.11 (to the nearest cent)

Sometimes it is easier to use the facts that 25% = $\frac{1}{4}$ and 50% = $\frac{1}{2}$ to work out the percentage of a quantity.

Example 13.2

Question

a Find 25% of $72. b Find 75% of $90.

Solution

a 25% of $72 = 72 ÷ 4
     = $18

b 50% of $90 = 90 ÷ 2
     = $45
   25% of $90 = $45 ÷ 2
     = $22.50
   So 75% of $90 = $45 + $22.50
     = $67.50

Using decimals

To calculate a percentage of an amount, multiply the amount by the decimal equivalent of the percentage.

Example 13.3

Question

a Find 37% of $48. b Find 2.4% of $36 000.

Solution

a 37% of $48 = $48 × 0.37
     = $17.76

b 2.4% of $36 000 = $36 000 × 0.024
     = $864

Expressing one quantity as a percentage of another

Exercise 13.2

1. Work out these.
   a. 10% of $200
   b. 10% of $6
   c. 10% of $7.20
   d. 20% of $30
   e. 30% of $60
   f. 20% of $5
   g. 20% of $300
   h. 30% of $18
   i. 30% of $8.60
   j. 5% of $40
   k. 5% of $300
   l. 5% of $6.80

2. A sales representative gets 15% commission on sales of $600.
   How much is his commission?

3. Which is more, 15% of $20 or 20% of $15?
   Show your working.

4. A shop increases its DVD prices by 5%.
   What is the increase on a DVD which originally cost $20?

5. A charity spends 5% of their income on administration.
   One year they receive $8000.
   How much is spent on administration?

6. 20% of the students in a school arrive by bus.
   There are 800 students in the school.
   How many arrive by bus?

7. 15% of the 720 students in a school are left-footed.
   How many students is this?

8. A school gave 30% of the money raised from a sponsored bike ride to a charity. The total raised was $260.
   How much did the charity receive?

9. In a year, a family spends 5% of their income on fuel.
   They have an income of $23 000.
   How much do they spend on fuel?

10. Selina paid 23% of her income in tax.
    How much tax did she pay in the month when she earned $1284?

11. Work out these.
    a. 63% of 475
    b. 75% of 27
    c. 32% of $720
    d. 37% of 950 m
    e. 46% of 246
    f. 99% of $172
    g. 13% of 156
    h. 6% of 46
    i. 8% of $32
    j. 3.5% of $60

12. Beardalls' Jewellers offered 37.5% off the price of all rings.
    How much did they offer off a ring priced at $420?

13. A bus company claim that over 85% of the seats in their buses are fitted with safety belts.
    What is the smallest number of seats that must have seat belts fitted in a 53-seater bus?

Expressing one quantity as a percentage of another

To express one quantity as a percentage of another, first write a fraction with the first quantity as the numerator of a fraction and the second quantity as the denominator.

Then multiply the fraction by 100.

13 PERCENTAGES

> **Example 13.4**
>
> **Question**
> a Express 4 as a percentage of 5.
> b Express $5 as a percentage of $30.
> c Express 70 cm as a percentage of 2.3 m.
>
> **Solution**
> a $\frac{4}{5}$ Write as a fraction.
>
> $\frac{4}{5} \times 100 = \frac{400}{5} = 80\%$ Multiply by 100.
>
>   Or: $\frac{4}{5} \times 100 = 0.8 \times 100 = 80\%$
>
> b $\frac{5}{30} = 0.166\ldots = 16.7\%$ to 1 decimal place
>
> c   2.3 m = 230 cm   Change 2.3 m to centimetres.
> $\frac{70}{230} = 0.304\ldots = 30.4\%$ to 1 decimal place

Exercise 13.3

1 Find these.
   a 12 as a percentage of 100
   b 4 as a percentage of 50
   c 80 as a percentage of 200
   d $5 as a percentage of $20
   e 4 m as a percentage of 10 m
   f 30c as percentage of $2

2 In each case, express the first quantity as a percentage of the second.

	a	b	c	d	e	f	g	h	i	j
First quantity	16	12	1 m	$3	73c	8c	$1.80	40 cm	50c	$2.60
Second quantity	100	50	4 m	$10	$1	$1	$2	2 m	$10	$2

3 In Class 7B, 14 of the 25 students are boys.
   What percentage is this?

4 A sailor has a rope 50 m long. He cuts off 12 m.
   What percentage of the rope has he cut off?

5 In a sales promotion, a company offers an extra 150 ml of cola free.
   The bottles usually hold 1 litre.
   What percentage of 1 litre is 150 ml?

6 Selma spends $4 of her $20 spending money on a visit to the cinema.
   What percentage is this?

7 At a keep-fit club, 8 of the 25 members are men.
   What percentage is this?

8 Derek scored 44 runs out of his team's total of 80.
   What percentage of the runs did he score?

9 Work out these.
   a 25 as a percentage of 200
   b 40 as a percentage of 150
   c 76c as a percentage of $1.60
   d $1.70 as a percentage of $2
   e 19 as a percentage of 24
   f 213 as a percentage of 321

10 Halima scores 17 out of 40 in a test.
What percentage is this?

11 Salma is given a discount of $2.50 on a meal that would have cost $17.
What percentage discount is she given?

12 During the 30 days of September in 2022, it rained on 28 of the days.
On what percentage of the days did it rain?

13 Last Saturday 6198 fans booked online for a festival and 4232 fans booked by phone. What percentage of the fans made online bookings?

14 A mathematics test has 45 questions. Abigail gets 42 right.
What percentage does Abigail get right?

15 A photo frame 20 cm by 30 cm contains a photo 14 cm by 18 cm.
What percentage of the whole area is the photo?

Percentage change

Percentage increases and decreases are worked out as percentages of the original amount, not the new amount.

Percentage profit or loss is worked out as a percentage of the cost price, not the selling price.

$$\text{Percentage change} = \frac{\text{change}}{\text{original amount}} \times 100$$

Example 13.5

Question

An art dealer buys a painting for $45 and sells it for $72.
What percentage profit is this?

Solution

Profit = $72 − $45 = $27

Percentage profit = $\frac{27}{45} \times 100 = 60\%$

Example 13.6

Question

The value of a computer drops from $1200 to $700 in a year.
What percentage decrease is this?

Solution

Decrease in value = $1200 − $700 = $500

Percentage decrease = $\frac{500}{1200} \times 100 = 41.7\%$ to 1 decimal place

Exercise 13.4

1. A shopkeeper buys an article for $10 and sells it for $12.
   What percentage profit does she make?
2. Adam earned $5 per hour. His pay is increased to $5.60 per hour.
   What is the percentage increase in Adam's pay?
3. Rahat bought a car for $1250. A year later he sold it for $600.
   What percentage of the value did he lose?
4. Umar buys a signed photo for $22 and sells it for $28.
   What is his percentage profit? Give your answer to the nearest 1%.
5. Hannah reduces the time it takes her to swim 30 lengths from 55 minutes to 47 minutes.
   What percentage reduction is this? Give your answer to the nearest 1%.

Percentage increase and decrease

To find the amount after a percentage increase, work out the increase and add it on to the original amount.

To find the amount after a percentage decrease, work out the decrease and subtract it from the original amount.

Example 13.7

Question

An engineer receives an increase of 3% in her annual salary.
Her salary was $24 000 before the increase.
What is her new salary?

Solution

3% = 0.03
Increase = $24 000 × 0.03 = $720   Find the increase.
New salary = $24 000 + $720 = $24 720   Add the increase to the original amount.

Example 13.8

Question

In a sale all prices are reduced by 15%.
Find the sale price of an article previously priced at $17.60.

Solution

15% = 0.15
17.60 × 0.15 = 2.64   Find the decrease.
$17.60 − 2.64 = $14.96   Subtract the decrease from the original amount.

Percentage increase and decrease using a multiplier

Exercise 13.5

1. Increase $400 by each of these percentages.
   a 20%   b 45%   c 6%   d 80%

2. Decrease $200 by each of these percentages.
   a 30%   b 15%   c 3%   d 60%

3. The rent for a shop was $25 000 a year and went up by 20%.
   What is the new rent?

4. Simon earns $12 000 per year.
   He receives a salary increase of 4%.
   Find his new salary.

5. Prices are reduced by 20% in a sale.
   The original price of a top was $13.
   What is the sale price?

6. Mike puts $1200 into a savings account.
   7.5% interest is added to the account at the end of the year.
   How much is in the account at the end of the year?

7. To test the strength of a piece of wire, it is stretched by 12% of its original length.
   The piece of wire was originally 1.5 m long.
   How long will it be after stretching?

8. Rushna pays 6% of her pay into a pension fund.
   She earns $185 per week.
   What will her pay be after taking off her pension payment?

9. A drum kit costs $280 before sales tax is added.
   What will it cost after sales tax at 17% has been added?

10. A savings account pays 6.2% per annum.
    Sohail put $2500 in this account and left it for a year.
    How much was in the account at the end of the year?

Percentage increase and decrease using a multiplier

You can find a quantity after a percentage increase or decrease in one step using a multiplier.

Example 13.9

Question

Amir's salary is $17 000 per year.
He receives a 3% increase.
Find his new salary.

Solution

Amir's new salary is 103% of his original salary.
So the multiplier is 1.03.
New salary = $17 000 × 1.03 = $17 510.

13 PERCENTAGES

Example 13.10

Question
In a sale all prices are reduced by 15%.
Rushna buys a tablet in the sale.
The original price was $155.
Calculate the sale price.

Solution
A 15% decrease is the same as 100% − 15% = 85%.
So the multiplier is 0.85.
Sale price = $155 × 0.85 = $131.75

Exercise 13.6

1 Write down the multiplier that will increase an amount by each of these percentages.
   a 13%   b 20%   c 68%   d 8%
   e 2%   f 17.5%   g 150%

2 Write down the multiplier that will decrease an amount by each of these percentages.
   a 14%   b 20%   c 45%   d 7%
   e 3%   f 23%   g 16.5%

3 Sanjay used to earn $4.60 per hour.
   He received a 4% increase.
   How much does he earn now? Give your answer to the nearest cent.

4 In a sale, everything is reduced by 30%.
   Ali buys a pair of shoes in the sale.
   The original price was $42.
   What is the sale price?

5 A shop increases its prices by 8%.
   What is the new price of a skirt which previously cost $30?

Finding the original quantity

To increase a quantity by 20%, you calculate

new amount = original amount × 1.20.

It follows that

original amount = new amount ÷ 1.20.

Example 13.11

Question
Jared received a salary increase of 20%.
After the increase his salary is $31 260.
What was his salary before the increase?

Solution
New salary = old salary × 1.20
Old salary = new salary ÷ 1.20
           = $31 260 ÷ 1.20
           = $26 050

Finding the original quantity

To decrease a quantity by 10%, you calculate

new amount = original amount × 0.90.

It follows that

original amount = new amount ÷ 0.90.

> ### Example 13.12
>
> **Question**
> In a sale, all prices are reduced by 10%.
> Selma paid $38.70 for a dress in the sale.
> What was the original price of the dress?
>
> **Solution**
> New price = original price × 0.90
> Original price = new price ÷ 0.90
> $\qquad\qquad$ = $38.70 ÷ 0.90
> $\qquad\qquad$ = $43

Exercise 13.7

1. After an increase of 12%, a quantity is 84 tonnes.
   What was it before the increase?
2. A quantity is decreased by 3%. It is now 38.8.
   What was it to start with?
3. In a sale everything is reduced by 5%.
   A pair of shoes costs $47.50 in the sale.
   How much did they cost before the sale?
4. A newspaper increases the average number of copies it sells in a day by 3%.
   The new number of copies sold is 58 195.
   What was it before the increase?
5. Mr Lee made a profit of $13 250 in the year 2023.
   This was an increase of 6% on his profit in 2022.
   What was his profit in 2022?
6. Santos sold his car for $8520.
   This was 40% less than he paid for it five years before.
   What did he pay for it?
7. A charity's income in 2023 was $8580.
   This was a decrease of $2\frac{1}{2}$% compared with 2022.
   What was its income in 2022?
8. Stephen was given a rise of 7%.
   His salary after the rise was $28 890.
   What was it before the rise?
9. Between 2012 and 2013 house prices increased by 12.5%.
   In 2013 the value of a house was $27 000.
   What was its value in 2012?
10. A television costs $564 including a sales tax of 17.5%.
    What is the cost without the sales tax?

Wages and salaries

A **wage** or **wages** is an amount you earn, usually according to how many hours or days you work each week or month. It may be paid weekly or monthly.

A **salary** is a fixed amount of money you earn each month or year from your job. It is paid monthly.

What you earn is called your **gross pay**.

Tax and other amounts may be deducted from your gross pay.

The amount actually received by the employee is often called the **take-home pay**.

Some employees, particularly sales staff, are paid a basic wage or salary plus a **commission**, which is usually a percentage of their sales.

Example 13.13

Question

a Jenny's annual salary is $8500.
   She is paid monthly.
   What is her monthly gross pay?

b Rob's basic annual pay is $9600.
   He also earns 5% commission on the sales he makes each month.
   In May his sales are $7500. What is his gross pay for May?

Solution

a Jenny's monthly gross pay = $8500 ÷ 12 = $708.333…
                              = $708.33

b Basic monthly pay = $9600 ÷ 12 = $800
   Commission = $7500 × 0.05 = $375   Use your preferred method to find a multiplier.
   Total pay for May = $800 + $375 = $1175

Discount, profit and loss

A **discount** is a reduction in the price of something.

This is often a percentage of the original price but can be a fixed amount.

Example 13.14

Question

A jacket is priced at $38. The shopkeeper offers Rashid a discount of 15%.
How much does Rashid pay for the jacket?

▶

Discount, profit and loss

Solution
Rashid will pay 100 − 15 = 85% of the original price

Rashid pays $38 × 0.85 = $32.30

Use your preferred method to find a percentage of an amount.

Percentage profit and loss are always worked out as a percentage of the price paid by the **seller**.

Example 13.15

Question
A shopkeeper buys potatoes at $1.40 per kilogram.
He sells them for $1.80 per kilogram.
Find his percentage profit.

Solution
Profit = $1.80 − $1.40 = $0.40
Percentage profit = $\frac{0.40}{1.40} \times 100$ = 28.6% to 1 decimal place

Example 13.16

Question
Françoise buys a car for $9500.
A year later she sells it for $8300.
Calculate her percentage loss.

Solution
Loss = $9500 − $8300 = $1200
Percentage loss = $\frac{1200}{9500} \times 100$ = 12.6% to 1 decimal place

Exercise 13.8

1. **a** Pali earns an annual salary of $10 980. How much is this a month?
   b Steve's wage is $195 per week. How much does he earn in a year?
2. In a special promotion, a health club offers 20% discount for membership.
   Membership usually costs $35.
   How much does it cost with the discount?
3. Laura earns £12 000 per year plus commission of 7.5% on her sales.
   Last year she sold $76 000 of goods.
   What was her average monthly pay last year?
4. A shopkeeper buys an article for $22 and sells it for $27.
   What is his percentage profit?
5. A1 Electrics buys washing machines for $255 and sells them for $310.
   Bob's Budget Bargains buys washing machines for $270 and sells them for $330.
   Which company makes the greater percentage profit and by how much?
6. A trader buys an item for $12.50 and sells it at a loss for $5.
   What is his percentage loss?
7. A shopkeeper buys 100 cups for $1.20 each.
   He sells 80 of them, making a profit of 15% on each.
   He sells the remaining 20 at 95c each.
   What is his overall percentage profit?

13 PERCENTAGES

Repeated percentage change

Earlier in this chapter, you were shown a single-step method for increasing a quantity by a percentage.

For example, to increase a quantity by 18%, you multiply by 1.18.

Similarly, to increase a quantity by 7%, you multiply by 1.07.

This method is particularly useful when calculating repeated increases.

Example 13.17

Question

Due to inflation, prices increase by 5% per year.
An item costs $12 now.
What will it cost in 2 years' time?

Solution

A 5% increase means that the new amount is 100 + 5 = 105% of the old amount each year.
To find 105%, you multiply by 1.05.
In 1 year the price will be $12 × 1.05 = $12.60.
In 2 years the price will be $12.60 × 1.05 = $13.23.
Alternatively, this calculation can be worked out as
$12 × 1.05 × 1.05 = $13.23
or $12 × 1.05^2 = $13.23

Example 13.18

Question

Selena invests $4000 and receives 5% interest a year which is added to the amount each year.
How much is her investment worth in total after
a   1 year   b   6 years?

Solution

a   After one year the investment is worth $4000 × 1.05 = £4200.
b   After 6 years the investment is worth
    4000 × 1.05 × 1.05 × 1.05 × 1.05 × 1.05 × 1.05
    = 4000 × $(1.05)^6$
    = $5360.38 (to the nearest cent)

Note

On a calculator, the calculation for part **b** can be done using the power key.
This is usually labelled $x^\blacksquare$ but may be labelled y^x or x^y or ^.

Compound and simple interest

When you invest money in a bank, you may receive interest.

There are two types of interest.

Simple interest is calculated as a percentage of the original amount.

Compound and simple interest

If you invest $4000 for 2 years in an account that pays 5% simple interest annually, you will receive

$4000 × 0.05 = $200 at the end of the first year and the same again each year.

By contrast, **compound interest** is added to the original amount so that the amount of interest paid each year increases.

Example 13.18 is an example of compound interest.

At the end of the first year the interest is

5% of $4000 = $200.

This is added to the original amount. At the end of the second year the interest is calculated as a percentage of this new amount.

5% of $4200 = $210

So each year the interest received increases.

Compound interest is just one example of repeated percentage change and you can calculate the new amount in the same way as you calculate other repeated percentage changes.

However, because it has such wide application in everyday life, it has its own formula for working out the value of the investment after a given time.

$$A = P\left(1 + \frac{r}{100}\right)^n$$

In this formula

- A is the value of the investment after the given time
- P is the amount invested and is called the **principal**
- r is the percentage rate of interest (usually per year but it can be per month)
- n is the time (usually in years) for which the money is invested.

You can use either the method used in Example 13.18 or the formula above.

Example 13.19

Question

Salim has $4000 to invest for 5 years.
The bank offers him two options.
- 3.5% simple interest or
- 3.25% compound interest

Which is the better option?
How much more will his investment be worth with the better option?

13 PERCENTAGES

Solution

3.5% simple interest
Interest for one year = 0.035 × 4000 = $140
Interest for 5 years = 5 × $140 = $700
Value of investment at the end of 5 years = $4000 + $700 = $4700
3.25% compound interest
The multiplier for a 3.25% increase is 1.0325.
Value of investment at the end of 5 years = $4000 × 1.0325^5 = $4693.65 (to the nearest cent).
Alternatively, using the formula

$$A = P\left(1 + \frac{r}{100}\right)^n$$
$$= 4000 \times \left(1 + \frac{3.25}{100}\right)^5$$
$$= \$4693.65 \text{ (to the nearest cent)}$$

So 3.5% simple interest is the better option.
Difference = $4700 − $4693.65 = $6.35 (to the nearest cent)

Exercise 13.9

1 Craig puts $240 into a savings account that earns compound interest of 6% annually.
   What will his savings be worth after 3 years?
   Give your answer to the nearest cent.

2 Calculate the amount each of these items will be worth if they reduce in value each year by the given percentage for the given number of years.

	Original value	% reduction	Number of years
a	$250	45	5
b	$3500	11	7
c	£1400	15	4
d	$10 500	12	10

3 $1500 is invested in an account that earns compound interest of 4% annually.
   What is the value of the investment after four years?

4 Calculate the final value of the investment when $3000 is invested with compound interest in these cases. Give your answers to the nearest cent.
   a 5% for 4 years   b 6% for 20 years   c 3.5% for 10 years

5 Martin had shares worth $8000.
   They increased in value by 7.5% each year.
   What was their value after 10 years?
   Give your answer to the nearest dollar.

6 Mr Costa was given an 8% pay rise every year.
   His starting salary was $28 500.
   How much does he earn after four rises? (Give your answer to the nearest dollar.)

7 Pulova has high inflation. In 2023, prices increased by 15% a month for the first six months and 12.5% for the next six months.
A car cost 78 000 dubs (their unit of currency) in January 2023.
How much did it cost
  a after six months?
  b in January 2024?
Give your answers to the nearest whole number.

8 Andrew invested $3500 in a six-year bond that paid 5% compound interest for the first three years and 7.5% compound interest for the next three years.
What is the amount in the bond, to the nearest cent
  a after three years?   b after six years?

9 Find the difference in interest earned by investing $500 for three years at 12% simple interest, or for three years at 10% compound interest.

10 Is it better to invest $1000 for five years at 8% compound interest or for four years at 9% compound interest?

11 Yassim invested $500 at 4% compound interest.
How many years will it take for the value of the investment to be over $600?

Key points

- A percentage is equivalent to a fraction with the percentage as the numerator and 100 as the denominator. To find the equivalent decimal, divide out the fraction.
- To find a percentage of a quantity, multiply the quantity by the fraction or decimal equivalent to the percentage.
- To find one quantity as a percentage of another, write a fraction with the first quantity as numerator and the second as denominator. Then, multiply by 100.
- Percentage change = $\dfrac{\text{change}}{\text{original amount}} \times 100$.
- To find a quantity after a percentage increase, add the percentage to 100% and use the equivalent decimal as a multiplier.
- To find a quantity after a percentage decrease, subtract the percentage from 100% and use the equivalent decimal as a multiplier.
- Simple interest paid on a sum of money is calculated on the original sum.
- Compound interest is added to the original investment.
  $A = P\left(1 + \dfrac{r}{100}\right)^n$
  A is the value of the investment at any time
  P is the amount invested
  r is the percentage rate of interest
  n is the number of years the money is invested.
- To find the original quantity, given the result of an increase or decrease, divide by the multiplier.

14 USING A CALCULATOR

BY THE END OF THIS CHAPTER YOU WILL BE ABLE TO:
- use a calculator efficiently
- enter values appropriately on a calculator
- interpret the calculator display appropriately.

CHECK YOU CAN:
- use the correct order of operations in calculations
- work with numbers in standard form
- use a calculator to find squares, square roots, cubes and cube roots
- round numbers to a given degree of accuracy
- estimate the answers to calculations by rounding to one significant figure.

Order of operations

You should know that this is the correct order of operations when carrying out any calculation:

- First work out anything in brackets.
- Then work out any powers (such as squares or square roots).
- Then do any multiplication or division.
- Finally do any addition or subtraction.

A scientific calculator should follow this order of operations.

Using the correct order of operations, $2 + 3 \times 4 = 14$.

Enter the calculation $\boxed{2}\ \boxed{+}\ \boxed{3}\ \boxed{\times}\ \boxed{4}$ on your calculator and check that you get the answer 14.

When a calculation is written as a fraction, you will need to use brackets to ensure that the calculator divides the result of the calculation in the numerator by the result of the calculation in the denominator.

Using the correct order of operations, $\frac{9+3}{4+2} = 2$.

On a calculator you will need to enter $\boxed{(}\ \boxed{9}\ \boxed{+}\ \boxed{3}\ \boxed{)}\ \boxed{\div}\ \boxed{(}\ \boxed{4}\ \boxed{+}\ \boxed{2}\ \boxed{)}$

for this calculation. Check that you get the answer 2 when you do this.

A scientific calculator has function keys that can be used in calculations.

You need to find out what these keys look like on your calculator.

Your calculator will have keys for squares, cubes, square roots and cube roots.

You should make sure that you can use these keys in calculations.

Example 14.1

Question

Use a calculator to work out these, giving each answer correct to 3 significant figures.

a $\dfrac{4.65 + 5.72}{0.651 \times 12.7}$

b $\sqrt{3.65^2 - 1.83^2}$

Solution

a In this calculation, the fraction line means that you must divide the result of the calculation in the numerator by the result of the calculation in the denominator.

You use brackets so that the calculator does this.

$(4.65 + 5.72) \div (0.651 \times 12.7)$ This calculation is entered on the calculator.
$= 1.254\ 278\ 699$ Round the result to 3 significant figures.
$= 1.25$
Check: $(5 + 6) \div (0.7 \times 10) = 11 \div 7 = 1.5…$
The estimate is close to the calculated answer, which suggests that the correct order of operations has been used.

b In this calculation, the complete subtraction is inside the square root sign.
$\sqrt{(3.65^2 - 1.83^2)}$ This calculation is entered on the calculator. Use the square key and the square root key. You may need to use brackets.
$= 3.158\ 100\ 695$ Round the result to 3 significant figures.
$= 3.16$
Check: $\sqrt{(4^2 - 2^2)} = \sqrt{(16-4)} = \sqrt{(12)}$; $\sqrt{(12)}$ is close to the calculated answer, which suggests that the correct order of operations has been used.

> **Note**
> It is a good idea to check that your answer is approximately correct by rounding the values to one significant figure and working out an estimate of the answer. If the estimated answer is not close to the calculated answer, this suggests that you have not used the correct order of operations.

Exercise 14.1

Use a calculator to work out these. Give your answers correct to 3 significant figures where appropriate.

1 $\dfrac{-4.7 + 2.6}{-5.7}$

2 $\dfrac{-4.72}{-1.4} \times \dfrac{8.61}{-7.21}$

3 $\dfrac{7.92 \times 1.71}{-4.2 + 3.6}$

4 $\dfrac{3.14 - 8.16}{-8.25 \times 3.18}$

5 $\dfrac{5.2 + 10.3}{3.1}$

6 $\dfrac{127 - 31}{25}$

7 $\dfrac{9.3 + 12.3}{8.2 - 3.4}$

8 $6.2 + \dfrac{7.2}{2.4}$

9 $2.8 \times (5.2 - 3.6)$

10 $\dfrac{5.3}{2.6 \times 1.7}$

11 $\dfrac{5.3}{2.6 + 1.7}$

12 $2.7^3 + 3.8^2$

13 $\sqrt{1.2^2 + 0.5^2}$

14 $\sqrt{15.7 - 3.8}$

15 $\dfrac{2.6^2}{1.7 + 0.82}$

16 $\dfrac{6.2 \times 3.8}{22.7 - 13.8}$

17 $\dfrac{5.3}{\sqrt{6.2 + 2.7}}$

18 $\dfrac{5 + \sqrt{25 + 12}}{6}$

19 $2.7^2 + 3.6^2 - 2 \times 2.7 \times 3.6 \times 0.146$

20 $\dfrac{6.2}{2.6} + \dfrac{5.4}{3.9}$

21 $\dfrac{2.6 + 4.2}{7.8 \times 3.6^2}$

22 $\sqrt[3]{2.5^2 - 1.8^2}$

> **Note**
> Some of these questions can be done using the fraction button on a calculator (see later in the chapter).

Standard form on your calculator

You can do calculations involving standard form on your calculator, using the ×10ˣ or EXP key.

14 USING A CALCULATOR

Check which of these your calculator has.

In following work $\boxed{\times 10^x}$ is used, but if your calculator has $\boxed{\text{EXP}}$ use that instead.

> ### Example 14.2
>
> **Question**
> Find $7 \times 10^7 \div 2 \times 10^{-3}$ using your calculator.
>
> **Note**
> If using the $\boxed{\text{EXP}}$ key, do not enter 10 as well.
>
> **Solution**
> These are the keys to press on your calculator.
>
> $\boxed{7}\ \boxed{\times 10^x}\ \boxed{7}\ \boxed{\div}\ \boxed{2}\ \boxed{\times 10^x}\ \boxed{(-)}\ \boxed{3}\ \boxed{=}$
>
> You should see 3.5×10^{10}.

Sometimes, your calculator will display an ordinary number which you will have to write in standard form, if necessary. Otherwise, your calculator will give you the answer in standard form.

Modern calculators usually give the correct version of standard form, for example, 3.5×10^{10}.

Older calculators often give a calculator version such as 3.5^{10}. You must write this in proper standard form, 3.5×10^{10}, for your answer.

Some graphical calculators display standard form as, for example, 3.5 E 10.

Again, you must write this in proper standard form for your answer.

Practise by checking Example 14.3 on your calculator.

> ### Example 14.3
>
> **Question**
> The radius of Neptune is 2.48×10^4 km.
> Assume that Neptune is a sphere.
> The surface area of a sphere is given by $4\pi r^2$.
> Calculate the surface area of Neptune.
> Give your answer in standard form, correct to 2 significant figures.
>
> **Solution**
> $A = 4\pi r^2 = 4 \times \pi \times (2.48 \times 10^4)^2$
>
> This is the sequence of keys to press.
>
> $\boxed{4}\ \boxed{\times}\ \boxed{\text{SHIFT}}\ \boxed{\pi}\ \boxed{\times}\ \boxed{(}\ \boxed{2}\ \boxed{.}\ \boxed{4}\ \boxed{8}\ \boxed{\times 10^x}\ \boxed{4}\ \boxed{)}\ \boxed{x^2}\ \boxed{=}$
>
> The result on your calculator should be 7 728 820 583.
> So the answer is 7.7×10^9 km² correct to 2 significant figures.

Exercise 14.2

1. Work out these. Give your answers in standard form correct to 3 significant figures.
   - a $6.21 \times 10^5 \times 3.78 \times 10^7$
   - b $8.34 \times 10^7 \div 1.78 \times 10^2$
   - c $5.92 \times 10^8 \div 3.16 \times 10^{-3}$
   - d $6.27 \times 10^{-4} \times 4.06 \times 10^{-3}$
   - e $9.46 \times 10^8 \div 3.63 \times 10^3$
   - f $7.3 \times 10^4 \times 3.78 \times 10^3$
   - g $(5.63 \times 10^5)^2$
   - h $(8.76 \times 10^{-4})^2$

2. The radius of Jupiter is 7.14×10^4 km.
   Assume that Jupiter is a sphere.
   The surface area of a sphere is given by $4\pi r^2$.
   Calculate the surface area of Jupiter.
   Give your answer in standard form, correct to 2 significant figures.

3. The population of the UK in 2011 was approximately 63 200 000.
   - a Write 63 200 000 in standard form.
   - b The area of the UK is approximately 2.44×10^5 km^2.
     Calculate the average number of people per km^2 in the UK in 2011.
     Give your answer to a suitable degree of accuracy.

4.
   - a The planet Neptune is 4.5×10^9 km from the Sun.
     Light travels at 3×10^5 km/s.
     How many seconds does it take light to travel from the Sun to Neptune?
     Give your answer in standard form.
   - b Earth is 1.5×10^8 km from the Sun.
     How much further from the Sun is Neptune than Earth?
     Give your answer in standard form.

5. The population of Asia in 2013 was estimated to be 4.299 billion.
   - a Write 4.299 billion in standard form.
   - b The area of Asia is approximately 4.34×10^7 km^2.
     Calculate the number of people per km^2 in Asia.
     Give your answer to a suitable degree of accuracy.

6. Light takes approximately 3.3×10^{-9} seconds to travel 1 metre.
   The distance from the Earth to the Sun is 150 000 000 km.
   - a Write 150 000 000 km in metres using standard form.
   - b How long does it take for light to reach the Earth from the Sun?

Checking accuracy

When you have solved a problem, it is useful to check whether the answer is sensible.

When the problem is set in a real-life context you may be able to use your experience to help you check. For example, if the question is about a shopping bill, think about whether the total you have found is realistic.

If the question is not set in context, you need to use other methods to check your answer.

You have checked answers by rounding the numbers in the calculation to one significant figure and finding an estimate for the result. There are also other methods that you can use.

14 USING A CALCULATOR

Using number facts

Look at these calculations.

$40 \times 5 = 200$ $40 \times 0.5 = 20$

$40 \times 2 = 80$ $40 \times 0.8 = 32$

$40 \times 1.1 = 44$ $40 \times 0.9 = 36$

You can see that when 40 is multiplied by a number greater than 1, the result is greater than 40.

When 40 is multiplied by a number less than 1, the result is smaller than 40.

Look at these calculations.

$40 \div 5 = 8$ $40 \div 0.5 = 80$

$40 \div 2 = 20$ $40 \div 0.8 = 50$

$40 \div 1.1 = 36.36\ldots$ $40 \div 0.9 = 44.44\ldots$

You can see that when 40 is divided by a number greater than 1, the result is smaller than 40.

When 40 is divided by a number less than 1, the result is greater than 40.

These results can be applied to any multiplication or division.

Starting with any positive number,

- multiplying by a number greater than 1 gives a result that is larger than the starting number
- multiplying by a number between 0 and 1 gives a result that is smaller than the starting number
- dividing by a number greater than 1 gives a result that is smaller than the starting number
- dividing by a number between 0 and 1 gives a result that is larger than the starting number.

Using inverse operations

Inverse operations can be used to check the result of a calculation.

You know that subtraction is the inverse of addition and that division is the inverse of multiplication.

You can use this to check the result of a calculation.

Work out $6.9 \div 75$ on your calculator. The answer is 0.092.

You can check this answer using the multiplication 0.092×75.

The answer to the multiplication is 6.9, so the answer to the division was correct.

Finding the square root is the inverse of finding the square of a number.

You can use these facts to check the result of a calculation.

Interpreting the calculator display

Example 14.4

Question

Explain how you can tell the answer to each of these calculations is wrong.
a $297 \div 2.8 = 10.6$ to 3 significant figures
b $752 \div 24 = 18\,048$
c $\sqrt{35} = 9.52$ to 2 decimal places

Solution

a Rounding to 1 significant figure gives $300 \div 3 = 100$, so the answer 10.6 is too small.
b 752 is divided by a number greater than 1, so the result should be less than 752 but 18 048 is greater than 752.
c $6^2 = 36$, so $\sqrt{35}$ must be less than 6, but 9.52 is greater than 6.
   Alternatively, $9^2 = 81$, so $\sqrt{35}$ must be much less than 9.

Exercise 14.3

1 Which of these answers might be correct and which are definitely wrong? Show how you decide.
   a $39.6 \times 18.1 = 716.76$
   b $175 \div 1.013 = 177.275$
   c $8400 \times 9 = 756\,000$
   d An elevator can hold 9 people, so a group of 110 people will need 12 trips.
   e Musa has $100 to spend. He thinks that he has enough money to buy 5 DVDs costing $17.99 each.

2 Look at these calculations. The answers are all wrong.

   For each calculation, show how you can tell this quickly, without using a calculator to work it out.

   a $-6.2 \div -2 = -3.1$
   b $12.4 \times 0.7 = 86.8$
   c $31.2 \times 40 = 124.8$
   d $\sqrt{72} = 9.49$ to 2 d.p.
   e $0.3^2 = 0.9$
   f $16.2 \div 8.1 = 20$
   g $125 \div 0.5 = 25$
   h $6.4 \times -4 = 25.6$
   i $24.7 + 6.2 = 30.8$
   j $76 \div 0.5 = 38$

Interpreting the calculator display

When a calculation involves units, you need to make sure that you use the same units for all of the quantities and that you give the correct units for your answer.

It is also important that you think about the accuracy needed when you give your answer.

If the answer is not exact, think about what degree of accuracy is sensible.

14 USING A CALCULATOR

If the question is set in context, the context may help you to decide on what accuracy is appropriate. For example, in a problem about money, the answer should be given to two decimal places if it is not exact.

If the problem is not set in context and the question does not tell you what accuracy to use, then you should give your answer correct to three significant figures.

If you need to use the result of a calculation in a second calculation, then you should use the unrounded value in the second calculation.

Example 14.5

Question

a $18 is shared equally between 10 people. How much does each person receive?
b $18 is shared equally between 13 people. How much does each person receive?

Solution

a $18 \div 10 = 1.8$ The calculator answer is 1.8.
   Each person receives $1.80. Money should be written with 2 decimal places, so add the final zero.
b $18 \div 13 = 1.384\,615\ldots$ The answer is not exact.
   Each person receives $1.38. Round the answer to 2 decimal places.

Exercise 14.4

1 These are the amounts of water Denise drinks each day for 4 days.
   1.5 litres   800 ml   2 litres   1 litre 250 ml
   How much water did she drink in total?

2 A piece of string 4 m long is cut into seven equal pieces.
   How long is each piece?

3 Work out how much each person receives in each case.
   a $250 is divided equally between 4 people.
   b $67 is divided equally between 8 people.
   c $137 is divided equally between 12 people.
   d $475 is divided equally between 17 people.

4 Naquila is paid $12.45 per hour.
   Work out how much she is paid for 8 hours' work.

Calculating with time

When a calculation involves time, you need to make sure that you use your calculator correctly.

Calculating with time

Entering time on a calculator

There are 60 minutes in an hour, so a time of 2 hours 30 minutes is the same as $2\frac{1}{2}$ hours.

You enter 2.5 on your calculator.

If you need to enter a time such as 1 hour 12 minutes, first convert the minutes to a fraction of an hour.

12 minutes is $\frac{12}{60}$ of an hour or, as a decimal, 0.2 hour.

So 1 hour 12 minutes is 1.2 hours.

You can use the same method to enter times given in minutes and seconds on your calculator.

Interpreting the result of a time calculation

Time calculations often require answers to be given in hours and minutes or minutes and seconds. If you have used a calculator, then the answer will be a decimal which you need to convert to the correct units.

Some conversions can be done mentally.

For example, you know that 1.5 minutes is the same as $1\frac{1}{2}$ minutes.

There are 60 seconds in a minute, so 1.5 minutes is the same as 1 minute 30 seconds.

Similarly, you know that 1.25 minutes is the same as $1\frac{1}{4}$ minutes.

There are 60 seconds in a minute, so 1.25 minutes is the same as 1 minute 15 seconds.

> ### Example 14.6
>
> **Question**
> a Convert 8 minutes 36 seconds into minutes.
> b Write 6.4 hours in hours and minutes.
>
> **Solution**
> a 8 minutes 36 seconds is $8\frac{36}{60}$ minutes.
> $36 \div 60 = 0.6$ Convert the fraction to a decimal using division.
>   8 minutes 36 seconds is 8.6 minutes.
> b $0.4 \times 60 = 24$ Multiply the decimal by 60 to convert to minutes.
>   0.4 hour is 24 minutes
>   6.4 hours is 6 hours 24 minutes.

Photocopying is prohibited

14 USING A CALCULATOR

Exercise 14.5

1. Convert each of these times into hours.
   - a 1 hour 45 minutes
   - b 2 hours 12 minutes
   - c 39 minutes
2. Convert each of these times into minutes.
   - a 5 minutes 15 seconds
   - b 3 minutes 27 seconds
   - c 42 seconds
3. Write each of these times in hours and minutes.
   - a 3.5 hours
   - b 1.3 hours
   - c 4.85 hours
4. Write each of these times in minutes and seconds.
   - a 3.75 minutes
   - b 2.4 minutes
   - c 1.9 minutes
5. Asma takes 8 minutes to make one birthday card.
   If she works at the same rate, how long will she take to make 20 cards?
   Give your answer in hours and minutes.

Hours, minutes and seconds on a calculator

Some calculators have a ° ' " key.

You can use this for working with time on your calculator.

When you do a calculation and the answer is a time in hours, but your calculator shows a decimal, you change it to hours and minutes by pressing the ° ' " key and then the = key.

To convert back to a decimal, press the ° ' " key again.

To enter a time of 8 hours 32 minutes on your calculator, press this sequence of keys:

[8] [° ' "] [3] [2] [° ' "] [=]

The display should look like this:

[8°32°0]

You may wish to experiment with this key and learn how to use it to enter and convert times.

Fractions on a calculator

You need to be able to calculate with fractions without a calculator.
However, when a calculator is allowed you can use the fraction key.

You may wish to experiment with calculations using the fraction key on your calculator. The fraction key will look like this ▤ or like this $a^b/_c$

The following instructions are for a calculator with a fraction key that looks like this ▤.

To enter a fraction such as $\frac{2}{5}$ into your calculator, you need to press this sequence of keys:

[2] [▤] [5] [=]

Fractions on a calculator

The display will look like this. $\frac{2}{5}$

This is the sequence of keys to press to do the calculation $\frac{2}{5} + \frac{1}{2}$:

The display should look like this $\frac{9}{10}$.

This is the sequence of keys to press to enter a mixed number, such as $2\frac{3}{5}$, into your calculator:

[2] [SHIFT] [▪] [3] [▼] [5] [=]

Your display will look like this $\frac{13}{5}$.

You can also cancel a fraction on your calculator.

When you press [8] [▪] [1] [2] [=], you should see: $\frac{8}{12}$.

When you press [=] again, the display changes to $\frac{2}{3}$.

When you do calculations with fractions on your calculator, it will automatically give the answer as a fraction in its simplest form.

Similarly, if you enter an improper fraction into your calculator and press the = key, the calculator will automatically change it to a mixed number.

Key points
- Use some method of estimating the answer as a check.
- Not all calculators are the same. Make sure you know how to use yours.
- Take care to correctly enter and interpret values representing time or money.

15 TIME

BY THE END OF THIS CHAPTER YOU WILL BE ABLE TO:
- calculate with time: seconds (s), minutes (min), hours (h), days, weeks, months, years, including the relationship between units
- calculate times in terms of the 24-hour and 12-hour clock
- read clocks and timetables.

CHECK YOU CAN:
- calculate speed.

The 24-hour clock

The time of the day can be given as a.m., p.m. or on the 24-hour clock.

Digital recorders use the 24-hour clock, so you are likely to be familiar with this.

You also need to be able to change from one to the other.

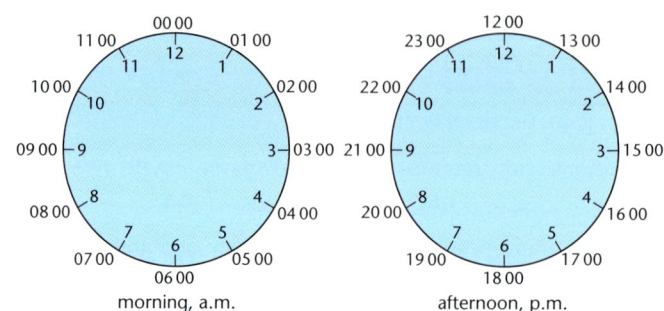

morning, a.m.   afternoon, p.m.

Example 15.1

Question

Write these times using the 24-hour clock.
- a 11.32 a.m.
- b 9.14 a.m.
- c 12.18 a.m.
- d 6.55 p.m.

Note
Times between midnight and 1 a.m. start with 00 on the 24-hour clock.

Solution
- a 11 32. Times using a.m. are the same as on the 24-hour clock.
- b 09 14. For times before 10 a.m. you need to add a zero at the start so that there are four figures.
- c 00 18. You need to add two zeros.
- d 18 55. To change times using p.m. to the 24-hour clock, add 12 to the hours.

Example 15.2

Question

Write these times using the 12-hour clock.
- a 11 46
- b 04 21
- c 14 22
- d 23 05

Note
Times between 12 noon (midday) and 1 p.m. do not have 12 subtracted from them for p.m. times.

Solution
- a 11.46 a.m.   Add a full stop and write a.m. 11:46 a.m. is also acceptable.
- b 4.21 a.m.   As it is in the morning, just add a.m. and drop the first zero.
- c 2.22 p.m.   As it is in the afternoon, subtract 12 hours and add p.m.
- d 11.05 p.m.

Exercise 15.1

1. Write these times using a.m. notation.
   - a 11 45
   - b 05 53
   - c 01 40
   - d 09 20
   - e 00 10
   - f 10 41
   - g 02 32
   - h 04 30
   - i 11 20
   - j 00 48

2. Write these times using p.m. notation.
   - a 13 45
   - b 15 53
   - c 21 40
   - d 22 59
   - e 12 10
   - f 14 40
   - g 17 23
   - h 19 40
   - i 20 19
   - j 12 03

3. Write these times using the 24-hour clock.
   - a 1.50 a.m.
   - b 2.40 p.m.
   - c 11.49 a.m.
   - d 6.30 p.m.
   - e 12.02 a.m.
   - f 3.20 a.m.
   - g 2.08 p.m.
   - h 12.49 a.m.
   - i 9.35 a.m.
   - j 11.02 p.m.

4. Write these times using the 12-hour clock.
   - a 03 45
   - b 14 56
   - c 23 40
   - d 11 59
   - e 12 55
   - f 04 35
   - g 15 16
   - h 21 40
   - i 01 59
   - j 14 52

Calculating with time

When calculating with time, remember that there are 60 minutes in an hour and 60 seconds in a minute.

Example 15.3

Question
Mavis left for school at 7.45 a.m. and arrived 40 minutes later.
At what time did she arrive?

Solution
7.45 + 40 minutes = 7 hours + 85 minutes   85 minutes is 1 hour and 25 minutes
                 = 8 hours 25 minutes
She arrived at school at 8.25 a.m.

Example 15.4

Question
Carla set her digital recorder to start recording at 15 20 and to stop at 17 10.
For how long was it recording?

Solution
Count to the next hour, 16 00. This is 40 minutes.
Then, to get to 17 10, a further 1 hour and 10 minutes are needed.
So the total time is 40 minutes + 1 hour 10 minutes = 1 hour 50 minutes.

15 TIME

Example 15.5

Question

It takes Nathee 35 minutes to prepare some food for cooking.

He is ready to start cooking at 6.15 p.m.

At what time did he start preparing the food?

Solution

Work backwards from 6.15 p.m.

15 minutes before 6.15 is 6.00 p.m.

That leaves 35 − 15 = 20 minutes still to take off.

20 minutes before 6.00 is 5.40 p.m.

So he started preparing the food at 5.40 p.m.

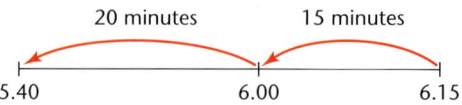

Timetables

Example 15.6

Question

This is the timetable for some trains from Delhi to Agra.

Delhi (depart)	06 00	08 35	10 05	15 55	20 55	21 15	22 30
Agra (arrive)	08 06	11 00	12 43	18 08	23 35	00 02	01 07

Work out how long these trains take to do the journey.

a   The 08 35 from Delhi

b   The 15 55 from Delhi

c   The 22 30 from Delhi

Solution

a   From 08 35 to 09 00 is 25 minutes.

 09 00 to 11 00 is 2 hours.

 So the 08 35 from Delhi takes 2 hours 25 minutes.

b   From 15 55 to 16 00 is 5 minutes.

 From 16 00 to 18 00 is 2 hours.

 From 18 00 to 18 08 is 8 minutes.

 So the 15 55 from Delhi takes 2 hours 13 minutes.

c   From 22 30 to 23 00 is 30 minutes.

 From 23 00 to 01 00 is 2 hours.

 From 01 00 to 01 07 is 7 minutes.

 So the 22 30 from Delhi takes 2 hours 37 minutes.

Exercise 15.2

1. A train due at 14 40 was 50 minutes late.
   What time did it arrive?
2. A ferry left Liverpool at 14 20 and arrived at Belfast at 21 05.
   How long did it take?
3. Kylie left home at 7.50 a.m. and took 2 hours 45 minutes to reach her friend's house.
   What time did she arrive at her friend's house?
4. Sam set his oven to switch on at 3.45 p.m. and switch off at 6.20 p.m.
   How long was the oven on?
5. Joe set his digital recorder to switch on at 8.35 p.m. and switch off at 10.20 p.m.
   How long was the recorder on?
6. A bus left Barnmouth at 15 47 and arrived in Manton at 18 20.
   How long did it take?
7. Here is part of the Howerdale railway timetable.

Stafford	08 15
Luker	09 05
Melpett	09 40
Haker	11 20
Golpath	11 35
Oldway	13 20

   a How long does it take to get from
     i   Stafford to Melpett
     ii Haker to Oldway
     iii Luker to Golpath
     iv Melpett to Oldway?
   b One day the train is delayed at Luker for 35 minutes.
     At what time did it arrive at
     i   Golpath
     ii Oldway?

8. Ciara's favourite film ended at 8.50 p.m.
   The film lasted for 1 hour and 45 minutes.
   What time did it start?
9. A train journey lasts 2 hours and 40 minutes.
   The train arrives at 10.35 a.m.
   What time did it start its journey?
10. A ferry leaves port at 08 35 on a 55-minute journey.
    The ferry arrives 20 minutes late due to bad weather.
    What time does it arrive at its destination?

Converting between hours and minutes and hours written as a decimal

When you are calculating with time, your calculator will often give times in hours using decimals, but many questions ask for answers in hours and minutes.

Similarly, a time in a question may be given in hours and minutes, but you need to change this to a decimal number to do the calculation.

Example 15.7

Question

Write 15.36 hours in hours and minutes, correct to the nearest minute.

Solution

0.36 hours = 0.36 × 60   To convert hours to minutes multiply by 60.
            = 21.6 minutes

So 15.36 hours = 15 hours 22 minutes correct to the nearest minute.

Example 15.8

Question

The Beijing to Shanghai train travels 1318 km in 5 hours 37 minutes.
Find the average speed of the train, correct to the nearest km/h.

Solution

37 minutes = 37 ÷ 60   To convert minutes to hours divide by 60.
           = 0.6166… hours

So the total time is 5 + 0.6166… = 5.6166… hours.

Speed = distance ÷ time

Speed = 1318 ÷ 5.6166… = 234.658…
      = 235 km/h correct to the nearest km/h

Time zones

The time of day varies in different countries, depending on their longitude.

In large countries, there may even be differences within the country.

Usually these differences are a whole number of hours.

For example, the local time in Sydney is 6 hours ahead of the local time in Islamabad.

This means that when it is 11 00 in Islamabad, it is 17 00 in Sydney.

The local time in Dubai is 5 hours behind the local time in Tokyo.

This means that when it is 21 00 in Tokyo, it is 16 00 in Dubai.

Example 15.9

Question

An aircraft leaves Dubai at 08 55 local time and flies to Tokyo.
The flight takes 9 hours 25 minutes.
Local time in Tokyo is 5 hours ahead of Dubai.
What is the local time in Tokyo when the aircraft lands?

> **Solution**
> 08 55 + 9 hours 25 minutes = 17 55 + 25 minutes = 18 20 (Dubai time)
> 18 20 + 5 hours = 23 20 (Tokyo time)

Exercise 15.3

1. Write these times in hours and minutes.
   Where necessary give the time to the nearest minute.
   a 5.4 hours   b 3.26 hours   c 2.84 hours   d 12.76 hours
2. Write these times in hours using decimals.
   Where necessary give the time to 2 decimal places.
   a 4 hours 15 minutes   b 1 hour 50 minutes   c 45 minutes
   d 10 hours 39 minutes   e 3 hours 38 minutes
3. Write these times in minutes and seconds.
   a 4.7 minutes   b 5.25 minutes   c 25.3 minutes   d 0.4 minutes
4. Assad leaves home at 13 48 and drives for 4.8 hours
   What time does he arrive?
5. Soraya drives for 3 hours 17 minutes and travels 140 km.
   What is her average speed?
6. A train travels at an average speed of 85 km/h.
   The journey distance is 495 km.
   The train leaves at 13 35.
   What time does it arrive? Give your answer to the nearest minute.
7. An aircraft leaves Karachi at 11 35, local time, to fly to London.
   The flight time is 8 hours 30 minutes.
   The time in London is 4 hours behind the time in Karachi.
   What is the local time in London when the aircraft arrives?
8. An aircraft leaves Dubai at 01 50 local time and arrives in Sydney at 22 30 local time.
   The flight time is 13 hours 40 minutes.
   How many hours ahead of Dubai time is Sydney time?

> **Key points**
> - Times of day can be given as a.m. (morning), p.m. (afternoon) or in the 24-hour clock.
> - Times in the morning have the same hours digits when converted to the 24-hour clock.
> - Times in the afternoon have 12 added to the hours digits when converted to the 24-hour clock.
> - 60 s = 1 min; 60 min = 1 h; 24 h = 1 day; 7 days = 1 week; 365 days = 1 year.
> - Local time varies in different countries, depending on longitude.

16 MONEY

BY THE END OF THIS CHAPTER YOU WILL BE ABLE TO:
- calculate with money
- convert from one currency to another.

CHECK YOU CAN:
- convert between metric units
- work with measures of rate.

Note
Whichever method you use, make sure that you make the right conclusion.

The better value is the one which costs less for each kilogram, gram, litre or millilitre or the one where you get more for each dollar or cent.

Value for money

When shopping, it is easy to assume that the biggest size is the best value for money, but this is not always the case.

Example 16.1

Question
A packet of Wheat-o-Flakes weighing 1.5 kg costs $4.99.
A packet of Wheat-o-Flakes weighing 500 g costs $1.70.
Which is better value for money?

Solution
Method 1
Find the price for each kilogram.
For the larger packet: $4.99 \div 1.5 = 3.326...$ $/kg
For the smaller packet: $1.70 \div 0.5 = 3.40$ $/kg
The larger packet costs slightly less for each kilogram so is better value for money.

Method 2
Find how many grams you get for each cent.
For the larger packet: $1500 \div 499 = 3.006...$ grams per cent
For the smaller packet: $500 \div 170 = 2.941...$ grams per cent
You get slightly more for each cent with the larger packet so it is better value for money.

Although the methods shown in Example 16.1 will always work, sometimes there is an obvious easier method.

For example, in Example 16.1, the larger packet weighs three times as much as the smaller packet.

When you buy three of the smaller packets, you get $0.5 \text{ kg} \times 3 = 1.5 \text{ kg}$ for $\$1.70 \times 3 = \$5.10$.

$4.99 is less than $5.10 so the larger packet is better value for money.

Example 16.2

Question

At Mega T, if you buy two T-shirts at $7.99 each, you get a third T-shirt free.
At T World, T-shirts cost $5.50 each if you buy two or more.
Explain which is the better offer if you want to buy

a three T-shirts   b four T-shirts.

Solution

a At Mega T, three T-shirts cost $2 \times \$7.99 = \$15.98$
   At T World, three T-shirts cost $3 \times \$5.50 = \$16.50$
   So Mega T has the better offer.

b At Mega T, you need to buy the fourth T-shirt at $7.99.
   Total cost = $15.98 + $7.99 = $23.97
   At T World, four T-shirts cost $4 \times \$5.50 = \$22.00$
   So T World has the better offer.

Exercise 16.1

1 Show which is the better value, 5 litres of water for $1.29 or 2 litres for 87c.
2 Here are some prices for bottles of cola.
   3 litres for $3.98   2 litres $2.70   1 litre $1.14   500 ml 65c
   Which size gives the best value?
3 A brand of shaving gel costs $2.38 for the 75 ml bottle and $5.78 for the 200 ml bottle. Show which is the better value.
4 Here are the prices of some packs of soda.
   12×150 ml cans for $5.90
   6×330 ml cans for $3.18
   12×330 ml cans for $5.98
   Which pack gives the most soda for your money?
5 An 800 g loaf of bread costs 96c.
   A 1.2 kg loaf costs $1.41.
   Show which is the better value for money.
6 A 420 g bag of chocolate bars costs $3.18.
   A 325 g bag of chocolate bars costs $2.18.
   Show which bag is the better value for money.
7 A 680 g block of cheese costs $5.18.
   A 500 g block of cheese costs $3.98.
   Show which block is the better value for money.
8 Li and Chan want to go out for a meal on Wednesday.
   Their two local restaurants have special offers.
   At restaurant A, if you buy a meal for $12.50, you get a second meal half price.
   At restaurant B, meals cost $9.50 on Monday to Thursday.
   Which restaurant will be cheaper for them?

16 MONEY

Currency conversion

Currency exchange rates tell you what one of the units of one currency is equal to in the other currency.

For example, if the exchange rate between US dollars ($) and euros (€) is $1 = €0.88, each US dollar is worth €0.88.

Exchange rates can change daily.

Example 16.3

Question

The exchange rate between dollars and euros is $1 = €0.88.

a Sam changes $135 into euros. How much will he receive in euros?

b Françoise changes €340 into dollars. How much will she receive in dollars?

Solution

a You know what each dollar is worth in euros, so you multiply.

$135 × 0.88 = €118.80

Note
The answer on the calculator was 118.8. When you write answers to money questions, you use two decimal places as the two decimal places give the number of the smaller unit, in this case, cents.

b You do not know how much in dollars each euro is worth.

You need to know how many 0.88s there are in $340, so you divide.

€340 ÷ 0.88 = $386.36 correct to the nearest cent.

Note
When you are given an exchange rate, the only decision is whether to multiply or divide by that rate.

You can check you have chosen correctly.

If $1 = €0.88 then the number of dollars will be greater than the number of euros.

Exercise 16.2

1 The exchange rate between US dollars (USD) and Pakistani rupees (PKR) is 1 USD = 101.42 PKR. Change
   a $240 to Pakistani rupees
   b 3500 PKR to US dollars.

2 The exchange rate between euros (EUR) and United Arab Emirates dirhams (AED) is 1 EUR = 4.16 AED. Change
   a €280 to dirhams
   b 850 AED to euros.

Currency conversion

3 Jackie goes on holiday from Australia to Japan.
   a She changes 1250 Australian dollars (AUD) into Japanese yen (JPY).
     The exchange rate is 1 AUD = 91.89 JPY.
     How many Japanese yen does she receive?
   b When she returns, she changes 15 000 Japanese yen back into Australian dollars.
     The exchange rate is now 1 AUD = 93.15 JPY.
     How many Australian dollars does she receive?
4 Brian travels from London to Mauritius. He changes £450 (GBP) into Mauritian rupees (MUR).
   The exchange rate is 1 GBP = 50.98 MUR.
   a Calculate how much he receives in Mauritian rupees.
   Whilst there, he buys presents worth 3500 MUR.
   b Calculate how much these are worth in British pounds.
5 Khalid is buying a laptop computer. He can buy it in Pakistan for 54 500 rupees. He sees an advert on the internet for the same laptop. If he buys it on the internet, it will cost $450 + $80 for delivery.
   The exchange rate is 1 USD = 101.50 PKR.
   a Is it cheaper for Khalid to buy the laptop in Pakistan or on the internet?
   b What is the difference in the cost? Give your answer in
      i Pakistani rupees   ii dollars.
6 Here are two exchange rates.
   £1 = €1.35   £1 = 2.08 Singapore dollars
   a Assuming proportional exchange rates,
      i how many Singapore dollars will you get for €1?
      ii how many euros will you get for 1 Singapore dollar?
   b A camera costs 240 Singapore dollars.
     Calculate its cost in euros.

> **Key points**
> - To compare which item or offer is better value, either:
>   ○ find the price for each unit (such as price per kilogram), or
>   ○ find the amount for each unit of money (such as grams per dollar).
> - To convert between currencies, multiply or divide by the exchange rate, as appropriate (for example, if $1 = €0.88, $100 = 100 × 0.88 = €88.00 and €100 = 100 ÷ 0.88 = $113.64).

17 EXPONENTIAL GROWTH AND DECAY

BY THE END OF THIS CHAPTER YOU WILL BE ABLE TO:
- use exponential growth and decay.

CHECK YOU CAN:
- use constant multipliers.

Exponential growth

In Chapter 13, you met compound interest. This involved a constant multiplier and is an example of exponential growth.

Consider $200 invested at 5% per year compound interest.

After 1 year, it will be worth

$200 \times 1.05 = \$210$.

After 2 years, it will be worth

$200 \times 1.05 \times 1.05 = 200 \times 1.05^2 = \$220.50$.

After 3 years, it will be worth

$200 \times 1.05 \times 1.05 \times 1.05 = 200 \times 1.05^3 = \$231.53$.

After 20 years, it will be worth 200×1.05^{20}.

On a calculator, this is done using the powers key. This is usually labelled $x^{\blacksquare}$

The calculation is 200×1.05 $x^{\blacksquare}$ $20 = \$530.66$.

The formula for this calculation is $\$A = 200 \times 1.05^n$, where $\$A$ is the amount of the investment and n is the number of years.

Example 17.1

Question
The number of bacteria present in a population doubles every hour. If there are 500 bacteria present at 12 noon, find the number present
a at 2 p.m.
b at midnight
c after n hours.

Solution
a $500 \times 2 \times 2 = 2000$
b $500 \times 2^{12} = 2\,048\,000$
c 500×2^n

Exponential decay

A population of bats is declining by 15% a year.

At the start, there are 140 bats.

$100\% - 15\% = 85\%$

After 1 year, there will be 140 × 0.85 = 119

After 2 years, there will be 140 × 0.85 × 0.85 = 140 × 0.85^2 = 101, and so on.

This calculation, where the constant multiplier is less than 1, is an example of exponential decay.

The calculations involved work just like those for exponential growth, except that the multiplier is less than 1, so the result gets smaller.

After 10 years, there will be 140 × 0.85^{10} = 28

The formula for this calculation is $A = 140 \times 0.85^n$, where A is the number of bats and n is the number of years.

Example 17.2

Question

Each year, a car decreases by 12% of its value at the beginning of the year.
At the beginning of the year, a car was valued at $9000.
Find its value after three years.

Solution

100% − 12% = 88%
Value after three years = $9000 × $(0.88)^3$ = $6133.25 (to the nearest cent).

Note
When the value of something decreases, as in Example 17.2, it is often called *depreciation*.

Exercise 17.1

1. The value of an old book increases by 10% a year.
   In 2023, it was worth $450.
   What will it be worth in 2030? Give your answer to the nearest dollar.
2. Each year, a car depreciates by 11% of its value at the start of the year.
   At the beginning of the year, it is worth $8000.
   What will it be worth after two years?
3. The size, y, of a population of bacteria is growing according to the rule $y = 25 \times 1.02^t$, where t minutes is the measured time.
   a. How many bacteria are there at time $t = 0$?
   b. What will the population be 5 hours after starting to measure the time? Give your answer to the nearest whole number.
4. A population of bacteria is estimated to increase by 12% every 24 hours.
   The population was 2000 at midnight on Friday.
   What was the population (to the nearest whole number) by midnight the following Wednesday?
5. A radioactive element has a mass of 50 g.
   Its mass reduces by 10% each year.
   a. Write down a formula for the mass, m, of the element after t years.
   b. Calculate the mass after
      i. three years
      ii. ten years.

17 EXPONENTIAL GROWTH AND DECAY

6 Tony says his boat is increasing in value by 6% a year.
  It is valued at $25 000.
  How much would it be worth, to the nearest hundred dollars, after six years if he is correct?

7 The mass (m grams) of a chemical t minutes after the start of a chemical reaction is given by $m = 100 \times 0.5^t$.
  a What was the mass at the start?
  b What was the mass after
     i 5 minutes
     ii 30 minutes?
  Give your answers to the nearest gram.

8 A painting was worth $15 000 in 2015.
  The painting increased in value by 15% every year for 6 years.
  How much was it worth at the end of the 6 years?
  Give your answer to the nearest dollar.

9 Elaine buys a car for $9000.
  It depreciates in value by 12% each year.
  a Write down a formula for the value, v, of the car after t years.
  b Calculate the value of the car after
     i three years
     ii eight years.

10 The population of a country is increasing at a rate of 5% a year.
   In 2020, the population was 60 million.
   a Write down a formula for the population size, P million, after t years.
   b What will the population be in
      i 2025
      ii 2100?

Key points
- Exponential growth can be calculated using a constant multiplier greater than 1.
- Exponential decay uses a constant multiplier less than 1.

18 SURDS

Surds are irrational numbers such as $\sqrt{2}$ or $5+\sqrt{3}$.

Numbers such as $\sqrt{36}$ are not surds, since 36 is a perfect square. $\sqrt{36} = 6$, which is a rational number.

Surds can be expressed in the form $a + b\sqrt{c}$, where a and b are rational numbers and c is an integer that is not a perfect square.

Simplifying surds

Surds can often be simplified by using the result $\sqrt{a \times b} = \sqrt{a} \times \sqrt{b}$.

This result can be demonstrated using $\sqrt{36}$.

$\sqrt{36} = 6 = 2 \times 3 = \sqrt{4} \times \sqrt{9}$

So $\sqrt{36} = \sqrt{4 \times 9} = \sqrt{4} \times \sqrt{9}$

> **Note**
> By definition of what we mean by a square root, $\sqrt{a} \times \sqrt{a} = a$.

Example 18.1

Question

Simplify $\sqrt{50}$.

Solution

$\sqrt{50} = \sqrt{25 \times 2} = \sqrt{25} \times \sqrt{2} = 5\sqrt{2}$

Example 18.2

Question

Simplify $\sqrt{72}$.

Solution

9 is a factor of 72, so $\sqrt{72} = \sqrt{9} \times \sqrt{8} = 3\sqrt{8}$

However, 4 is a factor of 8, so $3\sqrt{8} = 3 \times \sqrt{4} \times \sqrt{2}$
$= 3 \times 2 \times \sqrt{2}$
$= 6\sqrt{2}$

Or, if you spot straight away that 36 is a factor of 72, $\sqrt{72} = \sqrt{36 \times 2}$
$= \sqrt{36} \times \sqrt{2}$
$= 6\sqrt{2}$

> **BY THE END OF THIS CHAPTER YOU WILL BE ABLE TO:**
> - understand and use surds, including simplifying expressions
> - rationalise the denominator.

> **CHECK YOU CAN:**
> - multiply out and simplify brackets such as $(a + b)(c + d)$
> - recognise rational and irrational numbers.

> **Note**
> Look for as large a factor of the number as possible which has an exact square root. In Example 18.1, this is 25. In Example 18.2 it is 36.

18 SURDS

> **Example 18.3**
>
> **Question**
> Simplify $\sqrt{12} \times \sqrt{27}$.
>
> **Solution**
>
> **Method 1**
>
> $\sqrt{12} = \sqrt{4 \times 3}$ $\sqrt{27} = \sqrt{9 \times 3}$
> $\phantom{\sqrt{12}} = \sqrt{4} \times \sqrt{3}$ $\phantom{\sqrt{27}} = \sqrt{9} \times \sqrt{3}$
> $\phantom{\sqrt{12}} = 2 \times \sqrt{3}$ $\phantom{\sqrt{27}} = 3 \times \sqrt{3}$
>
> So $\sqrt{12} \times \sqrt{27} = 2 \times \sqrt{3} \times 3 \times \sqrt{3}$
> $\phantom{So \sqrt{12} \times \sqrt{27}} = 2 \times 3 \times \sqrt{3} \times \sqrt{3}$
> $\phantom{So \sqrt{12} \times \sqrt{27}} = 6 \times 3$
> $\phantom{So \sqrt{12} \times \sqrt{27}} = 18$
>
> **Method 2**
>
> $\sqrt{12} \times \sqrt{27} = \sqrt{12 \times 27}$
> $\phantom{\sqrt{12} \times \sqrt{27}} = \sqrt{324}$
> $\phantom{\sqrt{12} \times \sqrt{27}} = 18$

Manipulation of expressions of the form $a + b\sqrt{c}$

An expression such as $2 + \sqrt{3}$, which is the sum of a rational number and an irrational number, is irrational.

This is because $2 + 1.732\,050\,808\ldots = 3.732\,050\,808\ldots$ which is itself a decimal that goes on forever without recurring, and so is irrational.

> **Example 18.4**
>
> **Question**
> If $x = 5 + \sqrt{3}$ and $y = 3 - 2\sqrt{3}$, simplify these.
> **a** $x + y$ **b** $x - y$ **c** xy
>
> **Solution**
>
> **a** $x + y = 5 + \sqrt{3} + 3 - 2\sqrt{3}$
> $ = 5 + 3 + \sqrt{3} - 2\sqrt{3}$
> $ = 8 - \sqrt{3}$
>
> **b** $x - y = 5 + \sqrt{3} - (3 - 2\sqrt{3})$
> $ = 5 - 3 + \sqrt{3} + 2\sqrt{3}$
> $ = 2 + 3\sqrt{3}$
>
> These two results illustrate the fact that when adding and subtracting these numbers, you can deal with the rational and irrational parts separately.
>
> **c** $xy = (5 + \sqrt{3})(3 - 2\sqrt{3}) = 15 - 10\sqrt{3} + 3\sqrt{3} - 2\sqrt{3} \times \sqrt{3}$
> $ = 15 - 2 \times 3 - 10\sqrt{3} + 3\sqrt{3}$
> $ = 9 - 7\sqrt{3}$
>
> These expressions can be manipulated using the ordinary rules of arithmetic and algebra.

Manipulation of expressions of the form $a + b\sqrt{c}$

Example 18.5

Question

If $x = 5 + \sqrt{2}$ and $y = 3 - \sqrt{2}$, simplify these.

a $x + y$ 	b y^2

Solution

a $x + y = 5 + \sqrt{2} + 3 - \sqrt{2}$
$= 5 + 3 + \sqrt{2} - \sqrt{2}$
$= 8$

Note that this result indicates that it is possible for the sum of two irrational numbers to be rational.

b $y^2 = (3 - \sqrt{2})^2$
$= (3 - \sqrt{2})(3 - \sqrt{2})$
$= 9 - 3\sqrt{2} - 3\sqrt{2} + \sqrt{2} \times \sqrt{2}$
$= 9 + 2 - 6\sqrt{2}$
$= 11 - 6\sqrt{2}$

Note that this result is also an application of the algebraic result

$(a + b)^2 = a^2 + 2ab + b^2$

Exercise 18.1

 Calculators should **not** be used for any of the questions in this exercise.

1 Simplify the following, stating whether the result is rational or irrational.

a $\sqrt{12}$ 	b $\sqrt{1000}$ 	c $\sqrt{45}$
d $\sqrt{300}$ 	e $\sqrt{75}$ 	f $\sqrt{8} \times \sqrt{2}$
g $\sqrt{20} \times \sqrt{18}$ 	h $\sqrt{20} \div \sqrt{5}$ 	i $\sqrt{80} \times \sqrt{50}$
j $\sqrt{75} \times \sqrt{15}$ 	k $\sqrt{40}$ 	l $\sqrt{54}$
m $\sqrt{98}$ 	n $\sqrt{800}$ 	o $\sqrt{363}$
p $\sqrt{27} \times \sqrt{3}$ 	q $\sqrt{250} \times \sqrt{40}$ 	r $\sqrt{108} \div \sqrt{12}$
s $\sqrt{90} \times \sqrt{20}$ 	t $\dfrac{\sqrt{60} \times \sqrt{20}}{\sqrt{12}}$

2 If $x = 4 + \sqrt{3}$ and $y = 4 - \sqrt{3}$, simplify these.
a $x + y$ 	b $x - y$ 	c xy

3 If $x = 5 + \sqrt{7}$ and $y = 3 - \sqrt{7}$, simplify these.
a $x + y$ 	b $x - y$ 	c xy

4 If $x = 3 + \sqrt{5}$ and $y = 4 - 3\sqrt{5}$, simplify these.
a $x + y$ 	b $x - y$ 	c x^2

5 If $x = 4 + \sqrt{11}$ and $y = 9 - 2\sqrt{11}$, simplify these.
a $x + y$ 	b $x - y$ 	c x^2

6 If $x = 5 + 2\sqrt{3}$ and $y = 4 - 3\sqrt{2}$, simplify these.
   a $x\sqrt{3}$
   b x^2
   c y^2

7 If $x = 6 - 2\sqrt{5}$ and $y = 3 - 5\sqrt{3}$, simplify these.
   a $x\sqrt{5}$
   b $y\sqrt{3}$
   c x^2
   d y^2

8 Simplify $\sqrt{2}(5 + 3\sqrt{2})^2$.

9 Show that $(10 - 3\sqrt{7})(10 + 3\sqrt{7})$ is rational, finding its value.

Rationalising denominators

When dealing with fractions, it is easier if the denominator is a rational number. For this, you need to manipulate the fraction. This is called rationalising the denominator.

The rules of fractions are used when the denominator is of the form $\sqrt{b}$ or $a\sqrt{b}$. The next example shows the method.

Example 18.6

Question

Rationalise the denominator in these irrational fractions.

a $\dfrac{5}{\sqrt{2}}$
b $\dfrac{7}{\sqrt{12}}$

Solution

a Multiply the numerator and the denominator by $\sqrt{2}$. By the rules of fractions, since you have multiplied both the numerator and the denominator by the same quantity, you have not changed the value of the number.
This gives $\dfrac{5 \times \sqrt{2}}{\sqrt{2} \times \sqrt{2}} = \dfrac{5\sqrt{2}}{2}$ and the denominator is now a rational number.

b First, simplify the denominator and then repeat the process in part **a**, this time multiplying by $\sqrt{3}$.

$$\dfrac{7}{\sqrt{12}} = \dfrac{7}{\sqrt{4 \times 3}} = \dfrac{7}{2\sqrt{3}}$$

$$\dfrac{7\sqrt{3}}{2\sqrt{3} \times \sqrt{3}} = \dfrac{7\sqrt{3}}{2 \times 3}$$

$$= \dfrac{7\sqrt{3}}{6}$$

When the denominator is of the form $a + b\sqrt{c}$, you need to use the difference of squares as well as the rules of fractions.
Remember that $(x + y)(x - y) = x^2 - y^2$.

Rationalising denominators

Example 18.7

Question

Simplify

a $(5+\sqrt{3})(5-\sqrt{3})$

b $(7-2\sqrt{5})(7+2\sqrt{5})$

Note
If you remember that $(x + y)(x - y) = x^2 - y^2$, you can simplify the working in this example.

Solution

a $(5+\sqrt{3})(5-\sqrt{3}) = 25 - 5\sqrt{3} + 5\sqrt{3} - 3 = 25 - 3 = 22$

b $(7-2\sqrt{5})(7+2\sqrt{5}) = 49 + 14\sqrt{5} - 14\sqrt{5} - 4 \times 5 = 49 - 20 = 29$

When the denominator is of the form $a + b\sqrt{c}$, multiply both the numerator and the denominator by $a - b\sqrt{c}$ to rationalise the denominator.

Example 18.8

Question

Rationalise the denominator of the following, and simplify your answer.

a $\dfrac{6}{\sqrt{7}-2}$

b $\dfrac{1+\sqrt{3}}{5-2\sqrt{3}}$

Solution

a $\dfrac{6}{\sqrt{7}-2} = \dfrac{6}{\sqrt{7}-2} \times \dfrac{\sqrt{7}+2}{\sqrt{7}+2} = \dfrac{6(\sqrt{7}+2)}{7-4} = \dfrac{6(\sqrt{7}+2)}{3} = 2(\sqrt{7}+2)$

b $\dfrac{1+\sqrt{3}}{5-2\sqrt{3}} = \dfrac{1+\sqrt{3}}{5-2\sqrt{3}} \times \dfrac{5+2\sqrt{3}}{5+2\sqrt{3}} = \dfrac{5+2\sqrt{3}+5\sqrt{3}+2\times 3}{25-4\times 3} = \dfrac{11+7\sqrt{3}}{13}$

Exercise 18.2

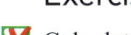 Calculators should **not** be used for any of the questions in this exercise.

1 Rationalise the denominator of each of these irrational fractions.

a $\dfrac{1}{\sqrt{2}}$ b $\dfrac{2}{\sqrt{5}}$ c $\dfrac{5}{\sqrt{7}}$

d $\dfrac{11}{\sqrt{18}}$ e $\dfrac{9}{\sqrt{20}}$ f $\dfrac{1}{\sqrt{7}}$

g $\dfrac{3}{\sqrt{2}}$ h $\dfrac{5}{\sqrt{11}}$ i $\dfrac{7}{\sqrt{50}}$

j $\dfrac{9}{\sqrt{32}}$

2 Rationalise the denominator and simplify each of these.

a $\dfrac{6}{\sqrt{8}}$ b $\dfrac{6}{\sqrt{300}}$ c $\dfrac{12}{\sqrt{75}}$

d $\dfrac{\sqrt{48}}{\sqrt{18}}$ e $\dfrac{10}{\sqrt{5}}$ f $\dfrac{15}{\sqrt{50}}$

g $\dfrac{20}{\sqrt{32}}$ h $\dfrac{12}{\sqrt{20}}$ i $\dfrac{10}{\sqrt{2}}$

18 SURDS

j $\dfrac{2}{\sqrt{10}}$ k $\dfrac{4}{3\sqrt{10}}$ l $\dfrac{14}{5\sqrt{8}}$

m $\dfrac{2\sqrt{3}}{3\sqrt{2}}$ n $\dfrac{12\sqrt{6}}{7\sqrt{15}}$

3 Rationalise the denominator and simplify each of these.

a $\dfrac{6+3\sqrt{2}}{\sqrt{2}}$ b $\dfrac{15+\sqrt{5}}{2\sqrt{5}}$

c $\dfrac{12+3\sqrt{2}}{2\sqrt{3}}$ d $\dfrac{5+2\sqrt{3}}{\sqrt{6}}$

4 Rationalise the denominator and simplify each of these.

a $\dfrac{1}{2-\sqrt{3}}$ b $\dfrac{10}{4-\sqrt{11}}$

c $\dfrac{12}{3+\sqrt{5}}$ d $\dfrac{1}{3+\sqrt{7}}$

5 Rationalise the denominator and simplify each of these.

a $\dfrac{2+3\sqrt{3}}{1+\sqrt{3}}$ b $\dfrac{5+\sqrt{3}}{3+\sqrt{3}}$

c $\dfrac{7-2\sqrt{5}}{2+\sqrt{5}}$ d $\dfrac{8+2\sqrt{7}}{3+\sqrt{7}}$

6 Express each of the following in the form $a+b\sqrt{5}$, where a and b are rational numbers.

a $\dfrac{7+3\sqrt{5}}{7-3\sqrt{5}}$ b $\sqrt{20}+\dfrac{3}{2+\sqrt{5}}$

c $4+2\sqrt{5}+\dfrac{1+2\sqrt{5}}{2+\sqrt{5}}$ d $12-\sqrt{45}+\dfrac{8-4\sqrt{5}}{3-\sqrt{5}}$

Key points
- Surds are irrational numbers, such as $\sqrt{2}$ or $5+\sqrt{3}$.
- Surds can be simplified using $\sqrt{a\times b}=\sqrt{a}\times\sqrt{b}$.
- Use the normal rules of algebra to add and multiply different surds of the form $a+b\sqrt{c}$.
- To rationalise the denominator of a fraction:
  - when the denominator is of the form $\sqrt{b}$ or $a\sqrt{b}$, multiply the numerator and the denominator by $\sqrt{b}$
  - when the denominator is of the form $a+b\sqrt{c}$, multiply both the numerator and the denominator by $a-b\sqrt{c}$.

REVIEW EXERCISE 3

Ch 13

1 **a** Daniel earns $760 each month.
He pays 15% of his earnings in tax.
Calculate the amount of money Daniel has each month after paying tax. [2]
b Daniel invests $1200 in a savings account.
The account pays simple interest at a rate of 2% per year.
Calculate the amount of money in the account after 6 years. [2]

Cambridge O Level Mathematics Syllabus D (4024) Paper 12 Q6, June 2019

Ch 13
Ch 16

2 Tanya owns a small business.
a The business made a profit of $25 700 in 2017 compared with $22 102 in 2018.
Calculate the percentage decrease in profit from 2017 to 2018. [2]
b Tanya invests $8500 in an account paying 3.1% per year compound interest.
At the end of 5 years she takes $9300 from the account to buy new equipment for the business.
Calculate how much money is left in the account after buying the new equipment. [3]

Cambridge O Level Mathematics Syllabus D (4024) Paper 22 Q1b & d, November 2019

Ch 17

3 The population of a type of bacteria grows exponentially.
The formula $P = 1200 \times 1.03^t$ shows how the population, P, changes with the number of years, t, after 1 January 2018.
a Find the population of the bacteria on 1 January 2018. [1]
b Calculate the population of the bacteria on 1 January 2022. [3]
c Work out the year in which the population is predicted to reach 1500. [2]

Ch 17

4 The value of a car decays exponentially, halving in value every 2 years.
At an age of 4 years a car has the value $42 000.
a Work out the value of the car when it was new. [2]
b Work out the value of a car at an age of 10 years. [2]
c David has saved $10 500.
Work out the age of the car he can afford to buy if he spends all his savings. [2]

Ch 18

5 Simplify the following.
a $\sqrt{63}$ [1]
b $6\sqrt{3} + 2\sqrt{5} - 5\sqrt{3} - 4\sqrt{5}$ [1]
c $\sqrt{75} + \sqrt{48}$ [3]

REVIEW EXERCISE 3

  6

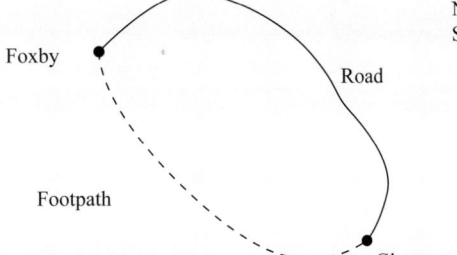

Two villages, Foxby and Glanton, are joined by a footpath and a road.
  a Abdul walks along the footpath from Foxby to Glanton.
     He walks for 2 hours 14 minutes and arrives at Glanton at 15 10.
     Calculate the time Abdul left Foxby. [1]
  b The distance, by road, between Foxby and Glanton is 15 km.
     A bus travels along the road between Foxby and Glanton.
     The bus journey takes 12 minutes.
     Calculate the average speed of the bus in kilometres per hour. [2]

Cambridge O Level Mathematics Syllabus D (4024) Paper 11 Q15a & b, June 2021

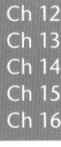 7 **a** Each week Leah works 5 days and is paid a total of $682.
     Each day she works from 08 45 until 12 15 and then from 13 15 until 17 30.
     Calculate Leah's hourly rate of pay. [2]
  b Carlos buys a new bicycle.
     After one year he sells it for $231.
     He makes a loss of 16% on the price he paid.
     Calculate the price Carlos paid for the bicycle. [2]
  c The exchange rate between dollars ($) and euros (€) is $1 = €0.44.
     Henry changes $850 to euros for his holiday.
     He spends €260 when he is on holiday.
     He changes the rest of the money back to dollars at the same exchange rate.
     Calculate how much money in dollars he receives.
     Give your answer correct to the nearest dollar. [3]

Cambridge O Level Mathematics Syllabus D (4024) Paper 22 Q1a, b & c, June 2018

  8 **a** Rationalise $\frac{5}{\sqrt{6}}$. [1]

  b Simplify $\frac{3}{\sqrt{7} - 2}$, giving your answer in the form $p + q\sqrt{7}$ where p and q are integers. [3]

19 INTRODUCTION TO ALGEBRA

Letters for unknowns

Imagine you had a job where you were paid by the hour. You would receive the same amount for each hour you worked. How could you work out how much you will earn in a week?

You would need to work it out as the number of hours you worked, multiplied by the amount you are paid for each hour.

This is a formula in words.

If you work 35 hours at $4.50 an hour, it is easy to work out $35 \times \$4.50$, but what if the numbers change?

The calculation '$35 \times \$4.50$' is only correct if you work 35 hours at $4.50 an hour. Suppose you move to another job where you are paid more for each hour?

You need a simple formula that always works. You can use symbols to stand for the numbers that can change.

You could use ? or □, but it is less confusing to use letters.

Let the number of hours worked be N

the amount you are paid each hour be $\$P$

the amount you earn in a week be $\$W$.

Then $W = N \times P$.

BY THE END OF THIS CHAPTER YOU WILL BE ABLE TO:
- know that letters can be used to represent generalised numbers
- substitute numbers into expressions and formulas.

CHECK YOU CAN
- add, subtract, multiply and divide whole numbers and decimals
- add, subtract, multiply and divide negative numbers
- add, subtract, multiply and divide fractions
- use the correct order of operations in arithmetic.

Example 19.1

Question

Use the formula $W = N \times P$ to find the amount earned in a week when $N = 40$ and $P = 5$.

Solution

$W = N \times P$

$= 40 \times 5$

$= 200$

The amount earned in a week is $200.

Some rules of algebra

You do not need to write the $\times$ sign.

$4 \times t$ is written $4t$.

In multiplications, the number is always written in front of the letter.

$p \times 6 - 30$ is written as $6p - 30$.

Photocopying is prohibited

19 INTRODUCTION TO ALGEBRA

> You often start a formula with the single letter you are finding.
>
> $2 \times l + 2 \times w = P$ is written as $P = 2l + 2w$.
>
> When there is a division in a formula, it is usually written as a fraction.
>
> $y = k \div 6$ can be written as $y = \frac{k}{6}$.

Substituting numbers into algebraic expressions

Numbers that can be substituted into algebraic expressions can be positive, negative, decimals or fractions.

Example 19.2

Question

a Find the value of $4x + 3$ when $x = 2$.
b Find the value of $3x^2 + 4$ when $x = 3$.
c Find the value of $2x^2 + 6$ when $x = -2$.

Solution

a $4x + 3 = 4 \times 2 + 3$ Work out each term separately and then collect together.
    $= 8 + 3$
    $= 11$

b $3x^2 + 4 = 3 \times 3^2 + 4$ Remember $3x^2$ means $3 \times x \times x$ not $3 \times x \times 3 \times x$.
    $= 3 \times 9 + 4$
    $= 27 + 4$
    $= 31$

c $2x^2 + 6 = 2 \times (-2)^2 + 6$ Take special care when negative numbers are involved.
    $= 2 \times 4 + 6$
    $= 8 + 6$
    $= 14$

Exercise 19.1

1 Find the value of these expressions when $a = 5$, $b = 4$ and $c = 2$.

a $a + b$
b $b + c$
c $a - c$
d $a + b + c$
e $2a$
f $3b$
g $5c$
h $3a + b$
i $3c - b$
j $a + 6c$
k $4a + 2b$
l $2b + 3c$
m $a - 2c$
n $8c - 2b$
o bc
p $4ac$
q abc
r $ac + bc$
s $ab - bc - ca$
t a^2
u $\frac{a+b}{3}$
v $\frac{ab}{c}$
w $b^2 + c^2$
x $3c^2$
y $a^2 b$
z c^3

Using harder numbers when substituting

2 Find the value of these expressions when $t = 3$.
  a $t + 2$ **b** $t - 4$ **c** $5t$ **d** $4t - 7$ **e** $2 + 3t$
  f $10 - 2t$ **g** t^2 **h** $10t^2$ **i** $t^2 + 2t$ **j** t^3

3 Use the formula $A = 5k + 4$ to find A for these values of k.
  a $k = 3$ **b** $k = 7$ **c** $k = 0$ **d** $k = \frac{1}{2}$ **e** $k = 2.1$

4 Use the formula $y = mx + c$ to find y in these.
  a $m = 3, x = 2, c = 4$
  b $m = \frac{1}{2}, x = 8, c = 6$
  c $m = 10, x = 15, c = 82$

5 Use the formula $D = \frac{m}{v}$ to find D in these.
  a $m = 24, v = 6$ **b** $m = 150, v = 25$
  c $m = 17, v = 2$ **d** $m = 4, v = \frac{1}{2}$

Using harder numbers when substituting

You need to be able to work with positive and negative integers, decimals or fractions, and to do so with or without a calculator.

Example 19.3

Question

Find the value of
a $a + 5b^3$ when $a = 2.6$ and $b = 3.2$.
b $5cd$ when $c = 2.4$ and $d = 3.2$.

Solution

a $a + 5b^3 = 2.6 + 5 \times 3.2^3$
  $= 2.6 + 5 \times 32.768$
  $= 2.6 + 163.84$
  $= 166.44$

Note
The working is shown here to remind you of the order of operations, but you should be able to do this calculation in one step on your calculator.

b $5cd = 5 \times 2.4 \times 3.2$
  $= 12 \times 3.2$
  $= 10 \times 3.2 + 2 \times 3.2$
  $= 32 + 6.4$
  $= 38.4$

Note
You could work out 2.4×3.2 first, but this would be harder – use your number facts to help you see the best way to do non-calculator calculations.

Example 19.4

Question

$P = ab + 4b^2$. Find P when $a = \frac{4}{5}$ and $b = \frac{3}{8}$, giving your answer as a fraction.

Solution

$P = \left(\frac{4}{5} \times \frac{3}{8}\right) + \left(4 \times \frac{3}{8} \times \frac{3}{8}\right) = \frac{3}{10} + \frac{9}{16} = \frac{24}{80} + \frac{45}{80} = \frac{69}{80}$

19 INTRODUCTION TO ALGEBRA

Exercise 19.2

 1 Work out these.
   a $V = ab - ac$ when $a = 3$, $b = -2$ and $c = 5$
   b $P = 2rv + 3r^2$ when $r = 5$ and $v = -2$
   c $T = 5s^2 - 2t^2$ when $s = -2$ and $t = 3$
   d $M = 2a(3b + 4c)$ when $a = 5$, $b = 3$ and $c = -2$
   e $R = \dfrac{2qv}{q+v}$ when $q = 3$ and $v = -4$
   f $L = 2n + m$ when $n = \tfrac{2}{3}$ and $m = \tfrac{5}{6}$
   g $D = a^2 - 2b^2$ when $a = \tfrac{4}{5}$ and $b = \tfrac{2}{5}$
   h $A = a^2 + b^2$ when $a = 5$ and $b = -3$
   i $P = 2c^2 - 3cd$ when $c = 2$ and $d = -5$
   j $B = p^2 - 3q^2$ when $p = -4$ and $q = -2$

2 Find the value of $M = \dfrac{ab}{(2a + b^2)}$ when $a = 2.75$ and $b = 3.12$.
Give your answer correct to 2 decimal places.

3 The distance S metres fallen by a pebble is given by the formula
$S = \tfrac{1}{2}gt^2$, where t is in seconds.
Find S in each of these cases.
   a $g = 10$ and $t = 12$
   b $g = 9.8$ and $t = 2.5$

4 The surface area (A cm^2) of a cuboid with sides x, y and z cm is given by the formula $A = 2xy + 2yz + 2xz$.
Find A when $x = 5$, $y = 4.5$ and $z = 3.5$.

5 The elastic energy of an elastic string is given by the formula $E = \dfrac{\lambda x^2}{2a}$.
Find E when $\lambda = 3.4$, $x = 5.7$ and $a = 2.5$.
Give your answer correct to 1 decimal place.

6 The focal length of a lens is given by the formula $f = \dfrac{uv}{u+v}$.
Find the focal length when $u = 6$ and $v = -7$.

Key points
- When you have substituted numbers in a formula and are working out the result, remember the order of operations – brackets and indices (powers), then divide and multiply, then add and subtract.

20 ALGEBRAIC MANIPULATION

Simplifying algebraic expressions

These methods can also be used in algebra to **simplify** an expression.

Collect all the positive terms and all the negative terms separately.

Then add up each set and find the difference.

The sign of the answer is the sign of the larger total.

Example 20.1

Question

Simplify $5b - 4b - 8b + 2b$.

Solution

$5b + 2b - 4b - 8b$ Collect all the positive terms and all the negative terms separately.

$= 7b - 12b$ Add up each set and find the difference.

$= -5b$

You can only add or subtract **like terms**.

Like terms use only the same letter or the same combination of letters.

Example 20.2

Question

Simplify $3a + 2b - 2a + 5b$.

Solution

$3a + 2b - 2a + 5b$

$= 3a - 2a + 2b + 5b$ Reorder with the like terms together.

$= a + 7b$

Algebraic terms can be multiplied together.

Remember that in algebra, you do not need to write the multiplication signs.

However, writing them in your working can help.

BY THE END OF THIS CHAPTER YOU WILL BE ABLE TO:

- simplify expressions by collecting like terms
- expand products of algebraic expressions
- factorise by extracting common factors
- factorise expressions of the form:
  $ax + bx + kay + kby$
  $a^2x^2 - b^2y^2$
  $a^2 + 2ab + b^2$
  $ax^2 + bx + c$
  $ax^3 + bx^2 + cx$
- complete the square for expressions in the form $ax^2 + bx + c$.

CHECK YOU CAN:

- add, subtract, multiply and divide natural numbers
- order integers on a number line
- understand that in algebra, letters stand for numbers and obey the rules of arithmetic
- express basic arithmetic processes algebraically
- understand and use index notation.

Photocopying is prohibited

20 ALGEBRAIC MANIPULATION

Example 20.3

Question

a Simplify $2a \times 3b$.

b Simplify $p \times q \times p^2 \times q$.

Solution

a $2a \times 3b$
   $= 2 \times a \times 3 \times b$
   $= 2 \times 3 \times a \times b$ Reorder with the numbers together.
   $= 6ab$

b $p \times q \times p^2 \times q$
   $= p \times p^2 \times q \times q$ Reorder with the like terms together.
   $= p^3 \times q^2$ $p \times p^2 = p \times p \times p = p^3$.
   $= p^3 q^2$ You say this as 'p cubed q squared'.

Note

$1a$ or $1 \times a$ is always written as just a.
Some people confuse $2a$ and a^2.
$2a = 2 \times a = a + a$, while $a^2 = a \times a$.

Example 20.4

Question

A rectangle has length $3a$ and width $2b$.
Find and simplify an expression for the perimeter of the rectangle.

Solution

Perimeter $= 3a + 2b + 3a + 2b$
$= 6a + 4b$

Exercise 20.1

Simplify the expressions in questions **1** to **16**.

1 $x + x + x + x + x$
2 $y + y + y + z + z$
3 $x + x + y + y$
4 $a + b + a + b + a$
5 $5 \times x$
6 $4p + 3p$
7 $b + 2b + 3b$
8 $p \times 3$
9 $s + 2s + s$
10 $a \times a + b \times b$
11 $a + 2b + 2a + b$
12 $2m + 3n - m - n$
13 $5x - 3x + 2y - y$
14 $x^2 + 3x - 5x - 15$
15 $3p \times 4q$
16 $a + 6b - 5a + 2b$

17 A square has sides $2a$ long.
    Find and simplify an expression for its perimeter.
18 A quadrilateral has sides of $2a$, $3b$, $4a$ and $6b$.
    Find and simplify an expression for its perimeter.
19 A pentagon has two sides x cm long and two sides $2x$ cm long.
    The perimeter of the pentagon is $9x$.
    What is the length of the fifth side?

20 Parvez walked c miles on Friday and d miles on Saturday.
Liz walked twice as far as Parvez on Friday and three times as far on Saturday.
Find and simplify an expression for the total distance they walked.

Simplifying more complex algebraic expressions

Terms such as ab^2, a^2b and a^3 cannot be collected together unless they are exactly the same type.

> **Note**
> Errors are often made by trying to go too far.
> For example, $2a + 3b$ cannot be simplified any further.
> Another common error is to work out $4a^2 - a^2$ as 4. The answer is $3a^2$.

Example 20.5

Question
Simplify each of these expressions by collecting together the like terms.

a $2x^2 - 3xy + 2yx + 3y^2$

b $3a^2 + 4ab - 2a^2 - 3b^2 - 2ab$

c $3 + 5a - 2b + 2 + 8a - 7b$

Solution
a $2x^2 - 3xy + 2yx + 3y^2$
$= 2x^2 - xy + 3y^2$

The middle two terms are like terms, because xy is the same as yx.

b $3a^2 + 4ab - 2a^2 - 3b^2 - 2ab$
$= 3a^2 - 2a^2 + 4ab - 2ab - 3b^2$
$= a^2 + 2ab - 3b^2$

Here the a^2 terms and the ab terms can be collected, but they cannot be combined with each other or with the single b^2 term.

c $3 + 5a - 2b + 2 + 8a - 7b$
$= 3 + 2 + 5a + 8a - 2b - 7b$
$= 5 + 13a - 9b$

Here there are three different types of terms, which can be collected separately, but not combined together.

Exercise 20.2

Simplify these where possible.

1 $3ab + 2ab - ab$
2 $5ac + 2ab - 3ac + 4ab$
3 $3abc + 2abc - 5abc$
4 $4ab - 3ac + 2ab - ac$
5 $a^2 + 3b^2 - 2a^2 - b^2$
6 $2x^2 - 3xy - xy + y^2$
7 $4b^2 + 3a^2 - 2b^2 - 4a^2$
8 $9a^2 - 3ab + 5ab - 6b^2$
9 $4ab + 2bc - 3ba - bc$
10 $2p^2 - 3pq + 4pq - 5p^2$
11 $3ab + 2ac + ad$
12 $9ab - 2bc + 3bc - 7ab$
13 $a^3 + 3a^3 - 6a^3$
14 $5a^3 + 4a^2 + 3a$
15 $a^3 + 3a^2 + 2a^3 + 4a^2$
16 $3ab^2 - 4ba^2 + 7a^2b$
17 $3x^3 - 2x^2 + 4x^2 - 3x^3$
18 $8a^3 - 4a^2 + 5a^3 - 2a^2$
19 $abc + cab - 3abc + 2bac$

20 Pali, Gemma and Saad all drew rectangles.
Pali's was a cm long and $2b$ cm wide.
Gemma's was $3a$ cm long and b cm wide.
Saad's was $2a$ cm long and $3b$ cm wide.
Find and simplify an expression for the total area of the three rectangles.

20 ALGEBRAIC MANIPULATION

Expanding a single bracket

What is $2 \times 3 + 4$? Is it 14? Is it 10?

The rule is 'do the multiplication first', so the answer is 10.

If you want the answer to be 14, you need to add 3 and 4 first.

You use a bracket to show this $2 \times (3 + 4)$.

Notice that this is equal to $2 \times 3 + 2 \times 4$.

It is the same in algebra.

$a(b + c)$ means 'add b and c and then multiply by a' and this is the same as 'multiply b by a, multiply c by a and then add the results'.

So $a(b + c) = ab + ac$.

This is called expanding the bracket.

> **Note**
> Remember to multiply each of the terms inside the bracket by the number or letter outside the brackets.

Example 20.6

Question

Expand these brackets.

a $5(2x + 3)$ b $4(2x - 1)$

Solution

a $5(2x + 3) = 5 \times 2x + 5 \times 3 = 10x + 15$
b $4(2x - 1) = 4 \times 2x + 4 \times -1 = 8x - 4$

> **Note**
> Remember about multiplying with negative numbers. For example,
> $-3 \times x = -3x$
> $-3 \times -x = 3x$.

Exercise 20.3

Expand these brackets.

1 $2(a + b)$ 2 $4(2x + 1)$ 3 $2(p + 3)$ 4 $3(3x - 1)$
5 $2(2x - 3)$ 6 $7(3y + z)$ 7 $4(3 - 8a)$ 8 $10(2a + 3b)$
9 $5(3e - 8f)$ 10 $2(1 - x)$ 11 $5(p - q)$ 12 $a(a + 2)$
13 $y(y - 1)$ 14 $y(3 - 2y)$ 15 $x(2 - x)$ 16 $-y(2 + y)$
17 $3c(c + 4)$ 18 $-2x(5x - 3)$ 19 $2(3i + 4j - 5k)$ 20 $4(5m - 3n + 2p)$

Factorising algebraic expressions

Factors are numbers or letters which will divide into an expression.

The factors of 6 are 1, 2, 3 and 6.

The factors of p^2 are 1, p and p^2.

To factorise an expression, look for common factors.

For example, the common factors of $2a^2$ and $6a$ are 2, a and $2a$.

Factorising algebraic expressions

Example 20.7

Question

Factorise these.

a $4p + 6$ **b** $2a^2 - 3a$ **c** $8x^2 - 12x$

Solution

a The only common factor is 2.
You divide each term by the common factor, 2.
$4p \div 2 = 2p$ and $6 \div 2 = 3$
$4p + 6 = 2(2p + 3)$

b The only common factor is a.
You divide each term by the common factor, a.
$2a^2 \div a = 2a$ and $3a \div a = 3$
$2a^2 - 3a = a(2a - 3)$

c Think about the numbers and the letters separately and then combine them.
The highest common factor of 8 and 12 is 4, and the common factor of x^2 and x is x. Therefore, the common factor of $8x^2$ and $12x$ is $4 \times x = 4x$.
$4x(\ldots)$ You write this factor outside the bracket.
You then divide each term by the common factor, $4x$.
$8x^2 \div 4x = 2x$ and $12x \div 4x = 3$
$8x^2 - 12x = 4x(2x - 3)$

Note

Factorise means 'factorise fully'. Make sure that you have found all the common factors. Check that the expression in the brackets will not factorise further.

Exercise 20.4

For questions **1** to **8**, copy and complete the factorisations.

1 $12a + 3 = 3(\square + 1)$ 2 $9a + 18 = 9(a + \square)$ 3 $5y - 30 = 5(y - \square)$
4 $6b - 4 = \square(3b - 2)$ 5 $4x + 16 = \square(x + 4)$ 6 $y^2 + 2y = y(\square + \square)$
7 $2b + 6b^2 = 2b(\square + \square)$ 8 $8a^2 + 20a = \square(2a + 5)$

For questions **9** to **31**, factorise each of the expressions.

9   $2x + 6$
10 $4x - 20$
11 $9 - 12x$
12 $3x^2 + 5x$
13 $5a^2 + 10b$
14 $24 + 36a^2$
15 $10x^2 - 100x$
16 $24x + 32y$
17 $15ab - 20ac$
18 $30f^2 - 18fg$
19 $42ab + 35a^2$
20 $5a^2b + 10ab^2$
21 $3ab - 2ac + 3ad$
22 $2x^2y^2 - 3x^3y$
23 $5x^2 - 15x + 15$
24 $12x - 6y + 8z$
25 $9ab + 6b^2$
26 $4a^2c - 2ac^2$
27 $12x^2y + 8xy - 4xy^2$
28 $3a^2b - 9a^3b^2$
29 $5a^2b^2c^2 - 10abc$
30 $2a^2b - 3a^2b^3 + 7a^4b$
31 $4abc - 3ac^2 + 2a^2b$

Factorising expressions of the form $ax + bx + kay + kby$

Example 20.8

Question

Factorise $x(a + b) + 2y(a + b)$.

Solution

The bracket $(a + b)$ is a common factor, so this can be factorised as $(a + b)(x + 2y)$.

Example 20.9

Question

Factorise $x^2 - 3x + 2ax - 6a$.

Solution

There is no common factor for all four terms, but the first pair of terms have a common factor and the second pair of terms also have a common factor.

Factorising in pairs gives $x(x - 3) + 2a(x - 3)$.

The bracket $(x - 3)$ is a common factor so
$x(x - 3) + 2a(x - 3) = (x - 3)(x + 2a)$

Sometimes you have to change the order of the terms, as in the next example.

Example 20.10

Question

Factorise $5ax - 9b + 3bx - 15a$.

Solution

The pairs of terms in this order do not have common factors.
Changing the order gives $5ax + 3bx - 15a - 9b$.
You can then factorise in pairs.
$x(5a + 3b) - 3(5a + 3b) = (5a + 3b)(x - 3)$

Note

Taking the factor -3 out means the sign of $3b$ is $+$, since $-3 \times +3b = -9b$.

Exercise 20.5

Factorise these.

1. $a(2x + 3) + b(2x + 3)$
2. $5x(a - 2b) - 3(a - 2b)$
3. $p^2 + pq + pr + qr$
4. $ax - ay + bx - by$
5. $a^2 + ab - ac - bc$
6. $a^2 - ab - ac + bc$
7. $4ab + 12a + 3b + 9$
8. $2ax - 4ay + 3x - 6y$
9. $5ax^2 - 5bx - 2ax + 2b$
10. $3a^3 + 6ab - 2a^2b - 4b^2$
11. $10ax - 5ay + 6bx - 3by$
12. $8ac - 6ad - 12bc + 9bd$
13. $8a^2 - 6ab - 15bc + 20ac$
14. $a^2 + ab + a + b$
15. $ac + 6bd + 3ad + 2bc$
16. $10ac + 3bd - 2ad - 15bc$
17. $10ab - 5ac - 3cd + 6bd$
18. $6x^2 + 3yz - 2xz - 9xy$
19. $10x^2 + 15y^2 - 25x^2y - 6y$
20. $2x^2 - 6x + 3 - x$

Expanding a pair of brackets

You expand a pair of brackets by multiplying one bracket by the other bracket.

Look at this rectangle.   Now split the rectangle up.

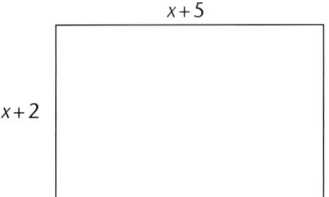 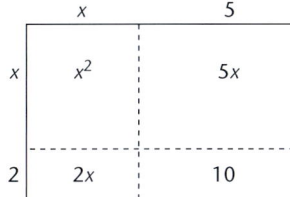

The area of the rectangle is $(x + 5)(x + 2)$.

So $(x + 5)(x + 2) = x^2 + 2x + 5x + 10 = x^2 + 7x + 10$

When multiplying out a pair of brackets you must multiply each term inside the second bracket by each term inside the first bracket.

The example which follows shows three methods for doing this.

Choose the method you prefer and stick to it.

Example 20.11

Question

Expand these.

a $(a+2)(a+5)$ b $(b+4)(b-1)$ c $(m-5)(m-4)$

Note
'Multiply out', 'simplify', 'expand' or 'remove' the brackets all mean the same thing.

Solution

a **Method 1**

Expand the first bracket and then multiply out.

$(a+2)(a+5) = a(a+5) + 2(a+5)$
$\qquad\qquad\quad = a^2 + 5a + 2a + 10$ Notice that the middle two terms are like terms and so can be collected.
$\qquad\qquad\quad = a^2 + 7a + 10$

b **Method 2**

×	b	+4
b	b^2	$+4b$
−1	$-1b$	-4

$(b+4)(b-1) = b^2 + 4b - 1b - 4$
$\qquad\qquad\quad = b^2 + 3b - 4$

Note
Most errors are made in expanding brackets when negative signs are involved.

c **Method 3**

Use the word FOIL to make sure you multiply each term in the second bracket by each term in the first.

If you draw arrows to show the multiplications, you can think of a smiley face.

$(m-5)(m-4) = m \times m + m \times -4 - 5 \times m - 5 \times -4$
$\qquad\qquad\quad\; = m^2 - 4m - 5m + 20$
$\qquad\qquad\quad\; = m^2 - 9m + 20$

Note
F: first × first
O: outer × outer
I: inner × inner
L: last × last

There are two special types of expansion of a pair of brackets that you need to know.

The first is when a bracket is squared, as in parts **a** and **b** of the next example. The important thing with this type of expansion is to make sure that you write the brackets separately and that you end up with three terms.

The second type is shown in part **c** of the example. With this type of expansion, you get only two terms because the middle terms cancel each other out. This type is known as **the difference of two squares**, because $(A-B)(A+B) = A^2 - B^2$.

Example 20.12

Question

Expand these.

a $(x + 3)^2$ b $(x - 3)^2$ c $(x + 3)(x - 3)$

Solution

a $(x + 3)^2 = (x + 3)(x + 3)$ The method shown here is method 1, but you can use any of the methods that were shown in Example 20.11. You will get the same answer.

$= x(x + 3) + 3(x + 3)$
$= x^2 + 3x + 3x + 9$
$= x^2 + 6x + 9$

b $(x - 3)^2 = (x - 3)(x - 3)$
$= x(x - 3) - 3(x - 3)$
$= x^2 - 3x - 3x + 9$
$= x^2 - 6x + 9$

Note
Take care with the negative signs.

c $(x + 3)(x - 3) = x(x - 3) + 3(x - 3)$
$= x^2 - 3x + 3x - 9$
$= x^2 - 9$

Exercise 20.6

Expand these.

1 $(x + 2)(x + 3)$ 2 $(a + 4)(a + 3)$ 3 $(a + 2)(a + 1)$
4 $(x + 5)(x - 2)$ 5 $(x + 7)(x - 3)$ 6 $(x - 5)(x - 6)$
7 $(x - 3)(x + 6)$ 8 $(x - 5)(x - 4)$ 9 $(x + 1)(x + 3)$
10 $(a + 3)(a + 3)$ 11 $(a + 2)(a + 1)$ 12 $(x - 2)(x + 1)$
13 $(p + 4)(p - 2)$ 14 $(a + 7)(a + 8)$ 15 $(x - 6)(x + 4)$
16 $(x - 9)(x - 3)$ 17 $(x + 10)(x - 1)$ 18 $(x + 3)^2$
19 $(a - 5)^2$ 20 $(b + 1)^2$ 21 $(x - 2)^2$
22 $(a + 2)^2$ 23 $(x - 10)^2$ 24 $(x + 8)^2$
25 $(b - 7)^2$ 26 $(x + 4)^2$ 27 $(x + 2)(x - 2)$
28 $(x + 6)(x - 6)$ 29 $(x - 4)(x + 4)$ 30 $(x + 1)(x - 1)$
31 $(x + 10)(x - 10)$ 32 $(x + 7)(x - 7)$

Expanding more complex brackets

The same methods can be used for brackets containing more complex expressions.

This is shown in the following example.

Example 20.13

Question

Expand and simplify these.

a $(3x + 2)(5x + 1)$ b $(5a - 2b)(3a - b)$ c $(y + 2)(2y - 5)(y - 1)$

> **Note**
> Expressions of the form $ax^2 + bx + c$ are called quadratic.

Solution

a $(3x + 2)(5x + 1) = 3x(5x + 1) + 2(5x + 1)$
$= 15x^2 + 3x + 10x + 2$
$= 15x^2 + 13x + 2$

b $(5a - 2b)(3a - b) = 5a \times 3a + 5a \times -b - 2b \times 3a - 2b \times -b$
$= 15a^2 - 5ab - 6ab + 2b^2$
$= 15a^2 - 11ab + 2b^2$

c $(y + 2)(2y - 5)(y - 1) = y(2y - 5)(y - 1) + 2(2y - 5)(y - 1)$
$= y(2y^2 - 7y + 5) + 2(2y^2 - 7y + 5)$
$= 2y^3 - 7y^2 + 5y + 4y^2 - 14y + 10$
$= 2y^3 - 3y^2 - 9y + 10$

Exercise 20.7

Expand these.

1. $(5 + x)(x - 6)$
2. $(5x - 1)(2x - 4)$
3. $(2x - 5)(3x - 2)$
4. $(5x + 6)(2x - 3)$
5. $(x + y)(2x + y)$
6. $(3x - 5y)(x - 4y)$
7. $(2x - 3y)(x - 2y)$
8. $(7x + 8y)(6x - 4y)$
9. $(2g - 3h)(2g - 7h)$
10. $(3j - 8m)(2j - 7m)$
11. $(2m + 7n)(5m - 6n)$
12. $(2r + 3n)(3r - 5n)$
13. $(2q + 7p)(2q - 9p)$
14. $(3r - 8s)(2r - 7s)$
15. $(2s - 3t)(2s - 7t)$
16. $(x + 1)(x + 2)(x + 3)$
17. $(2y + 5)(4 + y)(y - 1)$
18. $(3a + 2)(4a - 3)(a + 1)$
19. $(m + n)(2m - 3n)(m - n)$
20. $(2p - 3q)(p - q)(3p - 2q)$

Factorising expressions of the form $a^2x^2 - b^2y^2$

When expanding brackets earlier in this chapter, you met expressions of the form $a^2 - b^2$.

They are called **the difference of two squares**.

We have already seen that $(x - 3)(x + 3) = x^2 - 9$.

So, following this pattern, we see that factorising $x^2 - a^2$ gives $(x - a)(x + a)$.

> **Note**
> It is worth learning this so that you recognise it when you see it!

Factorising expressions of the form $x^2 + bx + c$

Example 20.14

Question
Factorise $x^2 - 16$.

Solution
$x^2 - 16 = x^2 - 4^2$
$= (x - 4)(x + 4)$

In fact *any* expression which can be written as the difference of two squares can be factorised in this way.

Example 20.15

Question
Factorise each of these expressions.
a $25x^2 - 1$
b $9x^2 - 4y^2$

Solution
a $25x^2 - 1 = 5^2 x^2 - 1$
   $= (5x - 1)(5x + 1)$
b $9x^2 - 4y^2 = 3^2 x^2 - 2^2 y^2$
   $= (3x - 2y)(3x + 2y)$

When a numerical term is not a square number, check to see whether a common factor can be extracted, leaving an expression in a form that allows you to use the difference of two squares method.

Example 20.16

Question
Factorise $3x^2 - 12$.

Solution
$3x^2 - 12 = 3(x^2 - 4)$ Take out the common factor 3.
$= 3(x^2 - 2^2)$
$= 3(x - 2)(x + 2)$

Exercise 20.8

Factorise each of these expressions.

1 $x^2 - 25$
2 $x^2 - 4$
3 $4a^2 - b^2$
4 $9 - 16y^2$
5 $25x^2 - 49y^2$
6 $9x^2 - 64$
7 $1 - 49t^2$
8 $100x^2 - 1$
9 $x^2 y^2 - 16a^2$
10 $y^2 - 169$
11 $121x^2 - 144y^2$
12 $8 - 2x^2$
13 $3x^2 - 192$
14 $45 - 20x^2$
15 $25x^2 y^2 - 100$
16 $3x^2 - 12$
17 $5x^2 - 45$
18 $7x^2 - 343$
19 $10x^2 - 4000$
20 $8x^2 - 200$

Factorising expressions of the form $x^2 + bx + c$

In the expression $x^2 + bx + c$, c is the numerical term, and b is called the **coefficient** of x.

20 ALGEBRAIC MANIPULATION

Factorising quadratic expressions with a positive numerical term

The expression $(x + 6)(x + 2)$ can be multiplied out and simplified to $x^2 + 8x + 12$.

Therefore $x^2 + 8x + 12$ can be factorised as a product of two brackets, by reversing the process.

The order in which you write the brackets does not matter.

Example 20.17

Question

Factorise $x^2 + 7x + 12$.

Solution

This will factorise into two brackets, with x as the first term in each.
$x^2 + 7x + 12 = (x \quad)(x \quad)$
As both the signs are positive, both the numbers will be positive.
You need to find two numbers that multiply to give 12 and add to give 7.
These are +3 and +4.
So $\quad x^2 + 7x + 12 = (x + 3)(x + 4)$
or $\quad x^2 + 7x + 12 = (x + 4)(x + 3)$.

Example 20.18

Question

Factorise $x^2 - 3x + 2$.

Solution

You need to find two negative numbers that multiply to give 2 and add to give −3. They are −2 and −1.
$x^2 - 3x + 2 = (x - 2)(x - 1)$.

Note

If the numerical term is positive, both the numbers in the brackets must have the same sign as the coefficient of x.

If the coefficient of x is negative and the numerical term is positive, both the numbers in the brackets will be negative.

Exercise 20.9

Factorise these expressions.

1. $x^2 + 5x + 6$
2. $x^2 + 6x + 5$
3. $x^2 + 4x + 3$
4. $x^2 + 6x + 8$
5. $x^2 + 5x + 4$
6. $x^2 + 9x + 20$
7. $x^2 + 2x + 1$
8. $x^2 - 7x + 6$
9. $x^2 - 9x + 18$
10. $x^2 - 7x + 10$
11. $x^2 - 4x + 3$
12. $a^2 - 2a + 1$
13. $y^2 - 9y + 14$
14. $x^2 - 6x + 8$
15. $a^2 + 8a + 12$
16. $a^2 - 6a + 9$
17. $b^2 - 12b + 32$
18. $x^2 + 11x + 24$
19. $x^2 - 9x + 20$
20. $x^2 - 15x + 56$

Factorising quadratic expressions with a negative numerical term

Example 20.19

Question
Factorise $x^2 - 3x - 10$.

Solution
As the numerical term is negative, you need two numbers, with opposite signs, that multiply to give −10 and add to give −3.

The numbers are −5 and +2.

$x^2 - 3x - 10 = (x - 5)(x + 2)$

Example 20.20

Question
Factorise $x^2 + 4x - 12$.

Solution
As the numerical term is negative, you need two numbers, with opposite signs, that multiply to give −12 and add to give +4.

The numbers are +6 and −2.

$x^2 + 4x - 12 = (x + 6)(x - 2)$

Exercise 20.10

Factorise these expressions.

1. $x^2 - 2x - 8$
2. $x^2 + 4x - 5$
3. $x^2 - x - 6$
4. $x^2 + 5x - 6$
5. $x^2 + 2x - 3$
6. $x^2 - 3x - 18$
7. $x^2 - 3x - 28$
8. $x^2 - 9x - 10$
9. $x^2 - 9x - 22$
10. $x^2 + 4x - 32$
11. $y^2 + 9y - 22$
12. $x^2 + x - 12$
13. $x^2 + x - 20$
14. $a^2 + 8a - 20$
15. $a^2 - 6a - 27$
16. $a^2 - 6a - 16$
17. $b^2 + 19b - 20$
18. $x^2 + 11x - 26$
19. $x^2 + 7x - 30$
20. $x^2 - 7x - 18$

Note

Remember that if the numerical term is negative, the two numbers in the brackets have different signs and the larger number has the sign of the coefficient of x.

It is easy to make a mistake when factorising. Always check by multiplying out the brackets.

Factorising expressions of the form $a^2 + 2ab + b^2$

We have found that $(x + 3)^2 = x^2 + 6x + 9$ and $(x - 3)^2 = x^2 - 6x + 9$.

These are examples of **perfect squares**.

In general,

$(a + b)^2 = (a + b)(a + b) = a^2 + ab + ab + b^2 = a^2 + 2ab + b^2$.

The result can be used in reverse to factorise expressions of the form $a^2 + 2ab + b^2$.

So $a^2 + 2ab + b^2 = (a + b)^2$.

Note

The first and third terms are squares and the middle term is twice the product of the quantities that are squared.

20 ALGEBRAIC MANIPULATION

Note
If the middle term is negative, the sign in the bracket will be negative.

Example 20.21

Question
Factorise $x^2 + 8x + 16$.

Solution
The 1st and 3rd terms are x^2 and 4^2 and the middle term is $8x = 2 \times 4 \times x$.
$x^2 + 8x + 16 = (x + 4)^2$

Example 20.22

Question
Factorise $9x^2 - 30xy + 25y^2$.

Solution
The first and third terms are $(3x)^2$ and $(5y)^2$ and the middle term $30xy = 2 \times 3x \times 5y$.
The middle term is negative so the sign in the bracket will be negative.
$9x^2 - 30xy + 25y^2 = (3x - 5y)^2$

Exercise 20.11

Factorise these expressions.

1. $x^2 + 2x + 1$
2. $x^2 - 4x + 4$
3. $x^2 - 10x + 25$
4. $a^2 + 20a + 100$
5. $9 - 12y + 4y^2$
6. $4x^2 + 4xy + y^2$
7. $49a^2 + 28a + 4$
8. $36x^2 - 60xy + 25y^2$
9. $16a^2 - 24ab + 9b^2$
10. $4a^2b^2 + 4abc + c^2$

Factorising quadratic expressions of the form $ax^2 + bx + c$

You have already learned how to factorise simple quadratic expressions such as $x^2 + bx + c$.

- If c is positive, you find two numbers with the same sign that multiply to give c and add to give b.
- If c is negative, you find two numbers with different signs that multiply to give c and add to give b.

The expression $ax^2 + bx + c$, when factorised, will be
$(px + q)(rx + s) = prx^2 + (ps + qr)x + qs$.

So $pr = a$, $qs = c$ and $ps + qr = b$.

It is easiest to look at examples.

Factorising quadratic expressions of the form $ax^2 + bx + c$

Factorising more complex quadratic expressions with a positive numerical term

Example 20.23

Question

Factorise $3x^2 + 11x + 6$.

Solution

As the numerical term is positive, both numerical terms in the brackets have the same sign, and as the coefficient of x is positive, they are both positive.

The only numbers that multiply to give 3 are 3 and 1.

So, as a start, $(3x + ...)(x + ...)$.

The numbers that multiply to give 6 are either 3 and 2 or 6 and 1.

So the possible answers are

$(3x + 2)(x + 3)$ or $(3x + 3)(x + 2)$, or $(3x + 6)(x + 1)$ or $(3x + 1)(x + 6)$.

By expanding the brackets, it can be seen that the first one is correct.

$3x^2 + 11x + 6 = (3x + 2)(x + 3)$

The coefficient of x is $3 \times 3 + 2 \times 1 = 9 + 2 = 11$.

It is very useful to check completely by multiplying out the two brackets.

Writing out all the possible brackets can be a long process, and it is quicker to test the possibilities until you find the correct coefficient of x and then multiply out the brackets to check.

Example 20.24

Question

Factorise $4x^2 - 14x + 6$.

Solution

First, check whether there is a common factor.

Here the common factor is 2.

$4x^2 - 14x + 6 = 2(2x^2 - 7x + 3)$

Now look at the quadratic expression.

As the numerical term is positive and the coefficient of x is negative, both the numerical terms in the brackets will be negative.

The possibilities for the coefficients of the x-terms in the brackets are 2 and 1.

The possibilities for the numerical terms in the brackets are -1 and -3 or -3 and -1.

The coefficient of x is thus $2 \times -1 + 1 \times -3 = -5$ or $2 \times -3 + 1 \times -1 = -7$.

The second is correct, so $2x^2 - 7x + 3 = (2x - 1)(x - 3)$.

So $4x^2 - 14x + 6 = 2(2x - 1)(x - 3)$.

20 ALGEBRAIC MANIPULATION

> **Note**
> First, look for any common factor, then try the most obvious pairs first. Remember, if the sign of c (the numerical term) is positive, both the numerical terms in the brackets have the same sign as b (the coefficient of x).

Exercise 20.12

Factorise each of these expressions.

1. $2x^2 + 6x + 4$
2. $3x^2 + 7x + 2$
3. $2x^2 + 9x + 4$
4. $2x^2 + 7x + 6$
5. $6x^2 - 15x + 6$
6. $3x^2 - 12x + 12$
7. $3x^2 - 11x + 6$
8. $3x^2 - 13x + 10$
9. $3x^2 - 11x + 10$
10. $4x^2 - 16x + 15$
11. $4x^2 + 8x + 3$
12. $7x^2 + 10x + 3$
13. $5x^2 - 13x + 6$
14. $5x^2 - 22x + 8$
15. $6x^2 - 19x + 10$
16. $8x^2 - 18x + 9$
17. $3x^2 + 17x + 20$
18. $2x^2 + 7x + 6$
19. $3x^2 + 13x + 4$
20. $5x^2 + 18x + 9$
21. $4x^2 + 6x + 2$
22. $3x^2 + 11x + 10$
23. $2x^2 + 5x + 2$
24. $4x^2 + 17x + 15$

Factorising more complex quadratic expressions with a negative numerical term

The examples looked at so far had a positive numerical term.

When c (the numerical term) is negative, the signs of the numerical terms in the brackets are different.

These are, again, best shown by examples.

> ### Example 20.25
>
> **Question**
> Factorise $3x^2 - 7x - 6$.
>
> **Solution**
> There is no common factor.
> The signs in the numerical terms in the brackets are different.
> The coefficients of the x-terms in the brackets are 3 and 1.
> The possibilities for the numerical terms are 1 and -6 or -1 and 6, 6 and -1 or -6 and 1, 3 and -2 or -3 and 2, or 2 and -3 or -2 and 3.
> Test the possibilities until you find the correct coefficient of x.
> $3 \times 1 + 1 \times -6 = -3$
> $3 \times -1 + 1 \times 6 = 3$
> $3 \times 6 + 1 \times -1 = 17$
> $3 \times -6 + 1 \times 1 = -17$ You don't need to test this pair as you can see that the result will be -17.
> $3 \times 3 + 1 \times -2 = 7$
> $3 \times -3 + 1 \times 2 = -7$ You can see from the previous calculation that this pair will give the required result.
> So, $3x^2 - 7x - 6 = (3x + 2)(x - 3)$.
> Multiply out to check.

Factorising expressions of the form $ax^3 + bx^2 + cx$

Example 20.26

Question

Factorise $6x^3 + 3x^2 - 30x$.

Solution

There is a common factor of $3x$.

$6x^3 + 3x^2 - 30x = 3x(2x^2 + x - 10)$

Look at the quadratic expression.

The signs in the numerical terms in the brackets are different.

The coefficients of the x-terms in the brackets are 2 and 1.

The possibilities for the numerical terms are 1 and −10 or −1 and 10, 10 and −1 or −10 and 1, 5 and −2 or −5 and 2, or 2 and −5 or −2 and 5.

Try the combinations you think may give the correct result first.

$2 \times 5 + 1 \times -2 = 8$ So neither 5 and −2 nor −5 and 2 is correct.

$2 \times 2 + 1 \times -5 = -1$ Which is the correct number but the wrong sign.

$2 \times -2 + 1 \times 5 = 1$ So the numerical terms in the brackets are −2 and 5.

So, $6x^3 + 3x^2 - 30x = 3x(2x + 5)(x - 2)$.

Check by multiplying out.

Exercise 20.13

Factorise each of these expressions.

1 $3x^2 + x - 10$
2 $2x^2 + 5x - 3$
3 $3x^2 - 2x - 8$
4 $3x^2 - 11x - 4$
5 $2x^2 + 9x - 5$
6 $3x^2 + 4x - 15$
7 $5x^2 - 15x - 50$
8 $5x^2 + 13x - 6$
9 $4x^2 - 4x - 3$
10 $7x^2 + 10x - 8$
11 $3x^2 - x - 14$
12 $3x^2 - 11x - 20$
13 $2x^2 - x - 21$
14 $2x^2 - 15x - 8$
15 $6x^2 - 17x - 14$
16 $6x^2 - 13x - 15$
17 $2x^2 - x - 15$
18 $3x^2 + x - 14$
19 $5x^2 - 17x - 12$
20 $3x^2 - 5x - 12$
21 $4x^3 - 3x^2 - 10x$
22 $2x^3 - 7x^2 - 15x$
23 $4x^3 - 7x^2 - 2x$
24 $3x^3 - 16x^2 - 12x$

Note

First, look for any common factor. If the sign of c (the numerical term) is negative, the numerical terms in the brackets will have different signs. Always check by multiplying out.

Completing the square

Another way to manipulate quadratic expressions is to arrange the x-terms in a square.

The expression $x^2 + 2x + 5$ can be written as $x^2 + 2x + 1 + 4 = (x + 1)^2 + 4$.

This is called completing the square.

It can be used in factorising, finding maximum and minimum values and solving equations (see Chapter 23).

20 ALGEBRAIC MANIPULATION

> ### Example 20.27
>
> **Question**
>
> a Complete the square for $x^2 - 6x + 5$.
> b Use the result to factorise $x^2 - 6x + 5$.
> c What is the minimum value of the expression $x^2 - 6x + 5$?
>
> **Solution**
>
> a $x^2 - 6x$ are terms in the expansion of $(x - 3)^2$
>
>   (The number in the bracket is half the coefficient of x in the expression.)
>
> $x^2 - 6x + 5 = x^2 - 6x + 9 - 4 = (x - 3)^2 - 4$
>
> b $(x - 3)^2 - 4$ is a difference of two squares
>
> $(x - 3)^2 - 4 = (x - 3)^2 - 2^2 = (x - 3 - 2)(x - 3 + 2) = (x - 5)(x - 1)$
>
> c $x^2 - 6x + 5 = (x - 3)^2 - 4$
>
>   Minimum value of expression when $x - 3 = 0$ (a square number cannot be negative)
>
>   Minimum $= -4$ when $x = 3$.

Exercise 20.14

Complete the square for these expressions and state the minimum value.

1 $x^2 + 2x - 3$
2 $x^2 - 4x - 2$
3 $x^2 + 10x + 100$
4 $x^2 - 20x - 10$
5 $4x^2 - 4x + 5$

> ### Key points
>
> - When simplifying expressions, you can only add or subtract like terms.
> - $a(b + c) = ab + ac$.
> - When factorising, always take out common factors first.
> - When multiplying out brackets, each term in one bracket must be multiplied by each term in the other bracket.
> - Factorising $x^2 - a^2$ gives $(x - a)(x + a)$ which is a difference of two squares.
> - $a^2 + 2ab + b^2 = (a + b)^2$ which is a perfect square.
> - The expression $ax^2 + bx + c$ can be factorised into $(px + q)(rx + s)$ where $pr = a$, $qs = c$ and $ps + qr = b$.
> - Completing the square for a quadratic expression involves arranging the x terms as a square and adjusting the value of the constant.
> $x^2 + 2bx + c = (x + b)^2 + c - b^2$

21 ALGEBRAIC FRACTIONS

Simplifying algebraic fractions

To simplify a numerical fraction, you cancel any common factors of both the numerator and the denominator to write it in its simplest terms.

In the same way, to simplify an algebraic fraction, you cancel any common factors of both the numerator and the denominator.

Factors can be numbers, letters or brackets.

> **BY THE END OF THIS CHAPTER YOU WILL BE ABLE TO:**
> - manipulate algebraic fractions
> - factorise and simplify rational expressions.

Example 21.1

Question

Simplify $\dfrac{4ab^2}{3c^2} \times \dfrac{9c^2}{2a^2b}$.

Solution

$\dfrac{4ab^2}{3c^2} \times \dfrac{9c^2}{2a^2b} = \dfrac{\overset{2}{\cancel{4}}a\overset{b}{\cancel{b^2}} \times \overset{3}{\cancel{9}}\cancel{c^2}}{\cancel{3c^2} \times \underset{a}{\cancel{2}\cancel{a^2}\cancel{b}}} = \dfrac{6b}{a}$

2, 3, a, b and c^2 are all factors of both the numerator and the denominator, so these cancel.

> **CHECK YOU CAN:**
> - simplify algebraic expressions
> - expand brackets
> - factorise algebraic expressions.

Example 21.2

Question

Simplify $\dfrac{x^2 + x}{x^2 - 2x - 3}$.

Solution

As it stands, this fraction cannot be cancelled.

First, both the numerator and denominator must be factorised.

$\dfrac{x^2 + x}{x^2 - 2x - 3} = \dfrac{x\cancel{(x+1)}}{(x-3)\cancel{(x+1)}}$ $(x + 1)$ is a factor of both the numerator and the denominator so it cancels.

$= \dfrac{x}{(x-3)}$

> **Note**
> Errors often occur by cancelling individual terms. Only factors, which can be individual numbers, letters or brackets, can be cancelled.

Photocopying is prohibited

21 ALGEBRAIC FRACTIONS

Exercise 21.1

Simplify these expressions.

1. $\dfrac{15a^2}{6} \times \dfrac{b^2}{a}$
2. $\dfrac{12abc}{a^2b} \times \dfrac{a^3b}{4c}$
3. $\dfrac{x^3y^2}{10x} \times \dfrac{15xy}{10y^2}$
4. $\dfrac{2x^2y^2}{xy} \times \dfrac{3xy^3}{4x^2}$
5. $\dfrac{5x^2 - 20x}{10x^2}$
6. $\dfrac{3x^2 - 6x}{x^2 + x - 6}$
7. $\dfrac{x^2 + 2x + 1}{x^2 - 1}$
8. $\dfrac{3x^2 - x - 4}{5x^2 - 5}$
9. $\dfrac{3x + 15}{x^2 + 3x - 10}$
10. $\dfrac{6x - 18}{x^2 - x - 6}$
11. $\dfrac{x^2 - 5x + 6}{x^2 - 4x + 3}$
12. $\dfrac{x^2 - 3x - 4}{x^2 - 4x - 5}$
13. $\dfrac{x^2 - 2x - 3}{x^2 - 9}$
14. $\dfrac{3x^2 - 12}{x^2 + 2x - 8}$
15. $\dfrac{3x^2 + 5x + 2}{2x^2 - x - 3}$
16. $\dfrac{2x^2 + x - 6}{x^2 + x - 2}$
17. $\dfrac{6x^2 - 3x}{(2x - 1)^2}$
18. $\dfrac{5(x + 3)^2}{x^2 - 9}$
19. $\dfrac{(x - 3)(x + 2)^2}{x^2 - x - 6}$
20. $\dfrac{2x^2 + x - 6}{(2x - 3)^2}$

Adding and subtracting algebraic fractions

To add and subtract numerical fractions, you first write all the fractions with a common denominator.

In the same way, to add and subtract algebraic fractions, you first write all the fractions with a common denominator.

Example 21.3

Question

Simplify $\dfrac{x + 3}{2} - \dfrac{x - 3}{3}$.

Solution

The common denominator is 6.
So $(x + 3)$ is multiplied by $6 \div 2 = 3$ and $(x - 3)$ is multiplied by $6 \div 3 = 2$.

$$\dfrac{x+3}{2} - \dfrac{x-3}{3} = \dfrac{3(x+3) - 2(x-3)}{6}$$
$$= \dfrac{3x + 9 - 2x + 6}{6}$$
$$= \dfrac{x + 15}{6}$$

Note
Do not miss out the first step. Most errors occur because of expanding the brackets wrongly.

The procedure is still the same when the denominators involve algebraic expressions as shown in the next example.

Adding and subtracting algebraic fractions

Example 21.4

Question

Simplify $\dfrac{3}{x+1} - \dfrac{2}{x}$.

Solution

The common denominator is $x(x + 1)$.

So 3 is multiplied by $x(x + 1) \div (x + 1) = x$ and 2 is multiplied by $x(x + 1) \div x = (x + 1)$.

$$\dfrac{3}{x+1} - \dfrac{2}{x} = \dfrac{3x - 2(x+1)}{x(x+1)}$$

$$= \dfrac{3x - 2x - 2}{x(x+1)}$$

$$= \dfrac{x - 2}{x(x+1)}$$

Exercise 21.2

Simplify these.

1. $\dfrac{x}{2} + \dfrac{3x}{5}$

2. $\dfrac{2x}{3} - \dfrac{3x}{5}$

3. $\dfrac{x-1}{2} - \dfrac{x-3}{5}$

4. $\dfrac{x-3}{5} + \dfrac{2x}{3} - \dfrac{3x-2}{10}$

5. $\dfrac{2x-1}{6} + \dfrac{3x}{4} - \dfrac{x-2}{12}$

6. $\dfrac{1}{x} + \dfrac{2}{x-1}$

7. $\dfrac{3}{x} + \dfrac{2}{x+1}$

8. $\dfrac{2x}{x+1} - \dfrac{x-1}{x+3}$

9. $\dfrac{x+1}{x-1} + \dfrac{3x-1}{x+2}$

10. $\dfrac{2x}{x-1} - \dfrac{3x+2}{x+2}$

11. $\dfrac{x}{x+1} - \dfrac{3}{5} + \dfrac{x-2}{x}$

12. $\dfrac{x}{x+1} + \dfrac{3}{5} - \dfrac{x+3}{x}$

13. $\dfrac{2x}{x-3} + \dfrac{x-1}{x+2} - \dfrac{4}{9}$

14. $\dfrac{2}{2x+1} + \dfrac{3x+5}{x+2}$

15. $\dfrac{4x+17}{x+3} - \dfrac{2x-15}{x-3}$

Key points
- To simplify algebraic fractions, first factorise the numerator and denominator. Common factors can then be cancelled.
- To add or subtract algebraic fractions, they must all be written with a common denominator.

22 INDICES 2

BY THE END OF THIS CHAPTER YOU WILL BE ABLE TO:
- understand and use indices (positive, zero, negative and fractional)
- understand and use the rules of indices.

CHECK YOU CAN:
- use index notation with numbers
- simplify algebraic fractions.

Simplifying algebraic expressions using indices

Indices can be used with algebraic expressions, as well as with numerical expressions.

You know that $a \times a \times a = a^3$ and $a \times a \times a \times a \times a = a^5$

So, $a^3 \times a^5 = (a \times a \times a) \times (a \times a \times a \times a \times a) = a^8$

The indices are added $\quad a^3 \times a^5 = a^{3+5} = a^8$

Similarly $\quad a^5 \div a^3 = (a \times a \times a \times a \times a) \div (a \times a \times a)$

$$= \frac{\cancel{a} \times \cancel{a} \times \cancel{a} \times a \times a}{\cancel{a} \times \cancel{a} \times \cancel{a}}$$

$$= a \times a$$

$$= a^2$$

The indices are subtracted $\quad a^5 \div a^3 = a^{5-3} = a^2$

The rules for multiplying and dividing numbers in index form can also be applied to algebraic expressions.

These are the general laws for indices.

$$a^m \times a^n = a^{m+n}$$

$$a^m \div a^n = a^{m-n}$$

Using the first law, you can see that $\quad (a^2)^3 = a^2 \times a^2 \times a^2 = a^{2+2+2} = a^6$

This is the same as $\quad (a^2)^3 = a^{2 \times 3} = a^6$

This gives another general law.

$$(a^n)^m = a^{m \times n}$$

Using the second law, you can see that $\quad a^3 \div a^3 = a^{3-3} = a^0$

But $\quad a^3 \div a^3 = 1$

This gives a fourth general law.

$$a^0 = 1$$

These four laws can be used to simplify algebraic expressions involving indices.

Simplifying algebraic expressions using indices

Example 22.1

Question
Simplify these expressions where possible.

a $3a^2 \times 4a^3$ **b** $\frac{6a^5}{2a^3}$ **c** $(2x^3)^4$ **d** $\frac{12ab^3 \times 3a^2b}{2a^3b^2}$ **e** $4a^2 + 3a^3$

Solution

a $3a^2 \times 4a^3 = 12a^5$ The numbers are multiplied and the indices are added.

b $\frac{6a^5}{2a^3} = 3a^2$ The numbers are divided and the indices are subtracted.

c $(2x^3)^4 = 16x^{12}$ The number is raised to the power 4 and the indices are multiplied.

d $\frac{12ab^3 \times 3a^2b}{2a^3b^2} = \frac{36a^3b^4}{2a^3b^2}$ First, simplify the numerator by multiplying the numbers and adding the indices for the a terms and the b terms separately.

$= 18a^0b^2$ Then divide the numbers and subtract the indices for the a terms and the b terms separately.

$= 18b^2$ Remember that $a^0 = 1$.

e $4a^2 + 3a^3$ This expression cannot be simplified.
The terms are not like terms so they cannot be added.

Exercise 22.1

1 Write these as a single power of a.
  a $a^2 \times a^3$ **b** $a^4 \times a^5$ **c** $a^4 \times a^2$ **d** $a^3 \times a^6$

2 Write these as a single power of a.
  a $a^6 \div a^4$ **b** $a^7 \div a^3$ **c** $a^8 \div a^2$ **d** $a^5 \div a^2$

3 Simplify these.
  a $2a^2 \times 3a^3$ **b** $4a^4 \times 3a^5$ **c** $3a^4 \times 4a^2$ **d** $5a^3 \times 3a^6$

4 Simplify these.
  a $6a^6 \div 2a^4$ **b** $10a^7 \div 5a^3$ **c** $6a^8 \div 3a^2$ **d** $12a^5 \div 4a^2$

5 Simplify these.
  a $(3a^3)^2$ **b** $(2a)^3$ **c** $(5a^4)^2$ **d** $(2a^3)^5$

6 Write these as a single power of x.
  a $\frac{x^2 \times x^5}{x^3}$ **b** $\frac{x^7}{x^2 \times x^3}$ **c** $\frac{x \times x^8}{x^3 \times x^4}$ **d** $\left(\frac{x^5}{x^2}\right)^4$

7 Simplify these where possible.
  a $2a^2b \times 3a^3b^2$ **b** $3a^2b^3 \times 2a^3b^4$ **c** $4a^2b - 2ab^2$

8 Simplify these.
  a $\frac{15a^2b^3 \times 3a^2b}{9a^3b^2}$ **b** $\frac{4abc \times 3a^2bc^2}{6a^2b^2c^2}$ **c** $\frac{(3a^2b^2)^3}{(a^3b)^2}$

9 Simplify these.
  a $\frac{6a^4b}{5c^2} \times \frac{10c^3}{9a^2b^3}$ **b** $\frac{12x^5y^4}{7z^5} \times \frac{14z^3}{15x^3y^2}$ **c** $\frac{8a^7b^5}{15c^4} \div \frac{4b^2}{9c}$

22 INDICES 2

Using the laws of indices with numerical and algebraic expressions

You have met these laws of indices.

$a^m \times a^n = a^{m+n}$

$a^m \div a^n = a^{m-n}$

$(a^n)^m = a^{n \times m}$

$a^0 = 1$

$a^{-n} = \dfrac{1}{a^n}$

$a^{\frac{1}{n}} = \sqrt[n]{a}$

These laws can be used together to simplify expressions involving numbers or letters.

Example 22.2

Question

Write each of these as a single power of 2, where possible.

a $\quad 2\sqrt{2}$

b $\quad \left(\sqrt[3]{2}\right)^2$

c $\quad 2^3 \div 2^{\frac{1}{2}}$

d $\quad 2^3 + 2^4$

e $\quad 8^{\frac{3}{4}}$

f $\quad 2^3 \times 4^{\frac{3}{2}}$

g $\quad 2^n \times 4^3$

Solution

a $\quad 2\sqrt{2} = 2^1 \times 2^{\frac{1}{2}} = 2^{\frac{3}{2}}$

b $\quad \left(\sqrt[3]{2}\right)^2 = \left(2^{\frac{1}{3}}\right)^2 = 2^{\frac{2}{3}}$

c $\quad 2^3 \div 2^{\frac{1}{2}} = 2^{3-\frac{1}{2}} = 2^{\frac{5}{2}}$

d $\quad 2^3 + 2^4$ These powers cannot be added.

e $\quad 8^{\frac{3}{4}} = (2^3)^{\frac{3}{4}} = 2^{3 \times \frac{3}{4}} = 2^{\frac{9}{4}}$

f $\quad 2^3 \times 4^{\frac{3}{2}} = 2^3 \times (2^2)^{\frac{3}{2}} = 2^3 \times 2^3 = 2^6$

g $\quad 2^n \times 4^3 = 2^n \times (2^2)^3 = 2^n \times 2^6 = 2^{n+6}$

Exercise 22.2

1 Write each of these in the form 2^k.

a $\quad 2^2 \times 2^{\frac{1}{2}}$

b $\quad 2^{-2} \times 2^{\frac{1}{4}}$

c $\quad (2^{-3})^2$

d $\quad (2^0)^2 \times 2^{-3}$

e $\quad \left(\dfrac{1}{\sqrt[3]{2}}\right)^2$

f $\quad \dfrac{\left(\sqrt{2}\right)^3}{2}$

2 Write each of these numbers as a power of 3, as simply as possible.

a $\quad 27$

b $\quad \dfrac{1}{3}$

c $\quad 3 \times \sqrt{3}$

d $\quad 81^{\frac{3}{2}}$

e $\quad 3^4 \times 9^{-1}$

f $\quad 9^n \times 27^{3n}$

3 Write each of these numbers as a power of 2, as simply as possible.

a $\quad 32$

b $\quad 8^{\frac{2}{3}}$

c $\quad 2 \times \sqrt[3]{64}$

d $\quad 0.25$

e $\quad 2^{2n} \times 4^{\frac{n}{2}}$

f $\quad 2^{3n} \times 16^{-2}$

Using the laws of indices with numerical and algebraic expressions

4 Write each of these numbers as a power of 5, as simply as possible.
   a 625
   b $25^{-\frac{1}{2}} \times 5^3$
   c 0.2
   d $125^{\frac{3}{2}} \times 5^{-3} \div 25^2$
   e $5^4 \div 5^n$
   f $25^{3n} \times 125^{\frac{n}{3}}$

5 Write each of these in the form $2^a \times 3^b$.
   a 24
   b $6^2 \times 4^2$
   c $18^{\frac{1}{3}}$
   d $\frac{4}{9}$
   e $13\frac{1}{2}$
   f 12^{2n}

6 Write each of these expressions as a power of x, as simply as possible.
   a $x^3 \times x^2$
   b $x^n \times x^n$
   c $x^{\frac{3}{2}} \div x^{-\frac{1}{2}}$
   d $\left(x^{\frac{1}{2}}\right)^5$
   e $\dfrac{x^4 \times x \times x^{\frac{1}{2}}}{\sqrt[3]{x^2}}$
   f $\sqrt{\dfrac{\sqrt{x} \times x^3}{x}}$

7 Solve these equations.
   a $2^x = 16$
   b $64^x = 4$
   c $3^x = 81$
   d $\left(\frac{1}{9}\right)^x = 3$
   e $2^x = \frac{1}{8}$
   f $16^x = \frac{1}{4}$

   Note
   Solve means find the value of x that makes the statement true.

8 Simplify these.
   a $2a^{-3} \times 5a^2$
   b $\frac{3}{4}a^5 \times \frac{2}{3}a^{-2}$
   c $\frac{3}{10}a^{\frac{1}{2}} \times 5a^{-1}$
   d $\frac{4}{9}a^{-5} \times 6a^{\frac{3}{2}}$

9 Simplify these.
   a $\left(\dfrac{9ab^2}{a^3}\right)^{\frac{1}{2}}$
   b $\left(\dfrac{a^5b^2}{27a^2b^8}\right)^{\frac{1}{3}}$
   c $\left(\dfrac{5ab^2}{15a^2b}\right)^{-2}$
   d $\left(\dfrac{75a^8b}{3a^2b^5}\right)^{-\frac{1}{2}}$

10 Simplify.
   a $4y^2 \div 2y^{-2}$
   b $\frac{2}{5}y^{-3} \div \frac{1}{5}y^4$
   c $5y^{\frac{1}{2}} \div 10y^{-1}$
   d $\frac{3}{4}y^{-2} \div \frac{3}{5}y^{\frac{1}{2}}$

Key points
The laws of indices:
- $a^m \times a^n = a^{m+n}$
- $a^m \div a^n = a^{m-n}$
- $(a^n)^m = a^{n \times m}$
- $a^0 = 1$
- $a^{-n} = \dfrac{1}{a^n}$
- $a^{\frac{1}{n}} = \sqrt[n]{a}$

REVIEW EXERCISE 4

1 Work out these.
   a $b^2 - 4ac$ when $a = 2$, $b = -4$ and $c = -3$ [1]
   b $\frac{3V^2}{m-c}$ when $c = -1$, $m = 8$ and $V = -6$ [1]

Ch 19

2 On Monday, Ravi goes on a 20 km run.
   a His average speed for the first 12 km is x km/h.
   Write down an expression, in terms of x, for the time taken for the first 12 km.
   Give your answer in minutes. [1]
   b His average speed for the final 8 km of the run is 1.5 km/h slower than for the first 12 km.
   Write an expression, in terms of x, for the time taken for the final 8 km of the run.
   Give your answer in minutes. [1]
   c Ravi takes 110 minutes to complete the full 20 km.
   Form an equation in x and show that it simplifies to $22x^2 - 273x + 216 = 0$. [4]

Cambridge O Level Mathematics Syllabus D (4024) Paper 22 Q9ai, ii & iii, June 2018

Ch 12
Ch 19

3 a $Z = (ab)^2 - 4ab^2$.
   Work out Z when $a = -5$ and $b = 3$. [1]

   b $W = \frac{V^2 k}{k + V^3}$

   Work out W when $V = -1$ and $k = 3$. [1]

Ch 19

4 Simplify these expressions.
   a $a + 2a + 3b - 4a + 5b$ [1]
   b $3a \times 2b \times 4a$ [1]

Ch 20

5 Factorise these expressions.
   a $3xy + 9x^2$ [2]
   b $x^2 - 6x + 5$ [1]
   c $16 - a^2$ [1]

Ch 20

6 a Simplify $4c - 3(2c - 5)$. [1]
   b Factorise $8 - 10y + 12x - 15xy$. [2]

Cambridge O Level Mathematics Syllabus D (4024) Paper 12 Q5, November 2018

Ch 20

7 Expand and simplify these brackets.
   a $(2x + 1)(4x - 1)$ [2]
   b $(3x - 2)(x^2 - 4x + 3)$ [2]
   c $(2x - 1)(x + 2)(x - 6)$ [3]

Ch 20

Review exercise 4

Ch 21 **8** Simplify this expression.
$$\frac{4c^4}{3d^3} \div \frac{2c}{6d^2}$$ [2]

9 Simplify fully $\frac{4x^2 - 1}{2x^2 - 9x - 5}$. [3]

Ch 21

Cambridge O Level Mathematics Syllabus D (4024) Paper 12 Q26 b, June 2016

10 a Expand and simplify.
$(x + 5)(x - 2)$ [2]

Ch 20
Ch 21 **b** Write as a single fraction in its simplest form.
$$\frac{3}{x - 4} + \frac{2}{x + 5}$$ [3]

Cambridge O Level Mathematics Syllabus D (4024) Paper 11 Q23, November 2021

11 a Evaluate $9^1 + 9^0$. [1]
Ch 22 **b** Find n, where $4^n = 2^{n-1}$. [2]

Cambridge O Level Mathematics Syllabus D (4024) Paper 12 Q21, November 2018

12 Evaluate
Ch 22 **a** 3^{-2}, [1]
b $125^{\frac{2}{3}}$. [1]

Cambridge O Level Mathematics Syllabus D (4024) Paper 11 Q18a, June 2017

13 Simplify

Ch 22 **a** $\frac{5x^7 y}{15x^3 y^4}$ [1]

b $\left(\frac{4t^2}{v^4}\right)^{-\frac{1}{2}}$ [1]

Cambridge O Level Mathematics Syllabus D (4024) Paper 12 Q11, June 2016

Ch 21 **14** Simplify $\frac{x^2 + 5x - 14}{2x^2 - 4x}$. [3]

Photocopying is prohibited

23 EQUATIONS

BY THE END OF THIS CHAPTER YOU WILL BE ABLE TO:
- construct expressions, equations and formulas
- solve linear equations in one unknown
- solve fractional equations with numerical and linear algebraic denominators
- solve simultaneous linear equations in two unknowns
- solve quadratic equations by factorisation, completing the square and by use of the quadratic formula
- change the subject of formulas.

CHECK YOU CAN:
- add and subtract, multiply and divide numerical fractions
- simplify algebraic expressions by collecting like terms
- multiply out expressions such as $3(2x - 5)$
- simplify algebraic fractions
- factorise quadratic expressions when this is possible.

For changing the subject of formulas, you will also need to be able to:
- solve linear equations
- factorise expressions such as $ax + bx$.

Writing formulas

Example 23.1

Question

Write a formula for the mean height, M cm, of a group of n people whose heights total h cm.

Solution

To find the mean height, you divide the total of the heights by the number of people, so the formula for M is

$M = h \div n$ or $M = \frac{h}{n}$

Note
If you are not sure whether to multiply or divide, try an example with numbers first.

Example 23.2

Question

The cost ($\$C$) of hiring a car is a fixed charge ($\$f$), plus the number of days (n) multiplied by the daily rate ($\$d$).

Write a formula for C.

Solution

$C = f + n \times d$ or $C = f + nd$

Exercise 23.1

Write a formula for each of these, using the letters given.

1. The cost (C cents) of x pencils at y cents each.
2. The area (A cm^2) of a rectangle m cm long and n cm wide.
3. The height (h cm) of a stack of n tins each t cm high.
4. The temperature in °F (F) is 32, plus 1.8 times the temperature in °C (C).
5. The area (A cm^2) of a triangle is half the base (b cm) times the height (h cm).
6. The current in a circuit (I) is the voltage (V) divided by the resistance (R).
7. The cost of fuel ($\$C$) is the number of litres (n) multiplied by the price of fuel per litre ($\$p$).
8. The total of the wages ($\$w$) in a supermarket is the number of workers (n) multiplied by the weekly wage ($\$q$).
9. The number of books (N) that can fit on a shelf is the length (L cm) of the shelf divided by the thickness (t cm) of each book.
10. The number of posts (Q) for a fence is the length of the fence (R metres) divided by 2, plus 1.
11. Bethany is x years old, Ekene is y years old and Chine is $2x$ years old. Write an expression for the total of their ages in years.

12 The lengths of the sides of a triangle, in centimetres, are s, $s + 6$ and $2s - 1$.
   Write an expression for the total of these lengths in centimetres.
13 n is an integer.
   Write an expression for
   a   the next integer (larger than n)
   b   the product of these two consecutive integers.

Before you do Exercise 23.2, check your answers to Exercise 23.1.

Exercise 23.2

Use the formulas from Exercise 23.1 to find each of these.

1 C when $x = 15$ and $y = 12$
2 A when $m = 7$ and $n = 6$
3 h when $n = 20$ and $t = 17$
4 F when $C = 40$
5 A when $b = 5$ and $h = 6$
6 I when $V = 13.6$ and $R = 2.5$
7 C when $n = 50$ and $p = 70$
8 w when $n = 200$ and $q = 150$
9 N when $L = 90$ and $t = 3$
10 Q when $R = 36$

Writing equations

Some everyday problems can be solved by writing equations and solving them.

Example 23.3

Question
The length of a rectangle is a cm.
The width is 15 cm shorter.
The length is three times the width.
Write down an equation in a and solve it to find the length and width of the rectangle.

Solution
If the length = a, the width = $a - 15$.
Also the length = 3 × width = $3(a - 15)$.
The equation is $a = 3(a - 15)$.

$a = 3(a - 15)$
$a = 3a - 45$ Multiply out the bracket.
$a + 45 = 3a$ Add 45 to each side.
$45 = 2a$ Subtract a from each side.
$22.5 = a$ Divide each side by 2.
$a = 22.5$ Rewrite with the subject on the left-hand side.

So the length = 22.5 cm and the width = 7.5 cm.

Exercise 23.3

1. Eira is x years old and Jaan is three years older than Eira.
   Their ages add up to 23.
   Write an equation in x and solve it to find their ages.
2. Two angles of a triangle are the same and the other is 15° bigger.
   Call the two equal angles a.
   Write an equation and solve it to find all the angles.
3. Chan and Ali go shopping.
   Chan spends x and Ali spends twice as much.
   They spend $45 altogether.
   Write an equation and solve it to find how much each spends.
4. A pentagon has two angles of 150°, two of $x°$ and one of $(x + 30)°$.
   The sum of the angles in a pentagon is 540°.
   a. Write down an equation in x and solve it.
   b. State the size of each of the angles.
5. On a school trip to China, there are 15 more girls than boys.
   Altogether, 53 students go on the trip.
   a. If the number of boys is x, write an equation in x and solve it.
   b. How many boys and how many girls go on the trip?
6. A company employs 140 people, of whom x are men.
   There are ten fewer women than men.
   Use algebra to find how many men and how many women work for the company.
7. Two angles in a triangle are $x°$ and $(2x - 30)°$.
   The first angle is twice the size of the second.
   Write an equation and solve it to find the size of the two angles.
8. The width of a rectangle is 3 cm and the length is $x + 4$ cm.
   The area is 27 cm².
   Write an equation and solve it to find x.
9. Sabir thinks of a number.
   If he doubles the number and then subtracts 5, he gets the same answer as if he subtracts 2 from the number and then multiplies by 3.
   Let the number be n.
   Write an equation and solve it to find n.
10. On a bus trip, each child pays c and each adult pays $12 more than this.
    There are 28 children and 4 adults on the bus.
    The same amount of money is collected from all the children as from all the adults.
    Write an equation and solve it to find out how much each child and each adult pays.

Solving simple linear equations

When you solve an equation, you always do the same thing to both sides of the equation.

The following example will give you a reminder about solving simple linear equations.

Solving equations with a bracket

Example 23.4

Question

Solve $5x + 2 = 12$.

Solution

$5x + 2 = 12$

$\quad 5x = 10 \quad$ Subtract 2 from both sides.

$\quad\ x = 2 \quad$ Divide both sides by 5.

Exercise 23.4

Solve these equations.

1. $3x + 4 = 16$
2. $5x - 4 = 16$
3. $7x - 3 = 18$
4. $9a + 4 = 40$
5. $8y + 3 = 27$

Solving equations with a bracket

Sometimes the equation formed to solve a problem will involve a bracket.

Example 23.5

Question

Solve $4(x - 5) = 18$.

Solution

By expanding the bracket, you will have an equation like that shown in Example 23.4.

Method 1

$4(x - 5) = 18$

$\quad 4x - 20 = 18 \quad$ Expand the brackets.

$\quad\quad 4x = 38 \quad$ Add 20 to both sides.

$\quad\quad\ x = 9\frac{1}{2} \quad$ Divide both sides by 4.

Note
You may find it easier to use Method 1.

Method 2

$4(x - 5) = 18$

$\quad x - 5 = 4\frac{1}{2} \quad$ Divide both sides by 4.

$\quad\quad x = 9\frac{1}{2} \quad$ Add 5 to both sides.

Note
Method 2 is usually shorter.

23 EQUATIONS

Exercise 23.5

Solve these equations.

1. $2(x + 1) = 10$
2. $4(x - 1) = 12$
3. $5(x + 6) = 20$
4. $2(x + 4) = 8$
5. $2(x - 4) = 8$
6. $3(x + 7) = 9$
7. $2(x - 7) = 3$
8. $2(2x + 3) = 18$
9. $5(x - 1) = 12$
10. $2(5 + 2x) = 17$
11. $4(3x - 1) = 20$
12. $3(5x - 13) = 21$
13. $2(4x + 7) = 12$
14. $5(x - 6) = 20$
15. $7(x + 3) = 28$
16. $5(3x - 1) = 40$
17. $2(5x - 3) = 14$
18. $7(x - 4) = 28$
19. $3(5x - 12) = 24$
20. $2(2x - 5) = 12$

Solving equations with the unknown on both sides

Sometimes the equation formed to solve a problem will have the unknown on both sides.

Example 23.6

Question

Solve $2(3x - 1) = 3(x - 2)$.

Solution

$2(3x - 1) = 3(x - 2)$
$6x - 2 = 3x - 6$ Expand the brackets.
$6x = 3x - 4$ Add 2 to both sides.
$3x = -4$ Subtract $3x$ from both sides.
$x = -1\frac{1}{3}$ Divide both sides by 3.

Exercise 23.6

Solve these equations.

1. $2x - 1 = x + 3$
2. $5x - 6 = 3x$
3. $4x + 1 = x - 8$
4. $4x - 1 = 3x + 7$
5. $4x - 3 = x$
6. $2(x + 3) = x + 7$
7. $5(2x - 1) = 3x + 9$
8. $2(x - 1) = x + 2$
9. $2(2x + 3) = 3x - 7$
10. $2x - 3 = 7 - 3x$
11. $4x - 1 = 2 + x$
12. $3x - 2 = x + 7$
13. $x - 5 = 2x - 9$
14. $3x - 4 = 2 - 3x$
15. $5x - 6 = 16 - 6x$
16. $49 - 3x = x + 21$
17. $3(x - 1) = 2(x + 1)$
18. $3(4x - 3) = 10x - 1$
19. $3(4x - 1) = 5(2x + 3)$
20. $7(x - 2) = 3(2x - 7)$

Solving equations involving fractions

You may have to solve equations involving fractions.

The first step is to eliminate the fraction or fractions.

Example 23.7

Question

Solve $\frac{x}{3} = 2x - 3$.

Solution

$\frac{x}{3} = 2x - 3$

$x = 3(2x - 3)$ Multiply both sides by 3.

$x = 6x - 9$ Expand the bracket.

$x + 9 = 6x$ Add 9 to both sides.

$9 = 5x$ Subtract x from both sides.

Collecting the x-terms on the right-hand side of the equation makes the final x-term positive.

$x = \frac{9}{5}$ or $1\frac{4}{5}$ or 1.8   Divide both sides by 5.

Note

A common error when multiplying through by a number or letter is to multiply just the first term. Use a bracket to make sure.

Note

It is much easier to collect all the x-terms on to the side where the final x-term will be positive.

Note

Another common error in examples like this would be to give the answer as $\frac{5}{9}$ rather than $\frac{9}{5}$.

Example 23.8

Question

Solve the equation $\frac{400}{x} = 8$.

Solution

$\frac{400}{x} = 8$

$400 = 8x$ Multiply both sides by x.

$x = 50$ Divide both sides by 8.

Example 23.9

Question

Solve $\frac{x}{3} + \frac{x}{2} = 5$.

Solution

$\frac{x}{3} + \frac{x}{2} = 5$

$\frac{6x}{3} + \frac{6x}{2} = 5 \times 6$ Multiply every term by 6, because 6 is the common denominator of $\frac{x}{3}$ and $\frac{x}{2}$.

$\frac{^2 \cancel{6}x}{\cancel{3}_1} + \frac{^3 \cancel{6}x}{\cancel{2}_1} = 30$ Cancel the common factor in the numerator and denominator of each fraction.

$2x + 3x = 30$ This will eliminate both fractions.

$5x = 30$

$x = 6$ Divide both sides by 5.

Note

It is much easier to multiply both sides of the equation by a number that will eliminate the fraction than to expand the bracket.

Photocopying is prohibited

23 EQUATIONS

Exercise 23.7

Solve these equations.

1. $\frac{x}{2} = 3x - 10$
2. $\frac{2x}{3} = x - 2$
3. $\frac{2x}{5} = x - 3$
4. $\frac{3x}{5} = 4 - x$
5. $\frac{5x}{3} = 4x - 2$
6. $\frac{x}{3} + 5 = 9$
7. $\frac{2x}{7} - 2 = 5$
8. $\frac{300}{x} = 15$
9. $\frac{200}{x} = 4$
10. $\frac{39}{2x} = 3$
11. $\frac{1}{3}(x - 2) = 7$
12. $\frac{3x - 1}{5} = 4$
13. $\frac{5x - 12}{2} = x$
14. $\frac{5}{x} + 3 = 7$
15. $\frac{6}{x} - 5 = -1$
16. $\frac{7}{2x} - 2 = 3$
17. $\frac{5}{3x} + 7 = 10$
18. $\frac{x}{2} - \frac{x}{4} = 3$
19. $\frac{3x}{2} - \frac{x}{6} = 12$
20. $\frac{x}{2} - 4 = \frac{x}{3}$

Solving harder equations involving fractions

The following example illustrates how to solve harder equations involving fractions.

Example 23.10

Question

Solve $\frac{x - 2}{4} - \frac{x + 3}{5} = 1$.

Solution

$\frac{x - 2}{4} - \frac{x + 3}{5} = 1$ The common denominator of 4 and 5 is 20.

$\frac{20(x - 2)}{4} - \frac{20(x + 3)}{5} = 20$ Multiply every term by the common denominator.

$\frac{^5 \cancel{20}(x - 2)}{\cancel{4}_1} - \frac{^4 \cancel{20}(x + 3)}{\cancel{5}_1} = 20$ Cancel the common factors in the numerator and denominator of each fraction.

$5(x - 2) - 4(x + 3) = 20$ This will eliminate the fractions.

$5x - 10 - 4x - 12 = 20$ Expand the bracket. Take care with the signs.

$x - 22 = 20$ Collect like terms.

$x = 42$ Add 22 to both sides.

Note

The most common error is in the signs when the left-hand side is a subtraction. Make sure you put the brackets in.

Exercise 23.8

Solve these equations.

1. $\dfrac{x+2}{3} + \dfrac{x-3}{2} = 5$

2. $\dfrac{3x}{2} = \dfrac{3x-2}{5} + 4$

3. $\dfrac{2x-1}{5} - \dfrac{x-3}{3} = \dfrac{5}{2}$

4. $\dfrac{2x-1}{6} = \dfrac{x-3}{2} + \dfrac{2}{3}$

5. $\dfrac{x-2}{3} + \dfrac{2x-1}{2} = \dfrac{17}{6}$

6. $\dfrac{1}{3}(x+2) - \dfrac{1}{2}(x-3) = 2$

7. $\dfrac{2x-3}{3} - \dfrac{2x+1}{6} + \dfrac{3}{2} = 0$

8. $\dfrac{3x-2}{2} = \dfrac{x-3}{3} + \dfrac{7}{6}$

9. $\dfrac{x-1}{3} - \dfrac{2x}{5} = \dfrac{3x-1}{5}$

10. $\dfrac{1}{5}(x+2) - \dfrac{1}{3}(x-5) = 2$

11. $\dfrac{1}{2}(x-1) - \dfrac{1}{5}(x-3) = 1$

12. $\dfrac{2x-1}{6} + \dfrac{3x}{4} = \dfrac{x-2}{12}$

13. $\dfrac{1}{7}(x-5) + \dfrac{1}{2}(x-8) = 1$

14. $\dfrac{x-3}{2} + \dfrac{2x+3}{5} = \dfrac{4x-5}{4}$

15. $\dfrac{1}{5}(x+3) - \dfrac{1}{2}(2x-1) = \dfrac{1}{3}(11-x)$

Solving linear simultaneous equations

An equation in two unknowns does not have a unique solution.

For example, the graph of the equation $x + y = 4$ is a straight line. Every point on the line will have coordinates that satisfy the equation.

When you are given two equations in two unknowns, such as x and y, they usually have a common solution where the two lines meet at a point.

These are called **simultaneous equations**.

It is possible to solve simultaneous equations graphically. However, it can be time-consuming and you do not always obtain an accurate solution.

So using algebra to solve them accurately is often better.

If two quantities, A and B, are equal and two other quantities, C and D, are equal, it follows logically that

$A + C = B + D$ and

$A - C = B - D$

Therefore, you can add the left-hand sides of two simultaneous equations and the result will equal the two right-hand sides added together.

Similarly, you can subtract the left-hand sides of two simultaneous equations and the result will equal the two right-hand sides subtracted.

This is shown in the following examples.

23 EQUATIONS

Example 23.11

Question

Solve the simultaneous equations $2x + 5y = 9$ and $2x - y = 3$.

Solution

$2x + 5y = 9$ (1)   Write the two equations, one under the other, and label them.

$2x - y = 3$ (2)

This time $(+)2x$ appears in each equation, so subtract to eliminate the x-terms.

$2x - 2x + 5y - (-y) = 9 - 3$ (1) − (2). Take care with the signs.

           $6y = 6$ $5y - (-y) = 5y + y$.

           $y = 1$

$2x + 5 = 9$ Substitute $y = 1$ in (1). $5y$ is replaced by $5 \times 1 = 5$.

   $2x = 4$

     $x = 2$

The solution is $x = 2$, $y = 1$.

Check in equation (2): the left-hand side is $2x - y = 4 - 1 = 3$ which is correct.

> **Note**
> When eliminating, if the signs of the letter to be eliminated are the same, subtract. If they are different, add.

> **Note**
> When subtracting, take great care with the signs. If your check is wrong, see if you have made an error with any signs.

Example 23.12

Question

Solve the simultaneous equations $x + 3y = 10$ and $3x + 2y = 16$.

Solution

$x + 3y = 10$ (1)   Write the two equations, one under the other, and label them.

$3x + 2y = 16$ (2)

The coefficients of x and y are different in the two equations.

Multiply (1) by 3 to make the coefficient of x the same as in equation (2).

$3x + 9y = 30$ (3)   (1) × 3

$3x + 2y = 16$ (2)

Now $(+)3x$ appears in both equations, so subtract.

$3x - 3x + 9y - 2y = 30 - 16$ (3) − (2)

          $7y = 14$

           $y = 2$

$x + 6 = 10$ Substitute $y = 2$ in (1).

   $x = 4$

The solution is $x = 4$, $y = 2$.

Check in equation (2): the left-hand side is $3x + 2y = 12 + 4 = 16$ which is correct.

> **Note**
> When subtracting equations, you can do equation (1) − equation (2) or equation (2) − equation (1). It is better to make the letter positive. Always write down clearly what you are doing.

Solving harder simultaneous equations

Exercise 23.9

Solve these simultaneous equations.

1. $x + y = 5$
   $2x - y = 7$
2. $3x + y = 9$
   $2x + y = 7$
3. $2x + y = 6$
   $2x - y = 2$
4. $2x - y = 7$
   $3x + y = 13$
5. $2x + y = 7$
   $4x - y = 5$
6. $2x + 3y = 13$
   $x - y = 4$
7. $3x - y = 11$
   $3x - 5y = 7$
8. $2x + y = 6$
   $3x + 2y = 10$
9. $3x + y = 7$
   $2x + 3y = 7$
10. $2x - 3y = 0$
    $3x + y = 11$
11. $x + 2y = 19$
    $3x - y = 8$
12. $x + 2y = 6$
    $2x - 3y = 5$
13. $3x + 2y = 13$
    $x + 3y = 16$
14. $2x + 3y = 7$
    $3x - y = 5$
15. $2x + y = 3$
    $3x - 2y = 8$
16. $x + y = 4$
    $4x - 2y = 7$
17. $2x + 4y = 11$
    $x + 3y = 8$
18. $2x + 2y = 7$
    $4x - 3y = 7$
19. $2x - 3y = 9$
    $3x + y = 8$
20. $2x - y = -4$
    $x + 3y = -9$
21. $x + y = 0$
    $2x + 4y = 3$

Solving harder simultaneous equations

Sometimes the letters in the equations are not in the same order, so the first thing to do is to rearrange them.

Sometimes each of the equations needs to be multiplied by a different number.

Example 23.13

Question

Solve simultaneously the equations $3y = 4 - 4x$ and $6x + 2y = 11$.

Solution

$4x + 3y = 4$ (1)   Rearrange the first equation.
$6x + 2y = 11$ (2)

To eliminate x, multiply (1) by 3 and (2) by 2 and subtract,
or, to eliminate y, multiply (1) by 2 and (2) by 3 and subtract.

$12x + 9y = 12$ (3)   (1) × 3
$12x + 4y = 22$ (4)   (2) × 2
$5y = -10$ (3) − (4) Eliminate x.
$y = -2$
$4x - 6 = 4$ Substitute in (1). $3y$ is replaced by -6.
$\quad 4x = 10$
$x = \frac{10}{4} = \frac{5}{2} = 2\frac{1}{2}$

The solution is $x = 2\frac{1}{2}$, $y = -2$.

Check in equation (2): LHS = $6x + 2y = 15 - 4 = 11$, which is correct.

Note

If the equations are not already in the form $ax + by = c$, rearrange them so that they are.

23 EQUATIONS

Solving simultaneous equations using substitution

If x or y is the subject of one of the equations, there is an easier method to eliminate one of them.

This method is called substitution and is shown in Example 23.14.

Example 23.14

Question

Solve simultaneously the equations $3x - 2y = 6$ and $y = 2x - 5$.

Solution

Since y is the subject of the second equation, we can substitute it in the first.

$3x - 2y = 6$ (1)
$y = 2x - 5$ (2)
$3x - 2(2x - 5) = 6$ Substitute (2) in (1). Replace y in equation (1) by $2x - 5$.
$3x - 2(2x - 5) = 6$ Solve the resulting equation.
$3x - 4x + 10 = 6$ Expand the bracket.
$-x = -4$
$x = 4$
$y = 2 \times 4 - 5$ Now substitute $x = 4$ into equation (2).
$y = 3$

So the solution is $x = 4$, $y = 3$.

Note

Always substitute the answer you have found first into the equation with x or y as the subject to find the other value.

The most common error is to forget the brackets when making the substitution.

Exercise 23.10

Note

If there is a choice whether to add or subtract, it is usually easier to add.

Solve these simultaneous equations.

1. $y = 2x - 1$
   $x + 2y = 8$
2. $y = 3 - 2x$
   $3x - 3y = 0$
3. $5y = x + 1$
   $2x + 2y = 10$
4. $y = 3x - 3$
   $2x + 3y = 13$
5. $3x + 2y = 7$
   $2x - 3y = -4$
6. $3x - 2y = 3$
   $2x - y = 4$
7. $3x - 2y = 11$
   $2x + 3y = 16$
8. $2x - 3y = 5$
   $3x + 4y = 16$
9. $3x + 4y = 5$
   $2x + 3y = 4$
10. $4x - 3y = 1$
    $5x + 2y = -16$
11. $4x + 3y = 1$
    $3x + 2y = 0$
12. $y = x + 2$
    $2x - 4y = -9$
13. $4x - 2y = 3$
    $5y = 23 - 3x$
14. $x = 2y + 9$
    $3x - 2y = 19$
15. $4x + 2y = 18$
    $y = 3x - 4$

Solving quadratic equations of the form $x^2 + bx + c = 0$ by factorisation

For any two numbers, if $A \times B = 0$, then either $A = 0$ or $B = 0$.

If $(x - 3)(x - 2) = 0$ then either $(x - 3) = 0$ or $(x - 2) = 0$.

To solve a quadratic equation, factorise it into two brackets and then use this fact.

Remember, to factorise $x^2 + bx + c$:

- if c is positive, you find two numbers with the same sign as b that multiply to give c and add to give b
- if c is negative, you find two numbers with different signs that multiply to give c and add to give b.

If an equation is written as $x^2 + bx = c$ or $x^2 = bx + c$, first rearrange it so that all the terms are on one side, leaving zero on the other side.

Example 23.15

Question
Solve the equation $x^2 = 4x - 4$ by factorisation.

Solution
$x^2 - 4x + 4 = 0$ Rearrange so that all the terms are on the same side.
$(x - 2)(x - 2) = 0$ Factorising: c is positive and b is negative, $-2 \times -2 = 4$, $-2 + -2 = -4$.
$x - 2 = 0$ or $x - 2 = 0$
The solution is $x = 2$ (repeated).

Note
There are always two solutions, so if they are both the same, write 'repeated'.

Example 23.16

Question
Solve the equation $x^2 + 5x = 0$.

Solution
$x(x + 5) = 0$ Factorising with x as a common factor.
$x = 0$ or $x + 5 = 0$
The solution is $x = 0$ or $x = -5$.

23 EQUATIONS

Example 23.17

Question

Solve the equation $x^2 - 49 = 0$.

Solution

$(x + 7)(x - 7) = 0$ — Factorising by 'difference of two squares'.

$x + 7 = 0$ or $x - 7 = 0$

The solution is $x = -7$ or $x = 7$. This can be written as $x = \pm 7$.

You can use an alternative method for equations like this.

$x^2 - 49 = 0$

$x^2 = 49$ — Add 49 to both sides.

$x = \pm 7$ — Take the square root of both sides, remembering that this can give 7 or −7.

This method is perhaps simpler, but it is easy to forget the negative solution.

Exercise 23.11

Solve these equations by factorisation.

1. $x^2 - 5x + 6 = 0$
2. $x^2 - 6x + 5 = 0$
3. $x^2 - 4x + 3 = 0$
4. $x^2 - 100 = 0$
5. $x^2 + 6x + 8 = 0$
6. $x^2 + 5x + 4 = 0$
7. $x^2 + 9x + 20 = 0$
8. $x^2 - 25 = 0$
9. $x^2 + 2x + 1 = 0$
10. $x^2 - 7x + 6 = 0$
11. $x^2 - 9x + 18 = 0$
12. $x^2 - 8x = 0$
13. $x^2 + 7x + 12 = 0$
14. $x^2 + 3x = 0$
15. $x^2 + 6x = 0$
16. $x^2 - 10x + 24 = 0$
17. $x^2 - 6x + 8 = 0$
18. $x^2 - 169 = 0$
19. $x^2 - 225 = 0$
20. $x^2 + 2x - 3 = 0$
21. $x^2 + 4x - 5 = 0$
22. $x^2 = 10x$
23. $x^2 - x - 12 = 0$
24. $x^2 + 5x - 6 = 0$
25. $x^2 = x$
26. $x^2 - 2x - 15 = 0$
27. $x^2 - 3x = 18$
28. $x^2 - 9x = 10$
29. $x^2 - 17x + 30 = 0$
30. $x^2 + 4x = 32$

Solving quadratic equations of the form $ax^2 + bx + c = 0$ by factorisation

The following example has a question where the coefficient of x^2 is not 1.

Example 23.18

Question

Solve the equation $3x^2 + 11x + 6 = 0$.

Solution

$3x^2 + 11x + 6 = 0$

$(3x + 2)(x + 3) = 0$

$(3x + 2) = 0$ or $(x + 3) = 0$

$3x = -2$ or $x = -3$

$x = -\frac{2}{3}$ or $x = -3$

Exercise 23.12

Solve these equations.

1. $2x^2 - 5x - 12 = 0$
2. $3x^2 - 10x - 8 = 0$
3. $2x^2 + 5x + 3 = 0$
4. $2x^2 - 3x - 5 = 0$
5. $3x^2 + 2x - 1 = 0$
6. $2x^2 + 11x + 5 = 0$
7. $2x^2 - 13x + 15 = 0$
8. $12x^2 + 10x - 8 = 0$
9. $2x^2 + 2x - 60 = 0$
10. $3x^2 - 12x + 9 = 0$
11. $2x^2 - 2x - 12 = 0$
12. $3x^2 - 14x - 24 = 0$
13. $2x^2 - 8 = 0$
14. $3x^2 - 27 = 0$
15. $5x^2 - 125 = 0$

Solving quadratic equations by completing the square

This quadratic equation factorises.

$$x^2 - 4x + 3 = 0$$

This one does not.

$$x^2 - 4x + 1 = 0$$

$x^2 - 4x$ is part of the expansion $(x - 2)^2 = x^2 - 4x + 4$.

In this method for solving a quadratic equation, you 'complete the square' so that the left-hand side of the equation is a square $(mx + k)^2$.

When the coefficient of x^2 is 1,

$(x + k)^2 = x^2 + 2kx + k^2$,

so to find the number to add to complete the square, you halve the coefficient of x and square it.

Example 23.19

Question

Solve $x^2 - 4x + 1 = 0$.

Give your answer correct to 2 decimal places.

Solution

$x^2 - 4x + 1 = 0$

$x^2 - 4x + 1 + 3 = 0 + 3$ Add a number so the left-hand side is a complete square.

$x^2 - 4x + 4 = 3$

$(x - 2)^2 = 3$ Factorise the left-hand side.

$x - 2 = \sqrt{3}$ or $-\sqrt{3}$ Take the square root of both sides.

$x = 2 + \sqrt{3}$ or $2 - \sqrt{3}$ This is usually written as $x = 2 \pm \sqrt{3}$.

$x = 3.73$ or 0.27 to 2 decimal places

Note

Look carefully at the question. If it asks for an exact answer then you must give $2 + \sqrt{3}$ and $2 - \sqrt{3}$. If it asks for the answer to a given number of decimal places or significant figures, then you must give the decimal correctly rounded.

23 EQUATIONS

Completing the square may also be used to write quadratic expressions in completed square form.

For example,

$x^2 - 4x + 1 = (x - 2)^2 - 3.$

Since the least value that a square can be is zero, this gives the information that the least value of the expression is -3 and this occurs when $x = 2$.

Example 23.20

Question

Write these quadratic expressions in completed square form, $a(x + b)^2 + c$.

a $x^2 + 6x - 3$
b $3x^2 + 15x + 14$

Solution

a $x^2 + 6x - 3 = (x + 3)^2 - 9 - 3$ $x^2 + 6x$ is part of the expansion of $(x + 3)^2$, but without the +9 term, so subtract 9.

   $= (x + 3)^2 - 12$

b $3x^2 + 15x + 14$ Take out a factor of 3 from the terms involving x.

   $= 3(x^2 + 5x) + 14$

   $= 3\left[\left(x + \frac{5}{2}\right)^2 - \frac{25}{4}\right] + 14$ Recognise that $(x^2 + 5x)$ is part of the expansion of $\left(x + \frac{5}{2}\right)^2$ but without the term $\left(\frac{5}{2}\right)^2$.

Remember that you need to multiply the $-\frac{25}{4}$ term by 3.

   $= 3\left(x + \frac{5}{2}\right)^2 - \frac{75}{4} + 14$

   $= 3\left(x + \frac{5}{2}\right)^2 - \frac{19}{4}$

Exercise 23.13

Solve the equations in questions **1** to **14** by completing the square.

1 $x^2 - 6x + 4 = 0$
2 $x^2 + 10x + 5 = 0$
3 $x^2 - 7x + 2 = 0$
4 $x^2 + 6x - 4 = 0$
5 $x^2 - 10x + 6 = 0$
6 $x^2 - 5x + 2 = 0$
7 $4x^2 - 6x + 1 = 0$
8 $3x^2 - 4x - 5 = 0$
9 $2x^2 + 12x + 3 = 0$
10 $3x^2 + 2x - 2 = 0$
11 $5x^2 - 6x - 4 = 0$
12 $2x^2 - 5x - 2 = 0$
13 $5x^2 - x - 1 = 0$
14 $16x^2 + 12x + 1 = 0$

15 a Write $x^2 + 12x + 12$ in the form $(x + m)^2 - n$.
  b Hence state the minimum value of $y = x^2 + 12x + 12$.
  c Using your answer to part **a**, or otherwise, solve the equation $x^2 + 12x + 12 = 0$.
  Give your answer in the form $p \pm q\sqrt{6}$.

16 Write these quadratic expressions in the form $a(x + b)^2 + c$.
  a $2x^2 + 8x + 5$ **b** $5x^2 + 5x - 2$ **c** $2x^2 + 18x + 7$

Solving quadratic equations using the quadratic formula

Using the method of completing the square it may be shown that the equation $ax^2 + bx + c = 0$ has roots $x = \dfrac{-b \pm \sqrt{b^2 - 4ac}}{2a}$.

Note
You do not need to memorise the quadratic formula but you should be able to use it correctly.

Example 23.21

Question
Solve the equation $3x^2 + 4x - 2 = 0$.
Give your answers to 2 decimal places.

Solution
In the equation $3x^2 + 4x - 2 = 0$, $a = 3$, $b = 4$, $c = -2$.

$x = \dfrac{-b \pm \sqrt{b^2 - 4ac}}{2a}$

$x = \dfrac{-4 \pm \sqrt{16 - 4 \times 3 \times (-2)}}{2 \times 3}$

$= \dfrac{-4 \pm \sqrt{16 + 24}}{6}$

$= \dfrac{-4 \pm \sqrt{40}}{6}$

$= \dfrac{-4 + 6.324...}{6}$

$= \dfrac{-4 + 6.324}{6}$ or $\dfrac{-4 - 6.324}{6}$

$x = 0.39$ or -1.72 to 2 decimal places

Note
The main errors that occur in using the formula are
- errors with the signs, especially with $-4ac$
- failure to divide the whole expression by $2a$.

23 EQUATIONS

Exercise 23.14

Use the formula to solve the equations in questions **1** to **12**.

Give your answers correct to 2 decimal places.

1. $x^2 + 8x + 6 = 0$
2. $2x^2 - 2x - 3 = 0$
3. $3x^2 + 5x - 1 = 0$
4. $5x^2 - 12x + 5 = 0$
5. $5x^2 + 9x - 6 = 0$
6. $x^2 - 5x - 1 = 0$
7. $3x^2 + 9x + 5 = 0$
8. $x^2 + 7x + 4 = 0$
9. $2x^2 - 3x - 4 = 0$
10. $3x^2 + 2x - 2 = 0$
11. $5x^2 - 13x + 7 = 0$
12. $5x^2 + 9x + 3 = 0$

13. A garden is 8 m longer than it is wide and it has an area of 25 m².

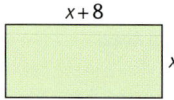

Write down an equation and solve it to find the dimensions correct to the nearest centimetre.

14. A rectangular pen is constructed along an existing wall, using 20 m of fencing.

a. The width of the pen is x m.
   Write an expression for the area enclosed.
b. Write an equation and solve it to find the dimensions to give an area of 40 m².
c. By completing the square, find the maximum possible area of a pen constructed with 20 m of fencing.

15. A rectangular lawn measuring 22 m by 15 m is surrounded by a path x m wide.

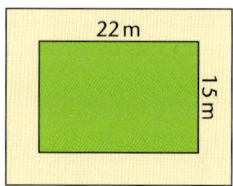

a. Form and simplify an expression for the total area of lawn and path.
b. Write an equation and solve it to find the width of the path correct to the nearest centimetre, when the total area is 400 m².

Changing the subject of formulas

When solving equations, you simplify and collect terms if necessary, then use inverse operations to get the unknown on its own.

Formulas can be treated in the same way as equations. This means they can be rearranged to change the subject. Use the same steps as you would if the formula was an equation.

Example 23.22

Question

Rearrange each of these formulas to make the letter in the bracket the subject.

a $a = b + c$ (b)   **b** $a = bx + c$ (b)   **c** $n = m - 3s$ (s)   **d** $p = \dfrac{q+r}{s}$ (r)

Solution

a
$a = b + c$
$a - c = b$ Subtract c from both sides.
$b = a - c$ Reverse to get b on the left.

b
$a = bx + c$
$a - c = bx$ Subtract c from both sides.
$\dfrac{a-c}{x} = b$ Divide both sides by x.
$b = \dfrac{a-c}{x}$ Reverse to get b on the left.

c
$n = m - 3s$
$n + 3s = m$ Add $3s$ to both sides.
$3s = m - n$ Subtract n from both sides.
$s = \dfrac{m-n}{3}$ Divide both sides by 3.

d
$p = \dfrac{q+r}{s}$
$sp = q + r$ Multiply both sides by s.
$sp - q = r$ Subtract q from both sides.
$r = sp - q$ Reverse to get r on the left.

Note

If you are not sure whether to multiply or divide, try an example with numbers first.

23 EQUATIONS

Example 23.23

Question

The formula for the total cost, T, of entry to the cinema for three adults and three children is given by the formula $T = 3(a + c)$, where a is the price of an adult ticket and c is the price of a child ticket.

a Make c the subject of the formula.
b Find the cost for a child when the cost for an adult is $5 and the total cost is $24.

Solution

a $T = 3(a + c)$
   $T = 3a + 3c$ Multiply out the bracket.
   $T - 3a = 3c$ Subtract $3a$ from both sides.
   $\frac{T - 3a}{3} = c$ Divide both sides by 3.
   $c = \frac{T - 3a}{3}$ Reverse to get c on the left.

b $c = \frac{24 - 3 \times 5}{3}$
   $= \frac{24 - 15}{3}$
   $= \frac{9}{3} = 3$
   The cost for a child is $3.

Exercise 23.15

1 Rearrange each formula to make the letter in the bracket the subject.
   a $a = b - c$ (b)
   b $3a = wx + y$ (x)
   c $v = u + at$ (t)
   d $A = \frac{T}{H}$ (T)
   e $C = P - 3T$ (T)
   f $P = \frac{u + v}{2}$ (u)
   g $C = 2\pi r$ (r)
   h $A = p(q + r)$ (q)
   i $p = q + 2r$ (q)
   j $B = s + 5r$ (r)
   k $s = 2u - t$ (t)
   l $m = \frac{pqr}{s}$ (q)
   m $L = 2G - 2F$ (G)
   n $F = \frac{m + 4n}{t}$ (n)
   o $T = \frac{S}{2a}$ (S)
   p $A = t(x - 2y)$ (y)

2 The formula for finding the perimeter P of a rectangle of length l and width w is $P = 2(l + w)$.
   a Rearrange the formula to make l the subject.
   b Find the length of a rectangle of width 8 metres and perimeter 44 metres.

3 The cost ($$C$) of catering for a conference is given by the formula $C = A + 32n$, where A is the cost of the room and n is the number of delegates.
   a Rearrange the formula to make n the subject.
   b Work out the number of delegates when A is $120 and the total cost C is $1912.

4 The cooking time, T minutes, for w kg of meat is given by $T = 45w + 40$.
   a Make w the subject of this formula.
   b What is the value of w when the cooking time is 2 hours and 28 minutes?

5 The curved surface area (S cm²) of a cylinder of radius r cm and height h cm is given by $S = 2\pi rh$.
   a Rearrange this formula to make r the subject.
   b Find the radius of a cylinder of height 4 cm which has a curved surface area of 60.3 cm².

6 The formula for the volume V of a cone of height h and base radius r is $V = \frac{1}{3}\pi r^2 h$.
   a Rearrange the formula to make h the subject.
   b Find the height of a cone with base radius 9 cm and volume 2290 cm³.
7 The cost (C) of booking a bus for n people is $C = 40 + 5n$.
   a Make n the subject of this formula.
   b Find the number of people when the cost is $235.
8 The total surface area (S cm²) of a cylinder of radius r cm and height h cm is given by $S = 2\pi rh + 2\pi r^2$.
   a Rearrange this formula to make h the subject.
   b Find the height of a cylinder of radius 6 cm which has a total surface area of 500 cm².

Changing the subject of harder formulas

All the formulas that you have rearranged up to this point have contained the new subject only once, and also the subject has not been raised to a power. However, this is now extended in the following examples.

Example 23.24

Question
Rearrange the formula $V = \frac{4}{3}\pi r^3$ to make r the subject.

Solution
$V = \frac{4}{3}\pi r^3$

$3V = 4\pi r^3$ Multiply both sides by 3.

$4\pi r^3 = 3V$ Rearrange to get all terms involving r on the left.

$r^3 = \frac{3V}{4\pi}$ Divide both sides by 4π.

$r = \sqrt[3]{\frac{3V}{4\pi}}$ Take the cube root of both sides.

Example 23.25

Question
Rearrange the formula $a = x + \frac{cx}{d}$ to make x the subject.

Solution
$a = x + \frac{cx}{d}$

$ad = dx + cx$ Multiply both sides by d.

$dx + cx = ad$ Rearrange to get all terms involving x (the subject) on the left.

$x(d + c) = ad$ Factorise the left-hand side, taking out the factor x.

$x = \dfrac{ad}{d + c}$ Divide both sides by the bracket $(d + c)$.

23 EQUATIONS

Example 23.26

Question

Rearrange the equation $ax + by = cy - ad$ to make a the subject.

Solution

$ax + by = cy - ad$

$ax + ad = cy - by$ — Rearrange to get all terms involving a on the left and all the other terms on the right. This is done by adding ad to both sides and subtracting by from both sides.

$a(x + d) = cy - by$ — Factorise the left-hand side, taking out the factor a.

$a = \dfrac{cy - by}{x + d}$ — Divide both sides by the bracket $(x + d)$.

Example 23.27

Question

Rearrange the formula $a = \dfrac{1}{p} + \dfrac{1}{q}$ to make p the subject.

Solution

$a = \dfrac{1}{p} + \dfrac{1}{q}$

$apq = q + p$ — Multiply through by pq.

$apq - p = q$ — Collect all the terms in p on the left.

$p(aq - 1) = q$ — Factorise.

$p = \dfrac{q}{aq - 1}$ — Divide both sides by the bracket $(aq - 1)$.

Exercise 23.16

For questions **1–25** rearrange the formula to make the letter in the bracket the subject.

1. $s = at + 2bt$ (t)
2. $s = ab - bc$ (b)
3. $P = t - \dfrac{at}{b}$ (t)
4. $v^2 = u^2 + 2as$ (u)
5. $A = 4\pi r^2$ (r)
6. $3(a + y) = by + 7$ (y)
7. $ab - cd = ac$ (a)
8. $2(a - 1) = b(1 - 2a)$ (a)
9. $s = 2r^2 - 1$ (r)
10. $s - 2ax = b(x - s)$ (x)
11. $a(b + d) = c(b - d)$ (d)
12. $a = \dfrac{t}{b} - st$ (t)
13. $V = 5ab^2 + 3c^3$ (c)
14. $A = P + \dfrac{PRT}{100}$ (P)
15. $s = \dfrac{uv}{u + v}$ (v)
16. $s = \dfrac{1}{a} + b$ (a)
17. $a = \dfrac{1}{b + c}$ (c)
18. $a = b + \dfrac{c}{d + 1}$ (d)

19 $m = \dfrac{100(a-b)}{b}$ (b) 20 $\dfrac{a}{p} = \dfrac{1}{1+p}$ (p)

21 $\dfrac{a}{x+1} = \dfrac{b}{2x+1}$ (x) 22 $T = 2\pi\sqrt{\dfrac{L}{g}}$ (L)

23 $y = 3x^2 - 4$ (x) 24 $V = \dfrac{1}{3}\pi r^2 h$ (r)

25 The formula for finding the length, d, of the diagonal of a cuboid whose dimensions are x, y and z is $d = \sqrt{x^2 + y^2 + z^2}$.

  a Find d when $x = 2$, $y = 3$ and $z = 4$.
  b How long is the diagonal of a cuboid block of concrete with dimensions 2 m, 3 m and 75 cm?
  c Rearrange the formula to make x the subject.
  d Find x when $d = 0.86$ m, $y = 0.25$ m, and $z = 0.41$ m.

Solving equations involving algebraic fractions

You need to be able to solve equations involving fractions with algebraic expressions in the denominator.

You deal with these in the same way as you dealt with fractions with numerical denominators.

You multiply every term by the common denominator to eliminate the fractions.

The resulting equation may be linear or quadratic.

Example 23.28

Question

Solve $\dfrac{3}{x+1} - \dfrac{2}{x} = \dfrac{1}{x-2}$.

Solution

$\dfrac{3}{x+1} - \dfrac{2}{x} = \dfrac{1}{x-2}$ The common denominator of $x+1$, x and $x-2$ is $x(x+1)(x-2)$.

$3x(x-2) - 2(x+1)(x-2) = x(x+1)$ Multiply by $x(x+1)(x-2)$. Make sure you multiply every expression on both sides.

$3x^2 - 6x - 2(x^2 - x - 2) = x^2 + x$ Expand the bracket and collect on one side.
$3x^2 - 6x - 2x^2 + 2x + 4 - x^2 - x = 0$
$-5x + 4 = 0$ Collect like terms.
$4 = 5x$
$x = \dfrac{4}{5}$

Note

In equations with algebraic fractions, multiply through by the common denominator. This eliminates the fractions.

23 EQUATIONS

Exercise 23.17

Solve these.

1. $\dfrac{2x}{3} - \dfrac{3x}{5} = \dfrac{1}{3}$
2. $\dfrac{x-1}{2} - \dfrac{x-3}{5} = 1$
3. $\dfrac{2x-1}{6} + \dfrac{3x}{4} = \dfrac{x-2}{12}$
4. $\dfrac{x-1}{3} + \dfrac{2x}{5} = \dfrac{3x+1}{5}$
5. $\dfrac{3}{x} - \dfrac{2}{x+1} = 0$
6. $\dfrac{5}{x} - \dfrac{2}{x-3} = 0$
7. $\dfrac{5}{6x} - \dfrac{1}{x+1} = \dfrac{1}{3x}$
8. $\dfrac{4}{x-2} - \dfrac{1}{x+1} = \dfrac{3}{x}$
9. $4x = \dfrac{3}{x} - 1$
10. $4x + \dfrac{3}{x} = 7$
11. $2x^2 - \dfrac{x}{3} = 5$
12. $2x + \dfrac{4}{x} = 9$
13. $\dfrac{1}{x-1} - \dfrac{3}{x+2} = \dfrac{1}{4}$
14. $\dfrac{2x}{3x+1} - \dfrac{5}{x+3} = 0$
15. $\dfrac{2x}{x-3} - \dfrac{x}{x-2} = 3$
16. $\dfrac{x}{x-2} - 2x = 3$

Key points

- The two sides of an equation must always be kept equal. Operations used to simplify or solve an equation must always be the same for each side.
- To solve equations with x on both sides, collect the x terms together on one side of the equation.
- To solve equations with brackets, multiply out the brackets first.
- To solve equations with fractions, first multiply every term in the equation by the lowest common denominator of the fractions. Then, multiply out any brackets.
- To write a formula or equation, think first what operations you would use for the first line of a solution if the letters were numbers. Then, write those operations using the letters instead.
- Some problems can be solved by writing an equation and then solving it.
- To solve two linear simultaneous equations using algebra, you can use the elimination method or the substitution method.
- To solve a quadratic equation by factorisation, first rearrange the equation, if necessary, so that all the terms are on the left-hand side, with zero on the right-hand side. Next, factorise the left-hand side. Lastly, set each bracket to zero to find the two solutions.
- Quadratic equations which will not factorise may be solved by completing the square.
  - Rearrange the equation, if needed, so that all the terms with x or x^2 are on the left-hand side.
  - Multiply, if necessary, so that the coefficient of x^2 is a perfect square.
  - Add a number to the left-hand side so that the left-hand side is a complete square which will factorise to $(ax + b)^2$.
  - Take the square root of both sides.
  - Solve the two resulting linear equations.
- The roots of the quadratic equation $ax^2 + bx + c = 0$ may be found by substituting into this formula:
  $x = \dfrac{-b \pm \sqrt{b^2 - 4ac}}{2a}$.
- When cancelling algebraic fractions, factorise if necessary. Only cancel factors.
- Rearrange a formula using the same steps as you would if it were an equation with numbers instead of letters.
- To rearrange a formula where the new subject occurs twice, first rearrange the formula so that all the terms containing the new subject are on one side and the remaining terms are on the other side. Then, take the new subject out as a common factor. Lastly, divide by the other factor.
- When a formula contains a power or root of a new subject, rearrange the formula to get that term by itself. Then, use the inverse operation. For example, if the formula has $\sqrt{a}$, then to find a, square the other side of the formula.

24 INEQUALITIES

Solving inequalities

$a < b$ means 'a is less than b'
$a \leqslant b$ means 'a is less than or equal to b'
$a > b$ means 'a is greater than b'
$a \geqslant b$ means 'a is greater than or equal to b'

Expressions involving these signs are called **inequalities**.

Example 24.1

Question

Find the integer values of x that satisfy each of these inequalities.

a $-3 < x \leqslant -1$ b $1 \leqslant x < 4$

Solution

a If $-3 < x \leqslant -1$, then $x = -2$ or -1. Note that -3 is not included, but -1 is.

b If $1 \leqslant x < 4$, then $x = 1, 2$ or 3. Note that 1 is included, but 4 is not.

In equations, if you always do the same thing to both sides, the equation still has the same solution.

The same is usually true for inequalities, but there is one important exception.

If you multiply or divide an inequality by a negative number, you must reverse the inequality sign.

For example, $5 < 7$ but, if you multiply by -2, -10 is greater than -14 (see number line on the next page).

Otherwise inequalities can be treated in the same way as equations.

Example 24.2

Question

Solve each of these inequalities and show the solution on a number line.

a $3x + 4 < 10$
b $2x - 5 \leqslant 4 - 3x$
c $x + 4 < 3x - 2$

> **BY THE END OF THIS CHAPTER YOU WILL BE ABLE TO:**
> - represent and interpret inequalities, including on a number line
> - construct, solve and interpret linear inequalities
> - represent and interpret linear inequalities in two variables graphically
> - list inequalities that define a given region.

> **CHECK YOU CAN:**
> - use the inequality signs, $<, >, \leqslant$ and $\geqslant$, correctly
> - draw straight-line graphs.

Photocopying is prohibited

24 INEQUALITIES

Solution

a $3x + 4 < 10$

$3x < 6$ — Subtract 4 from both sides.

$x < 2$ — Divide both sides by 3.

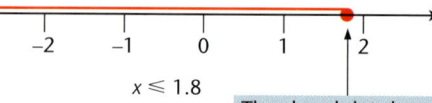

$x < 2$

The open dot shows that 2 is not included.

b $2x - 5 \leqslant 4 - 3x$

$2x \leqslant 9 - 3x$ — Add 5 to both sides.

$5x \leqslant 9$ — Add $3x$ to both sides.

$x \leqslant 1.8$ — Divide both sides by 5.

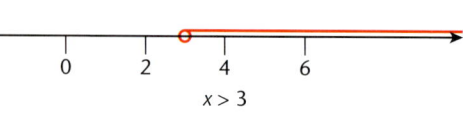

$x \leqslant 1.8$

The closed dot shows that 1.8 is included.

c $x + 4 < 3x - 2$

$x < 3x - 6$ — Subtract 4 from both sides.

$-2x < -6$ — Subtract $3x$ from both sides.

$x > 3$ — Divide both sides by -2 and change < to > (when dividing by -2).

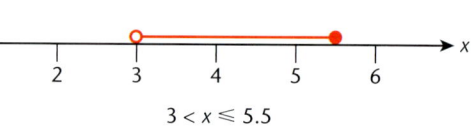

$x > 3$

Example 24.3

Question

Solve this inequality and show the solution on a number line.

$5 < 2x - 1 \leqslant 10$

Solution

$5 < 2x - 1 \leqslant 10$

$6 < 2x \leqslant 11$ — Add 1 to each part of the inequality.

$3 < x \leqslant 5.5$ — Divide each part of the inequality by 2.

$3 < x \leqslant 5.5$

Exercise 24.1

1. Write down two values of x that satisfy the inequality $x < -2$.
2. Write down the integer values of x that satisfy each of these inequalities.
   - **a** $-4 \leqslant x < 0$
   - **b** $1 < x \leqslant 5$
   - **c** $1 < x \leqslant 4$
   - **d** $-5 < x \leqslant -1$

Solve the inequalities in questions **3** to **29**.

For questions **3** to **7**, **28** and **29**, represent your solution on a number line.

3. $x - 3 \leqslant 4$
4. $2x - 3 < 5$
5. $3x + 4 \leqslant 7$
6. $2x + 3 > 6$
7. $3x - 4 \leqslant 8$
8. $2x \geqslant x + 5$
9. $5x > 3 - x$
10. $4x > 2x + 5$
11. $3x - 6 \geqslant x + 2$
12. $5a - 3 > 2a$
13. $2x - 3 < x + 1$
14. $3x + 7 < x + 3$
15. $8x - 10 > 3x + 25$
16. $6x + 11 \leqslant 18 - x$
17. $2x + 9 > 4x + 5$
18. $3(2x - 1) \geqslant 11 - x$
19. $x + 4 > 2x$
20. $2x - 5 < 4x + 1$
21. $3(x - 4) > 5(x + 1)$
22. $x - 2 < 2x + 4$
23. $2x - 1 > x - 4$

24 $3(x+3) \geq 2x - 1$ **25** $3(2x-4) < 5(x-6)$ **26** $\frac{2x+7}{3} \leq 5$

27 $\frac{x+5}{2} > \frac{2x-1}{3}$ **28** $6 < 3x < 15$ **29** $-2 < 5x + 3 \leq 15$

30 Find the integer values of x that satisfy each of these inequalities.
   a $3 \leq 2x - 1 \leq 5$ **b** $-4 \leq 3x + 2 \leq 11$

31 A rectangle has width a cm and length $(2a - 5)$ cm.
   The perimeter of the rectangle is greater than 20 cm but less than 38 cm.
   Write an inequality in a and solve it.

32 Sajid's age is y years. His brother Ali's age is $(2y - 4)$ years.
   The total of Sajid and Ali's ages is greater than 11 and less than or equal to 23.
   a Form an inequality and solve it.
   b Hence state the greatest number that Ali's age can be.

Showing regions on graphs

It is often possible to show the region on a graph that satisfies an inequality.

> An inequality involving the signs < or > is represented on a graph using a dashed line.
>
> An inequality involving the signs $\leq$ or $\geq$ is represented by a solid line.

Example 24.4

Question
Write down the inequality that describes the region shaded in each graph.

a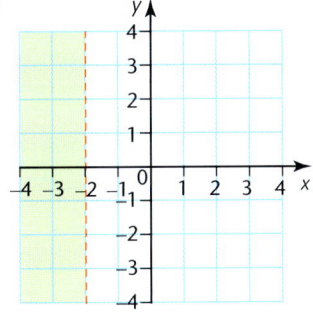

b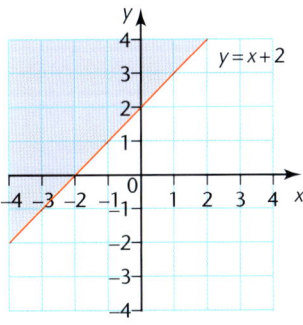

Solution
a $x < -2$

The line is dashed, so the inequality is either $x < -2$ or $x > -2$.

Choose any point in the shaded region, for example, $(-3, 0)$.

Substitute for x in the inequality.

$-3 < -2$, so the region represents $x < -2$.

b $y \geq x + 2$

The line is solid, so the inequality is either $y \geq x + 2$ or $y \leq x + 2$.

Choose any point in the shaded region, for example, $(0, 3)$.

Substitute for x and y.

$3 \geq 0 + 2$, so the region represents $y \geq x + 2$

24 INEQUALITIES

Example 24.5

Question

On separate grids, shade each of these regions.

a $y \geq 2$ 　　　　b $y < 2x - 3$

Solution

a Draw the solid line $y = 2$.

Choose a point on one side of the line, for example, (0, 0).

Substitute into the inequality.

$0 \geq 2$

Since this is *not* a true statement, (0, 0) is not in the required region.

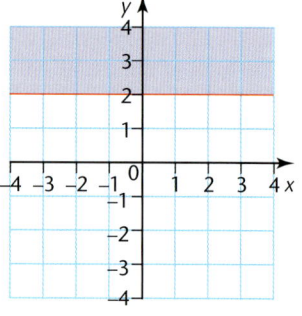

b Draw the dashed line $y = 2x - 3$.

Choose a point on one side of the line, for example, (3, 0).

Substitute into the inequality.

$0 < 2 \times 3 - 3$

$0 < 3$

Since this *is* a true statement, (3, 0) is in the required region.

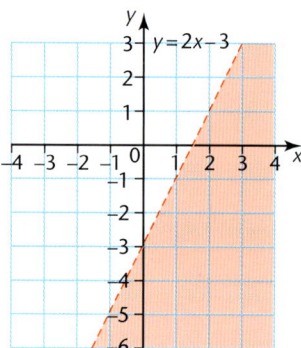

> **Note**
> When testing a point, if possible use (0, 0).
> If the line goes through (0, 0), choose a point with positive coordinates, for example (1, 0), to test in the inequality.

Exercise 24.2

1 Write down the inequality that describes the shaded region in each of these diagrams.

a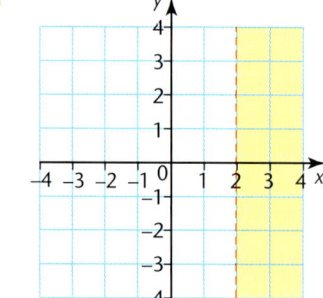

b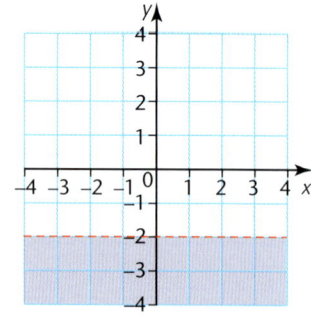

Representing regions satisfying more than one inequality

c 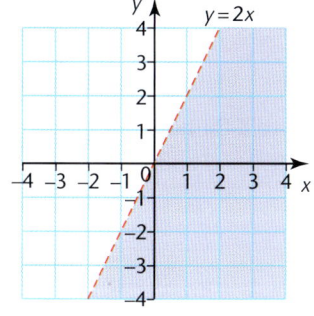 d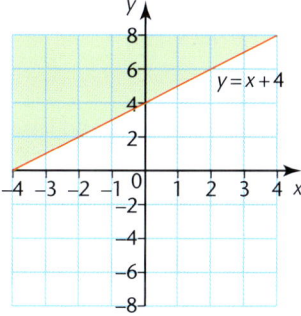

e 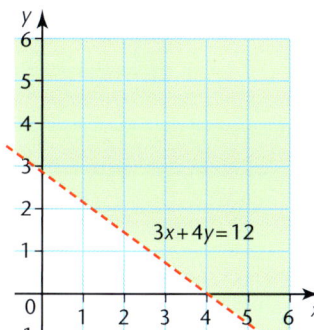 f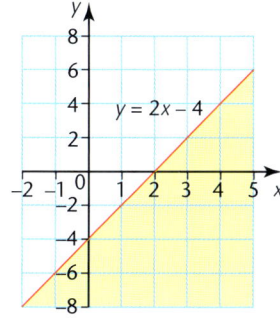

2. Draw a set of axes and label them from −4 to 4 for x and y.
   Shade the region $y > -3$.
3. Draw a set of axes and label them from −4 to 4 for x and y.
   Shade the region $x \geqslant -1$.
4. Draw a set of axes and label them from −4 to 4 for x and y.
   Shade the region $y < 3 - x$.
5. Draw a set of axes and label them from −4 to 4 for x and y.
   Shade the region $y \geqslant 2x + 1$.
6. Draw a set of axes and label them from −3 to 6 for x and from −3 to 5 for y.
   Shade the region $2x + 5y < 10$.
7. Draw a set of axes and label them from −1 to 6 for x and from −2 to 5 for y.
   Shade the region $4x + 5y \leqslant 20$.
8. Draw a set of axes and label them from 0 to 5 for x and y.
   Shade the region $3x + 5y > 15$.

Note
When plotting a line like $3x + 4y = 12$, choose $x = 0$ to give $4y = 12$ and $y = 3$.
Then choose $y = 0$ to give $3x = 12$ and $x = 4$.
This gives the points where the line crosses the axes, (0, 3) and (4, 0).

Representing regions satisfying more than one inequality

When you represent a region that satisfies more than one inequality, it is often better to shade the region that *does not* satisfy the inequality; the region that *does* satisfy the inequality is left unshaded.

Several inequalities can be represented on the same axes and the region where the values of x and y satisfy all of them can be found.

24 INEQUALITIES

Example 24.6

Question

Draw a set of axes and label them from 0 to 8 for both x and y.

Show, by shading, the region where $x \geq 0$, $y \geq 0$ and $x + 2y \leq 8$.

Solution

First, draw the line $x + 2y = 8$.

Then, shade the regions $x < 0$, $y < 0$ and $x + 2y > 8$.

These are the regions where the values of x and y do not satisfy the inequalities.

The required region, where the values of x and y satisfy all three inequalities, is left unshaded. It is labelled R.

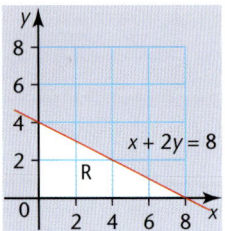

Example 24.7

Question

List the three inequalities that define the shaded region in the diagram.

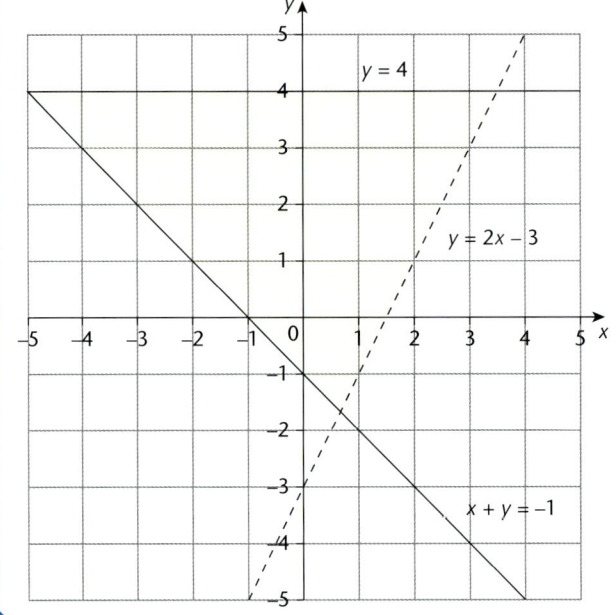

Representing regions satisfying more than one inequality

> **Solution**
> First, choose any point inside the region, such as (0, 0).
> Substitute the x- and y-values into each equation to determine the direction of the inequality sign.
>
> | | | | | |
> |---|---|---|---|---|
> | $y = 4$ | $0 \leq 4$ | solid line, so equals with the inequality | so, | $y \leq 4$ |
> | $y = 2x - 3$ | $0 > 0 - 3$ | dotted line, so no equals with the inequality | so, | $y > 2x - 3$ |
> | $x + y = -1$ | $0 + 0 \geq -1$ | solid line, so equals with the inequality | so, | $x + y \geq -1$ |
>
> The three inequalities are $y \leq 4$, $y > 2x - 3$, $x + y \geq -1$

Exercise 24.3

1. Draw a set of axes and label them from 0 to 6 for both x and y.
   Show, by shading, the region where $x \geq 0$, $y \geq 0$ and $x + 2y \leq 6$.
2. Draw a set of axes and label them from 0 to 12 for x and from 0 to 8 for y.
   Show, by shading, the region where $y > 0$, $x > 0$ and $3x + 5y < 30$.
3. Draw a set of axes. Label them from -1 to 4 for x and from -3 to 4 for y.
   Show, by shading, the region where $x \geq 0$, $y \leq 3$ and $y \geq 2x - 3$.
4. Draw a set of axes and label them from -4 to 4 for both x and y.
   Show, by shading, the region where $x > -2$, $y < 3$ and $y > 2x$.
5. Draw a set of axes and label them from 0 to 6 for both x and y.
   Show, by shading, the region where $x < 4$, $y < 3$ and $3x + 4y < 12$.
6. Draw a set of axes and label them -1 to 5 for both x and y.
   Show, by shading, the region where $y \geq 0$, $y \leq x + 1$ and $3x + 5y < 15$.
7. Draw a set of axes and label them from -1 to 8 for x and from -1 to 7 for y.
   Show, by shading, the region where $x > 0$, $3x + 8y > 24$ and $5x + 4y < 20$.
8. List the three inequalities that define the shaded region in the diagram.

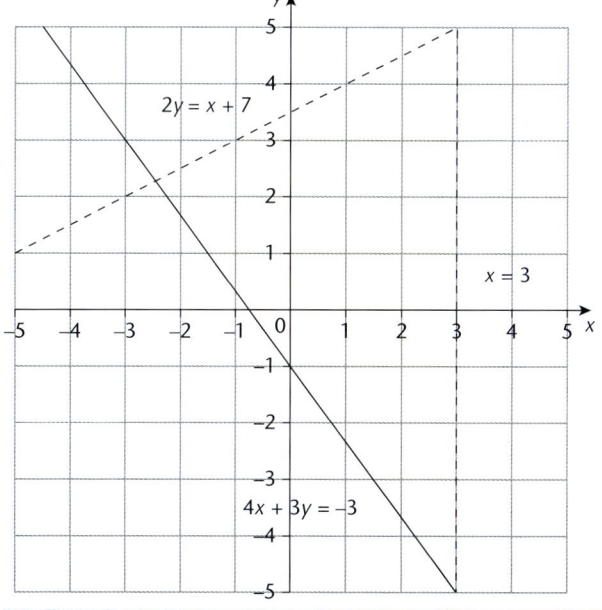

24 INEQUALITIES

9 List the four inequalities that define the shaded region in the diagram.

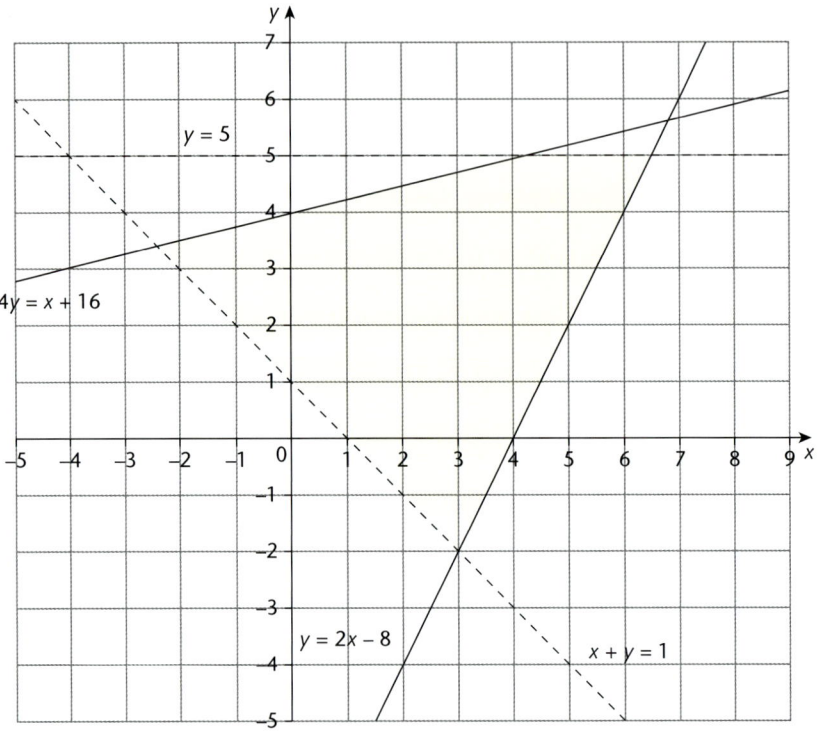

10 List the four inequalities that define the shaded region in the diagram.

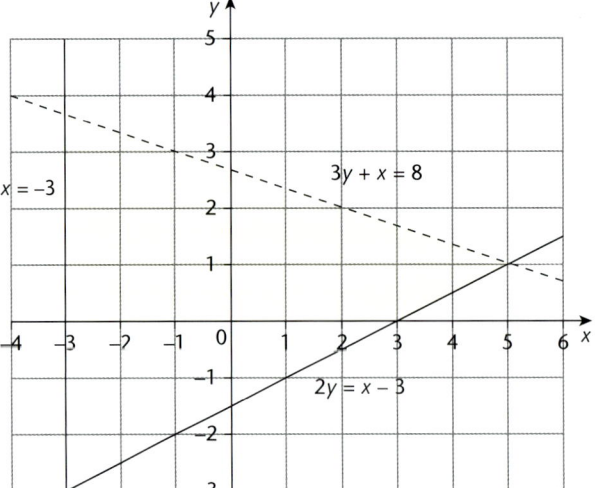

Key points

- '$a < b$' means 'a is less than b'.
  '$a \leq b$' means 'a is less than or equal to b'.
  '$a > b$' means 'a is greater than b'.
  '$a \geq b$' means 'a is greater than or equal to b'.
- Inequalities can be treated like equations, except when multiplying or dividing by a negative number. Then, you must reverse the symbol, since, for example, $3 < 5$ but $-3 > -5$.
- An inequality in y and/or x, involving $\leq$ or $\geq$ is represented by a region on a coordinate grid bounded by a solid line.
- An inequality in y and/or x, involving $<$ or $>$ is represented by a region on a coordinate grid bounded by a dashed line.
- Given an inequality, the region it represents can be found by:
  - drawing the appropriate straight line, solid or dashed
  - using a coordinate point on one side of the line to determine which is the required region.
- Given a region on a coordinate grid, the inequality it represents can be found by:
  - finding the equation of the line which is the boundary
  - using a coordinate point on one side of the line to determine the direction of the inequality
  - including an equals sign in the inequality if the line is solid.
- Given more than one inequality, the region which satisfies them all can be found by shading the region **not** satisfied by each inequality. The part of the grid left unshaded is the required region.
- Given a region on a coordinate grid bounded by straight lines, the inequalities it represents can be found by finding the inequality represented by each line separately.

25 SEQUENCES

BY THE END OF THIS CHAPTER YOU WILL BE ABLE TO:
- continue a given number sequence or pattern
- recognise patterns in sequences, including the term-to-term rule, and relationships between different sequences
- find and use the nth term of sequences.

CHECK YOU CAN:
- add, subtract, multiply and divide whole numbers
- substitute numbers into a formula
- solve simultaneous equations.

Number patterns

Here is a number pattern.

6   11   16   21   26   31   …

You can see that, to get from one number to the next, you add 5.

Here is another number pattern.

1   2   4   8   16   32   …

To get from one number to the next this time you multiply by 2.

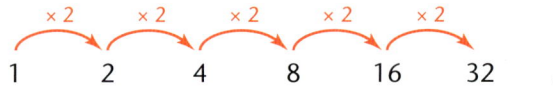

Number patterns like these are called **sequences**.

The rule for finding the next number in a sequence from the previous one is called the **term-to-term rule**.

> ### Example 25.1
>
> **Question**
>
> Find the term-to-term rule for each of these sequences and give the next number.
>
> a 1   4   7   10   13   16   …
> b 63   56   49   42   35   28   …
> c 1 000 000   100 000   10 000   1000   100   …
>
> **Solution**
>
> a To get from one number to the next you add 3.
>
>
>
> The next number is 16 + 3 = 19.
>
> b To get from one number to the next you subtract 7.
>
>
>
> The next number is 28 − 7 = 21.
>
> c To get from one number to the next you divide by 10.
>
>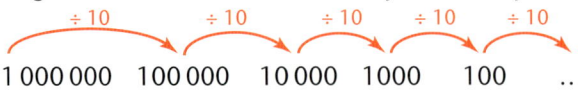
>
> The next number is 100 ÷ 10 = 10.

Number patterns

In the sequences in the next example, some of the numbers are missing.

It is possible to work out what they should be by looking at the numbers in the sequence and working out how to get from one number to the next.

Example 25.2

Question

Find the missing numbers in each of these sequences.

a 10   17   ☐   ☐   38   45
b 35   ☐   23   17   ☐   5
c ☐   15   ☐   23   27   31

Solution

a The term-to-term rule is 'add 7'.

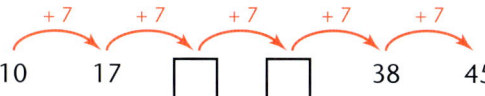

So the first missing number is $17 + 7 = 24$.
The second missing number is $24 + 7 = 31$.

b The term-to-term rule is 'subtract 6'.

So the first missing number is $35 - 6 = 29$.
The second missing number is $17 - 6 = 11$.

c The term-to-term rule is 'add 4'.

To get to the previous number, you must subtract 4.
So the first missing number is $15 - 4 = 11$.
You find the second missing number by adding 4 in the usual way.
The second missing number is $15 + 4 = 19$.

Exercise 25.1

1 Find the term-to-term rule for each of these sequences and give the next number.

   a 2   6   10   14   18   22   …
   b 3   11   19   27   35   43   …
   c 4   9   14   19   24   29   …

25 SEQUENCES

2 Find the term-to-term rule for each of these sequences and give the next number.
   a 1   3   9   27   81   243   …
   b 15 625   3125   625   125   25   …
   c 1   2   4   7   11   16   …

3 Find the missing numbers in each of these sequences.
   a 7   12   17   ☐   ☐   32   ☐
   b 25   ☐   19   16   ☐   10   ☐
   c 1   4   16   ☐   ☐   1024   4096

4 Explain how you know whether or not 77 is in each of these sequences.
   a 7   14   21   28   35   42   …
   b 14   18   22   26   30   34   …

5 Look at this sequence of numbers.
   1   4   9   16   25   36   …
   a Find the next number in the sequence.
   b Explain how you get each number of the sequence.

Linear sequences

Look at this sequence.

2, 5, 8, 11, 14, …

The terms of the sequence increase by 3 each time.

When the increase is constant like this, the sequence is called a **linear sequence**.

In the same way, the sequence 11, 9, 7, 5, 3, 1, −1, … is also linear, since the terms decrease by the same amount (2) each time.

Another way of thinking of this is that you are adding −2 each time.

This idea will help with finding the nth term.

Exercise 25.2

   a Which of these sequences are linear and which are not?
   b For each sequence, write down the next two terms.
1 3, 5, 7, 9, …
2 43, 40, 38, 35, 32, …
3 78, 75, 72, 69, …
4 1, 5, 9, 13, …
5 4, 9, 16, 25, …
6 1, 6, 15, 28, …
7 2, 9, 16, 23, …

The nth term

It is often possible to find a formula to give the terms of a sequence.

> **Example 25.3**
>
> **Question**
> The formula for the nth term of a sequence is $2n + 1$. Find each of these terms.
>
> a u_1 b u_2 c u_3
>
> **Solution**
> a $u_1 = 2 \times 1 + 1 = 3$
> b $u_2 = 2 \times 2 + 1 = 5$
> c $u_3 = 2 \times 3 + 1 = 7$

> **Note**
> You usually use n to stand for the number of a term and u_n for the nth term.

Exercise 25.3

Each of these is the formula for the nth term of a sequence.

Find the first four terms of each sequence.

1 $n + 1$	2 $2n$	3 $2n - 1$	4 $n + 5$	5 $3n$
6 $3n + 1$	7 $5n - 3$	8 $10n$	9 $7n - 7$	10 $2 - n$
11 n	12 $n + 3$	13 $4n$	14 $n - 1$	15 $2n + 1$
16 $3n - 1$	17 $6n + 5$	18 $2n - 3$	19 $5 - n$	20 $10 - 2n$
21 n^2	22 $n^2 + 2$	23 $n^2 - 5$	24 $3n^2$	25 n^3

Finding the formula for the nth term

Look back at the formulas in Exercise 25.3 questions **1–20** and the sequences in your answers, and notice how they are connected.

All the sequences in questions **1–20** are linear.

If the formula contains the expression $2n$, the terms increase by 2 each time.

If it contains the expression $5n$, the terms increase by 5 each time.

Similarly, if it contains the expression $-3n$, the terms increase by -3 (or decrease by 3) each time.

So to find a formula for the nth term of a given sequence, find how much more (or less) each term is than the one before it.

> **Note**
> This will always work if the differences are constant.
>
> The difference between terms in a linear sequence is always constant.

25 SEQUENCES

> ### Example 25.4
>
> **Question**
> Find the formula for the nth term of this sequence.
> 3, 5, 7, 9, ...
>
> **Solution**
> The difference between each term is 2, so the formula will include the expression $2n$.
> When $n = 1$, $2n = 2$.
> But the first term of the sequence is 3, which is 1 more.
> So the formula for the nth term of this sequence is $2n + 1$.

You can also use algebra to find the formula for the nth term.

The nth term can be written as T_n or u_n.

> ### Example 25.5
>
> **Question**
> Find the formula for the nth term of this sequence.
> 20, 17, 14, 11, ...
>
> **Solution**
> As the difference between the terms is constant, this is a linear sequence.
> The general formula for the nth term of a linear sequence is
> $T_n = an + b$.
> $T_1 = 20$ so $20 = a + b$
> $T_2 = 17$ so $17 = 2a + b$
> Solving the simultaneous equations gives
> $a = -3$ and $b = 23$
> So $T_n = 23 - 3n$
>
> **Note**
> Notice the similarity between this and the equation of a straight line, $y = mx + c$.

Exercise 25.4

Find the formula for the nth term of each of these sequences.

> **Note**
> Check that your formula for the nth term is correct by trying it out for the first few terms. Substitute $n = 1, 2, 3, ...,$ into your formula.

1. 1, 2, 3, 4, ...
2. 4, 6, 8, 10, ...
3. 4, 8, 12, 16, ...
4. 0, 2, 4, 6, ...
5. 7, 11, 15, 19, ...
6. 1, 7, 13, 19, ...
7. 11, 21, 31, 41, ...
8. 5, 8, 11, 14, ...
9. 101, 201, 301, 401, ...
10. 0, 1, 2, 3, ...
11. 2, 5, 8, 11, ...
12. 7, 9, 11, 13, ...
13. 4, 9, 14, 19, ...
14. 15, 20, 25, 30, ...
15. −1, 3, 7, 11, ...
16. 5, 7, 9, 11, ...
17. 101, 102, 103, 104, ...
18. 4, 3, 2, 1, ...
19. 7, 4, 1, −2, ...
20. 25, 23, 21, 19, ...

The nth term of quadratic, cubic and exponential sequences

This is the sequence of square numbers.

$1^2 \quad 2^2 \quad 3^2 \quad 4^2 \quad 5^2 \quad \ldots \quad n^2 \quad \ldots$
$\downarrow \quad \downarrow \quad \downarrow \quad \downarrow \quad \downarrow \quad \quad \downarrow$
$1 \quad 4 \quad 9 \quad 16 \quad 25 \quad \ldots \quad n^2 \quad \ldots$

This is the sequence of cube numbers.

$1^3 \quad 2^3 \quad 3^3 \quad 4^3 \quad 5^3 \quad \ldots \quad n^3 \quad \ldots$
$\downarrow \quad \downarrow \quad \downarrow \quad \downarrow \quad \downarrow \quad \ldots \quad \downarrow \quad \ldots$
$1 \quad 8 \quad 27 \quad 64 \quad 125 \quad \ldots \quad n^3 \quad \ldots$

This is an exponential sequence.

$3^1 \quad 3^2 \quad 3^3 \quad 3^4 \quad 3^5 \quad \ldots \quad 3^n$
$\downarrow \quad \downarrow \quad \downarrow \quad \downarrow \quad \downarrow \quad \ldots \quad \downarrow$
$3 \quad 9 \quad 27 \quad 81 \quad 243 \quad \ldots \quad 3^n$

You can find the formula for the nth term of a quadratic, cubic or exponential sequence by comparing the sequence with the sequence of square numbers, cube numbers or exponential numbers.

Example 25.6

Question

Find the formula for the nth term of this cubic sequence.

$-1 \quad 6 \quad 25 \quad 62 \quad 123 \quad \ldots$

Solution

Compare the given sequence with the sequence of cube numbers.

$1 \quad 8 \quad 27 \quad 64 \quad 125 \quad \ldots \quad n^3$
$\downarrow -2 \quad \downarrow -2 \quad \downarrow -2 \quad \downarrow -2 \quad \downarrow -2 \quad \ldots \quad \downarrow -2$
$-1 \quad 6 \quad 25 \quad 62 \quad 123 \quad \ldots \quad n^3 - 2$

Notice that every term is 2 smaller than the corresponding term in the sequence of cube numbers.

So the formula is $n^3 - 2$.

Example 25.7

Question

Find the formula for the nth term of this quadratic sequence.

$8 \quad 13 \quad 20 \quad 29 \quad 40 \quad \ldots$

Solution

Since this is a quadratic sequence, use $T_n = an^2 + bn + c$.

$n = 1 \quad\quad 8 = a + b + c$
$n = 2 \quad\quad 13 = 4a + 2b + c \quad\quad 5 = 3a + b$
$n = 3 \quad\quad 20 = 9a + 3b + c \quad\quad 7 = 5a + b \quad\quad 2 = 2a$
$\quad\quad\quad\quad\quad\quad\quad\quad\quad\quad\quad\quad\quad\quad\quad\quad\quad\quad\quad\quad\quad\quad a = 1$

Substitute $a = 1$
$\quad\quad\quad\quad\quad\quad\quad\quad\quad 5 = 3 + b$
$\quad\quad\quad\quad\quad\quad\quad\quad\quad b = 2$

Substitute $a = 1$, $b = 2$
$\quad\quad 8 = 1 + 2 + c$
$\quad\quad c = 5$

So that $T_n = n^2 + 2n + 5$

Example 25.8

Question

The nth term of a sequence is found using the formula $u_n = An^2 + B^n$.
$u_1 = 3$ and $u_2 = 8$.
Find u_3.

Solution

Substitute for $n = 1$ and $n = 2$ in the formula.
$3 = A + B \quad\quad (1)$
$8 = 4A + B^2 \quad\quad (2)$

Solve the simultaneous equations.
$12 = 4A + 4B \quad\quad (1) \times 4$
$8 = 4A + B^2$
$4 = 4B - B^2 \quad\quad$ Subtract.
$B^2 - 4B + 4 = 0 \quad\quad$ Solve the quadratic equation.
$(B - 2)^2 = 0$
So $\quad\quad B = 2$
Substituting in equation (1) gives $A = 1$.
So $\quad\quad u_n = n^2 + 2^n$
$u_3 = 3^2 + 2^3 = 9 + 8 = 17$

Note

Remember to check your answer by substituting in the formula.

Exercise 25.5

1 Find the formula for the nth term of each of these quadratic sequences.
  a 3, 6, 11, 18, 27, …
  b −4, −1, 4, 11, 20, …
  c 2, 6, 12, 20, 30, …
  d 2, 8, 18, 32, 50, …
  e 3, 9, 17, 27, 39, …
  f 3, 4, 7, 12, 19, …

2 Find the formula for the nth term of each of these cubic sequences.
  a 11, 18, 37, 74, 135, …
  b −8, −1, 18, 55, 116, …
  c 2, 16, 54, 128, 250, …
  d 2, 12, 36, 80, 150, …
  e 0, 6, 24, 60, 120, …

3 The formula for the nth term of the sequence 2, 4, 8, 16, 32, … is 2^n.
  Use this to find the formula for the nth term of each of these sequences.
  a 1, 3, 7, 15, 31, …   b 14, 16, 20, 28, 44, …
  c 3, 6, 11, 20, 37, …   d 0, 0, 2, 8, 22, …

4 The nth term of a sequence is found using the formula $T_n = An^3 + Bn^2$.
  $T_1 = 1$ and $T_2 = 12$.
  a Show that $A + B = 1$ and $8A + 4B = 12$.
  b Solve the simultaneous equations to find A and B.
  c Find T_3 and T_4.

5 The first five terms of a sequence are
  $\frac{2}{3}, \frac{8}{5}, \frac{18}{7}, \frac{32}{9}, \frac{50}{11}$.
  Find a formula for the nth term.

6 Look at this sequence of diagrams made from lines and dots.

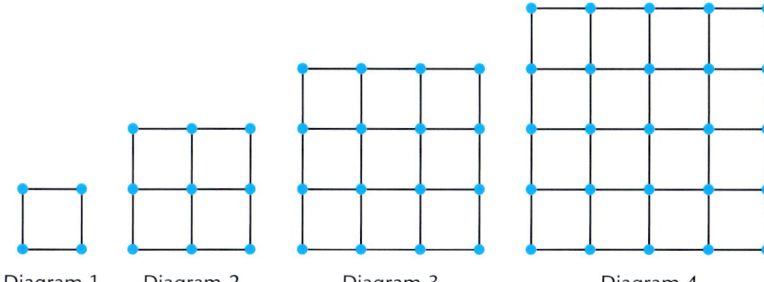

Diagram 1   Diagram 2   Diagram 3   Diagram 4

Copy and complete the table.

Diagram (n)	1	2	3	4	n
Number of small squares	1	4	9		
Number of dots	4	9			
Number of lines	4	12			

Key points
- The term-to-term rule for a sequence of numbers can be used to find the terms of a sequence.
- A linear sequence increases or decreases by a fixed value.
- The formula for the nth term of a sequence can be used to find every term in the sequence by substituting values for n. The first term is found using $n = 1$, the second term using $n = 2$, and so on.
- The formula for the nth term of a linear sequence has the form $an + b$, where a is the fixed value by which the sequence is changing from term to term and b is an integer to be found.
- Other sequences include quadratic (for example, n^2), cubic (for example, n^3) and exponential (for example, 2^n) sequences.
- The formula for the nth term of a non-linear sequence can be found by using a general formula for that sequence. For a quadratic sequence, use $T_n = an^2 + bn + c$ and substitute $n = 1$ and T_1, $n = 2$ and T_2, $n = 3$ and T_3. Solve the simultaneous equations to find a, b and c.

26 PROPORTION

BY THE END OF THIS CHAPTER YOU WILL BE ABLE TO:
- express direct and inverse proportion in algebraic terms and use this form of expression to find unknown quantities.

CHECK YOU CAN:
- find and use multipliers
- manipulate simple formulas
- substitute numbers into simple formulas.

Note
You can easily tell whether the proportion is direct or inverse – in direct proportion, both variables change in the same way, either up or down; in inverse proportion, when one variable goes up, the other will go down.

Proportion

In Chapter 11 you solved problems involving proportion. When two quantities are in **direct proportion**, as one quantity increases, so does the other.

For example, a truck travels x km using y litres of fuel.

In this case, y varies in the same way as x. If you double x, you also double y.

It is possible to express this in symbols as $y \propto x$, which is read as 'y varies as x' or 'y is proportional to x'.

When two quantities are in **indirect proportion**, as one quantity increases, the other decreases.

For example, x workers can dig a hole in y days.

In this case, y varies inversely as x. If you double x, you halve y.

This is the same as 'y varies as $\frac{1}{x}$' which is written in symbols as $y \propto \frac{1}{x}$.

Exercise 26.1

1 Describe the proportion in each of these situations, using the symbol $\propto$.
   a The time the journey takes, t, and the speed of the train, s.
   b The number of pages in a book, p, and the number of words, w.
   c The depth of water in a rectangular tank, d, and the length of time it has been filling, t.
   d The number of buses, b, needed to carry 2000 people, and the number of seats on a bus, s.
   e The time a journey takes, t, at a fixed speed, and the distance covered, d.
   f The number of ice creams you can buy, c, and the amount of money you have, m.

2 Describe the proportion shown in each of these tables of values. Use the symbol $\propto$.

a
x	3	15
y	1	5

b
x	4	15
y	28	105

c
x	8	20
y	10	4

d
x	10	12
y	50	60

e
x	24	4.8
y	16	3.2

f
x	3	15
y	5	1

g
x	3	15
y	2	10

h
x	8	20
y	10	25

Proportion as a formula

i

x	1	0.1
y	50	500

j

x	16	56
y	6.4	22.4

Proportion as a formula

Example 26.1

Question

Find the formula connecting x and y.

x	5	15
y	3	9

Solution

You can see that $y \propto x$.
So the formula will be $y = kx$.
When $x = 5$, $y = 3$.
Substituting these values in the equation gives $3 = k \times 5$, so $k = \frac{3}{5}$.
This gives the formula $y = \frac{3}{5}x$.

Example 26.2

Question

Find the formula connecting x and y.

x	4	20
y	10	2

Solution

$y \propto \frac{1}{x}$
So $y = k \times \frac{1}{x}$ or $y = \frac{k}{x}$.
When $x = 4$, $y = 10$.
Substituting these values in the formula gives $10 = \frac{k}{4}$ so $k = 10 \times 4 = 40$.
This gives the formula $y = \frac{40}{x}$.
Check the formula works for the other pair of values in the table: $y = \frac{40}{x} = \frac{40}{20} = 2$, which is correct.

Exercise 26.2

1 Find a formula for each proportion in Exercise 26.1, question 2.
2 Ohm's law states that if a voltage, V volts, is applied across the terminals of an electrical component, and if the current flowing in the component is I amperes, then $I \propto V$.
   When $V = 15$, $I = 2.5$.
   a Find the formula for the proportion.
   b Find I when $V = 60$ volts.

26 PROPORTION

3 If w m is the wavelength of a sound wave and f hertz is the frequency of the wave, then $w \propto \frac{1}{f}$.
   When $f = 300$ hertz, $w = 1.1$ m.
   a Find the formula for the proportion.
   b Find w when $f = 660$ hertz.

4 The variables A and B are such that $B = 50$ when $A = 20$.
   a i Write down a formula where A is directly proportional to B.
      ii Write down a formula where A is inversely proportional to B.
   b Calculate B when $A = 25$ for each of the two formulas you have found.

Other types of proportion

The examples you have met in this chapter so far have been for proportion: $y \propto x$ and $y \propto \frac{1}{x}$.

Sometimes two variables can be related in more complicated ways.

For example, $y \propto x^2$, $y \propto \frac{1}{x^2}$, $y \propto \sqrt{x}$, $y \propto \frac{1}{\sqrt{x}}$, $y \propto x^3$, $y \propto \sqrt[3]{x}$ and $y \propto \frac{1}{x^3}$

This is the method you use for finding the formula for the proportion. It is the same method you used previously.

- First write the proportionality using the $\propto$ sign. For example, $y \propto x^2$.
- Replace the $\propto$ sign with = and multiply the right-hand side by k. For example, $y = kx^2$.
- Substitute a known pair of values for x and y.
- Solve the equation to find k.
- Rewrite the formula using the found value of k.

Example 26.3

Question

y is proportional to x^2 and when $x = 5$, $y = 10$.
a Find a formula connecting x and y.
b Find the value of y when $x = 15$.

Solution

a $y \propto x^2$
   So $y = kx^2$
   $10 = k \times 5^2$ Substitute $x = 5$, $y = 10$.
   $k = \frac{10}{25} = \frac{2}{5}$ Solve for k.
   $y = \frac{2}{5}x^2$ Substitute for k in the formula.
b $y = \frac{2}{5} \times 15^2 = 90$ Substitute $x = 15$ into the formula.

Note

There are different ways of talking about proportion.
'y varies directly as x^2' means the same as 'y is proportional to x^2'.

Other types of proportion

Example 26.4

Question

y is inversely proportional to x^2 and when $x = 5$, $y = 10$.

a Find a formula connecting x and y.
b Find the value of y when $x = 10$.
c Find the value of x when $y = 1000$.

Solution

a $y \propto \dfrac{1}{x^2}$

So $y = k \times \dfrac{1}{x^2} = \dfrac{k}{x^2}$

Note
For inverse proportion you can go straight to the fraction version with k as the numerator.

$10 = \dfrac{k}{5^2}$ Substitute $x = 5$, $y = 10$.

$k = 10 \times 25 = 250$ Solve for k.

$y = \dfrac{250}{x^2}$ Substitute for k in the formula.

b $y = \dfrac{250}{10^2} = 2.5$ Substitute $x = 10$ into the formula.

c $1000 = \dfrac{250}{x^2}$ Substitute $y = 1000$ into the formula.

$1000x^2 = 250$

$x^2 = \dfrac{250}{1000}$

$x^2 = \dfrac{1}{4}$

$x = \pm\sqrt{\dfrac{1}{4}} = \pm\dfrac{1}{2}$

Example 26.5

Question

$y \propto \dfrac{1}{x^2}$ and $y = 7$ when $x = 7$.
Find y when $x = 14$.

Solution

Although you can use the formula method, this is quicker.
x has been **multiplied** by 2.
Since y is **inversely** proportional to x^2, y will be **divided** by 2^2.
$y = \dfrac{7}{2^2} = 1.75$

Photocopying is prohibited

26 PROPORTION

Example 26.6

Question

y varies inversely as the square root of x.
When $x = 4$, $y = 12$.
Find the value of y when $x = 9$.

Solution

First, find the formula connecting x and y.

$$y \propto \frac{1}{\sqrt{x}}$$

So $y = k \times \frac{1}{\sqrt{x}} = \frac{k}{\sqrt{x}}$

$12 = \frac{k}{\sqrt{4}}$ Substitute $x = 4$, $y = 12$.

$k = 12 \times 2 = 24$ Solve for k.

$y = \frac{24}{\sqrt{x}}$ Substitute for k in the formula.

$y = \frac{24}{\sqrt{9}} = 8$ Substitute $x = 9$ into the formula.

Example 26.7

Question

y is inversely proportional to x^2.
When x takes a certain value, y is 18.
Find the value of y when x is multiplied by 3.

Solution

x is **multiplied** by 3.
Since y is **inversely** proportional to x^2, y will be **divided** by 3^2.
$y = \frac{18}{3^2} = 2$

Exercise 26.3

1. $y \propto x^2$ and $y = 3$ when $x = 6$.
   a. Find a formula connecting x and y.
   b. Find y when $x = 12$.
   c. Find x when $y = 48$.

2. $y \propto \frac{1}{x^2}$ and $y = 4$ when $x = 4$.
   a. Find a formula connecting x and y.
   b. Find y when $x = 8$.
   c. Find x when $y = 25$.

Other types of proportion

3 y is directly proportional to the square root of x and $y = 8$ when $x = 9$.
   a Find a formula connecting x and y.
   b Find y when $x = 36$.
   c Find x when $y = 4$.

4 y varies inversely as the cube of x and $y = 5$ when $x = 2$.
   a Find a formula connecting x and y.
   b Find y when $x = 5$.
   c Find x when $y = 320$.

5 y is directly proportional to x^3 and $y = 20$ when $x = 2$.
   a Find a formula connecting x and y.
   b Find y when $x = 4$.
   c Find x when $y = 540$.

6 $y \propto x^2$ and $y = 9$ when $x = 7.5$.
Find y when $x = 5$.

7 $y \propto \frac{1}{x^2}$ and $y = 10$ when $x = 3$.
Find y when $x = 12$.

8 $y \propto \sqrt{x}$ and $y = 6$ when $x = 25$.
Find y when $x = 16$.

9 $y \propto \frac{1}{x^3}$ and $y = \frac{1}{2}$ when $x = 2$.
Find y when $x = 1$.

10 $y \propto \sqrt[3]{x}$ and $y = 6$ when $x = 27$.
Find y when $x = 1$.

11 y is inversely proportional to x^2.
$y = 4$ when $x = 3$.
Find y when $x = 10$.

12 A car skids to a halt in an accident.
Investigators measure the length, l metres, of the skid to estimate the speed, v m/s, at which the car had been travelling before the accident. The speed is proportional to the square root of the length of the skid.
   a Given that a car travelling at 20 m/s skids 25 m, find a formula for v in terms of l.
   b Use your formula to find the speed of a car which skids 100 m.

13 Seven people can paint a bridge in 15 days.
   a How long will it take 3 people to paint the bridge?
   b The bridge was painted in t days.
Write down an expression, in terms of t, for the number of people needed to paint the bridge.

14 The force, F, between two particles is inversely proportional to the square of the distance between them.
The force is 36 newtons when the distance between the particles is r metres.
Find the force when the distance between the particles is $3r$ metres.

26 PROPORTION

15 The braking distance of a car is directly proportional to the square of its speed.
When the speed is p metres per second, the braking distance is 6 m.
When the speed is increased by 300%, find
  a an expression for the speed of the car
  b the braking distance
  c the percentage increase in the braking distance.

16 Describe the variation shown in each of these tables of values.
Use the symbol $\propto$.

a
x	5	25
y	5	125

b
x	5	10
y	5	1.25

c
x	5	2.5
y	5	10

d
x	4	6
y	18	8

e
x	2	6
y	7	63

f
x	1	0.25
y	1	4

g
x	54	21.6
y	33	13.2

h
x	24	48
y	4	1

Key points
- The symbol $\propto$ means 'is proportional to'.
- For direct proportion, $y \propto x$, $y \propto x^2$, $y \propto x^3$, $y \propto \sqrt{x}$, $y \propto \sqrt[3]{x}$.
- For indirect proportion, $y \propto \frac{1}{x^2}$, $y \propto \frac{1}{x}$, and so on.
- Proportion can be expressed as a formula. For instance, $y \propto x^2$, has the formula $y = kx^2$. The value of k can be found by substituting a pair of corresponding values for x and y.

REVIEW EXERCISE 5

Ch 23 **1** Solve these equations.
   a $\frac{x}{3} - 5 = 1$ [2]
   b $3(2x + 4) = 2x$ [2]
   c $\frac{2}{x+1} + \frac{x}{x-3} = 1$ [4]

Ch 23 **2** Rearrange each formula to make a the subject.
   a $d = 4a + 3f$ [2]
   b $c = \frac{t}{a} - 3b$ [2]
   c $av = 3(2a + w)$ [3]

3 Make p the subject of the formula $t = \frac{p+3}{p-4}$. [3]

Ch 23
Cambridge O Level Mathematics Syllabus D (4024) Paper 12 Q26a, June 2016

Ch 24 **4** Solve these inequalities.
   a $4x - 15 > 3$ [2]
   b $3 < \frac{x}{2} + 4 \leq 10$ [2]

Ch 23 **5 a** Solve these simultaneous equations.
   Show your working.
   $2x - 4y = 11$
   $3x + 3y = -6$ [4]
   b Solve the equation $2x^2 = 3(8 - x)$.
   Show all your working and give your answers correct to 2 decimal places. [4]

Cambridge O Level Mathematics Syllabus D (4024) Paper 22 Q5a & b, June 2020

Ch 20 **6 a** Solve $5x^2 + 13x - 6$ by factorisation. [3]
Ch 23 **b i** Write $x^2 - 6x - 5$ in completed square form, $(x + a)^2 + b$. [2]
   ii Hence solve the equation $x^2 - 6x - 5 = 0$ and give your answers in a form which is exact. [2]

7 Solve $\frac{6}{x+1} = \frac{5}{x-3}$. [2]

Ch 23
Cambridge O Level Mathematics Syllabus D (4024) Paper 11 Q24a, June 2017

8 y is proportional to $(x - 1)^2$.
Given that $y = 18$ when $x = 4$, find y when $x = 6$. [2]

Ch 26
Cambridge O Level Mathematics Syllabus D (4024) Paper 11 Q19, June 2021

Photocopying is prohibited

215

REVIEW EXERCISE 5

9 y varies inversely as the cube root of x and $y = -3$ when $x = 64$.
   a Find a formula connecting x and y. [2]
   b Find x when $y = 4$. [1]

10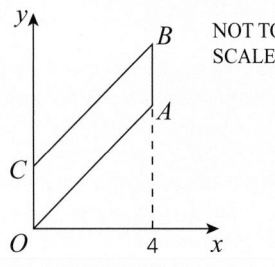

In the diagram, $OABC$ is a parallelogram.
The equation of the line CB is $y = x + 2$.
The region **inside** the parallelogram is defined by four inequalities.
One of these is $y < x + 2$.
Write down the other three inequalities. [2]

Cambridge O Level Mathematics Syllabus D (4024) Paper 11 Q16a, November 2020

11 Here are the first three patterns in a sequence made using dots and lines.

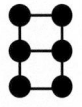

Pattern 1   Pattern 2   Pattern 3

a Complete the table for the first five patterns in this sequence.

Pattern number	1	2	3	4	5
Number of dots	3	6			
Number of lines	2	7			

[2]

b Find an expression, in terms of n, for the number of lines in Pattern n. [2]
c Anwar makes one of these patterns using 92 lines.
   Find the number of dots in Anwar's pattern. [2]

Cambridge O Level Mathematics Syllabus D (4024) Paper 12 Q22, June 2019

27 GRAPHS IN PRACTICAL SITUATIONS

Conversion graphs

A conversion graph can be used to convert quantities from one unit to another.

They are often used to convert between different currencies.

Note
Conversion graphs will only give approximate values. If you know the conversion rate, then you can find a more accurate answer by calculation.

BY THE END OF THIS CHAPTER YOU WILL BE ABLE TO:
- use and interpret graphs in practical situations including travel graphs and conversion graphs
- draw graphs from given data
- apply the idea of rate of change to simple kinematics involving distance–time and speed–time graphs, acceleration and deceleration
- calculate distance travelled as area under a speed–time graph.

Example 27.1

Question

The exchange rate between pounds and dollars is £1 = $1.60.
a Draw a conversion graph for pounds to dollars.
b Use your graph to convert
  i £30 to dollars   ii $150 to pounds.

Solution

a As £1 is equivalent to $1.60, then £100 is equivalent to $160.
Draw a straight line from (0, 0) to (100, 160).
Label each axis with the currency.

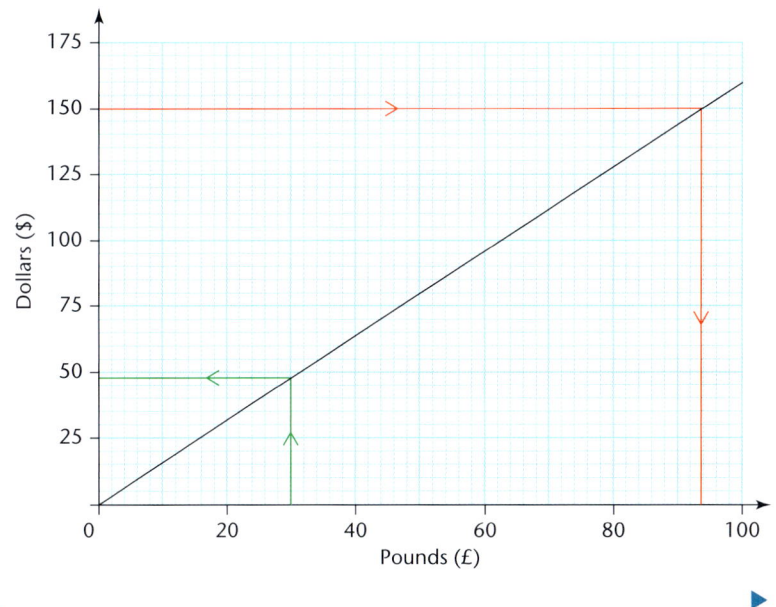

CHECK YOU CAN:
- plot points and read values from a graph
- work with the formula for speed
- find the area of a rectangle and a triangle.

27 GRAPHS IN PRACTICAL SITUATIONS

> **b i** Reading from the graph, £30 = $48
> This is an estimated value.
> The line meets the axis between $45 and $50, closer to $50 so it is estimated at $48.
> **ii** Reading from the graph, $150 = £94
> This is an estimated value.
> The line meets the axis between £92 and £94, but very close to £94, so the answer is given correct to the nearest pound.

Exercise 27.1

1 The exchange rate between dollars and euros (€) is $1 = €0.70.
   a Draw a conversion graph between dollars and euros, for up to $100.
   b Use your graph to convert
      i $35 to euros   ii €60 to dollars.
2 This conversion graph is for pounds (£) to New Zealand dollars (NZ$) for amounts up to £100.

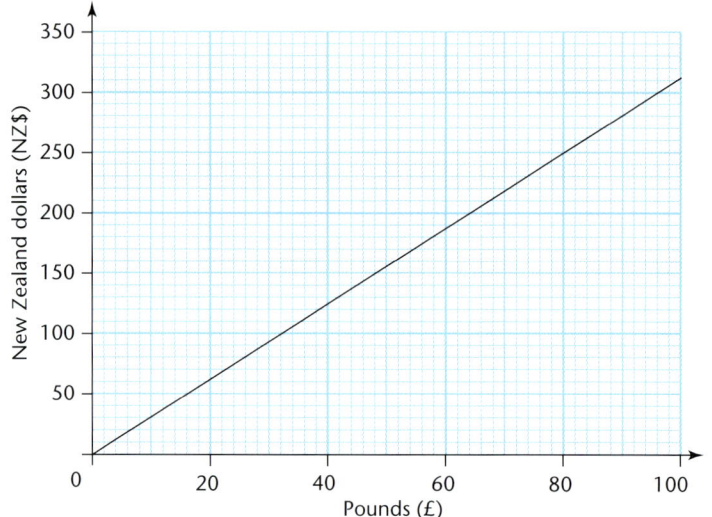

   a Use the graph to find the number of New Zealand dollars equal to
      i £20   ii £85.
   b Use the graph to find the number of pounds equal to
      i NZ$100   ii NZ$250.
   c How many pounds are equal to NZ$600?

Conversion graphs

 3 This conversion graph is for the number of kilometres (km) equal to miles up to 100 km.

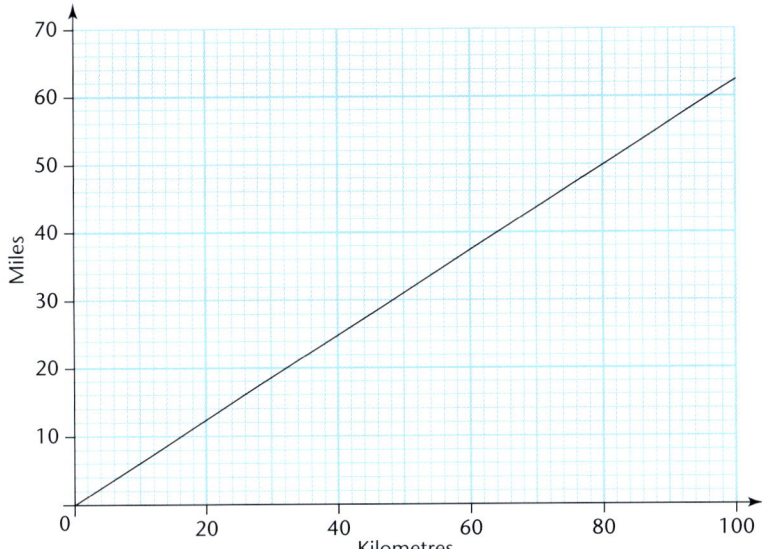

a Use the graph to find the number of miles equal to
   i 40 km
   ii 65 km.
b Use the graph to find the number of kilometres equal to
   i 10 miles
   ii 56 miles.
c How many kilometres are equal to 150 miles?

 4 This conversion graph is for temperatures in °C to °F, from 0 to 100 °C.

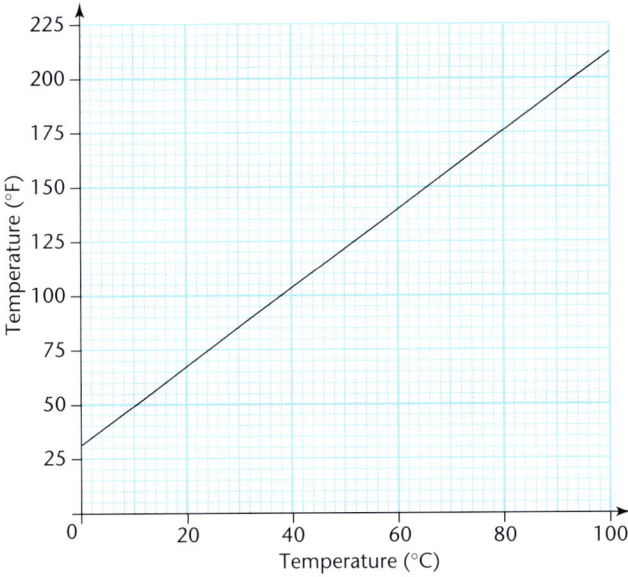

a Use the graph to find the temperature in °F equal to
   i 0 °C
   ii 85 °C.
b Use the graph to find the temperature in °C equal to
   i 100 °F
   ii 170 °F.

27 GRAPHS IN PRACTICAL SITUATIONS

Travel graphs

A travel graph is a graph showing a journey.

Time is plotted on the horizontal axis and distance is plotted on the vertical axis.

This graph shows the distance, in kilometres, of a truck from point A at a given time.

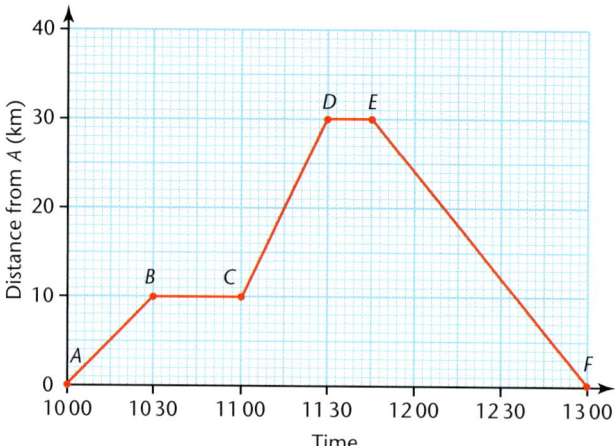

Note
Take care when reading the time scale. There are 60 minutes in an hour, so each small square on this graph represents 3 minutes.

The graph can be used to describe the journey.

A to B The truck travels 10 km at constant speed away from A. It takes 30 minutes.

B to C The truck is stationary for 30 minutes.

C to D The truck travels 20 km at constant speed away from A. It takes 30 minutes. The speed is greater than the speed from A to B.

D to E The truck is stationary for 15 minutes.

E to F The truck travels 30 km at constant speed back to A. It takes 1 hour 15 minutes.

Travel graphs

Example 27.2

Question
Lalith ran the first 3 kilometres to school at a speed of 12 km/h.

She then waited 5 minutes for her friend.

They walked the last 2 kilometres to school together, taking 20 minutes.

Draw a graph to show Lalith's journey.

Solution
It takes 15 minutes to run 3 kilometres at 12 km/h, so the first part of the graph is a straight line from the origin to (15, 3).

The second part of the graph is horizontal from (15, 3) to (20, 3) as Lalith is not moving for 5 minutes.

The final part of the journey covers a further 2 kilometres in 20 minutes, so is a straight line from (20, 3) to (40, 5). This assumes that they walk at a constant speed.

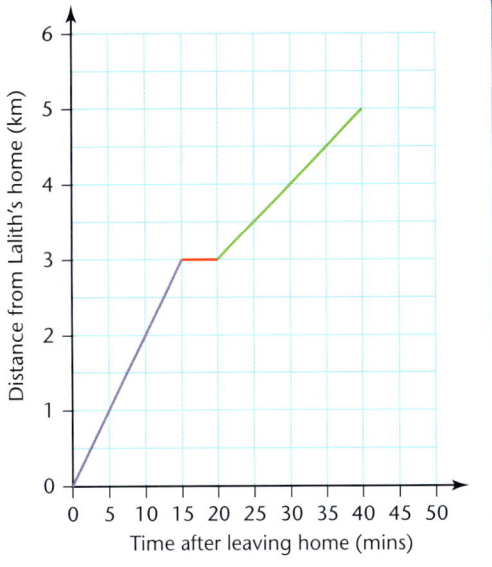

Exercise 27.2

1. Amrit left home and went for a walk, stopping at the supermarket on the way and ending at his brother's house.
   This graph shows his journey.

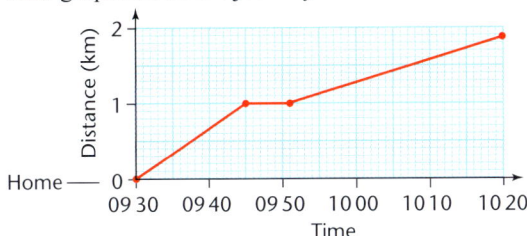

   a. At what time did Amrit reach the supermarket?
   b. How long did he stop at the supermarket?
   c. How far is it from his house to his brother's house?
   d. On which part of the walk did he walk faster?
   e. What was his speed on the first part of his journey?

Note
The slope of the line represents the speed.

If the line is horizontal, the person is not moving.

If the line is straight, the speed is constant.

The steeper the slope of the line, the faster the person is moving.

2 This graph shows Nasreen's walk to town from home one day and her return on the bus.

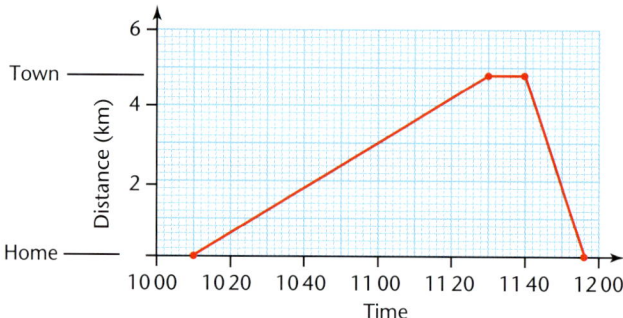

   a At what time did Nasreen leave home?
   b How far did she walk?
   c How long did she stop in town for?
   d How long did the bus journey take?

3 This graph shows the journeys of two men, Yousef and Usman, who both left Lahore. Yousef left first.

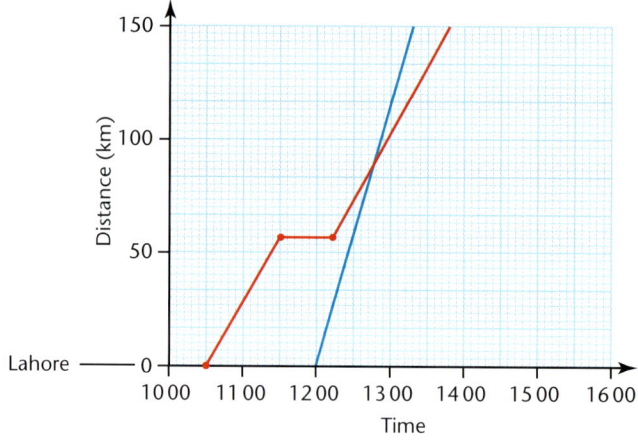

   a How much later than Yousef did Usman leave?
   b How long did Yousef stop for?
   c How far had they travelled when Usman passed Yousef?
   d At what time did Usman pass Yousef?

4 Mr Khan left home at 7.15 a.m. and walked the 500 m to the station in 10 minutes.
   He then waited for the train which left at 7.46 a.m.
   The train took 10 minutes to cover the 12 km to the next station, where Mr Khan got off.
   He then took a taxi for 2 km to his office, taking 8 minutes.
   a Draw a distance–time graph to show this journey.
      Show time from 7 a.m. to 8.40 a.m. and use a scale of 2 cm for 20 minutes.
      Show distance from 0 to 15 km and use a scale of 2 cm for 5 km.
   b How long did Mr Khan wait at the station?
   c At what time did he arrive at his office?

Rate of change on a distance–time graph

The **gradient** of a graph is a measure of its steepness or rate of change.

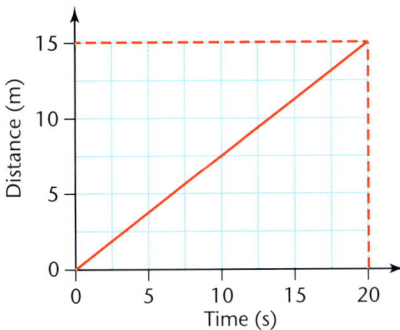

For this distance–time graph

$$\text{gradient} = \frac{\text{change in distance}}{\text{change in time}}$$

$$= \frac{15 \text{ metres}}{20 \text{ secounds}}$$

$$= 0.75 \text{ metres per second}$$

The units of the gradient in this case are metres per second or m/s.

The gradient gives the speed of the object.

The graph is a straight line, so the object is moving with a constant speed of 0.75 m/s.

If the distance–time graph is not a straight line, then the speed of the object is changing.

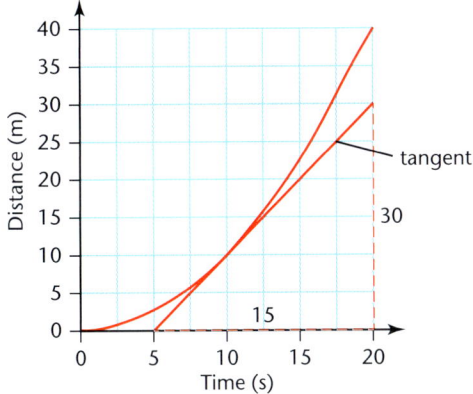

The graph is curving upwards, so the speed is increasing.

You can estimate the speed of the object at a given time by drawing a line which touches the curve at that point. This is called a **tangent**.

You then calculate the gradient of the tangent to the curve at that point.

27 GRAPHS IN PRACTICAL SITUATIONS

The gradient of the tangent to the curve at time 10 seconds gives the speed of the object after 10 seconds.

$$\text{gradient} = \frac{\text{change in distance}}{\text{change in time}}$$

$$= \frac{30 \text{ metres}}{15 \text{ seconds}}$$

$$= 2 \text{ metres per second}$$

The speed of the object after 10 seconds is 2 m/s.

Example 27.3

Question

This distance–time graph shows Manjit's journey from his home to a meeting.

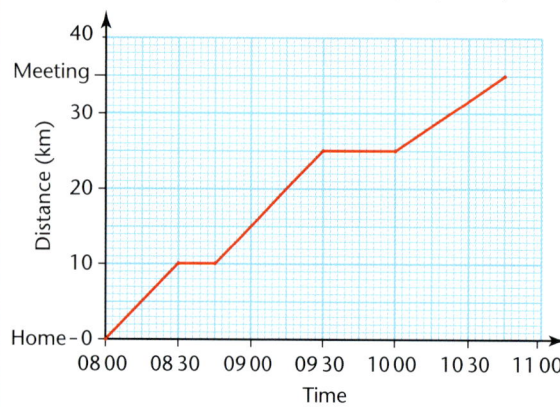

a Work out Manjit's speed, in kilometres per hour, for each stage of the journey.
b Sketch a speed–time graph for Manjit's journey.

Solution

a The graph is made up of five straight-line segments.
Each of these lines represents a stage with a constant speed.
To work out the speed, find the gradient of the line.

From 08 00 to 08 30, speed $= \dfrac{10 \text{ km}}{0.5 \text{ hours}} = 20 \text{ km/h}$

From 08 30 to 08 45, the line is horizontal, so Manjit is not moving.

From 08 45 to 09 30, speed $= \dfrac{(25-10) \text{ km}}{0.75 \text{ hours}} = 20 \text{ km/h}$

From 09 30 to 10 00, the line is horizontal, so Manjit is not moving.

From 10 00 to 10 45, speed $= \dfrac{(35-25) \text{ km}}{0.75 \text{ hours}} = 13.33\ldots \text{ km/h}$

Note

The question asks for the speed in kilometres per hour so you must use the correct units in the gradient calculation.
You must convert 30 minutes to 0.5 hours and 45 minutes to 0.75 hours.

b The speed–time graph is made up of five horizontal line segments.

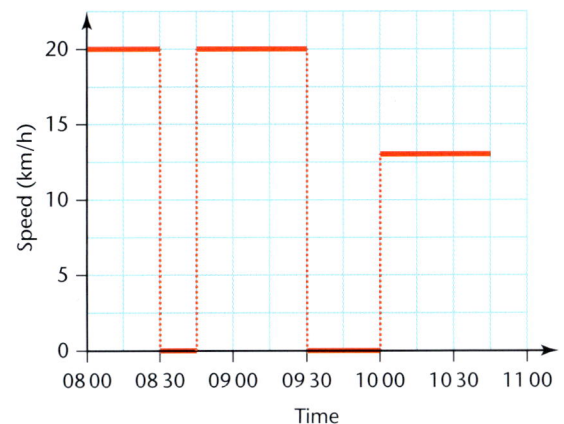

Note
This is a sketch of the situation, because the actual speed would not change instantaneously from 20 km/h to 0 km/h, for example.

Exercise 27.3

1 This distance–time graph shows Baljit's cycle ride.

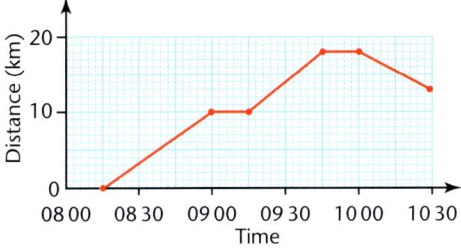

Work out Baljit's speed, in km/h, for each stage of his journey.

2 Work out the speed, in m/s, shown on each of these distance–time graphs.

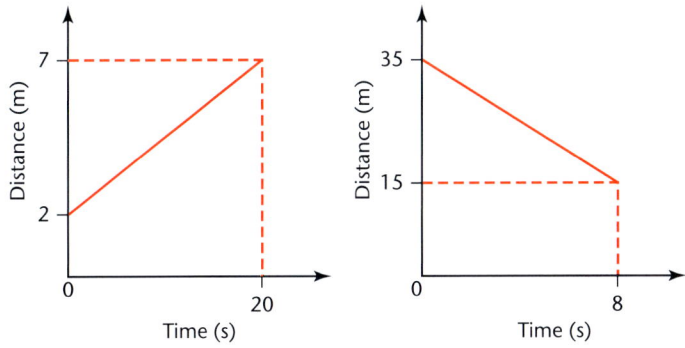

27 GRAPHS IN PRACTICAL SITUATIONS

3 Use the graphs below to estimate the speed, in m/s, of each object when $t = 10$.

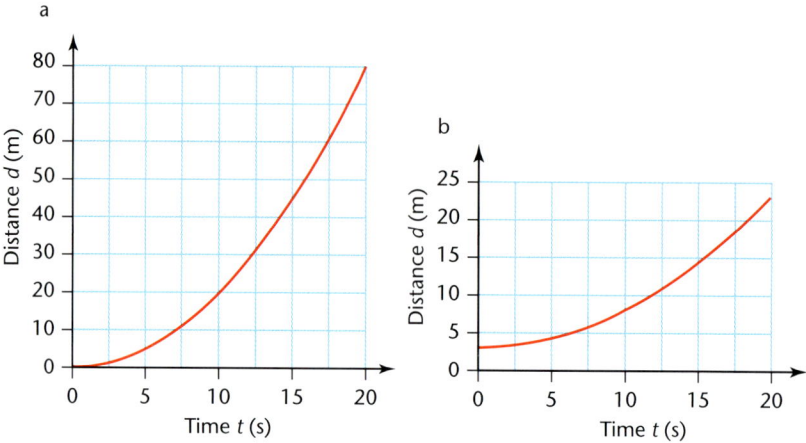

Rate of change on a speed–time graph

This speed–time graph shows the speed of an object during a given time period.

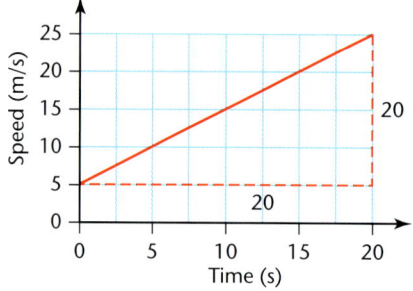

The gradient of the graph represents the rate of change of speed with time.

The graph is a straight line, so the speed is changing at a constant rate.

$$\text{gradient} = \frac{20 \text{ m/s}}{20 \text{ s}} = 1 \text{ m/s}^2$$

The rate of change of speed with time is the **acceleration** of the object.

The units of acceleration in this case are metres per second squared or m/s^2.

If the speed–time graph is not a straight line, then the acceleration is changing.

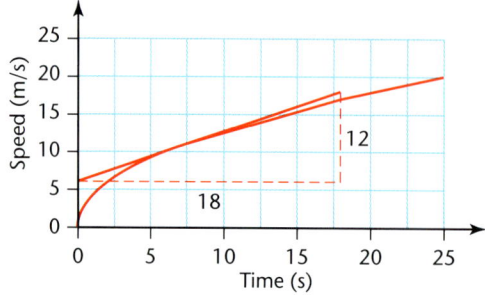

Rate of change on a speed–time graph

The graph is curving downwards, so the acceleration is decreasing.

You can estimate the acceleration of the object at a given time by calculating the gradient of the tangent to the curve at that point.

The gradient of the tangent to the curve at time 7.5 seconds gives the speed of the object after 7.5 seconds.

$$\text{gradient} = \frac{\text{change in speed}}{\text{change in time}}$$

$$= \frac{12\,\text{m/s}}{18\,\text{seconds}}$$

$$= 0.67\,\text{m/s}^2 \text{ correct to 2 decimal places}$$

The acceleration of the object after 7.5 seconds is $0.67\,\text{m/s}^2$.

Example 27.4

Question

The speed–time graph shows part of a car's journey.

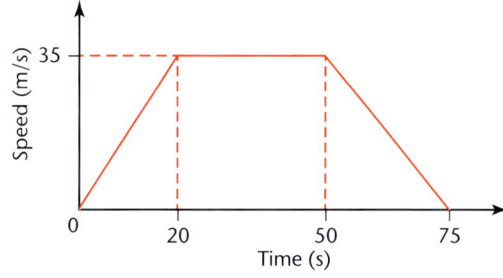

a Work out the acceleration, in m/s², for each stage of this journey.
b Calculate the speed of the car after 60 seconds.

Solution

a The graph is made up of three straight-line segments.

Each of these lines represents a stage with constant acceleration.

To work out the acceleration, find the gradient of the line.

From 0 to 20 seconds, acceleration = $\frac{35\,\text{m/s}}{20\,\text{s}} = 1.75\,\text{m/s}^2$.

From 20 to 50 seconds, the speed is constant, so acceleration = $0\,\text{m/s}^2$.

From 50 to 75 seconds, acceleration = $\frac{-35\,\text{m/s}}{25\,\text{s}} = -1.4\,\text{m/s}^2$.

b You know that the speed of the car is 35 m/s after 50 seconds.

After this the car is decelerating at $1.4\,\text{m/s}^2$.

In 10 seconds, the speed will decrease by $10 \times 1.4 = 14\,\text{m/s}$.

After 60 seconds, the speed is $35 - 14 = 21\,\text{m/s}$.

Note

When the acceleration is negative, the car is slowing down.

Negative acceleration is known as **deceleration**.

You can say that from 50 to 75 seconds, the car has a deceleration of $1.4\,\text{m/s}^2$.

27 GRAPHS IN PRACTICAL SITUATIONS

Exercise 27.4

1 Calculate the acceleration, in m/s², for each of these speed–time graphs.

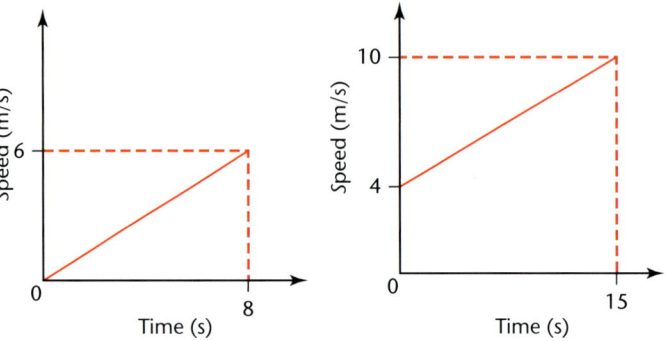

2 Calculate the deceleration, in m/s², for each of these speed–time graphs.

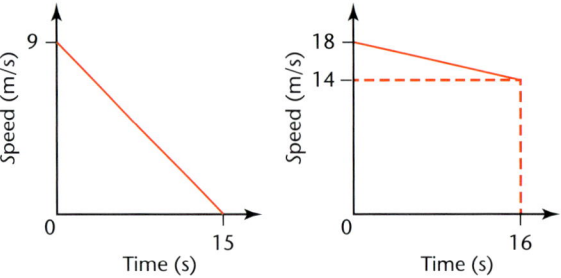

3 Use the graphs to estimate the acceleration of each object when $t = 5$.

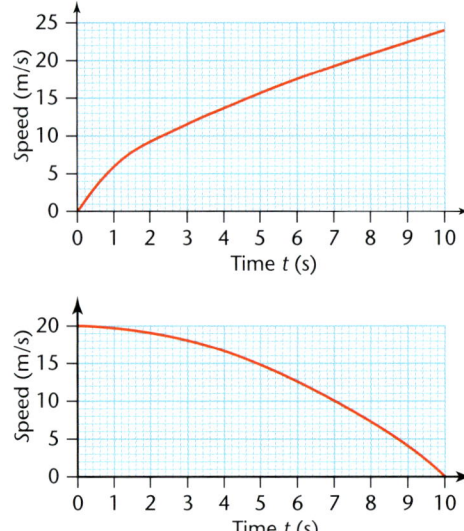

4 This is the speed–time graph for part of a car's journey.

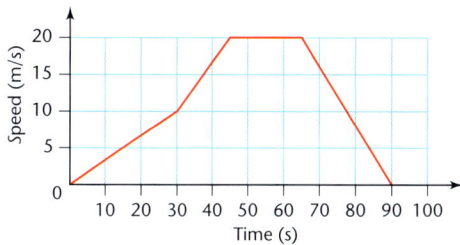

a Calculate the acceleration, in m/s², for each stage of the journey.
b Calculate the speed of the car when
   i $t = 20$ ii $t = 75$.
5 A train slows down uniformly from a speed of 8 m/s to rest in 24 seconds.
a Sketch a speed–time graph to show this information.
b Calculate the train's deceleration in m/s².
c Calculate the speed of the train after 9 seconds.
6 a An object moves from rest with constant acceleration of 1.2 m/s².
   Calculate the speed of the object after 15 seconds.
b An object slows down from a speed of 16 m/s with a constant deceleration of 0.8 m/s².
   Calculate the time taken for the object to stop.

Area under a speed–time graph

The distance travelled by an object moving at 30 m/s for 20 seconds can be found using the formula distance = speed × time.

Distance = 30 × 20 = 600 metres

This is the speed–time graph for an object moving with a constant speed of 30 m/s for 20 seconds.

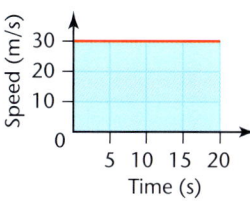

The blue rectangle is the area under the graph.

The area of the rectangle is 30 × 20 = 600.

The units of the area in this case are metres per second × seconds = metres.

So the area under the graph represents 600 metres.

This is the distance travelled by the object.

> The area under a speed–time graph gives the distance travelled by the object in that time.

27 GRAPHS IN PRACTICAL SITUATIONS

Example 27.5

Question

The speed–time graph shows part of a car's journey.

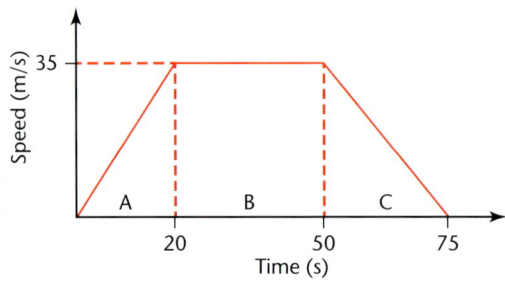

Calculate the distance travelled by the car in these 75 seconds.

Solution

The area under the graph gives the distance travelled.

The area can be found by adding the areas of two triangles and a rectangle.

Distance travelled = area of triangle A + area of rectangle B + area of triangle C

$= \frac{1}{2} \times 20 \times 35 + 30 \times 35 + \frac{1}{2} \times 25 \times 35$

$= 1837.5$ metres

Exercise 27.5

1. Calculate the distance travelled in 20 seconds from each of these speed–time graphs.

   a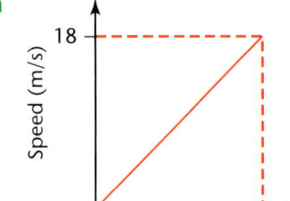

   b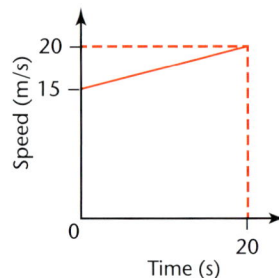

   c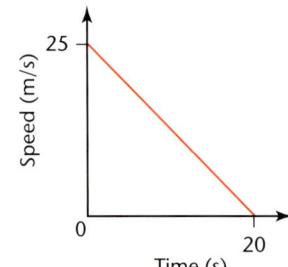

   d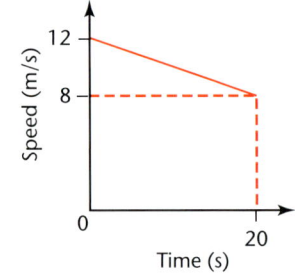

Area under a speed–time graph

2 The speed–time graph shows 100 seconds of a car's journey.

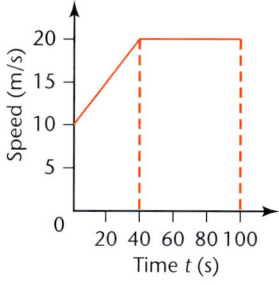

a Calculate the acceleration of the car when $t = 20$.
b i Calculate the distance travelled by the car in the first 40 seconds.
  ii Calculate the total distance travelled in 100 seconds.

3 The speed–time graph shows 50 seconds of a car's journey.

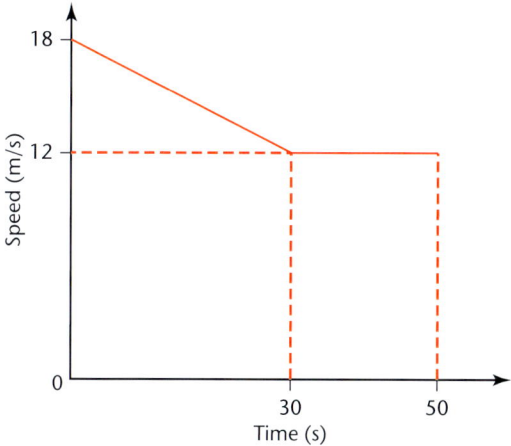

a Calculate the distance travelled by the car in these 50 seconds.
b After 50 s, the car slows down with a constant deceleration of $0.4 \, \text{m/s}^2$.
Work out how many more seconds it takes for the car to come to rest.

4 The speed–time graph shows part of a train's journey.

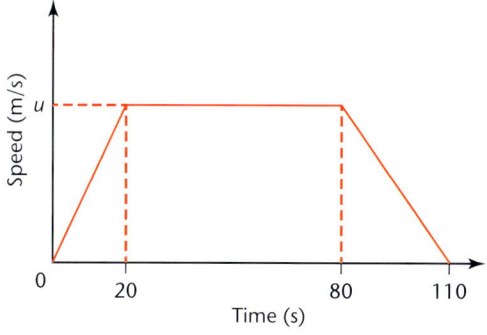

The train travels a total of 680 m in 110 seconds.
Work out the value of u.

27 GRAPHS IN PRACTICAL SITUATIONS

5 The speed–time graph shows the first 30 s of a car's journey.

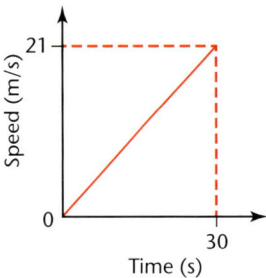

   a i Work out the speed of the car after 15 seconds.
      ii Work out the distance travelled by the car in the first 15 seconds.
   b Sketch a distance–time graph for the car's journey.

> ### Key points
> - Conversion graphs can be used to convert from one unit to another.
> - Travel graphs represent distance against time and show a journey. The rate of change or slope of the line represents the speed. If the line is not straight, the rate of change can be estimated by drawing a tangent at the point.
> - The rate of change on a speed–time graph represents the acceleration.
> - The area under a speed–time graph gives the distance travelled.

28 GRAPHS OF FUNCTIONS

Quadratic graphs

Quadratic graphs are graphs of equations of the form $y = ax^2 + bx + c$, where a, b and c are constants, and b and c may be zero.

When the graph of a quadratic equation is drawn, it produces a curve called a **parabola**.

BY THE END OF THIS CHAPTER YOU WILL BE ABLE TO:
- construct tables of values, and draw, recognise and interpret graphs for functions of the following forms: ax^n (includes sums of no more than three of these) and $ab^x + c$ where $n = -2, -1, -\frac{1}{2}, 0, \frac{1}{2}, 1, 2, 3$; a and c are rational numbers; and b is a positive integer
- solve associated equations graphically, including finding and interpreting roots by graphical methods
- draw and interpret graphs representing exponential growth and decay problems
- estimate gradients of curves by drawing tangents.

CHECK YOU CAN:
- draw the graph of a straight line, given its equation
- find the gradient of a straight line
- use percentages, including growth and decay.

Example 28.1

Question

a Draw the graph of $y = x^2 - 2x - 3$ for values of x from -2 to 4.
Label the axes from -2 to 4 for x and from -5 to 5 for y.

b Find the values of x for which $y = 0$.

Note
Sometimes a table will be given with only the two rows for the x and y values.
You may find it useful to include all the rows and then just add the correct rows to get the values of y.

Solution

a

x	-2	-1	0	1	2	3	4
x^2	4	1	0	1	4	9	16
$-2x$	4	2	0	-2	-4	-6	-8
-3	-3	-3	-3	-3	-3	-3	-3
$y = x^2 - 2x - 3$	5	0	-3	-4	-3	0	5

To find the value of y, add together the values in rows 2, 3 and 4.
Now plot the points $(-2, 5)$, $(-1, 0)$, and so on. Join them with a smooth curve.
Label the curve.

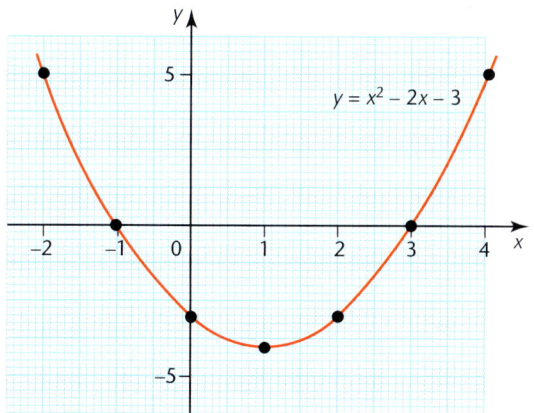

Note
Practise drawing smooth curves. Keep your pencil on the paper. Make the curve a thin single line. Look ahead to the next point as you draw the line. Keep your hand inside the curve by turning the paper.

Note
Complete a table of values to work out the points.
Make your own table if needed.
A common error is to include the value in the x-row when adding to find y, so separate this row off clearly.

b $y = 0$ when the curve crosses the x-axis.
This is when $x = -1$ or $x = 3$.

28 GRAPHS OF FUNCTIONS

> **Note**
> If you find that you have two equal lowest (or highest) values for y, the curve will go below (or above) that value.
> To find the y-value between the equal values, find the value of x halfway between the two points. Substitute to find the corresponding value of y.

> **Note**
> In question **2**, to find the value of y, rows 3 and 4 are added.

Exercise 28.1

1 Copy and complete this table for $y = x^2 + 3x - 7$. Do not draw the graph. Notice the extra value at $x = -1.5$.

x	−4	−3	−2	−1	0	1	2	−1.5
x^2			4					
+3x			−6					
−7			−7					
$y = x^2 + 3x - 7$			−9					

2 Copy and complete this table for $y = 2x^2 - 8$. Do not draw the graph.

x	−3	−2	−1	0	1	2	3
x^2	9						
$2x^2$	18						
−8	−8						
$y = 2x^2 - 8$	10						

3 Copy and complete this table for $y = -x^2 - 5x + 6$. Do not draw the graph. Remember that $-x^2$ is always negative. Notice the extra value at −2.5.

x	−6	−5	−4	−3	−2	−1	0	1	2	−2.5
$-x^2$			−16						−4	
−5x			20						−10	
+6			6						6	
$y = -x^2 - 5x + 6$			10						−8	

Use 2 mm graph paper to draw the graphs in questions **4** to **10**.

4 a Copy and complete this table for $y = x^2 - 3x$.

x	−1	0	1	2	3	4	5	1.5
x^2								
−3x								
$y = x^2 - 3x$								

b Draw the graph of $y = x^2 - 3x$ for values of x from −1 to 5.
Label the x-axis from −1 to 5 and the y-axis from −5 to 10.
Use a scale of 1 cm to 1 unit on the x-axis and 2 cm to 5 units on the y-axis.

5 Draw the graph of $y = x^2 - 6x + 5$ for values of x from −1 to 6.

6 Draw the graph of $y = -x^2 + 4x - 3$ for values of x from −1 to 5.

7 a Draw the graph of $y = x^2 - 5x + 2$ for values of x from −1 to 6.
b Find the values of x on your graph when $y = 0$.
Give your answers correct to 1 decimal place.

8 a Draw the graph of $y = 2x^2 - 5x + 1$ for values of x from −2 to 4.
b Find the value of x on your graph where $y = 0$.
Give your answers correct to 1 decimal place.

> **Note**
> An extra point at $x = 2.5$ might be useful for question **7**.

9 a Draw the graph of $y = 2x^2 - 12x$ for values of x from -1 to 7.
   b Write down the values of x where the curve crosses $y = 5$.
10 When a stone is dropped from the edge of a cliff, the distance, d metres, it falls is given by the formula $d = 5t^2$, where t is the time in seconds.
   a Work out the values of d for values of t from 0 to 5.
   b Draw the graph for $t = 0$ to 5.
   c The cliff is 65 metres high.
     How long does it take the stone to reach the bottom of the cliff? Give your answer correct to 1 decimal place.

Note
Note that in some questions, the values of y are not in pairs, so the lowest point is not exactly halfway between two of the given points.

Using graphs to solve equations

One way of finding solutions to quadratic equations is to draw and use a graph.

Example 28.2

Question

a Draw the graph of $y = x^2 - 2x - 8$ for values of x from -3 to 5.
b Solve the equation $x^2 - 2x - 8 = 0$.
c Solve the equation $x^2 - 2x - 8 = 5$.

Note
The solutions of an equation are also called the roots of the equation.

Solution

a

x	-3	-2	-1	0	1	2	3	4	5
x^2	9	4	1	0	1	4	9	16	25
$-2x$	6	4	2	0	-2	-4	-6	-8	-10
-8	-8	-8	-8	-8	-8	-8	-8	-8	-8
$y = x^2 - 2x - 8$	7	0	-5	-8	-9	-8	-5	0	7

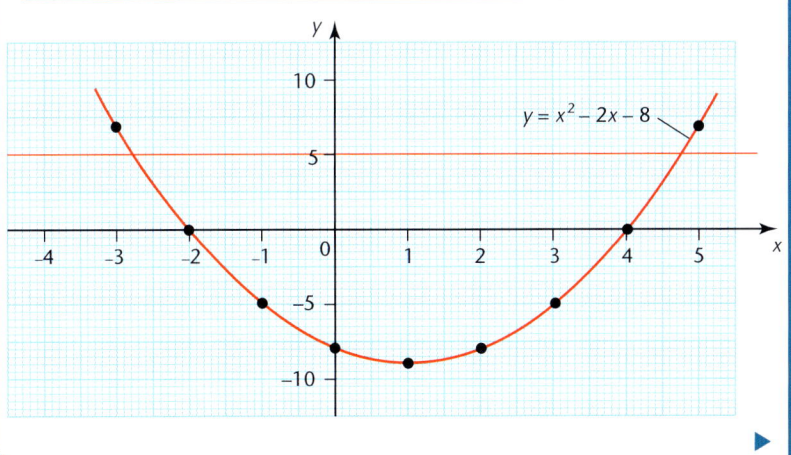

28 GRAPHS OF FUNCTIONS

b The solution of $x^2 - 2x - 8 = 0$ is when $y = 0$, where the curve cuts the x-axis. The solution is $x = -2$ or $x = 4$.

c The solution of $x^2 - 2x - 8 = 5$ is when $y = 5$.

Draw the line $y = 5$ on your graph and read off the values of x where the curve cuts the line.

The solution is $x = -2.7$ or $x = 4.7$, to 1 decimal place.

Sometimes you may have drawn a quadratic graph and then need to solve an equation that is different from the graph you have drawn. Rather than drawing another graph, it may be possible to rearrange the equation to obtain the one you have drawn.

Example 28.3

Question

a Draw the graph of $y = 2x^2 - x - 3$ for values of x between -3 and 4.

b Use this graph to solve these equations.

   i $2x^2 - x - 3 = 6$ **ii** $2x^2 - x = x + 5$

Solution

a

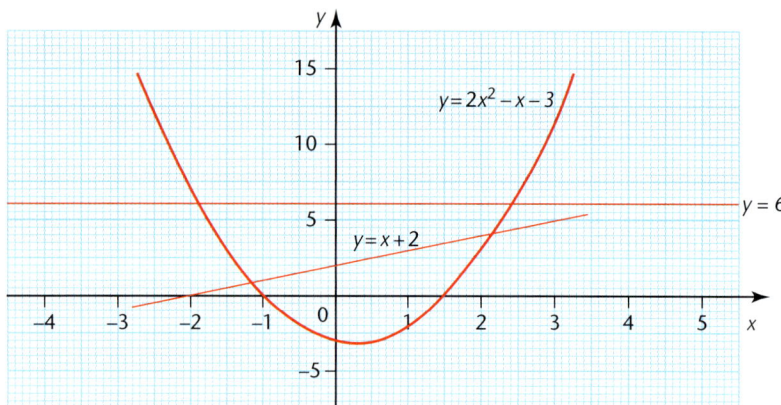

b **i** To solve $2x^2 - x - 3 = 6$ you draw the line $y = 6$ on your graph.

At the points of intersection of the line and the curve, $y = 6$ and $2x^2 - x - 3$ and therefore $2x^2 - x - 3 = 6$.

The curve and line cross at approximately $x = -1.9$ and $x = 2.4$.

ii To solve $2x^2 - x = x + 5$ you must rearrange the equation so that you have $2x^2 - x - 3$ on the left-hand side.

You do this by subtracting 3 from both sides.

$2x^2 - x = x + 5$

$2x^2 - x - 3 = x + 2$

Now draw the line $y = x + 2$ on your graph.

The points where the line and the curve intersect are the solution to $2x^2 - x = x + 5$.

The curve and line cross at approximately $x = -1.2$ and $x = 2.2$.

Exercise 28.2

Save the graphs you draw in this exercise. You will use some of them again in Exercise 28.5.

1. **a** Draw the graph of $y = x^2 - 4x + 3$ for values of x from -1 to 5.
   b Use your graph to find the roots of the equation $x^2 - 4x + 3 = 0$.
2. **a** Draw the graph of $y = x^2 - 3x$ for values of x from -2 to 5.
   b Use your graph to find the roots of the equation $x^2 - 3x = 0$.
3. **a** Draw the graph of $y = x^2 - 2x - 3$ for values of x from -2 to 4.
   b Use your graph to solve these equations.
     i $x^2 - 2x - 3 = 0$ **ii** $x^2 - 2x - 3 = -2$
     iii $x^2 - 2x - 3 = x$ **iv** $x^2 - 2x - 5 = 0$
4. **a** Draw the graph of $y = x^2 - 2x + 2$ for values of x from -2 to 4.
   b Use your graph to solve these equations.
     i $x^2 - 2x + 2 = 8$ **ii** $x^2 - 2x + 2 = 5 - x$
     iii $x^2 - 2x - 5 = 0$
5. **a** Draw the graph of $y = 2x^2 + 3x - 9$ for values of x from -3 to 2.
   b Use your graph to solve these equations.
     i $2x^2 + 3x - 9 = -1$ **ii** $2x^2 + 3x - 4 = 0$
6. **a** Draw the graph of $y = x^2 - 5x + 3$ for values of x from -2 to 8.
   b Use your graph to solve these equations.
     i $x^2 - 5x + 3 = 0$ **ii** $x^2 - 5x + 3 = 5$
     iii $x^2 - 7x + 3 = 0$
7. An archer shoots an arrow from ground level towards a castle 45 m high. The height of the arrow (h metres) above the ground after t seconds is given by
   $$h = 30t - 5t^2$$
   a Draw the graph for values of t from 0 to 6.
   b Can the arrow reach the top of the castle? Explain.
   c If the arrow does not hit anything, after how long will it be at ground level again?

   For the remaining questions, do not draw the graphs.
8. The graph of $y = x^2 - 8x + 2$ has been drawn.
   What other line needs to be drawn to solve the equation $x^2 - 8x + 6 = 0$?
9. Find and simplify the equation in x that is solved by the intersections of the graph of $y = 2x^2 - 5x + 1$ and the graph of $y = 4x - 3$.
10. Find and simplify the equation in x that is solved by the intersections of the graph of $y = 3x^2 + 7x - 2$ and the graph of $y = 5x + 4$.

Cubic graphs

Cubic graphs are graphs of equations of the form

$$y = ax^3 + bx^2 + cx + d$$

where a, b, c and d are constants, and b, c and d may be zero.

28 GRAPHS OF FUNCTIONS

Example 28.4

Question

a Draw the graph of $y = x^3$ for values of x from -3 to 3.
  Label the x-axis from -3 to 3 and the y-axis from -30 to 30.
b Use your graph to solve the equation $x^3 = 12$. Give your answer to 1 decimal place.

Solution

a

x	−3	−2.5	−2	−1	0	1	2	2.5	3
y = x³	−27	−15.625	−8	−1	0	1	8	15.625	27

The outside points are a long way apart, so plotting the values for $x = 2.5$ and $x = -2.5$ helps with the drawing of the curve.

b To solve $x^3 = 12$, you need to draw the line $y = 12$ on your graph and find where it intersects the curve. The solution is $x = 2.3$, to 1 decimal place.

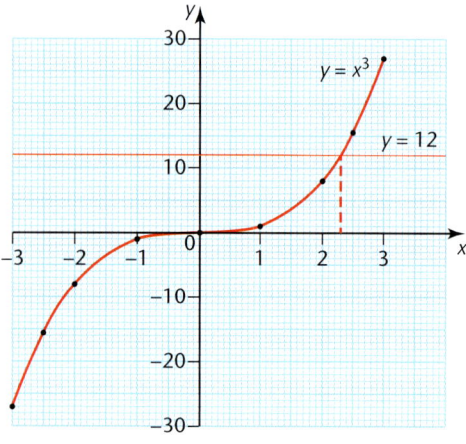

The curve goes from bottom left to top right and has a 'double bend' in the middle.

All cubic curves with a positive x^3 term have a similar shape.

If the x^3 term is negative the curve goes from top left to bottom right.

Example 28.5

Question

a Draw the graph of $y = x^3 - 2x$ for values of x from -2 to 2.
  Label the x-axis from -2 to 2 and the y-axis from -4 to 4.
b Use the graph to find the roots of the equation $x^3 - 2x = 0$.
  Give your answers correct to 1 decimal place.

Reciprocal graphs

Solution

a

x	−2	−1	−0.5	0	0.5	1	2
x^3	−8	−1	−0.125	0	0.125	1	8
$-2x$	4	2	1	0	−1	−2	−4
$y = x^3 - 2x$	−4	1	0.875	0	−0.875	−1	4

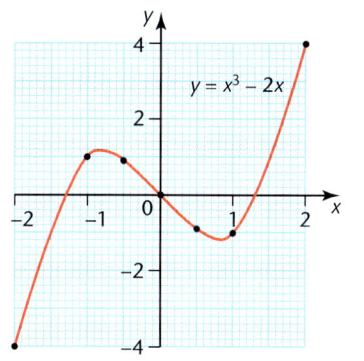

It helps to see more clearly where the curve is highest and lowest if you work out the values of y for x = −0.5 and x = 0.5.

b To find the roots of $x^3 - 2x = 0$, you need to find where $y = x^3 - 2x$ crosses $y = 0$, which is the x-axis.

The roots are x = −1.4 (to 1 d.p.), 0 and 1.4 (to 1 d.p.).

The shape of the curve is similar to the previous one. The −2x term makes the 'double bend' more pronounced.

Reciprocal graphs

Reciprocal graphs are graphs of equations of the form

$$y = \frac{a}{x}$$

where a is a non-zero constant.

Example 28.6

Question

Draw the graph of $y = \frac{4}{x}$.

Solution

x	−4	−3	−2	−1.5	−1	1	1.5	2	3	4
$y = \frac{4}{x}$	−1	−1.33…	−2	−2.66…	−4	4	2.66…	2	1.33…	1

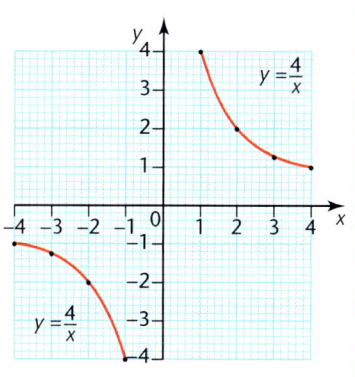

You cannot use 0 as a point in this type of graph, since you cannot divide 4 (or any other number) by 0.

Again, it is helpful to work out extra points to give a better curve. In this case you might work out the value of y when x = −1.5 and 1.5. It is also useful to have the same scale for both x and y in this case.

All equations of this type have graphs of the same shape, with two separate branches in opposite quadrants.

Plotting points for x = 0.5 and x = 0.1 would help to show that the curve gets closer to the y-axis without ever meeting it.

28 GRAPHS OF FUNCTIONS

Plotting points for $x = 5$ and $x = 6$ would help to show that the curve gets closer to the x-axis without ever meeting it.

> ## Example 28.7
>
> ### Question
> Draw the graph of $y = \frac{2}{x^2}$.
>
> ### Solution
>
> | x | −4 | −3 | −2 | −1.5 | −1 | 1 | 1.5 | 2 | 3 | 4 |
> |---|---|---|---|---|---|---|---|---|---|---|
> | $y = \frac{2}{x^2}$ | 0.125 | 0.22… | 0.5 | 0.88… | 2 | 2 | 0.88… | 0.5 | 0.22… | 0.125 |
>
>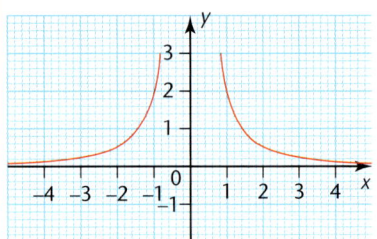

All equations of this type have graphs of the same shape, with two separate branches side by side.

Plotting points for $x = 0.5$ and $x = 0.1$ would help to show that the curve gets closer to the y-axis without ever meeting it.

Plotting points for $x = 5$ and $x = 6$ would help to show that the curve gets closer to the x-axis without ever meeting it.

Other graphs

In reciprocal graphs, you have learned that x cannot be zero. Here is another type of graph where the value of x is restricted:

$y = a\sqrt{x}$, or $y = ax^{\frac{1}{2}}$, where a is a number.

Here, since the square of a number cannot be negative, x cannot be negative.

Other graphs

Example 28.8

Question
Draw the graph of $y = 2\sqrt{x}$.

Solution

x	0	1	2	3	4	5	6	0.25	0.5	0.75
y	0	2	2.82…	3.46…	4	4.47…	4.89…	1	1.41…	1.73…

Some extra points have been added to the table to help draw the correct shape on the steep part of the curve.

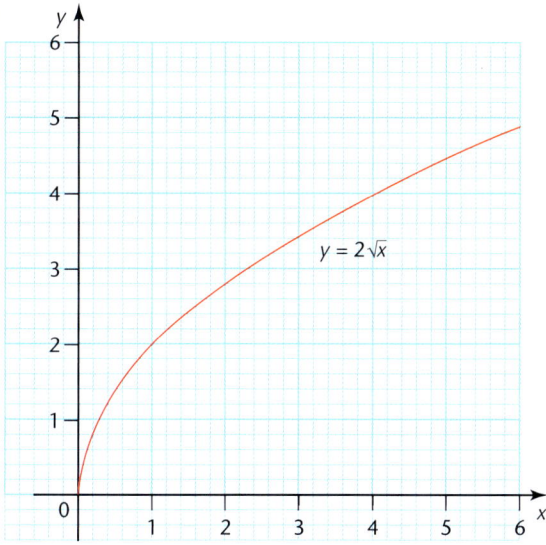

Sometimes, different graphs that you have met may be combined in a function. You will find that the shape can change a lot. This happens when the function includes reciprocals or roots as well as positive integer powers of x.

Example 28.9

Question
a Complete this table for $y = 5x - \frac{2}{x}$. Give values correct to 1 d.p. when needed.

x	−3	−2	−1	−0.5	−0.1	0.1	0.5	1	2	3
y										

b Draw the graph of $y = 5x - \frac{2}{x}$.

28 GRAPHS OF FUNCTIONS

Solution

a

x	−3	−2	−1	−0.5	−0.1	0.1	0.5	1	2	3
y	−14.3	−9	−3	1.5	19.5	−19.5	−1.5	3	9	14.3

b

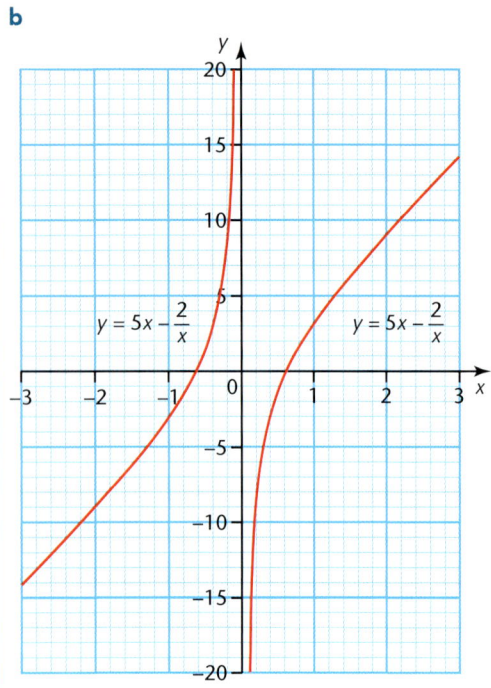

Note

When x is near zero, the graph is very different from that of $y = 5x$. When x is far from zero, the reciprocal part of the equation has much less effect.

Exercise 28.3

Save the graphs you draw in this exercise. You will use some of them again in Exercise 28.5.

1 a Copy and complete this table for $y = x^3 + 5$.

x	−3	−2	−1	0	1	2	3
x^3	−27			0		8	
+5	5						
y	−22						

b Draw the graph of $y = x^3 + 5$.
c Use your graph to solve the equation $x^3 + 5 = 0$.

2 a Copy and complete this table for $y = -x^3$.

x	−3	−2	−1	0	1	2	3
x^3	−27			0		8	
y	27						

b Draw the graph of $y = -x^3$.
c Use your graph to solve the equation $-x^3 = 6$.

3 a Copy and complete this table for $y = \frac{1}{x}$.

x	−10	−5	−2	−1	−0.5	−0.1	0.1	0.5	1	2	5	10
y						−10					0.2	

b Draw the graph of $y = \frac{1}{x}$.
  Use a scale of 1 cm to 2 units on both axes.
c Use your graph to solve each of these equations.
  i $\frac{1}{x} = 0.3$ **ii** $\frac{1}{x} = -5$

4 a Make a table of values for $y = x^3 - 12x + 2$ for $x = -3$ to 4.
b Draw the graph of $y = x^3 - 12x + 2$.
c Find the roots of the equation $x^3 - 12x + 2 = 0$.

5 a Make a table of values for $y = x^3 - x^2 - 6x$ for $x = -3$ to 4.
b Draw the graph of $y = x^3 - x^2 - 6x$.
c Use the graph to find the roots of the equation $x^3 - x^2 - 6x = 0$.

6 a Make a table of values for $y = \frac{8}{x}$ for $x = -8, -4, -2, -1, 1, 2, 4, 8$.
b Draw the graph of $y = \frac{8}{x}$.
  Use a scale of 1 cm to 1 unit on both axes.

7 a Make a table of values for $y = \frac{12}{x^2}$ for $x = -12, -8, -6, -4, -3, -2, -1, 1, 2, 3, 4, 6, 8, 12$.
b Draw the graph of $y = \frac{12}{x^2}$.

8 a Make a table of values for $y = \frac{5}{x}$ for $x = -5, -4, -2.5, -2, -1, 1, 2, 2.5, 4, 5$.
b Draw the graph of $y = \frac{5}{x}$.
  Use a scale of 1 cm to 1 unit on both axes.
c On the same grid, draw the graph of $y = x$.
d Use your graph to solve $x^2 = 5$, giving your answers to 1 decimal place.

9 The temperature ($y\,°C$) of a cup of coffee t minutes after it was made is given by the equation $y = 20 + \frac{50}{t}$.
a Plot the equation for values of t between 1 and 10.
b Estimate how long it takes for the coffee to cool to 40 °C.
c Estimate the temperature after 30 minutes.

10 The graph of $y = x^3 - 2x^2$ has been drawn.
What other curve needs to be drawn to solve the equation $x^3 - x^2 - 4x + 3 = 0$?

11 a Draw the graph of $y = \sqrt{x}$ for values of x from 0 to 6.
b On the same axes, draw the graph of $y = \frac{1}{\sqrt{x}}$ for $0 < x < 6$.

12 a Complete a table of values for $y = x^2 - \frac{1}{x^2}$ for $x = -3, -2, -1, -0.5, -0.2, 0.2, 0.5, 1, 2, 3$.
b Draw the graph of $y = 5x^2 - \frac{1}{x^2}$.

> **Note**
> Remember: always make a table of values.

28 GRAPHS OF FUNCTIONS

Exponential graphs

Exponential graphs are graphs of equations of the form

$y = ka^x$.

For the graphs that you will meet in this chapter, a is a positive integer.

Example 28.10

Question

Plot a graph of $y = 3^x$ for values of x from −2 to 3.

Use your graph to estimate

a the value of y when $x = 2.4$ **b** the solution to the equation $3^x = 20$.

Solution

Make a table of values and plot the graph.

x	−2	−1	0	1	2	3
y	0.111…	0.333…	1	3	9	27

a Draw the line $x = 2.4$ and find where it intersects the curve.

$y = 14.0$

b Draw the line $y = 20$ and find where it intersects the curve.

$x = 2.7$

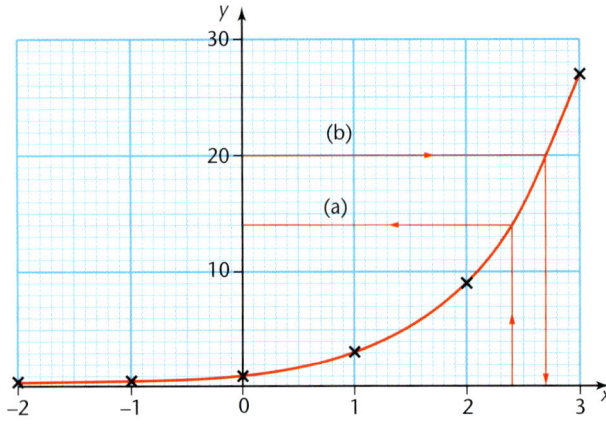

Growth and decay

You learned about this in Chapter 17.

In the function $y = a \times b^x$, the value a is repeatedly multiplied by the multiplier b.

When the multiplier is larger than 1, there is growth.

When the multiplier is less than 1, there is decay.

So exponential graphs can be used to represent and interpret growth and decay.

Growth and decay

Example 28.11

Question

The population of a village was 700 at the beginning of 2020.
After x years, the population P is given by this equation
$P = 700 \times 1.05^x$.

a What does the number 1.05 represent in this equation?
b Draw the graph of $P = 700 \times 1.05^x$ for $x = 0$ to 10.
c Use your graph to estimate the population at the end of June 2025.

Solution

a $1.05 = 105\%$. The population is growing by 5% each year.

b

x	0	1	2	3	4	5	6	7	8	9	10
P	700	735	772	810	851	893	938	985	1034	1086	1140

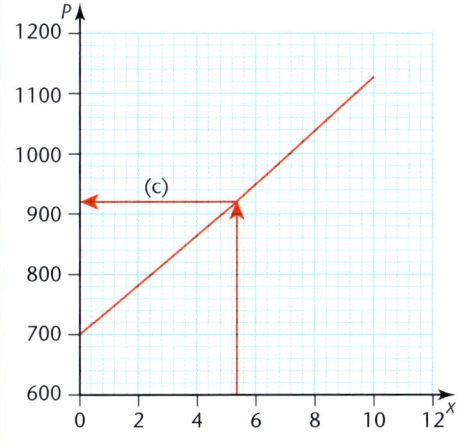

Note
The P-axis has been drawn starting from 600, not 0, so that the graph can be to a larger scale. This makes values easier to read.

c Reading from the graph, when $x = 5.5$, $P = 920$.

Exercise 28.4

1 Plot a graph of $y = 2^x$ for values of x from -2 to 5.
 Use a scale of 2 cm to 1 unit on the x-axis and 2 cm to 5 units on the y-axis.
 Use your graph to estimate
 a the value of y when $x = 3.2$
 b the solution to the equation $2^x = 20$.
2 Plot the graph of $y = 3^x + 5$, for values of x from -2 to 3.
 a How do the values of your graph compare with the graph in Example 28.10?
 b Use your graph to estimate the value of x when $y = 20$.
3 Plot a graph of $y = 1.5^x$ for values of x from -3 to 5.
 Use a scale of 2 cm to 1 unit on both axes.
 Use your graph to estimate
 a the value of y when $x = 2.4$
 b the solution to the equation $1.5^x = 6$.

28 GRAPHS OF FUNCTIONS

4 Without drawing the graph, write down how the graph of $y = 3 \times 2^x$ would look compared to the graph of $y = 2^x$.

5 Zari buys a car for $5000.
Its value, $V, t years after she buys it, is given by this formula
$V = 5000 \times 0.85^t$
   a By what percentage does the value decrease each year?
   b Draw the graph of V against t for $t = 0$ to 7.
   c How long after Zari buys it is the value of the car $2500?

6 Vivek puts some money into a bank account earning compound interest and leaves it there.
After t years, the amount of money in the account, $D, is given by this formula
$D = 500 \times 1.09^t$
   a How much money did Vivek put into the account?
   b Draw the graph of D against t for $t = 0$ to 6.
   c Vivek withdraws the money when it reaches $800.
How many years does he leave it in the account?

7 Irina makes a cup of coffee and leaves it to cool.
The temperature T °C after x minutes is given by this formula
$D = 60 \times 0.9^x + 20$
   a What is the temperature of the coffee when Irina makes it?
   b Draw the graph of T against x for $x = 0$ to 30, plotting at every 5 minutes.
   c Irina leaves her coffee for 4 minutes before drinking it.
What is its temperature then?
   d Estimate the temperature of the coffee if it is left for an hour without being drunk.

Estimating the gradient to a curve

The gradient of a straight line is constant, and gradient $= \dfrac{\text{increase in } y}{\text{increase in } x}$.

The gradient of a curve varies. At any point, the gradient of the curve is the gradient of the tangent to the curve at that point.

So to find the gradient of a curve at a given point, first draw the tangent at the point, and then work out its gradient. Since you cannot know whether the tangent you have drawn is completely accurate, this method only gives an estimate of the gradient of the curve.

Estimating the gradient to a curve

Example 28.12

Question

Estimate the gradient of the curve $y = x^2$ at the point (2, 4).

Solution

Plot the graph and draw a tangent to the curve at (2,4).

$$\text{Gradient} = \frac{\text{increase in } y}{\text{increase in } x}$$
$$= \frac{8-0}{3-1}$$
$$= 4$$

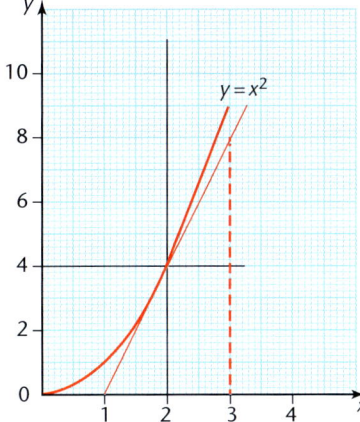

Note

To draw as good a tangent as possible, try to move your ruler so you have an equal amount of space between the curve and the tangent on each side of the given point.

When choosing two points to work out the gradient of the tangent, it is easiest to make the x increase an easy number.

Exercise 28.5

1. Use the graph of $y = x^2 - 4x + 3$ that you drew in Exercise 28.2 question **1**.
   Draw tangents to the curve at $x = 4$ and $x = 1$.
   Use the tangents to estimate the gradient of the curve at $x = 4$ and $x = 1$.

2. Use the graph of $y = x^2 - 3x$ that you drew in Exercise 28.2 question **2**.
   Draw tangents to the curve at $x = 4$ and $x = 1$.
   Use the tangents to estimate the gradient of the curve at $x = 4$ and $x = 1$.

3. Use the graph of $y = x^3 + 5$ that you drew in Exercise 28.3 question **1**.
   Draw tangents to the curve at $x = 2$ and $x = -1$.
   Use the tangents to estimate the gradient of the curve at $x = 2$ and $x = -1$.

4. Use the graph of $y = x^3 - 12x + 2$ that you drew in Exercise 28.3 question **4**.
   Draw tangents to the curve at $x = 0$, $x = 2$ and $x = 3$.
   Use the tangents to estimate the gradient of the curve at $x = 0$, $x = 2$ and $x = 3$.

5. Look at the graph of $y = \frac{1}{x}$ that you drew in Exercise 28.3 question **3**.
   a Identify the x values of the points on the curve where the gradient is -1.
   b Draw tangents to estimate the gradient of the curve at the points where $x = 2$ and $x = -\frac{1}{2}$.

Key points

- When drawing a graph, complete a table of values to work out the points.
- Draw a smooth curve. Look ahead to the next point to help you do this.
- Graphs may be used to solve equations. Rearrange the equation if necessary so that the function you have plotted is on the left-hand side.
- Reciprocal functions have graphs with two separate branches. This is because you cannot divide any number by zero.
- Exponential curves may be used to represent growth and decay problems.
- The gradient of a curve at a given point can be estimated by drawing a tangent to the curve at that point. The gradient of the tangent is the gradient of the curve.

29 SKETCHING CURVES

BY THE END OF THIS CHAPTER YOU WILL BE ABLE TO:

- recognise, sketch and interpret graphs of the following functions: linear, quadratic, cubic, reciprocal and exponential.

CHECK YOU CAN:

- plot graphs
- solve linear equations
- solve quadratic equations by factorising, by using the quadratic formula, and by completing the square
- extract common factors in algebraic expressions
- recognise symmetry.

Families of graphs

You learned about these in the chapter about graphs of functions. In this chapter, you will bring together your knowledge about graphs and solving equations. By analysing a function, you can find information to help you to sketch its graph.

When you sketch a graph, you need to draw the correct shape. You also need to give other information, such as where it crosses the axes.

To find where the graph crosses the y-axis, substitute $x = 0$.

To find where the graph crosses the x-axis, substitute $y = 0$.

Then, solve the resulting equation.

Linear graphs

Example 29.1

Question

Sketch the graph of $2x + 3y = 12$.

Solution

When $x = 0$, $3y = 12$, so $y = 4$. The graph crosses the y-axis at $(0, 4)$.
When $y = 0$, $2x = 12$, so $x = 6$. The graph crosses the x-axis at $(6, 0)$.

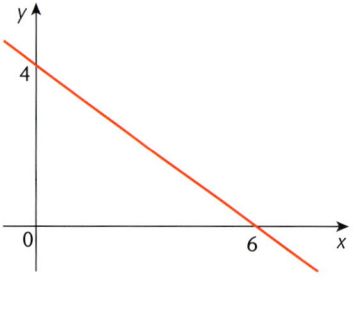

Note
When sketching lines or curves, do not scale the axes. Just label them and put on the sketch the values where the graph crosses the axes. Add the coordinates of any turning points that you find. You should rule straight lines. For curved graphs, aim for the correct general shape.

You will learn more about linear graphs in Chapter 31.

Quadratic graphs

These are graphs of the functions $y = ax^2 + bx + c$.

Their shape is

For $a > 0$

For $a < 0$

They are symmetrical. The line of symmetry goes through the turning point.

Example 29.2

Question

Sketch the graph of $y = 3x - x^2$. Find also the coordinates of its turning point.

Solution

The coefficient of x^2 is -1. So, the turning point is at the top of the graph.
When $x = 0$, $y = 0$.
When $y = 0$, $3x - x^2 = 0$.
Factorising, $x(3 - x) = 0$.
So, $x = 0$ or 3.
Sketch:

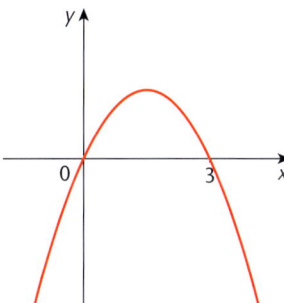

Using symmetry, the x-value of the turning point will be halfway between 0 and 3. This is 1.5.
Substituting, $y = 3 \times 1.5 - 1.5^2 = 2.25$
The turning point is (1.5, 2.25).

Completing the square can be used to find the turning point of a quadratic graph.

29 SKETCHING CURVES

Example 29.3

Question

Express $y = x^2 - 6x + 10$ in completed square form.
Hence, find the coordinates of the turning point of the graph of this function.
Then, sketch the graph.

Solution

$y = x^2 - 6x + 10$
$ = (x - 3)^2 - 9 + 10$
$ = (x - 3)^2 + 1$

The minimum value of a perfect square is 0.
So, the minimum value of $y = 0 + 1 = 1$.
This occurs when $x - 3 = 0$, i.e. when $x = 3$.
The turning point is at (3, 1) and is a minimum.
This means that the graph does not cross the x-axis.
When $x = 0$, $y = 10$.
The line of symmetry goes through the turning point. Its equation is $x = 3$.
Sketch:

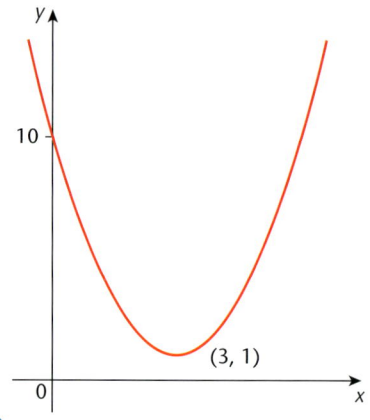

Exercise 29.1

1. Sketch the graph of $2x - 5y = 10$.
2. **a** Sketch the graph of $y = x^2$.
   b On the same pair of axes, sketch the graph of $y = x^2 + 3$.
3. **a** Sketch the graph of $y = x^2 - x - 6$.
   b Use the line of symmetry to find the turning point of this graph.
4. **a** Write $x^2 - 8x + 12$ in the form $(x - a)^2 - b$.
   b Solve $x^2 - 8x + 12 = 0$.
   c Sketch the graph of $y = x^2 - 8x + 12$. Show the coordinates of its turning point.
   d State the equation of the line of symmetry of this graph.

Cubic graphs

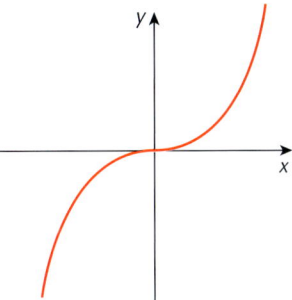

This is a sketch of $y = ax^3$, when a is positive.

Example 29.4

Question
Sketch the graph of $y = 8 - x^3$.

Solution
When $x = 0$, $y = 8$.
When $y = 0$, $x^3 = 8$, so $x = 2$.
When x is negative, y is positive.
When x is greater than 2, y is negative.
Sketch:

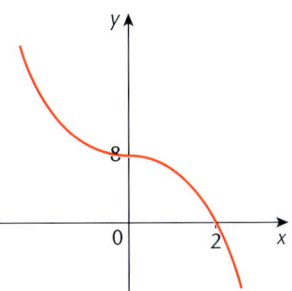

Note
Comparing this with the graph of $y = x^3$, shows that the 8 has moved the graph of $y = x^3$ up 8 units, and the negative x^3 term means that the graph of $y = x^3$ has been reflected in the y-axis.

As you know from Chapter 28, where a cubic graph crosses the axis more than once, the 'double bend' is more pronounced.

Example 29.5

Question
Find where the graph of $y = x^3 + 5x^2 - 6x$ crosses the axes.
Use your result to help you to sketch this graph.

29 SKETCHING CURVES

Solution

When $x = 0$, $y = 0$.
When $y = 0$, $x^3 + 5x^2 - 6x = 0$.
x is a factor of the left-hand side, so
$x(x^2 + 5x - 6) = 0$
$x(x + 6)(x - 1) = 0$
$x = 0, -6$ or 1.
Sketch:

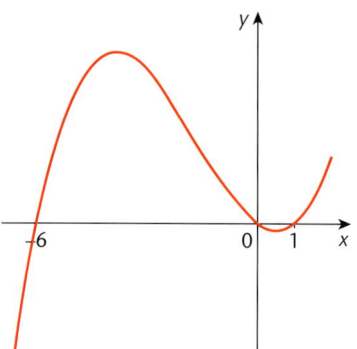

Note
Learn to find the factors of cubic expressions where x is a factor, as in this example.

Reciprocal graphs

You met the shape of these in Chapter 28.

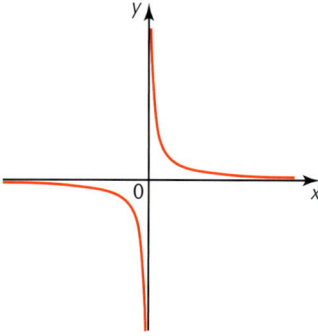

$y = \frac{a}{x}$, where $a > 0$.

In these graphs, the curves do not meet the axes but get closer and closer to them.

Lines where this happens are called asymptotes.

These graphs are symmetrical.

For quadratic graphs, the line of symmetry is vertical.

For graphs of the form $y = \frac{a}{x}$, there are two lines of symmetry. These are the diagonal lines, $y = x$ and $y = -x$.

Example 29.6

Question
Sketch the graph of $y = \frac{-2}{x}$.

Solution
You cannot divide −2 by zero, so $x = 0$ is an asymptote.

As x gets very large, $\frac{-2}{x}$ gets close to zero, so $y = 0$ is an asymptote.

When x is negative, y is positive.
When x is positive, y is negative.
Sketch:

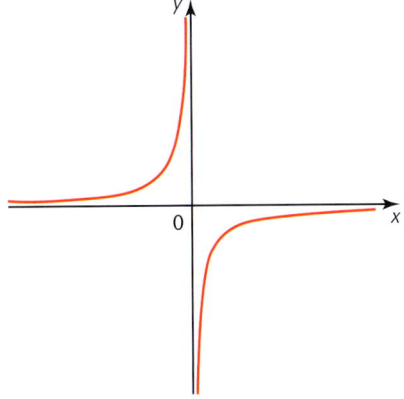

Example 29.7

Question
Sketch the graph of $y = \frac{12}{x} + 4$.

Solution
You cannot divide 12 by zero, so $x = 0$ is an asymptote.
When $y = 0$, $\frac{12}{x} = -4$
So, $12 = -4x$
$x = -3$

As x gets very large, $\frac{12}{x}$ gets close to zero, so y approaches 4.

So, $y = 4$ is the other asymptote.

Sketch:

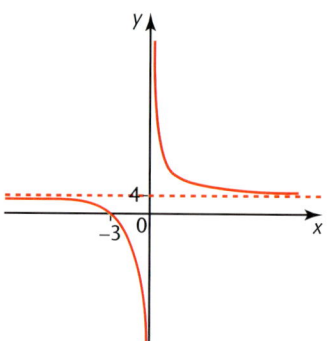

Note
Where an asymptote is not an axis, you should show it on the sketch as a dashed line.

The effect of the $+4$ in the equation of the graph has been to move it up by 4.

Exponential graphs

You met the shape of these in Chapter 28.

This is a sketch of $y = a^x$, where a is a positive integer.

The graph crosses the y-axis at $(0, 1)$.

The x-axis is an asymptote.

Example 29.8

Question
Sketch the graph of $y = 2^x - 8$.

Solution
When $x = 0$, $y = 1 - 8 = -7$.
When $y = 0$, $2^x = 8$. You know that $2^3 = 8$, so $x = 3$.
For $y = 2^x$, the asymptote is $y = 0$. So, for $y = 2^x - 8$, it is $y = -8$.
Sketch:

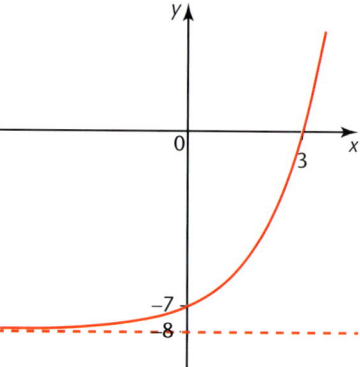

Exponential graphs

Exercise 29.2

1. Sketch the graph of $y = 2x^3 + 3$.
2. a Factorise $x^3 - 7x^2 + 10x$.
   b Sketch the graph of $y = x^3 - 7x^2 + 10x$.
3. a Sketch the graph of $y = \frac{-4}{x}$.
   b State the equations of the lines of symmetry of this graph.
4. Sketch the graph of $y = \frac{6}{x} + 2$.
5. a Sketch these curves on the same axes.
      $y = 2^x$, $y = 3.5^x$ and $y = 5^x$
   b The graph shows three curves, A, B and C.
      Match each curve to its equation
      $y = 6^x$ $y = 1.5^x$ $y = 3^x$

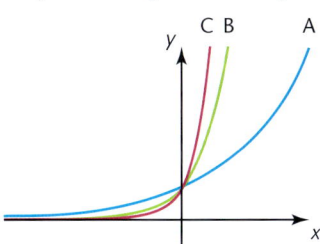

6. Without sketching the graph,
   a write down how the graph of $y = 2^x - 4$ would look compared with the graph of $y = 2^x$
   b find where the graph of $y = 2^x - 4$ crosses the axes.

> ### Key points
> - When you sketch a graph, you need to draw the correct shape. You also need to show where it crosses the axes.
> - To find where the graph crosses the y-axis, substitute $x = 0$. To find where the graph crosses the x-axis, substitute $y = 0$. Then, solve the resulting equation.
> - For quadratic graphs, you may need to use the completed square form $y = (x - a)^2 + b$. This tells you that the turning point is (a, b) and the line of symmetry is $x = a$.
> - An asymptote is a line that a curve does not meet but gets closer and closer to. Reciprocal curves have two asymptotes. Exponential curves have one asymptote.
> - Learn the shapes for the families of graphs in this chapter, so that you can recognise them and sketch them.

30 FUNCTIONS

BY THE END OF THIS CHAPTER YOU WILL BE ABLE TO:
- understand functions, domain and range, and use function notation
- understand and find inverse functions $f^{-1}(x)$
- form composite functions as defined by $gf(x) = g(f(x))$.

CHECK YOU CAN:
- substitute numbers into algebraic expressions
- simplify algebraic expressions
- solve equations
- change the subject of a formula.

Function notation

$f: x \mapsto 2x - 3$ means the **function** that maps x onto $2x - 3$.

This is often written instead as $f(x) = 2x - 3$.

$f(5)$ means the value of the function f when $x = 5$.

To find this, substitute 5 in the expression.

In the example above, this will be $f(5) = 2 \times 5 - 3 = 7$.

Sometimes other letters, such as g and h, are used for functions.

Example 30.1

Question

You are given that $f: x \mapsto 3(x - 2)$.

a Find the value of $f(6)$.
b Solve the equation $f(x) = 6$.
c Find and simplify an expression for $f(x + 4)$.

Solution

a $f(6) = 3(6 - 2)$
$= 3 \times 4$
$= 12$

b $3(x - 2) = 6$
$3x - 6 = 6$
$3x = 12$
$x = 4$

Note
You could start by dividing both sides by 3, giving $x - 2 = 2$.

c To find $f(x + 4)$, substitute $x + 4$ instead of x in the expression for f.
$f(x + 4) = 3(x + 4 - 2)$
$= 3(x + 2)$

Domain and range

In the two previous functions, the starting value of x can be any real number. The set of starting values is called the **domain**. The set of finishing values of $f(x)$ is called the **range**.

Sometimes, we want to look at the effect of using a function on a limited domain. Mapping diagrams are a useful way of representing this.

Domain and range

Example 30.2

Question
Draw a mapping diagram for g(x) = 5x + 2 for the domain {−1, 0, 1, 2, 3}.

Solution
g(−1) = 5 × (−1) + 2 = −3.
Similarly, g(0) = 2, g(1) = 7,
g(2) = 12, g(3) = 17.
The range is {−3, 7, 12, 17}
Mapping diagram:

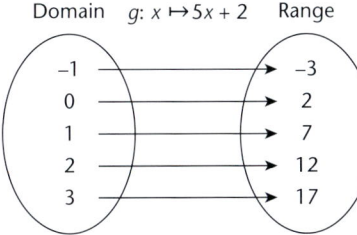

Note
Finding the values of g(x) = 5x + 2 for a mapping diagram is like completing a table for y = 5x + 2.

Example 30.3

Question
Draw a mapping diagram for the function h(x) = 2x² + 3 with domain {−5, −2, 0, 2, 5}.

Solution

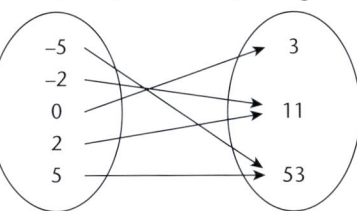

Example 30.3 shows that, for some functions, more than one number in the domain may map onto a number in the range. Functions can be 'one-to-one' or 'many-to-one' but never 'one-to-many'.

Exercise 30.1

1 When f: $x \mapsto 4x + 3$, find
   a f(1)   **b** f(−2)   **c** f(0).

2 When g(x) = 6 − 2x, find
   a g(2)   **b** g(−3)   **c** g(0.5).

30 FUNCTIONS

3 When $h(x) = \frac{x+6}{x}$, find
   a h(3)   b h(−4)   c $h\left(\frac{1}{2}\right)$.

4 When $f(x) = x^2 + 7$, find
   a f(4)   b f(−3)   c $f\left(\frac{1}{2}\right)$.

5 Draw a mapping diagram for the function $f(x) = 3(x + 2)$ with domain {0, 1, 2, 3}.

6 a Draw a mapping diagram for the function $g(x) = \frac{2}{x-4}$ with domain {0, 2, 6}.
   b Explain why the domain for this function cannot include $x = 4$.

7 When $g: x \mapsto 5(2 + 3x)$,
   a find g(−4)
   b solve g(x) = 0
   c find an expression for g(−x).

8 When $h(x) = \frac{12}{x+1}$,
   a find h(2)
   b solve h(x) = −6
   c find and simplify an expression for h(2x − 1).

9 When $f(x) = 2x − 3$ and $g(x) = 7 − 2x$,
   a find f(4)
   b solve g(x) = 0
   c solve f(x) = g(x).

10 When $h(x) = x^2 − 5$,
   a find h(4)
   b find and simplify an expression for h(2x)
   c find the two values of x for which h(x) = 31.

11 When $f(x) = x(x − 5)$,
   a find f(4)
   b solve f(x) = 0
   c solve f(x) = 14.

12 When $g(x) = \frac{2x+3}{5}$ and $h(x) = 4x − 1$,
   a find g(2)
   b solve h(x) = 6
   c solve g(x) = h(x).

Inverse functions

The **inverse** of a function f is the function that maps all the values of f(x) back to their original x values.

The inverse of a function is written as $f^{-1}(x)$.

For example, when $f(x) = x + 2$, $f^{-1}(x) = x − 2$.

To find the inverse function, write y in place of f(x).	$y = x + 2$
Rearrange the formula to make x the subject.	$x = y − 2$
Replace x with $f^{-1}(x)$ and y with x to give the expression for the inverse function.	$f^{-1}(x) = x − 2$

Inverse functions

Example 30.4

Question
When $f(x) = 3x - 5$,
a find $f^{-1}(x)$
b find $f^{-1}(7)$.

Solution

a $y = 3x - 5$ Write y in place of $f(x)$.
  $y + 5 = 3x$ Rearrange the formula to make x the subject.
  $x = \dfrac{y+5}{3}$
  $f^{-1}(x) = \dfrac{x+5}{3}$ Replace x with $f^{-1}(x)$ and y with x.

b $f^{-1}(7) = \dfrac{7+5}{3} = 4$

Note
Finding $f^{-1}(7)$ is the same as finding the value of x when $f(x) = 7$. We could solve the equation $3x - 5 = 7$ if we had not found $f^{-1}(x)$ first.

Mapping diagrams may be used to illustrate inverse functions. The range of a function is the domain of its inverse.

Example 30.5

Question
a Draw a mapping diagram for the function $h(x) = 3(x + 4)$ and domain $\{-1, 0, 1, 2, 3\}$.
b Find $h^{-1}(x)$.
c Draw the mapping diagram for this inverse function, using as the domain the range in part **a**.

Solution

a Domain $h: x \mapsto 3(x+4)$ Range

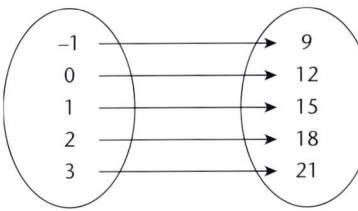

b $h(x) = 3(x + 4)$
  $y = 3(x + 4)$
  $\dfrac{y}{3} = x + 4$
  $x = \dfrac{y}{3} - 4$
  $h^{-1}(x) = \dfrac{x}{3} - 4$

c Domain $h^{-1}: x \mapsto \dfrac{x}{3} - 4$ Range

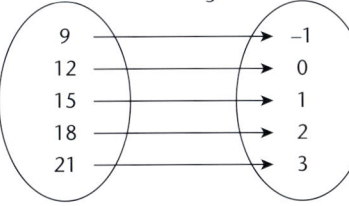

Note
Check that you have rearranged correctly by substituting a number from the new domain into $h^{-1}(x)$. Make sure that the result matches with its original value from $h(x)$.

30 FUNCTIONS

In Example 30.3 you looked at the function h(x) = 2x² + 3, where both 2 and −2 in the domain had the same value in the range. The inverse function must have only one outcome for each member of its domain. So, we need to use the positive square root only when finding $h^{-1}(x)$ for functions like this one:

$h(x) = 2x^2 + 3$

$y = 2x^2 + 3$

$y - 3 = 2x^2$

$\frac{y-3}{2} = x^2$

$x = \sqrt{\frac{y-3}{2}}$

$h^{-1}(x) = \sqrt{\frac{x-3}{2}}$

The domain for this inverse function is $\{x: x \geqslant 3\}$. Its range is $\{y: y \geqslant 0\}$.

Exercise 30.2

1. When f: $x \mapsto 4x + 3$,
   a. solve f(x) = 5
   b. find $f^{-1}(x)$
   c. use $f^{-1}(x)$ to find $f^{-1}(5)$.

2. When g(x) = 6 − 2x,
   a. find $g^{-1}(x)$
   b. use $g^{-1}(x)$ to find $g^{-1}(4)$ and check your work by substituting your answer into g(x).

3. When h: $x \mapsto 3(x − 6)$, find
   a. $h^{-1}(x)$
   b. $h^{-1}(9)$
   c. $h^{-1}(0)$.

4. When $f(x) = \frac{2x+1}{5}$, find
   a. $f^{-1}(x)$
   b. $f^{-1}(3)$
   c. $f^{-1}\left(\frac{4}{5}\right)$.

5. When $g(x) = \frac{x}{2} - 7$, find
   a. $g^{-1}(x)$
   b. $g^{-1}(-4)$
   c. $g^{-1}(12)$.

6. When $h(x) = \frac{12}{x+1}$, find
   a. $h^{-1}(x)$
   b. $h^{-1}(-4)$
   c. $h^{-1}(2)$.

7. When f(x) = 3(2x − 5), find
   a. $f^{-1}(9)$
   b. $f^{-1}(0)$
   c. $f^{-1}(-18)$.

8. When g(x) = 7 − 3x, find
   a. $g^{-1}(1)$
   b. $g^{-1}(-5)$
   c. $g^{-1}(-293)$.

9. Find the inverse of each of these functions.
   a. $f(x) = \frac{6x-1}{3}$
   b. g(x) = 2 − 5x
   c. $h(x) = \frac{6}{x}$
   d. $f(x) = \frac{3x}{4} - 7$
   e. g(x) = 4(2x + 1)
   f. $h(x) = \frac{2x+1}{x}$

Composite functions

Functions can be combined.

gf(4) means first finding f(4) and then g(the result).

The range for f(x) becomes the domain for g(x) when working out gf(x).

f: $x \mapsto f(x)$ g: $x \mapsto g(x)$

$x \longrightarrow f(x) \longrightarrow gf(x)$

Example 30.6

Question

$f(x) = 2x + 1$ and $g(x) = 3x - 2$

a Find gf(5).
b Find and simplify gf(x).

Solution

a $f(5) = 2 \times 5 + 1 = 11$
   $gf(5) = g(11) = 3 \times 11 - 2 = 31$
b $f(x) = 2x + 1$
   $gf(x) = g(2x + 1)$
   $\qquad = 3(2x + 1) - 2$ use (2x + 1) instead of x in the function g(x)
   $\qquad = 6x + 3 - 2$
   $\qquad = 6x + 1$

Note
You can substitute $x = 5$ into the answer in **b** to check the answer to **a**.

Exercise 30.3

1 When $f(x) = 3x + 4$ and $g(x) = (x + 2)^2$,
   a find
      i gf(2)   ii gf(−1)   iii fg(2)   iv fg(−1).
   b find and simplify
      i gf(x)   ii fg(x).

2 When $f(x) = \dfrac{2}{x+1}$ and $g(x) = 3x + 2$,
   a find
      i gf(1)   ii gf(5)   iii fg(1)   iv fg(5).
   b find and simplify
      i gf(x)   ii fg(x).

3 You are given that $g(x) = 3(x + 5)$ and $h(x) = x^2$.
   a Find
      i gh(2)   ii hg(5).
   b Solve the equation hg(x) = 36.

4 You are given that $f : x \mapsto 3x^2 - 2$ and $g : x \mapsto 4x + 5$.
  a Find
    i gf(2)   ii fg(−2).
  b i Find and simplify gf(x).
    ii Solve the equation gf(x) = 105.

5 When $h(x) = 2x + 1$ and $f(x) = \dfrac{3}{x - 2}$
  a find
    i hf(6).   ii fh(2).
  b find hf(x). Give your answer as a fraction in its simplest form.

6 You are given that $f(x) = \dfrac{12}{2x - 1}$ and $g(x) = x^2 + 4$.
  a Find
    i gf(2)   ii fg(−1).
  b Find fg(x). Give your answer as a fraction in its simplest form.

7 When $f(x) = \dfrac{2x + 1}{5}$ and $g(x) = 10x - 3$,
  a find
    i gf(x).   ii fg(x).
  b solve the equations
    i gf(x) = 19   ii fg(x) = 3.

8 When $h(x) = 2(x + 1)^2$ and $f(x) = \dfrac{6}{x + 2}$
  a find
    i hf(1).   ii fh(2).
  b i find fh(x), simplifying your answer
    ii solve the equation fh(x) = $\dfrac{3}{2}$.

> **Key points**
> - f: = x → 3x + 1 means the function that maps x onto 3x + 1. It may also be written as f(x) = 3x + 1.
>   f(2) means the value of the function when x = 2.
> - The domain of a function f is the set of starting values of x.
>   The range of a function is the set of values of f(x).
> - Mapping diagrams may be used to represent the effect of a function.
> - The inverse of the function f is written as f⁻¹(x). The inverse maps all the values of x back to their original values.
> - To find the inverse function, write y in place of f(x).
>   Rearrange the formula to make x the subject.
>   Replace x with f⁻¹(x) and y with x to give the expression for the inverse function.
> - Functions can be combined. gf(4) means finding first f(4) then g(the result). The range for f(x) becomes the domain for g(x) when working out gf(x).

31 COORDINATE GEOMETRY

The gradient of a straight-line graph

You have already found the **gradient** of a straight line as

$$\text{Gradient} = \frac{\text{increase in } y}{\text{increase in } x}$$

Remember that lines with a positive gradient slope up to the right and lines with a negative gradient slope down to the right.

If you know two points on the line then you do not have to draw it to work out the gradient.

For a line through two points (x_1, y_1) and (x_2, y_2), gradient $= \frac{y_2 - y_1}{x_2 - x_1}$.

Example 31.1

Question
Find the gradient of the line joining the points (3, 5) and (8, 7).

Solution
Gradient $= \frac{7-5}{8-3} = \frac{2}{5}$

Note
You can use the points the other way round in the formula. The answer will be the same.

Exercise 31.1

1 Find the gradient of these lines.

a (7, 4)

b passes through (0, 5) and (2, 0)

c (10, 5), passes through (0, 3)

BY THE END OF THIS CHAPTER YOU WILL BE ABLE TO:

- use and interpret Cartesian coordinates in two dimensions
- draw straight-line graphs for linear equations
- find the gradient of a straight line
- calculate the gradient of a straight line from the coordinates of two points on it
- calculate the length of a line segment
- find the coordinates of the midpoint of a line segment
- interpret and obtain the equation of a straight-line graph
- find the gradient and equation of a straight line parallel to a given line
- find the gradient and equation of a straight line perpendicular to a given line.

CHECK YOU CAN:

- use coordinates in two dimensions
- draw a straight–line graph, given its equation
- use Pythagoras' theorem to calculate lengths
- substitute into and rearrange simple formulas
- understand the meaning of the terms *parallel* and *perpendicular*.

31 COORDINATE GEOMETRY

2 Calculate the gradient of the line joining each of these pairs of points.
   a (1, 5) and (3, 5)   b (−1, 1) and (3, 2)   c (−2, 6) and (0, 4)
3 Calculate the gradient of the line joining each of these pairs of points.
   a (0.6, 3) and (3.6, −9)   b (2.5, 4) and (3.7, 4.9)
   c (2.5, 7) and (4, 2.2)

Line segments

A line can be extended in either direction forever. It is infinite.

A **line segment** is the part of a line between two points.

The midpoint of a line segment

The coordinates of the midpoint of a line segment are the mean of the coordinates of the two end points.

> The midpoint of the line joining the points (x_1, y_1) and (x_2, y_2)
> is $\left(\dfrac{x_1 + x_2}{2}, \dfrac{y_1 + y_2}{2}\right)$.

Example 31.2

Question

Find the coordinates of the midpoint of the line segment with these end points.
   a A(2, 1) and B(6, 7)
   b C(−2, 1) and D(2, 5)

Solution

a Midpoint = $\left(\dfrac{2+6}{2}, \dfrac{1+7}{2}\right)$

   = (4, 4)

b Midpoint = $\left(\dfrac{-2+2}{2}, \dfrac{1+5}{2}\right)$

   = (0, 3)

The length of a line segment

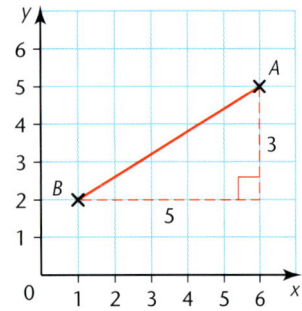

You can use Pythagoras' theorem to find the length of a line segment.

The length of the horizontal side is 6 − 1 = 5.

The length of the vertical side is 5 − 2 = 3.

You can then use Pythagoras' theorem to work out the length of AB.

$AB^2 = 5^2 + 3^2$

$AB^2 = 25 + 9 = 34$

$AB = \sqrt{34}$

$AB = 5.83$ units to 2 decimal places.

The general form of the equation of a straight line

The length of a line segment with end points (x_1, y_1) and (x_2, y_2) is $\sqrt{(x_1-x_2)^2 + (y_1-y_2)^2}$.

Example 31.3

Question

A is the point $(-5, 4)$ and B is the point $(3, 2)$.
Find the length AB.

Solution

$AB = \sqrt{(-5-3)^2 + (4-2)^2} = \sqrt{(-8^2) + 2^2} = \sqrt{64+4} = \sqrt{68}$

$= 8.25$ units to 2 decimal places.

Exercise 31.2

1. For each of these line segments
   i   find the coordinates of the midpoint
   ii calculate the length.
   You can draw a diagram to help you.
   a $A(1, 4)$ and $B(1, 8)$ b $C(1, 5)$ and $D(7, 3)$
   c $E(2, 3)$ and $F(8, 6)$ d $G(3, 7)$ and $H(8, 2)$
   e $I(-2, 3)$ and $J(4, 1)$ f $K(-4, -3)$ and $L(-6, -11)$

2. A is the point $(-2, -5)$.
   The midpoint of AB is $(3, 1)$.
   Find the coordinates of B.

3. A is the point $(-1, 5)$, B is the point $(7, 3)$ and C is the point $(1, -7)$.
   a Find the length of AC.
   b Find the length of the line joining the midpoints of BA and BC.
   c What do you notice?

The general form of the equation of a straight line

This line has gradient m and crosses the y-axis at the point $(0, c)$.

c is called the **y-intercept** because it is where the line intercepts, or crosses, the y-axis.

The equation of the line is $y = mx + c$.

The equation of any straight line is of the form $ax + by = k$.

This can be rearranged to give $y = mx + c$.

From this, the gradient and y-intercept can be found.

31 COORDINATE GEOMETRY

Example 31.4

Question

a The equation of a straight line is $y = 7$.
Find its gradient and y-intercept.

b The equation of a straight line is $5x + 2y = 10$.
Find its gradient and y-intercept.

Solution

a This can be written, $y = 0x + 7$.
So, gradient is 0 and y-intercept is 7.
[$y = 7$ is a horizontal straight line]

b $5x + 2y = 10$
$2y = -5x + 10$ Rearrange the equation into the form $y = mx + c$.
$y = -2.5x + 5$
So the gradient is -2.5 and the y-intercept is 5.

Finding the equation of a straight line

You can also use this method in reverse to find the equation of a line from its graph.

Example 31.5

Question

Find the equation of this straight line.

Solution

The gradient of a line, $m = \frac{\text{increase in } y}{\text{increase in } x}$

$= \frac{6}{2} = 3$

The line passes through $(0, -1)$, so the y-intercept, c, is -1.
So the equation of the line is $y = 3x - 1$.

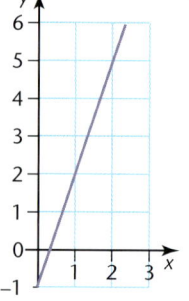

Example 31.6

Question

Find the equation of the line that passes through the points $(4, 6)$ and $(6, 2)$.

Solution

First find the gradient of the line.

Gradient, $m = \frac{y_2 - y_1}{x_2 - x_1} = \frac{2-6}{6-4} = -\frac{4}{2} = -2$

To find the y-intercept, c, substitute the coordinates of either of the given points and the gradient into the general equation of a line.

$y = mx + c$
$y = -2x + c$ The gradient is -2.
$6 = -2 \times 4 + c$ Substitute the coordinates of one of the given points.
$6 = -8 + c$
$6 + 8 = c$
$c = 14$ Solve to find c.
So $y = -2x + 14$
Or $2x + y = 14$

Parallel and perpendicular lines

Exercise 31.3

1. Find the gradient of each of these lines and their *y*-intercept.
   - a $y = 3$
   - b $y = 2x$
   - c $y = 3x - 2$
   - d $y = 5x - 3$
   - e $y = 2 + 5x$
   - f $y = 7 + 2x$
   - g $y = 7 - 2x$
   - h $y = 9 - 3x$

2. Find the gradient of each of these lines and their *y*-intercept.
   - a $y + 4 = 0$
   - b $y + 3x = 0$
   - c $y + 2x = 5$
   - d $y - 5x = 1$
   - e $4x + 2y = 7$
   - f $3x + 2y = 8$
   - g $6x + 5y = 10$
   - h $2x + 5y = 15$

3. Write down the equation of each of these straight lines.
   - a Gradient 3 and passing through (0, 2)
   - b Gradient −1 and passing through (0, 4)
   - c Gradient 5 and passing through (0, 0)
   - d Gradient 4 and passing through (0, −1)
   - e Gradient −2 and passing through (0, 5)
   - f Gradient 3 and passing through the origin

4. Find the equation of each of these lines.

   a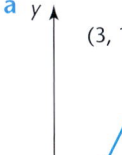
   b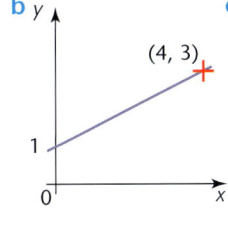
   c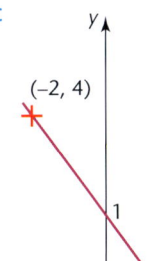
   d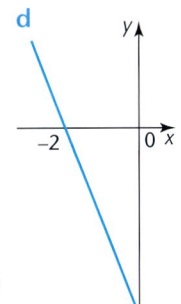

5. Find the equation of each of these lines.
   - a A line with gradient $-\frac{3}{4}$, passing through the point (3, 0).
   - b A line passing through (1, 4) and (4, 7).
   - c A line passing through (2, 3) and (5, 9).
   - d A line passing through (−1, 5) and (3, −7).
   - e A line passing through (3, 1) and (6, −1).

Parallel and perpendicular lines

Parallel lines have the same gradient.

There is also a connection between the gradients of **perpendicular** lines.

If a line has gradient m, a line perpendicular to this has gradient $-\frac{1}{m}$.

The product of the two gradients is −1.

> **Note**
> $m \times \dfrac{-1}{m} = -1$

31 COORDINATE GEOMETRY

Example 31.7

Question

Find the equation of the line that is parallel to the line $2y = 3x + 4$ and passes through the point (3, 2).

Solution

The gradient of the line $2y = 3x + 4$ is 1.5.
So the equation of a line parallel to this is $y = 1.5x + c$.

$y = 1.5x + c$ — To find the y-intercept, c, substitute the coordinates of the given point into the equation.

$2 = 1.5 \times 3 + c$
$2 = 4.5 + c$
$c = 2 - 4.5$
$c = -2.5$
So, $y = 1.5x - 2.5$
Or, $2y = 3x - 5$ — Multiply both sides of the equation by 2 to eliminate the decimals.

Example 31.8

Question

Find the equation of the line that crosses the line $y = 2x - 5$ at right angles at the point (3, 1).

Solution

The gradient of the line $y = 2x - 5$ is 2.
The gradient of a line perpendicular to this is $-\frac{1}{m}$.
So the gradient of the perpendicular line is $-\frac{1}{2}$.
So the equation of the perpendicular line is $y = -\frac{1}{2}x + c$.

$y = -\frac{1}{2}x + c$ — To find the y-intercept, c, substitute the coordinates of the given point into the equation.

$1 = -\frac{1}{2} \times 3 + c$
$1 = -\frac{3}{2} + c$
$c = 1 + \frac{3}{2}$
$c = 2\frac{1}{2}$
So $y = -\frac{1}{2}x + 2\frac{1}{2}$
$x + 2y = 5$ — Multiply both sides of the equation by 2 to eliminate the fractions and rearrange to give a positive x term.

> A line parallel to $y = mx + c$ will be of the form $y = mx + d$, where c and d are constants.
>
> A line perpendicular to $y = mx + c$ will be of the form $y = -\frac{1}{m}x + e$, where c and e are constants.

Exercise 31.4

1. In the diagram, *AB* and *BC* are two sides of a square, *ABCD*.
   a. Find the equation of the line *AD*.
   b. Find the equation of the line *DC*.
   c. Find the coordinates of *D*.
2. Which of these lines are
   a. parallel
   b. perpendicular?
   $y = 4x + 3$ $2y - 3x = 5$ $6y + 4x = 1$ $4x - y = 5$
3. Find the equation of the line that passes through (1, 5) and is parallel to $y = 3x - 3$.
4. Find the equation of the line that passes through (0, 3) and is parallel to $3x + 2y = 7$.
5. Find the equation of the line that passes through (1, 5) and is perpendicular to $y = 3x - 1$.
6. Find the equation of the line that passes through (0, 3) and is perpendicular to $3x + 2y = 7$.
7. Find the gradient of the perpendicular bisector of the line joining each of these pairs of points.
   a. (1, 2) and (5, −6)
   b. (2, 4) and (3, 8)
   c. (−2, 1) and (2, −4)
8. Two lines cross at right angles at the point (5, 3). One passes through (6, 0). What is the equation of the other line?

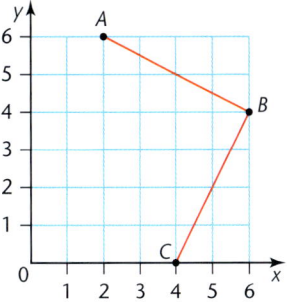

Note
A line that cuts another line in half at 90° is called its **perpendicular bisector**. You will meet perpendicular bisectors again in Chapter 32.

Key points

- For a line segment joining the points (x_1, y_1) and (x_2, y_2)

  Gradient $= \dfrac{y_1 - y_2}{x_1 - x_2}$

  Midpoint $= \left(\dfrac{x_1 + x_2}{2}, \dfrac{y_1 + y_2}{2} \right)$

  Length $= \sqrt{(x_1 - x_2)^2 + (y_1 - y_2)^2}$

- The general form of the equation of a straight line in a coordinate plane is $y = mx + c$, where m is the gradient and c is the intercept on the y-axis.
- When the gradient of the line is known and the y-intercept is not, it can be found by substituting the coordinates of a point on the line into the equation and rearranging.
- Parallel lines have the same gradient.
- A line perpendicular to a line with gradient m has gradient $-\dfrac{1}{m}$.

REVIEW EXERCISE 6

1 It is given that 100 dollars ($) is equivalent to 56 pounds (£).

Ch 27

   a Use this information to draw a conversion graph between pounds and dollars on the grid below.

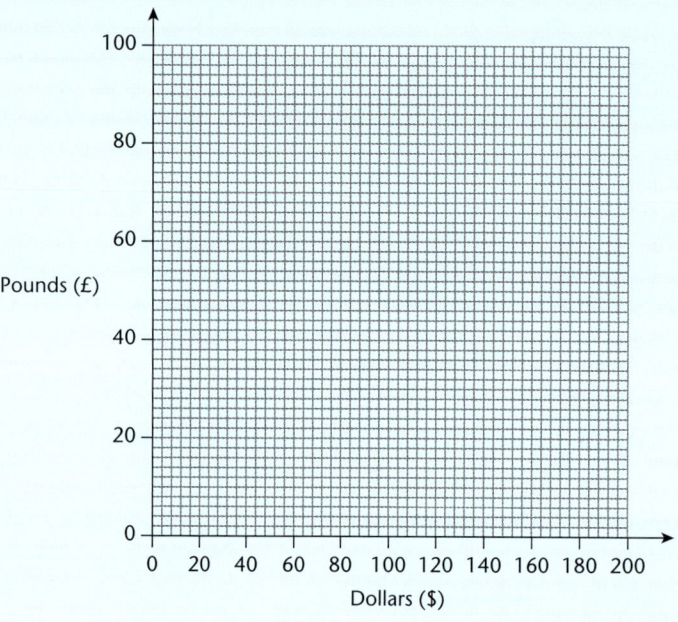

[1]

   b Use your graph to convert $64 to pounds. [1]

Cambridge O Level Mathematics Syllabus D (4024) Paper 12 Q3, June 2016

2 $f(x) = x - 3 \quad g(x) = x^2 + 1$

Ch 30

   a Find $f(-5)$. [1]

   b Find m given that $g(m - 3) = 17$. [3]

Cambridge O Level Mathematics Syllabus D (4024) Paper 11 Q24b, June 2017

Ch 30   **3** It is given that $f(x) = 3x - 5$ and $g(x) = \frac{x+2}{3}$.

   a Find $f^{-1}(x)$. [3]

   b Find and simplify $fg(x)$. [2]

Ch 31   **4** A is the point $(3, -2)$ and B is the point $(-5, 8)$.

   a Find the midpoint of the line joining A and B. [1]

   b Find the length of the line segment AB. [2]

5 Line L is perpendicular to the line $y = -\frac{1}{2}x + 3$.

Ch 31

   Line L passes through the point $(8, 9)$.

   Find the equation of line L. [3]

Cambridge O Level Mathematics Syllabus D (4024) Paper 11 Q21, November 2020

6 The diagrams are sketches of:

A $y = 2^x$ **B** $y = x^3 - x^2$ **C** $y = -x^2 + 2$ **D** $y = -2x + 2$ **E** $y = -\dfrac{2}{x}$

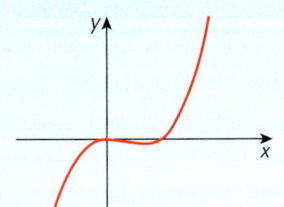

Sketch 1

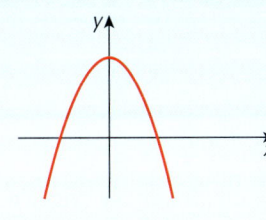

Sketch 2

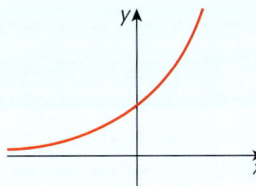

Sketch 3

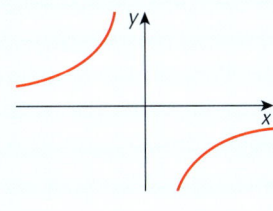

Sketch 4

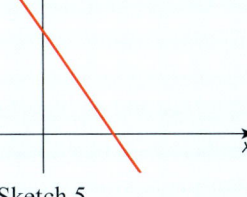

Sketch 5

Match each sketch with its correct equation. [3]

7 a i The points $(4, -3)$ and $(0, 5)$ lie on the line L.
Find the equation of line L. [2]

ii The line M is parallel to line L and passes through the point $(-2, 3)$.
Find the equation of line M. [2]

b The table below shows some values of x and the corresponding values of y for $y = x + \dfrac{3}{x} - 3$.

x	0.5	1	1.5	2	3	4	5	6
y	3.5	1	0.5	0.5	1	1.75	2.6	

i Complete the table. [1]

ii Using a scale of 2 cm to 1 unit on both axes, draw a horizontal x-axis for $0 \leq x \leq 7$ and a vertical y-axis for $0 \leq y \leq 4$.

Draw the graph of $y = x + \dfrac{3}{x} - 3$ for $0.5 \leq x \leq 6$. [3]

iii By drawing a tangent, estimate the gradient of the curve at $(1, 1)$. [2]

iv Use your graph to solve the equation $x + \dfrac{3}{x} = 5$. [2]

Cambridge O Level Mathematics Syllabus D (4024) Paper 21 Q7, November 2017

32 GEOMETRICAL TERMS

BY THE END OF THIS CHAPTER YOU WILL BE ABLE TO:

- use and interpret the following geometrical terms: point; vertex; line; plane; parallel; perpendicular; perpendicular bisector; bearing; right angle; acute, obtuse and reflex angles; interior and exterior angles; similar; congruent; scale factor
- use and interpret the vocabulary of: triangles; special quadrilaterals; polygons; nets; solids
- use and interpret the vocabulary of a circle.

CHECK YOU CAN:

- read scales
- use a ruler and compasses
- draw diagrams accurately on a square grid.

Dimensions

A **point** has no dimensions.

A **line** has length but no width. It has one dimension.

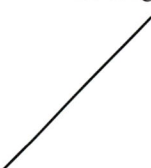

A **plane** has length and width but no height.

It is a flat **surface** with two dimensions.

A shape which has length, width and height has three dimensions.

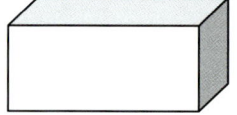

Angles

To describe and measure angles, we use a scale marked in degrees.

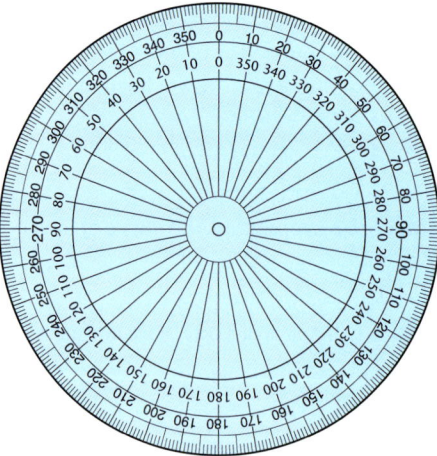

On this scale, one whole turn is equal to 360 degrees. This is written as 360°.

So a half turn is equal to 180° and a quarter turn is equal to 90°.

An angle of a half turn (180°) is a straight line.

Angles of a quarter turn (90°) are called **right angles**.

Angles of less than 90° are called **acute angles**.

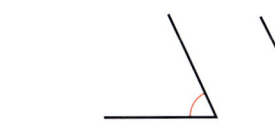

Angles of between 90° and 180° are called **obtuse angles**.

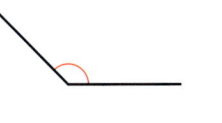 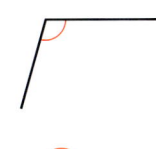

Angles of more than 180° are called **reflex angles**.

Lines

Two lines that never meet are **parallel** to each other.

Arrows are used to show that two lines are parallel.

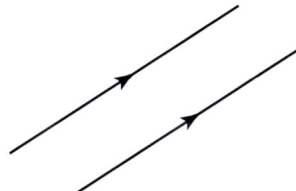

Two lines that meet or cross at 90° are **perpendicular** to each other.

A line that cuts another line in half at 90° is called its **perpendicular bisector**.

Exercise 32.1

1 For each of these angles, say whether it is acute, obtuse, reflex or a right angle.

a 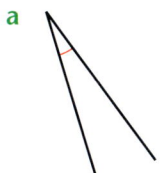   b    c

32 GEOMETRICAL TERMS

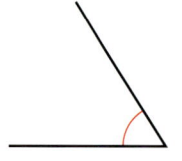 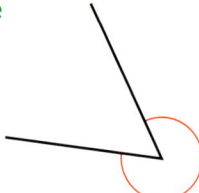 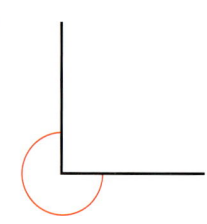

d   e   f

2 What sort of angle is each of these sizes?
   a 145°   **b** 86°   **c** 350°   **d** 190°
   e 126°   **f** 226°   **g** 26°   **h** 90°

3 *ABCD* is a rectangle.

Which side is

   a perpendicular to *AB*   **b** parallel to *DC*?

Bearings

Bearings are
- used to describe a direction
- measured in degrees from north clockwise
- made up of three digits.

Example 32.1

Question

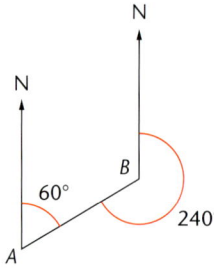

Find the bearing of
  a *B* from *A*   **b** *A* from *B*.

Solution

a The angle from the north line at *A* to the point *B* is 60° so the bearing of *B* from *A* is 060°.

b The angle from the north line at *B* to the point *A* is 240° so the bearing of *A* from *B* is 240°.

Triangles

Exercise 32.2

1. Write down the bearings for these directions.
   - **a** due east
   - **b** due south
   - **c** due west
   - **d** north-east
   - **e** south-east

2. Write these angles as bearings.
   - **a** 24°
   - **b** 101°
   - **c** 3°

3. Find the bearings from O of the points A, B, C, D.

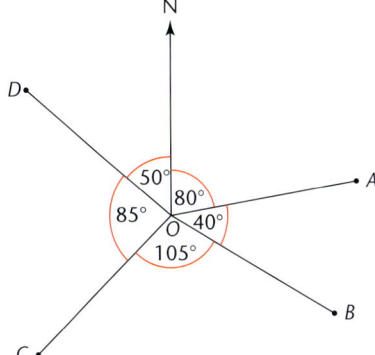

Triangles

A **triangle** is a three-sided shape.

These are **right-angled triangles**.

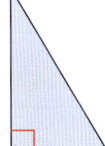   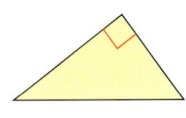

These are **isosceles triangles**.

They have two sides the same length.

They have two angles the same size.

The equal sides are marked with lines.

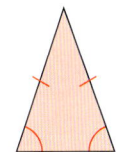   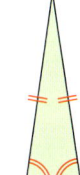

These are **equilateral triangles**.

All their sides are the same length.

All their angles are the same size.

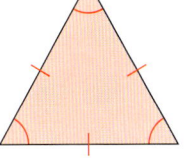   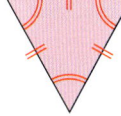

These triangles are **scalene triangles**.

In scalene triangles all the sides and all the angles are different.

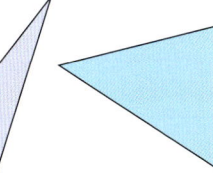   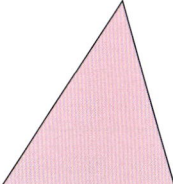

Quadrilaterals

A **quadrilateral** is a four-sided shape.

Here are seven different types.

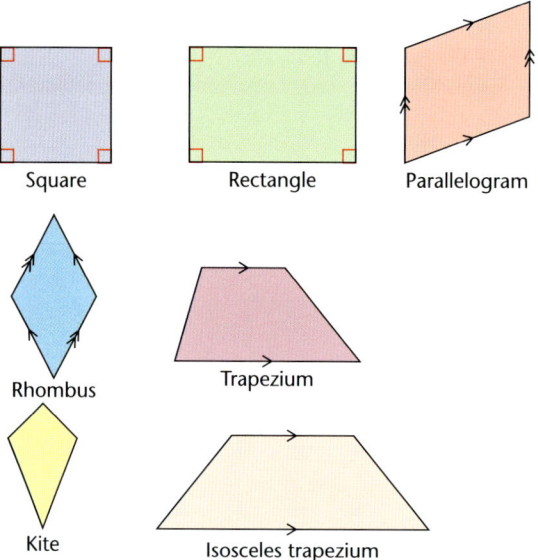

The table shows the names and geometrical properties of the seven types of quadrilaterals shown in the diagram.

Name	Angles	Sides		Diagonals
		Lengths	Parallel	
Square	All 90°	All equal	Opposite sides parallel	Equal length Bisect at 90°
Rectangle	All 90°	Opposite sides equal	Opposite sides parallel	Equal length Bisect but not at 90°
Parallelogram	Opposite angles equal	Opposite sides equal	Opposite sides parallel	Not equal Bisect but not at 90°
Rhombus	Opposite angles equal	All equal	Opposite sides parallel	Not equal Bisect at 90°
Trapezium	Can be different	Can be different	One pair of sides parallel	Nothing special
Kite	One pair of opposite angles equal	Two pairs of adjacent sides equal	None parallel	Not equal Only one bisected by the other They cross at 90°
Isosceles trapezium	Two pairs of adjacent angles equal	One pair of opposite sides equal	Other pair of opposite sides parallel	Equal length Do not bisect or cross at 90°

> **Example 32.2**
>
> **Question**
> Which quadrilaterals have
> a both pairs of opposite sides equal
> b just one pair of opposite angles equal?
>
> **Solution**
> a Square, rectangle, parallelogram, rhombus
> b Kite

Polygons

A **polygon** is a many-sided shape.

Here are some common ones, apart from the triangles and quadrilaterals you have met already.

Shape	Pentagon	Hexagon	Octagon	Decagon
Number of sides	5	6	8	10

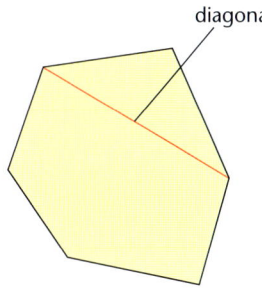

A line joining two corners of a polygon is called a **diagonal**.

When the sides of a polygon are all the same length and the angles of the polygon are all the same, it is called a **regular polygon**. Otherwise, the polygon is an **irregular polygon**.

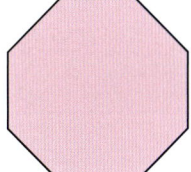 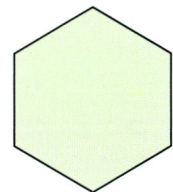 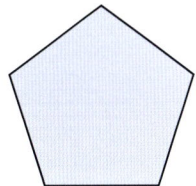

The angles inside a polygon are called **interior angles**.

If a side of the shape is continued outside the shape, the angle made is called an **exterior angle**.

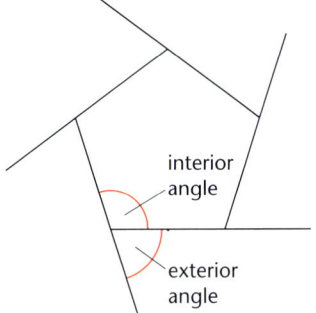

Exercise 32.3

1. Draw any pentagon.
   Draw diagonals across the pentagon from each vertex (corner of the pentagon) to another vertex.
   How many diagonals can you draw in the pentagon altogether?
2. Which quadrilaterals have all four angles the same?
3. Which quadrilaterals have opposite sides the same, but not all four sides the same?
4. A rectangle can be described as a special type of parallelogram. What else is true about a rectangle?
5. Which quadrilaterals have all four sides the same?
6. Which quadrilaterals have opposite angles the same, but not 90°?
7. A rhombus can be described as a special type of parallelogram. What else is true about a rhombus?
8. Which quadrilaterals have diagonals that cross at 90°?
9. A quadrilateral has diagonals the same length.
   What types of quadrilateral could it be?
10. List all the quadrilaterals that have both pairs of opposite sides parallel and equal.

Solids

Here is a **cuboid**.

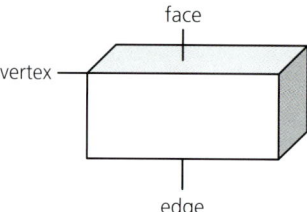

On this diagram of the cuboid, there are three hidden **edges**, three hidden **faces** and one hidden **vertex**.

On this diagram of a **cube**, the hidden edges have been shown with dashed lines. All the faces of a cube are squares.

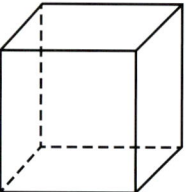

The cube and cuboid are examples of three-dimensional **solids** where all the faces are **planes**.

Some solids have curved **surfaces**.

A **sphere** has only a curved surface. A **hemisphere** (half a sphere) has both a flat and a curved surface, as do a **cylinder** and a **cone**.

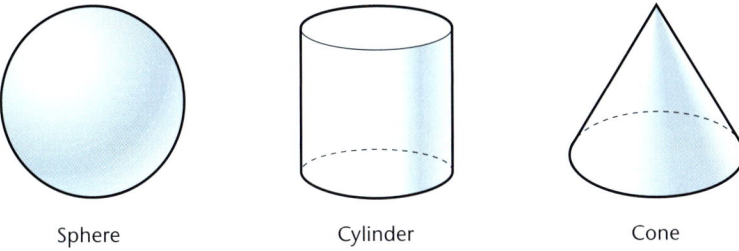

Sphere   Cylinder   Cone

Prisms and **pyramids** are also three-dimensional solid shapes.

Here are two examples of **prisms**.

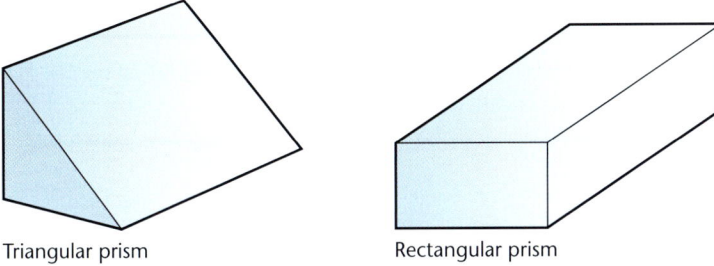

Triangular prism   Rectangular prism

Look at the shapes of the faces.

The shape on each end of the triangular prism is a triangle.

All the other faces are rectangles.

If you cut through the triangular prism at right angles to one of the rectangular faces you will see its cross-section. It is the same shape as the triangle at either end of the prism.

> The cross-section of a prism is the same all the way through.

The cross-section of a prism can be any shape.

The cross-section of the rectangular prism shown above is a rectangle. A rectangular prism is called a cuboid.

A **pyramid** has a base and triangular faces. The base can have any number of sides.

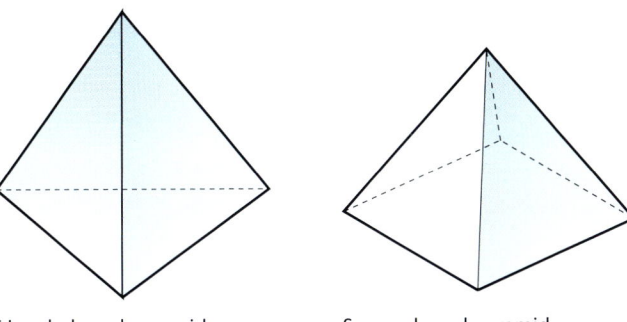

Triangle-based pyramid   Square-based pyramid

32 GEOMETRICAL TERMS

A **frustum** is a cut-off shape, cut parallel to the base of a cone or pyramid.
The frustum of a cone has two flat surfaces and a curved surface.
The surfaces of a frustum of a pyramid are all flat.

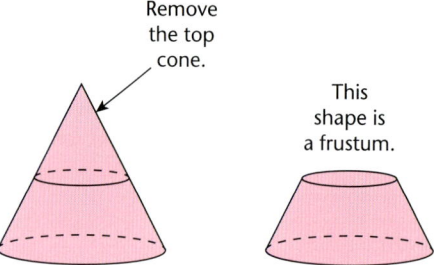

Nets of 3-D shapes

Three-dimensional objects with flat sides can be made by folding up a suitable flat object. This is called a **net**.

Example 32.3

Question

Which of these can be made from a net?
a   a cuboid
b   a sphere
c   a square-based pyramid
d   a cylinder.

Solution

a and **c**. The others have curved surfaces.

Example 32.4

Question

Which 3-D shape does each of these nets make?

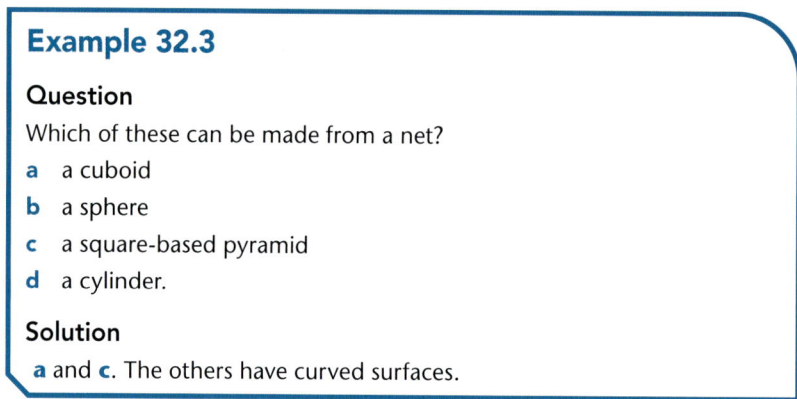

Solution

a   Triangular-based pyramid   b   Cube

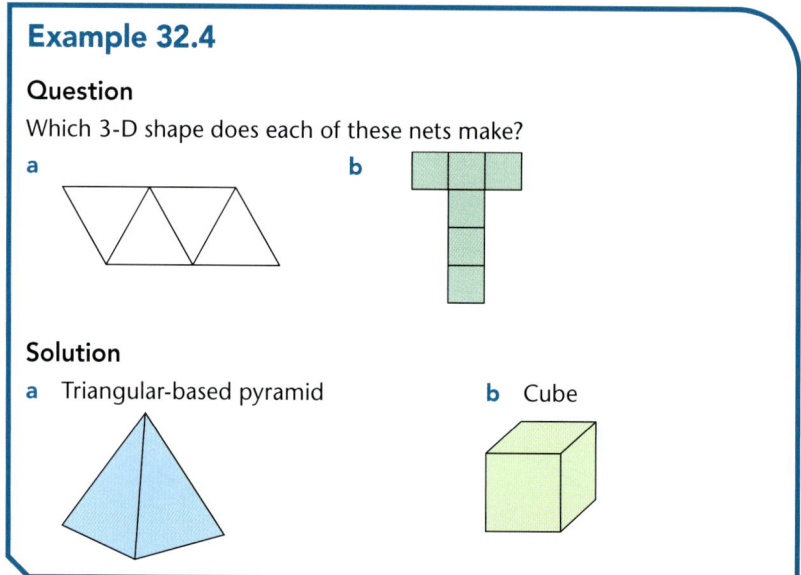

Exercise 32.4

1 Name each of these shapes.

a 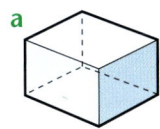 b 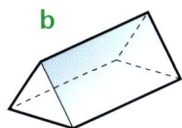 c 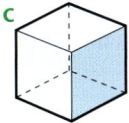 d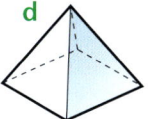

2 How many faces and how many vertices do each of the shapes in question **1** have?
3 Name each of these shapes.

a 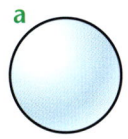 b 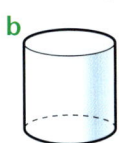 c  d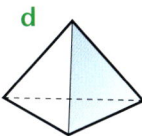

4 Which of the shapes in questions **1** and **3** are also prisms?

Exercise 32.5

1 Which of these shapes can have a net?

a 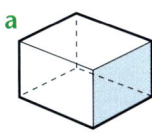 b 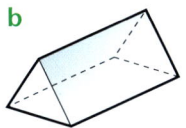 c 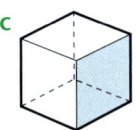 d

e 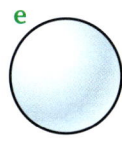 f 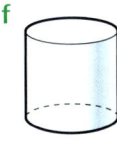 g 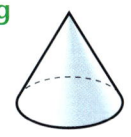 h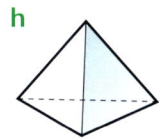

2 Which of the shapes **a**, **b** and **c** could be the net of this cuboid?

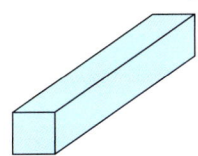

Note
There is more information about nets in Chapter 33.

a 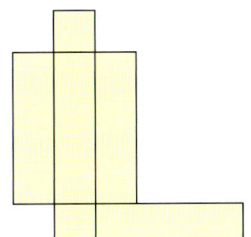 b c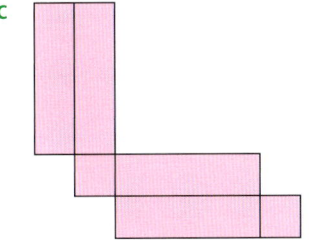

Congruence and similarity

Shapes that are exactly the same size and shape are said to be **congruent**.

If you cut out two congruent shapes, one shape would fit exactly on top of the other.

The corresponding sides are equal and the corresponding angles are equal.

Example 32.5

Question

Which of these triangles are congruent to triangle A?

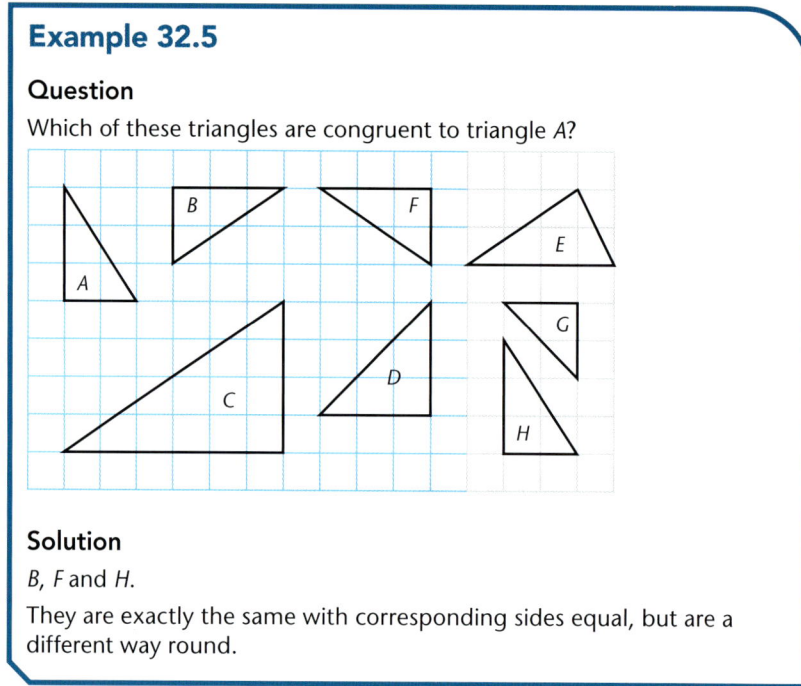

Solution

B, F and H.

They are exactly the same with corresponding sides equal, but are a different way round.

Every length in shape B is two times as long as the corresponding length in shape A.

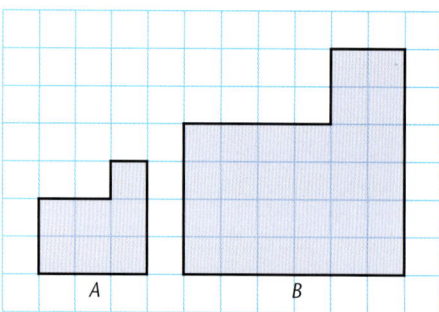

Shape B is an **enlargement** of shape A with scale factor 2.

We can also say that shapes A and B are **similar**.

Congruence and similarity

Shapes like these two quadrilaterals, which have the same angles and side lengths that are in proportion, are **similar**. One is an enlargement of the other.

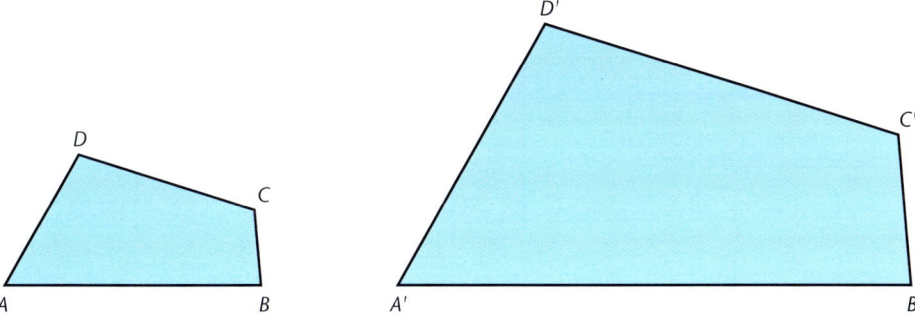

Exercise 32.6

1 In this question, sides that are equal are marked with corresponding lines, and angles that are equal are marked with corresponding arcs. Which of these pairs of triangles are congruent?

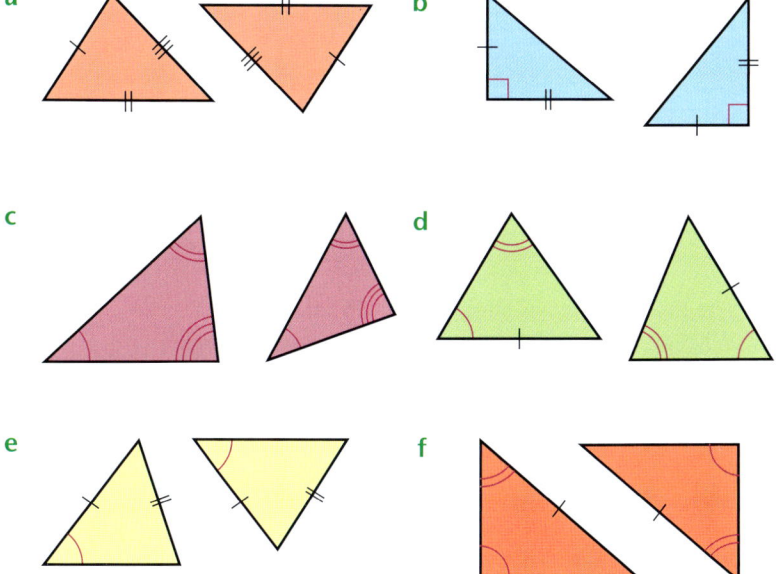

32 GEOMETRICAL TERMS

2 Which of these shapes are congruent to shape *A*?

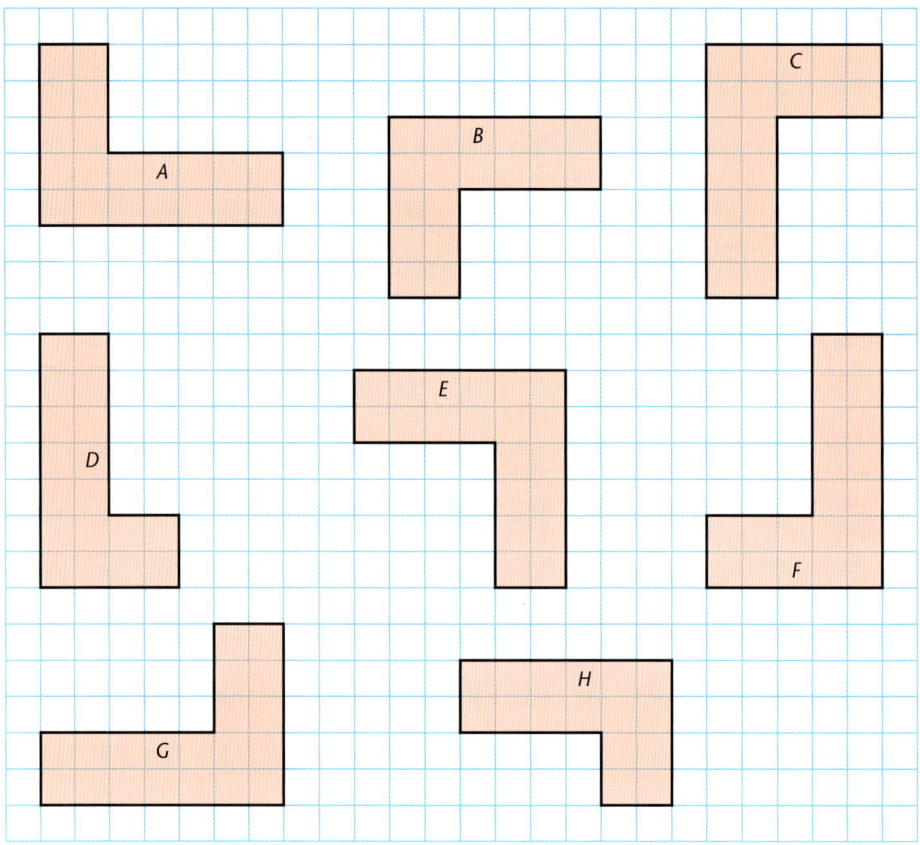

3 Two triangles are congruent.
The first triangle has angles of 35°, 75° and 70°.
What are the sizes of the angles in the second triangle?

4 Measure the lengths of the two shapes in each diagram.
Are the shapes similar? If they are, state the scale factor
of the enlargement.
  a

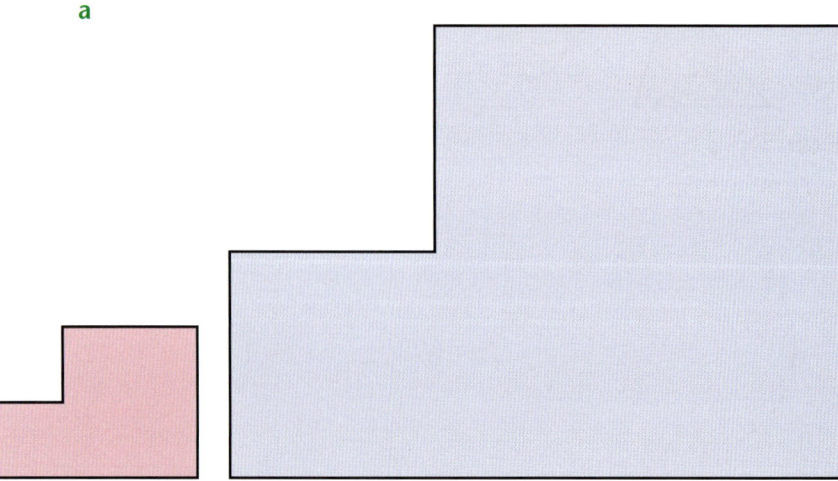

b

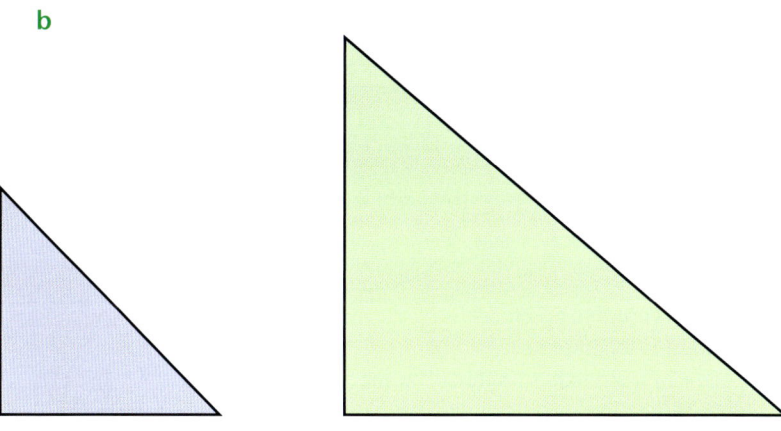

Circles

You need to be able to identify the parts of a **circle**.

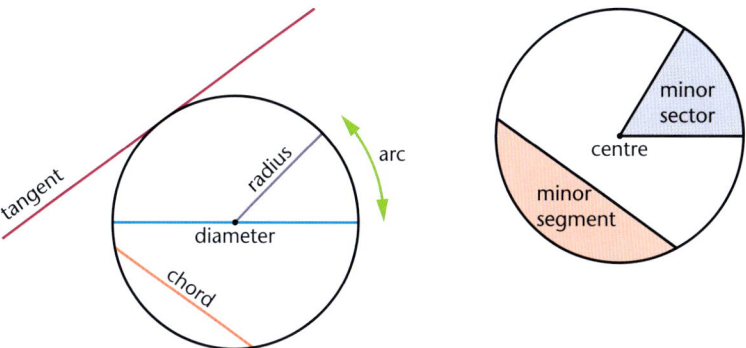

The **circumference** of a circle is the distance all the way round – the perimeter of the circle.

A chord divides a circle into two segments. The larger one is called the major segment and the smaller one is the minor segment.

Two radii divide a circle into two sectors. The larger one is called the major sector and the smaller one is the minor sector. The radii also divide the circumference into a major arc and a minor arc.

Exercise 32.7

1 Name each of these parts of a circle.

  a   b   c

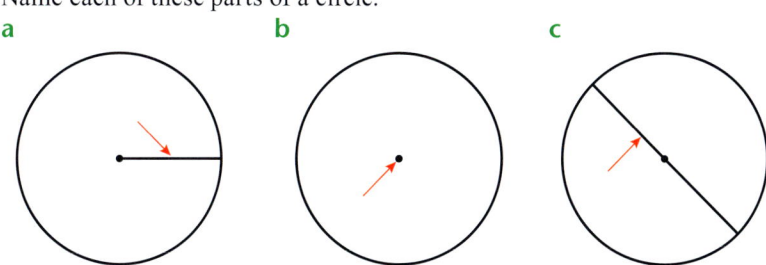

32 GEOMETRICAL TERMS

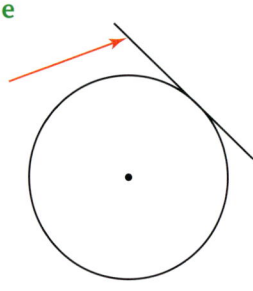

d   e

2 A circle has radius 5.7 cm.
   How long is its diameter?
3 A sector of a circle has angle 60° at the
   centre of the circle.
   The arc length of the sector is 4.2 cm.
   Use the fact that a full turn is 360° to find
   the length of the circumference of the circle.

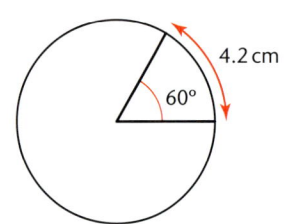

Key points
- A point has no dimensions. A line has one dimension (1-D). A flat shape, or plane, has two dimensions (2-D). A shape with length, width and height has three dimensions (3-D).
- Angles are measured in degrees. A full turn is 360°. A half-turn (which forms a straight line) is 180°. A quarter-turn is 90°, which is also called a right angle.
- Acute angles are less than 90°. Obtuse angles are between 90° and 180°. Reflex angles are more than 180°.
- Parallel lines never meet. Perpendicular lines meet or cross at 90°. A line that cuts another line in half at 90° is called its perpendicular bisector.
- Bearings are used to describe a direction. They are measured clockwise round from North and are always given with three digits.
- There are different types of triangle: right-angled, isosceles, equilateral and scalene.
- There are different types of quadrilateral: square, rectangle, parallelogram, rhombus, trapezium, kite, isosceles trapezium, as well as scalene.
- Pentagons, hexagons, octagons and decagons are all polygons.
- Regular polygons have all their sides the same length and all their angles the same size. Otherwise, a polygon is irregular.
- Angles inside a polygon are called interior angles. Exterior angles are made when a side of a shape is continued outside the shape.
- Congruent shapes are exactly the same shape and size.
- Similar shapes have the same angles but their sizes are different. The sides are in proportion to each other. So, when two shapes are similar, one is an enlargement of the other.
- 3-D solids that you should know about include: cube, cuboid, sphere, hemisphere, cylinder, cone, prism, pyramid and frustum.
- A 'corner' in a shape is called a vertex (the plural is 'vertices'). For 2-D shapes, a vertex is where two sides meet. For 3-D shapes, a vertex is where three or more edges meet.
- A flat surface on a 3-D shape is called a face.
- A net of a 3-D shape is a 2-D shape that can be folded to make the 3-D shape.
- Circle terms that you should know about include: centre, radius (plural radii), diameter, circumference, semi-circle, chord, tangent; also, major and minor arcs, sectors and segments.

33 GEOMETRICAL CONSTRUCTIONS

BY THE END OF THIS CHAPTER YOU WILL BE ABLE TO:
- measure and draw lines and angles
- construct a triangle, given the lengths of all sides, using a ruler and pair of compasses only
- draw, use and interpret nets.

CHECK YOU CAN:
- read the scales on a ruler
- read the scales on a protractor
- use a pair of compasses.

Measuring angles

The instrument used to measure an angle is called a protractor or an angle measurer.

Some protractors are full circles and can be used to measure angles up to 360°.

Most protractors are semi-circular in shape and can be used to measure angles up to 180°.

Example 33.1

Question
Measure angle PQR.

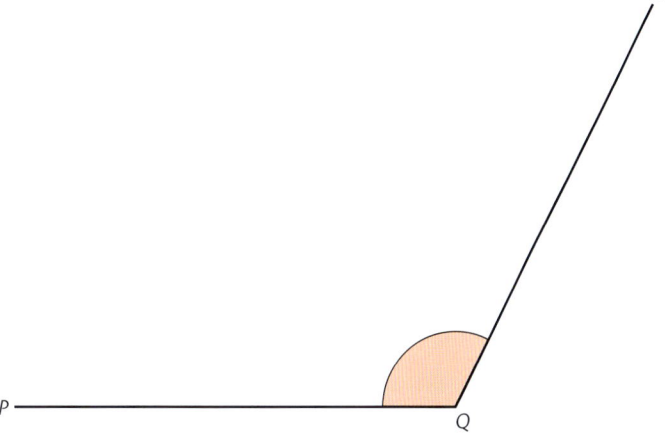

Solution
First make an estimate.
The angle is obtuse and so will be between 90° and 180°.
A rough estimate is about 120°.
Place your protractor so that the zero line is along one of the arms of the angle and the centre is at the point of the angle.

▶

Photocopying is prohibited

287

33 GEOMETRICAL CONSTRUCTIONS

> **Note**
> A common error in measuring angles is to use the wrong scale of the two on the protractor. Make sure you choose the one that starts at zero. A useful further check is to estimate the angle first. Knowing approximately what the angle is should prevent you using the wrong scale.

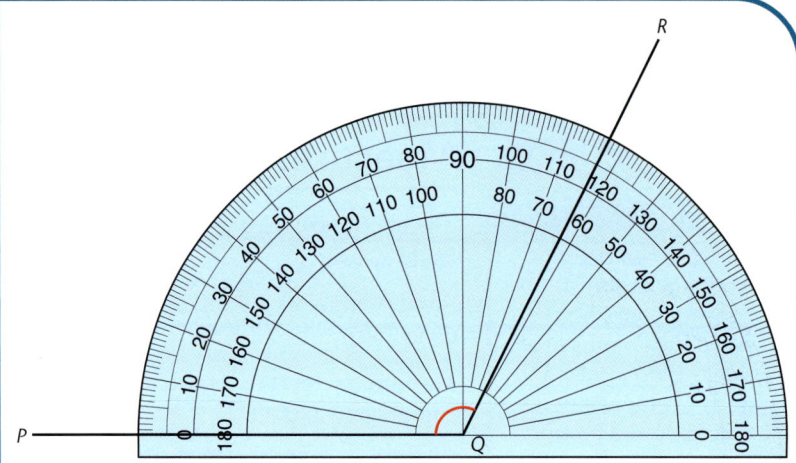

Start at zero.
Go round this scale until you reach the other arm of the angle.
Then read the size of the angle from the scale.
Angle PQR = 117°

Example 33.2

Question

Measure angle A.

Solution

A reflex angle is between 180° and 360°.
This reflex angle is over $\frac{3}{4}$ of a turn so it is bigger than 270°.
A rough estimate is 300°.
You can measure an angle of this size directly using a 360° circular angle measurer. However, the scale on a semi-circular protractor only goes up to 180°.

Measuring angles

You need to do a calculation as well as measure an angle.
Measure the acute angle first.
The acute angle is 53°.

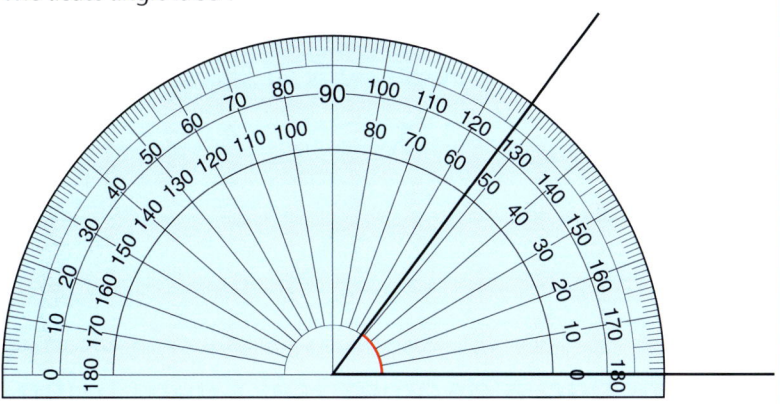

The acute angle and the reflex angle together make 360°.
Use the fact that the two angles add up to 360° to calculate the reflex angle.

Angle $A = 360° - 53°$
$= 307°$

Exercise 33.1

1 Copy and complete this table by
   i   estimating each angle
   ii measuring each angle.

	Estimated angle	Measured angle
a		
b		
c		
d		

a

b

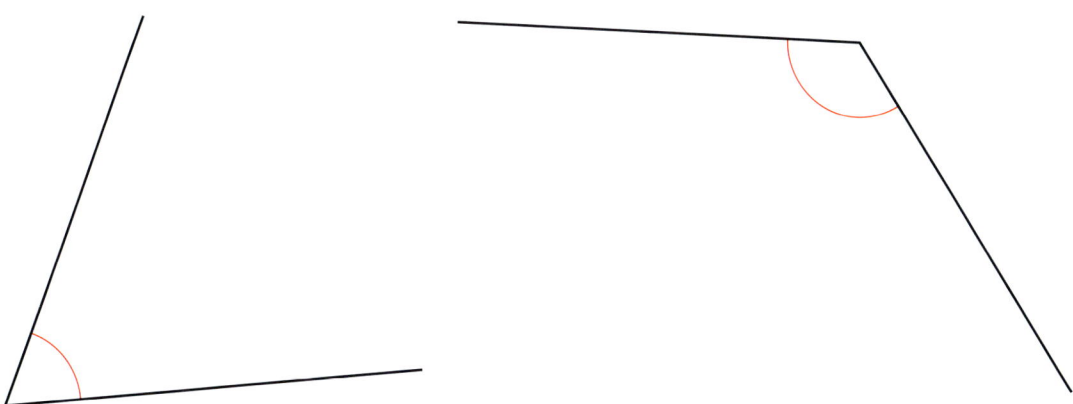

33 GEOMETRICAL CONSTRUCTIONS

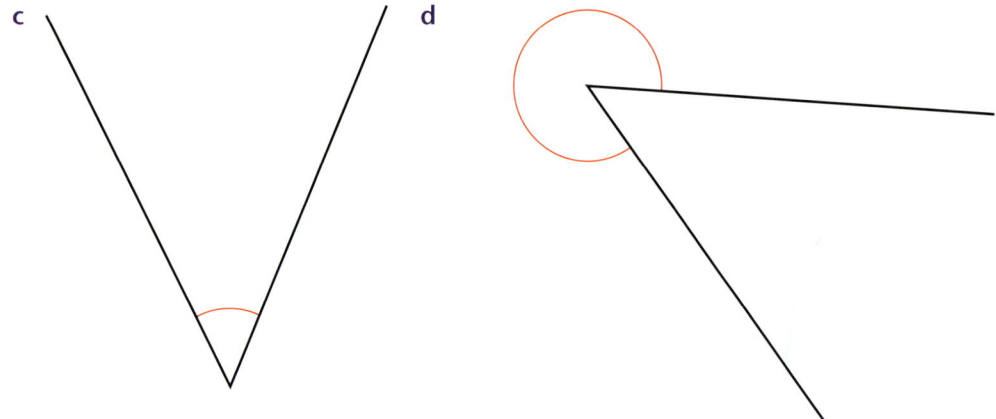

c   d

Constructing a geometrical figure using compasses

Constructing a triangle given three sides

Example 33.3

Question

Construct triangle *ABC*, where *AB* = 5 cm, *BC* = 4 cm and *AC* = 3 cm.

Solution

First draw a sketch.

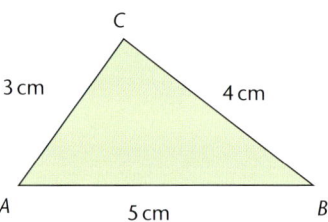

> **Note**
> Use a ruler for all straight lines. Leave in all your construction arcs.

Draw the line *AB*, 5 cm long. From *A*, with your compasses set to a radius of 3 cm, draw an arc above the line.

From *B*, with your compasses set to a radius of 4 cm, draw another arc to intersect the first. The point where the arcs meet is *C*.

Join points *A* and *B* to point *C* to complete the triangle.

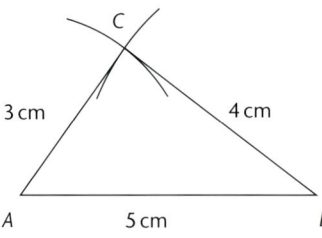

Exercise 33.2

1 Construct each of these triangles.

a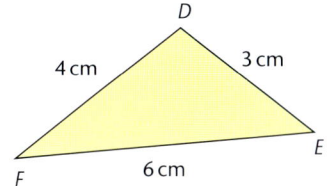

b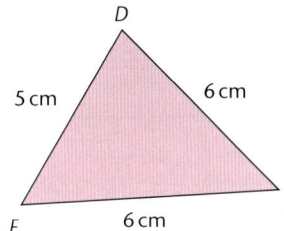

c Triangle *DEF* with *DE* = 4.3 cm, *EF* = 7.2 cm and *FD* = 6.5 cm.

2 For each triangle in question **1**, measure the angles on your drawing.
3 Construct an equilateral triangle with sides of 4.2 cm.
4 Draw an isosceles triangle with sides 6 cm and 5 cm.
  There are two possible triangles.
  Draw both of them.
  In each case, measure the angles in the triangle.
5 a Draw a rhombus *ABCD* with diagonal *AC* = 3.5 cm and sides of length 5.3 cm.
  b Measure the length of diagonal *BD*.

Nets of 3-D shapes

A **net** of a three-dimensional object is a flat shape that will fold to make that object.

This flat shape can be folded to make a cube – it is a net of the cube.

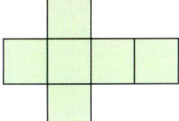 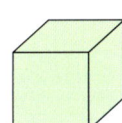

A regular triangular-based pyramid has edges that are all the same length.

Here is its net.

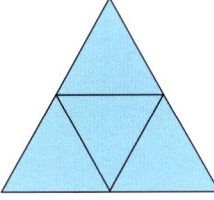 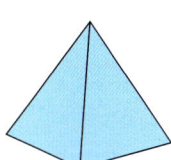

As a cuboid has six faces, its net needs six rectangles. But not every arrangement of the six rectangles will fold together to make the cuboid.

33 GEOMETRICAL CONSTRUCTIONS

Example 33.4

Question

This net can be folded to make a cuboid.

When it is folded, which point or points will meet with

a point A   b point D?

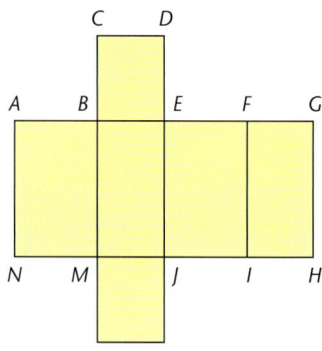

Solution

a Points C and G   b Point F

Example 33.5

Question

For each of these arrangements, say whether it is the net of a cuboid.
They are drawn on squared paper to size.

a
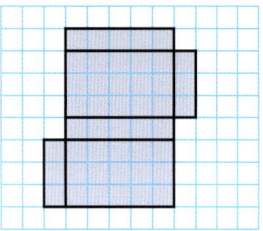

b
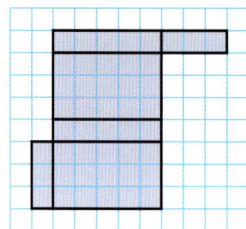

c
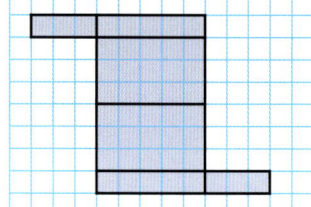

d
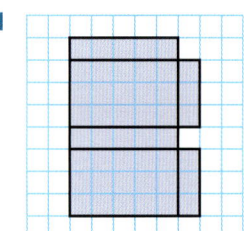

e
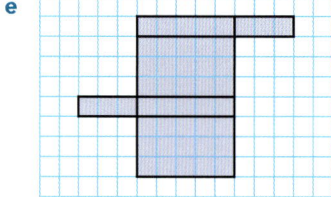

f
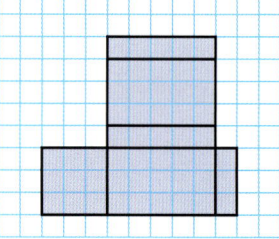

Solution

a Yes   b Yes   c No
d No   e Yes   f No

Nets of 3-D shapes

Example 33.6

Question

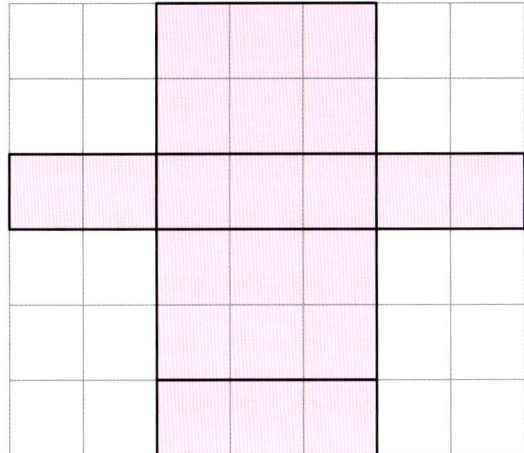

This net of a cuboid is drawn on a one-centimetre square grid.
a Write down the dimensions of the cuboid.
b By finding the area of the net, find the surface area of the cuboid.
c Find the volume of the cuboid.

Solution
a 3 cm by 2 cm by 1 cm
b Surface area = 2 × (3 × 2) + 2 × (3 × 1) + 2 × (2 × 1) = 22 cm² [or by counting squares]
c Volume = 3 × 2 × 1 = 6 cm³

Exercise 33.3

1 Look at each of these nets and say what shape it makes.

a 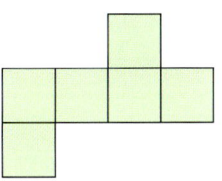   b    c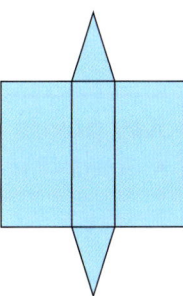

d    e 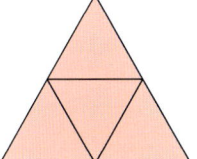   f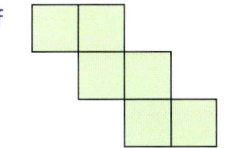

Photocopying is prohibited

33 GEOMETRICAL CONSTRUCTIONS

Note
If you are not sure whether a net is correct, cut it out and try folding it.

2 Draw a net for each of these cuboids on squared paper. All lengths are given in centimetres.

a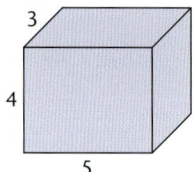

b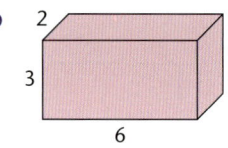

3 For a cuboid, state
   a the number of edges
   b the number of vertices
   c the number of faces.

4 a When the net is folded to make a cuboid, which point or points will meet with
      i Point A
      ii Point D?
   b Write down the dimensions of the cuboid.
   c Find the surface area of the cuboid.
   d Find the volume of the cuboid.

5 A box containing paper has a base 15 cm by 20 cm and it is 5 cm high. It has no top.
   a Draw a net for the box. Use a scale of 1 cm to 5 cm.
   b Find the surface area of the box. [Remember: each square on your net represents 25 cm².]
   c Find the volume of the box.

6 Construct a net of this triangular prism.

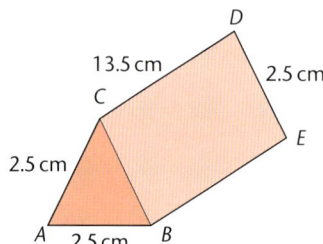

Key points
- Straight lines should be drawn and measured to an accuracy of 1 mm.
- Angles between lines should be drawn and measured to an accuracy of 1°.
- You can draw a triangle when you know the lengths of three sides. You need to use a pair of compasses to find the position of the third vertex.
- For all geometrical constructions, use a ruler for all straight lines and leave in any construction arcs.
- A net is a flat shape that will fold up to make a three-dimensional (3-D) object.

34 SCALE DRAWINGS

Scale drawings and maps

A **scale drawing** is exactly the same shape as the original drawing, but is different in size.

Large objects are scaled down in size so that, for example, they can fit on to the page of a book.

A map is a scale drawing of an area of land.

The scale of the drawing can be written as these examples.

1 cm to 2 m   This means that 1 cm on the scale drawing represents 2 m in real life.

2 cm to 5 km   This means that 2 cm on the scale drawing represents 5 km in real life.

> **BY THE END OF THIS CHAPTER YOU WILL BE ABLE TO:**
> - draw and interpret scale drawings
> - use and interpret three-figure bearings.

> **CHECK YOU CAN:**
> - use a ruler and a protractor
> - understand and use the terms north, south, east and west.

Example 34.1

Question

Here is a scale drawing of a truck.

The scale of the drawing is 1 cm to 2 m.

a How long is the truck?
b Will the truck go safely under a bridge 4 m high?
c The truck driver is 1.8 m tall.
  How high will he be on the scale drawing?

Solution

a Measure the length of the truck on the scale drawing.
  Length of truck on the drawing = 4 cm.
  As 1 cm represents 2 m, multiply the length on the drawing by 2 and change the units.
  Length of truck in real life = 4 × 2 = 8 m.
b Height of truck on the drawing = 2.5 cm.
  Height of truck in real life = 2.5 × 2 = 5 m.
  So the truck will not go under the bridge.
c To change from measurements in real life to measurements on the drawing, you have to divide by 2 and change the units.
  Height of driver in real life = 1.8 m.
  Height of driver on the drawing = 1.8 ÷ 2 = 0.9 cm.

Photocopying is prohibited

34 SCALE DRAWINGS

Exercise 34.1

1. Measure each of these lines as accurately as possible. Using a scale of 1 cm to 4 m, work out the length that each line represents.
   a _____
   b _____
   c _____
   d _____

2. Measure each of these lines as accurately as possible. Using a scale of 1 cm to 10 km, work out the length that each line represents.
   a _____
   b _____
   c _____
   d _____

3. Draw accurately the line to represent these actual lengths. Use the scale given.
   a 5 m   Scale: 1 cm to 1 m
   b 10 km   Scale: 1 cm to 2 km
   c 30 km   Scale: 2 cm to 5 km
   d 750 m   Scale: 1 cm to 100 m

4. Here is a plan of a house.
   The scale of the drawing is 1 cm to 2 m.

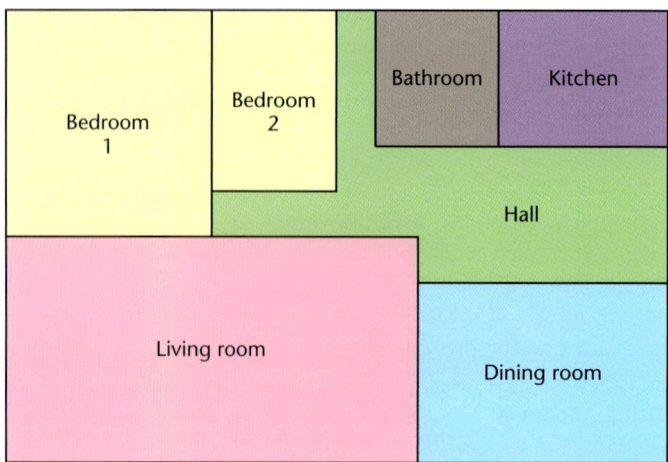

   a How long is the hall in real life?
   b Work out the length and width of each of the six rooms in real life.
   c The bungalow is on a plot of land measuring 26 m by 15 m. What will the measurements of the plot of land be on this scale drawing?

5 The map shows some towns and cities in the south-east of England.
The scale of the map is 1 cm to 20 km.

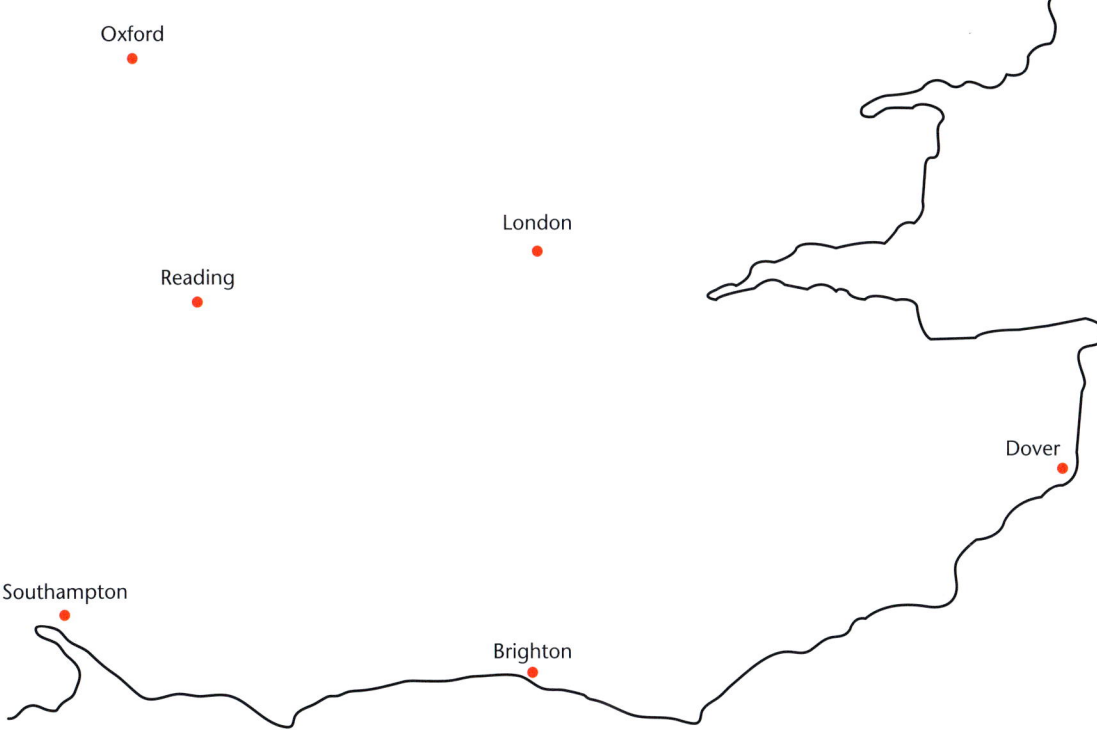

a What is the real-life distance, in kilometres, between these towns?
  i London and Reading
  ii Southampton and Dover
  iii London and Brighton
  iv Oxford and Reading
  v Brighton and Southampton
  vi Dover and Oxford
b It is 90 km from London to Cambridge.
How many centimetres will this be on the map?

34 SCALE DRAWINGS

Bearings

Bearings are used to describe direction.

They are measured clockwise from north.

Bearings are always given from some fixed point.

In the diagram, the bearing of B from A is $070°$.

The bearing of A from B is called the reverse bearing.

In this case,

the bearing of A from B = the bearing of B from A + $180°$
$= 070° + 180°$
$= 250°$

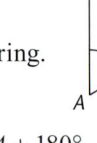

Note
If the bearing is greater than $180°$, the reverse bearing is found by subtracting $180°$.

Note
Bearings must have three figures, so if the angle is less than $100°$ you must put a zero in front of the figures.

Example 34.2

Question

Use a protractor to measure and write down each of these bearings.

a A from O
b B from O
c C from O

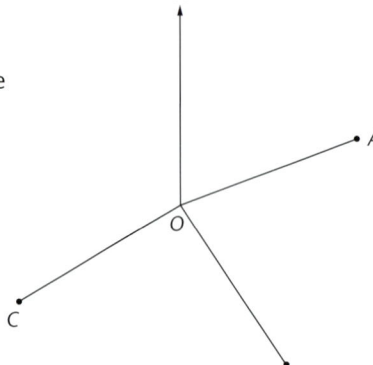

Note
Make sure you use the clockwise scale.

Solution

a $070°$
b $146°$
c $240°$

If you have a circular protractor you can measure all of these directly.

If you only have a semi-circular protractor then for part **c** you need to measure the obtuse angle and subtract it from $360°$.

Bearings

Example 34.3

Question

A, B and C are three towns.

B is 20 km from A, on a bearing of 085° from A.

C is 15 km from B, on a bearing of 150° from B.

a   Make a scale drawing showing the three towns. Use a scale of 1 cm to 5 km.

b   i   How far is A from C?

   ii   What is the bearing of A from C?

c   Town D is due west of C. What is the bearing of D from C?

Solution

a   Make a sketch and label the angles and lengths for the final diagram.

Then draw the diagram, starting far enough down the page to make sure it will all fit in.

b   i   CA measures 5.9 cm on the scale drawing, so the real distance is 5.9 × 5 = 29.5 km.

   ii   The bearing is 292°.

c   270°

Note
The directions north, east, south and west are at right angles to each other.

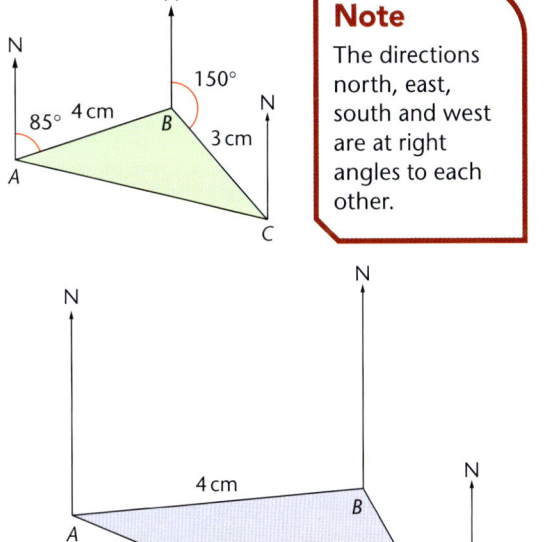

Exercise 34.2

1   Measure the bearings of A, B, C and D from O.

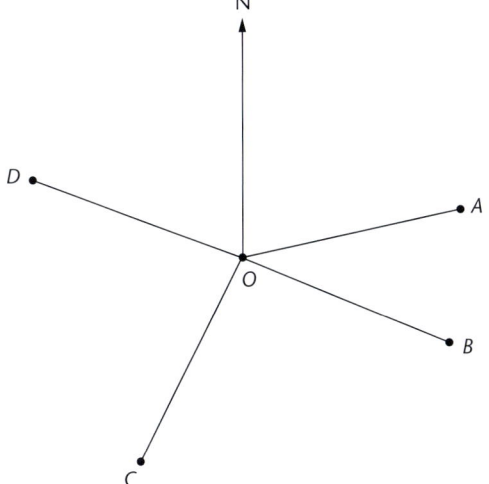

34 SCALE DRAWINGS

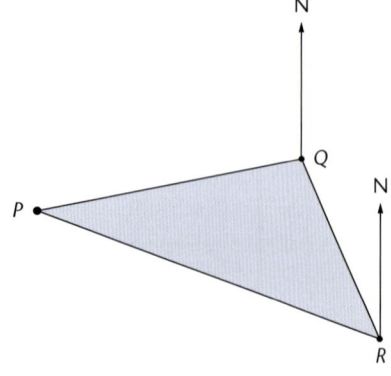

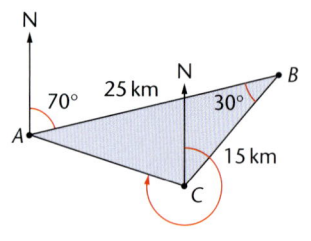

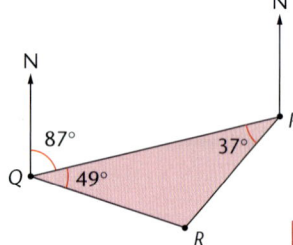

2. Mark a point *O*. Draw a north line through *O*. Mark points *A*, *B*, and *C*, all 5 cm from *O* on these bearings.
   A 055°   B 189°   C 295°

3. Point *A* is on a bearing of 124° from point *B*. What is the bearing of point *B* from point *A*?

4. Point *C* is on a bearing of 290° from point *D*. What is the bearing of point *D* from point *C*?

5. On this drawing, measure the bearing of
   a *P* from *Q*
   b *Q* from *R*.

6. A boat leaves port and sails for 10 km on a bearing of 285°. Make a sketch showing the direction in which the boat is sailing. Mark the distance and the angle clearly. Do not draw an accurate plan.

7. This sketch map shows the positions of three oil platforms – *A*, *B* and *C*.
   a Make an accurate drawing of the sketch. Use a scale of 1 cm to 5 km.
   b Measure the bearing of *A* from *C*.

8. This sketch map shows the positions of three hills – *P*, *Q* and *R*. Work out each of these bearings.
   a *R* from *Q*
   b *R* from *P*
   c *P* from *R*
   Do not make a scale drawing.

Key points
- The scale of a drawing or map tells you what every 1 unit represents; for example, every 1 cm represents 4 km.
- A scale drawing is exactly the same shape as the original object but is different in size.
- Bearings describe direction. A bearing is a three-figure angle measured clockwise from north.
- You can measure the bearing of one point from another using a protractor.
- You can make a scale drawing containing bearings and interpret it in context.

35 SIMILARITY

Similar shapes

In mathematics the word 'similar' has a very exact meaning. It does *not* mean 'roughly the same' or 'alike'.

For two shapes to be **similar**, each shape must be an exact enlargement of the other.

For example, look at these two rectangles.

The first rectangle is 2 cm wide by 4 cm long.

The second rectangle is 4 cm wide by 7 cm long.

The rectangles are *not* similar, because although the width of the large one is twice the width of the small one, the length of the large one is *not* twice the length of the small one.

If the length of the large one was 8 cm then the rectangles would be similar.

Now look at these two shapes.

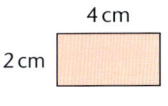 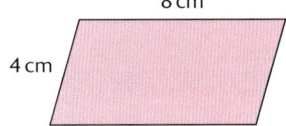

Although the scale factor for both pairs of sides is 2, the shapes are not similar because corresponding angles are not the same.

> For two shapes to be similar
> - all corresponding sides must have proportional lengths
> - all corresponding angles must be equal.

Because the lengths of three sides define a unique triangle, for two triangles to be similar, only one of the tests here needs to be made.

If you can establish that the angles are the same, you can conclude that the triangles are similar and carry out calculations to find the lengths of the sides.

BY THE END OF THIS CHAPTER YOU WILL BE ABLE TO:
- calculate lengths of similar shapes
- use the relationships between lengths and areas of similar shapes and lengths, surface areas and volumes of similar solids
- solve problems and give simple explanations involving similarity.

CHECK YOU CAN:
- use the angle properties of triangles and quadrilaterals
- use the angle properties of parallel lines
- understand the term scale factor.

Photocopying is prohibited

35 SIMILARITY

Calculating lengths in similar shapes

Example 35.1

Question

The rectangles ABCD and PQRS are similar.
Find the length of PQ.

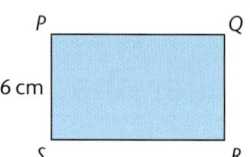

Solution

Since the widths of the rectangles are 6 cm and 4 cm, the scale factor is $6 \div 4 = 1.5$.

Length of $PQ = 7 \times 1.5$
$= 10.5$ cm

Example 35.2

Question

In the triangle, angle ABC = angle BDC = 90°, $AB = 6$ cm, $BC = 8$ cm and $BD = 4.8$ cm.

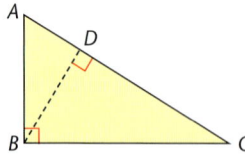

a Explain why triangles ABC and BDC are similar.
b Calculate the length of DC.

Note
When explaining why two triangles are similar, look for reasons why the angles are equal. Usual reasons include: 'opposite angles are equal', 'alternate angles are equal' and 'the angle is in both triangles (common angles)'.

Solution

a In the triangles ABC and BDC, angle ABC = angle BDC = 90° (given).
  The angle at C is in both triangles.
  Since the angle sum of a triangle is 180°, the third angles must be equal.
  So, since all the corresponding angles are equal, the triangles are similar.

b First redraw the triangles so they are the same way round as each other.

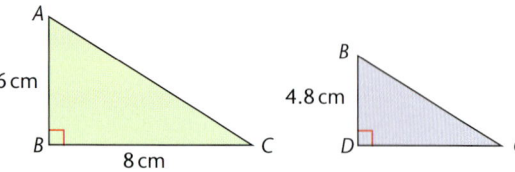

Note
It is always easier to spot the corresponding sides if the shapes are the same way round.
It is worth spending time redrawing the shapes separately and the same way round, and marking the lengths on the new diagram.

Since $AB = 6$ cm and $BD = 4.8$ cm the scale factor $= 4.8 \div 6 = 0.8$.
$CD = 8 \times 0.8$
$= 6.4$ cm

Exercise 35.1

1. The two rectangles in the diagram are similar.
   Find the length of the larger rectangle.

   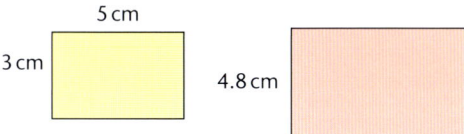

2. These two parallelograms are similar.
   Find the length of *PQ*.

   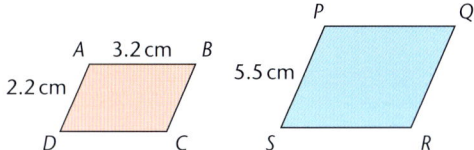

3. Josh bought a set of table mats.
   The set contained mats of three different sizes.
   They were all rectangular and similar in shape.
   The smallest one was 5 cm by 8 cm, the middle one was an enlargement of the smallest by scale factor 3, and the shorter side of the largest mat was 25 cm.
   Find the dimensions of the two larger mats.

4. The triangles *ABC* and *PQR* are similar.
   Calculate the lengths of *PQ* and *PR*.

   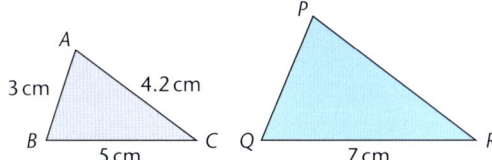

5. The triangles *ABC* and *PQR* are similar.
   Calculate the lengths of *PQ* and *QR*.

   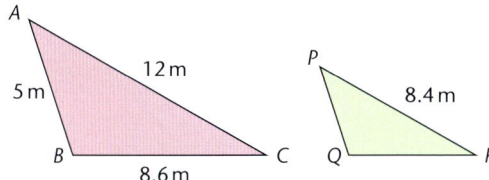

6. In the diagram, *DE* is parallel to *BC*, *AB* = 4.5 cm, *BC* = 6 cm and *DE* = 10 cm.
   Calculate the length of *BD*.

   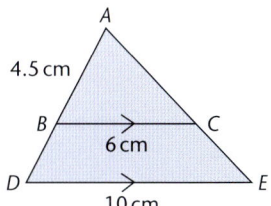

7 At noon a radio mast of height 12 m has a shadow length of 16 m. Calculate the height of a tower with a shadow length of 56 m at noon.

8 a Explain how you know that the two triangles in the diagram are similar.
  b Calculate the lengths of AC and CB.

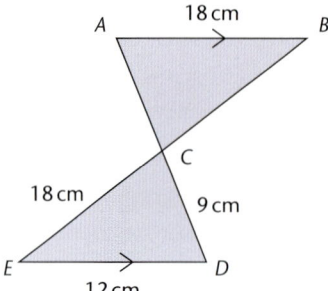

9 Explain why these quadrilaterals are *not* similar.

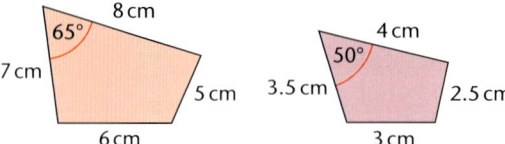

10 In the diagram, angle BAC = angle ADC = 90°, AD = 3 cm and DC = 5 cm.
  a Explain why triangles ADC and BDA are similar.
  b Calculate the length of BD.

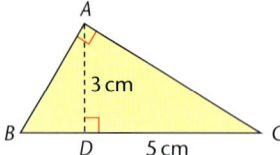

The areas and volumes of similar shapes

This cube has a volume of 8 cm³. This cube has a volume of 512 cm³.

 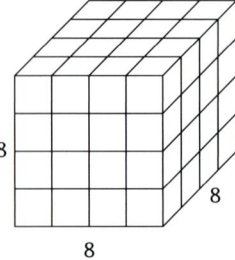

The lengths of the small cube have been enlarged by scale factor 4.

The volume has been enlarged by scale factor 64.

Since there are three dimensions for volume, and each dimension has been enlarged by scale factor 4, the volume scale factor = 4^3.

Similarly, consider the area of the face of each cube.

For the small cube, the area is $4\,\text{cm}^2$.

For the large cube, the area is $64\,\text{cm}^2$.

The area has been enlarged by scale factor 16.

There are two dimensions for area, so the area scale factor is 4^2.

> For mathematically similar shapes
> - area scale factor = (length scale factor)^2
> - volume scale factor = (length scale factor)^3.

Example 35.3

Question

A model aircraft is made to a scale of $1:50$.
The area of the wing on the model is $18\,\text{cm}^2$.
What is the area of the wing on the real aircraft?

Solution

Length scale factor = 50
Area scale factor = 50^2
Area of real wing = 18×50^2
$\qquad\qquad\qquad = 45\,000\,\text{cm}^2$
$\qquad\qquad\qquad = 4.5\,\text{m}^2$

Note
Remember that
$1\,\text{m} = 100\,\text{cm}$
$1\,\text{m}^2 = 10\,000\,\text{cm}^2$
$1\,\text{m}^3 = 1\,000\,000\,\text{cm}^3$

Example 35.4

Question

A jug holding $50\,\text{cl}$ is $12\,\text{cm}$ high.
A similar jug holds 2 litres.
What is its height?

Solution

$50\,\text{cl} = 0.5$ litres
Volume scale factor $= \dfrac{2}{0.5} = 4$
Length scale factor $= \sqrt[3]{4}$
Height of larger jug $= 12\,\text{cm} \times \sqrt[3]{4}$
$\qquad\qquad\qquad\qquad = 19.0\,\text{cm}$ to 1 decimal place

Exercise 35.2

1. State the area scale factor for each of these length scale factors.
   a 2   b 3   c 5   d 4   e 6   f 10
2. State the volume scale factor for each of these length scale factors.
   a 10   b 4   c 5   d 2   e 3   f 8

Photocopying is prohibited

3 State the length scale factor for each of these.
  a Area scale factor of 16
  b Volume scale factor of 216
  c Area scale factor of 64
  d Volume scale factor of 1000

4 A model of a building is made to a scale of 1 : 50.
A room in the model is a cuboid with dimensions 7.4 cm by 9.8 cm by 6.5 cm high.
Calculate the floor area of the room in
  a the model
  b the actual building.

5 A glass holds 15 cl.
The heights of this and a larger similar glass are in the ratio 1 : 1.2.
Calculate the capacity of the larger glass.

6 A tray has an area of 160 cm².
What is the area of a similar tray whose lengths are one and a half times as large?

7 To what scale is a model drawn if an area of 5 m² in real life is represented by 20 cm² on the model?

8 Three similar wooden boxes have heights in the ratio 3 : 4 : 5.
What is the ratio of their volumes?

9 The area of a rug is 2.4 times as large as the area of a similar rug.
The length of the smaller rug is 1.6 m.
Find the length of the larger rug.

10 An artist makes a model for a large sculpture.
It is 24 cm high.
The finished sculpture will be 3.6 m high.
  a Find the length scale factor.
  b Find the area scale factor.
  c The model has a volume of 1340 cm³.
  What will be the volume of the sculpture?

11 A glass is 12 cm high.
How tall is a similar glass that holds twice as much?

12 A model aircraft is made to a scale of 1 : 48.
The area of the wing of the real aircraft is 52 m².
What is the area of the wing of the model?

13 A square is enlarged by increasing the length of its sides by 10%.
The length of the sides was originally 8 cm.
What is the area of the enlarged square?

14 Two similar containers hold 1 litre and 2 litres of milk.
The surface area of the larger container is 1100 cm².
Calculate the surface area of the smaller one.

> ## Key points
> - For two shapes to be similar
>   - all corresponding sides must have proportional lengths, and
>   - all corresponding angles must be equal.
> - For mathematically similar shapes
>   - area scale factor = (length scale factor)²
>   - volume scale factor = (length scale factor)³.

36 SYMMETRY

Line symmetry

A **line of symmetry** divides a shape into two identical parts.

A shape may have more than one line of symmetry.

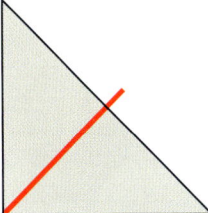

The triangle has one line of symmetry.

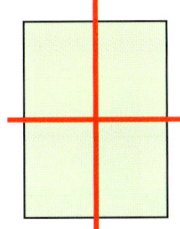

The rectangle has two lines of symmetry.

> **BY THE END OF THIS CHAPTER YOU WILL BE ABLE TO:**
> - recognise line symmetry and order of rotational symmetry in two dimensions
> - recognise symmetry properties of prisms, cylinders, pyramids and cones.
>
> **CHECK YOU CAN:**
> - use the vocabulary of shapes and solids
> - show that triangles are congruent
> - use the angle properties of a triangle.

Example 36.1

Question

Complete the pattern so that *L* is a line of symmetry.

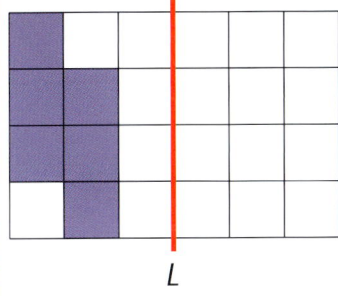

Solution

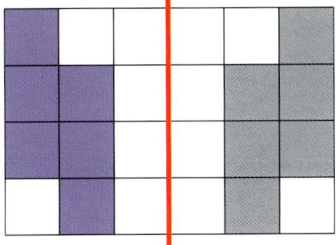

The image is the same distance from the line of symmetry as the original shape, but on the opposite side of the line.

Photocopying is prohibited

36 SYMMETRY

Exercise 36.1

1 For each shape, write down the number of lines of symmetry.

a  b c

2 Copy and complete each pattern so that L is a line of symmetry.

a 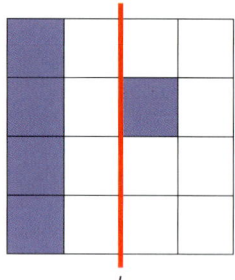 b c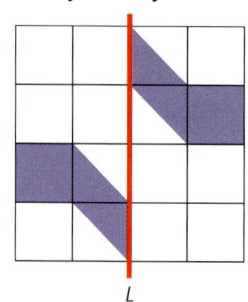

3 Copy and complete each pattern so that lines L and M are both lines of symmetry.

a 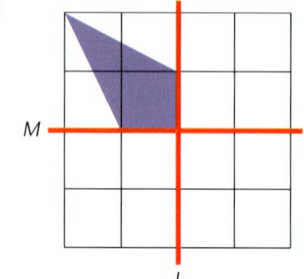 b 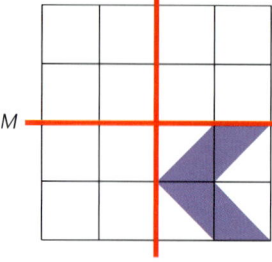 c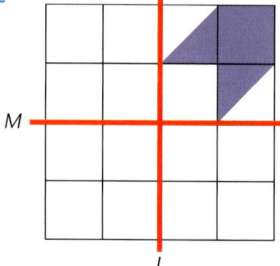

Rotational symmetry

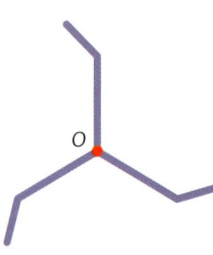

This shape has rotational symmetry of order 3 about centre O.

There are three positions where the shape fits onto the original shape in one complete turn about centre O.

The **order of rotational symmetry** is the number of times that a shape fits onto itself in one complete turn.

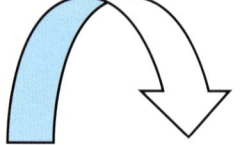

This shape has no rotational symmetry.

There is only one position where the shape fits onto the original in one complete turn.

It has rotational symmetry of order 1.

Rotational symmetry

Example 36.2

Question

Shade three more squares in this shape so that the pattern has rotational symmetry of order 2.

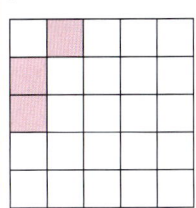

Solution

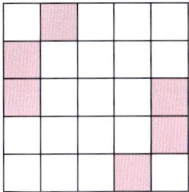

You can check that your diagram is correct by rotating it through a half turn.

Exercise 36.2

1. Copy this diagram and shade two more squares so that the pattern has rotational symmetry of order 2.

   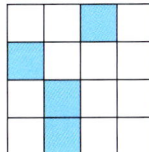

2. Copy this diagram and shade nine more squares so that the pattern has rotational symmetry of order 4.

   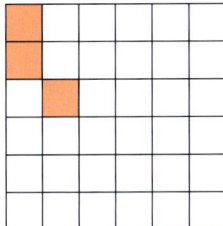

3. Shade squares on a 4 × 4 grid to make a pattern with rotational symmetry of order 2 but no lines of symmetry.

4. Copy and complete this diagram so that it has rotational symmetry of order 2.

36 SYMMETRY

Symmetry properties of shapes and solids

This is an isosceles triangle.

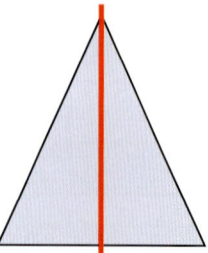

It has one line of symmetry and no rotational symmetry.

These symmetry properties are the same for any isosceles triangle.

This is a cuboid.

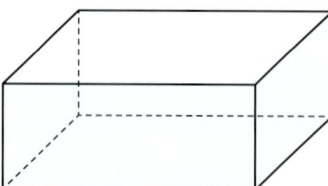

A **plane of symmetry** divides a solid into two identical parts.

The cuboid has three planes of symmetry.

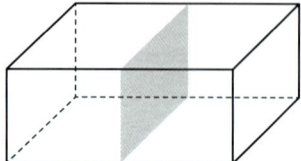 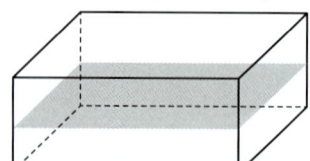 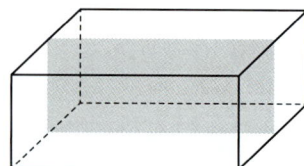

The cuboid has rotational symmetry of order 2 about the axis shown.

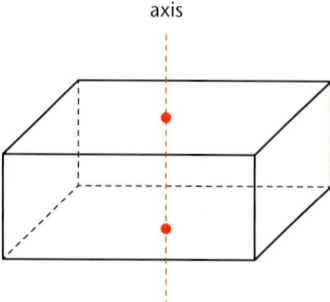

The cuboid has three axes of rotational symmetry.

When describing the symmetry of a shape, remember to describe both rotational and line symmetry.

Symmetry properties of shapes and solids

Example 36.3

Question

Describe the symmetry of
- **a** a parallelogram
- **b** a square-based pyramid
- **c** a regular nine-sided polygon. Use this to find the size of each interior angle of the polygon.

Solution
- **a** A parallelogram has no lines of symmetry and rotational symmetry of order 2.
- **b** A square-based pyramid has four planes of symmetry and one axis of rotational symmetry of order 4.
- **c** A regular nine-sided polygon has nine lines of symmetry and rotational symmetry of order 9.
  By symmetry: exterior angle = 360 ÷ 9 = 40°; interior angle = 180 − 40 = 140°.

Exercise 36.3

1. Describe the symmetry of each of these shapes.
   - **a** a rectangle
   - **b** an equilateral triangle
   - **c** a regular pentagon
2. Write down the number of planes of symmetry of each of these solids.
   - **a** a cube
   - **b** a cone
3. Describe the symmetry of a prism which has an isosceles triangle for its cross-section.
4. Sketch an octagon with two lines of symmetry and rotational symmetry of order 2.
5. **a** Which special quadrilateral has four lines of symmetry and rotational symmetry of order 4?
   b Which special quadrilateral has one line of symmetry and no rotational symmetry?
6. Which solid has an infinite number of planes of symmetry and an infinite number of axes of symmetry?
7. Use the symmetry of a regular hexagon to find the interior angle of the polygon.

Key points

- In two dimensions, a figure has line symmetry if it reflects onto itself in a line.
- A two-dimensional shape has rotational symmetry if it will rotate about a centre onto itself in more than one position. The number of ways it fits is the order of rotational symmetry.
- In three dimensions, a shape has plane symmetry if it reflects onto itself in a plane.
- A three-dimensional shape has rotational symmetry if it will rotate about an axis onto itself in more than one position. The number of ways it fits is the order of rotational symmetry.

37 ANGLES

BY THE END OF THIS CHAPTER YOU WILL BE ABLE TO:

- calculate unknown angles and give simple explanations using the following geometrical properties:
  - sum of angles at a point = 360°
  - sum of angles at a point on a straight line = 180°
  - vertically opposite angles are equal
  - angle sum of a triangle = 180° and angle sum of a quadrilateral = 360°
- calculate unknown angles and give geometric explanations for angles formed within parallel lines:
  - corresponding angles are equal
  - alternate angles are equal
  - co-interior (supplementary) angles sum to 180°
- know and use angle properties of regular and irregular polygons.

CHECK YOU CAN:

- distinguish between acute, obtuse, reflex and right angles
- distinguish between equilateral, isosceles, scalene and right-angled triangles
- understand the meaning of the term *congruent*
- understand and use geometrical terms.

Angles formed by straight lines

The sum of the angles at a point is 360°.

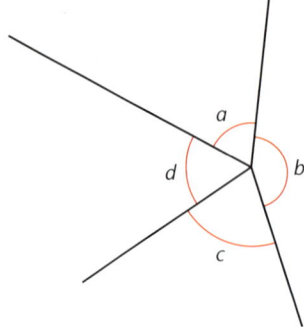

$a + b + c + d = 360°$

The sum of the angles on a straight line is 180°.

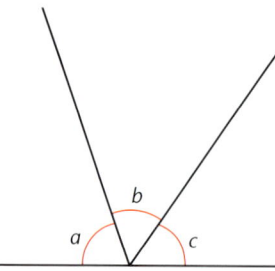

$a + b + c = 180°$

When two lines cross, vertically opposite angles are equal.

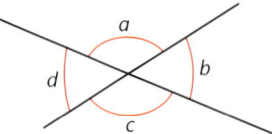

$a = c$ and $b = d$

Angles formed by straight lines

Example 37.1

Question
Work out the size of angles a, b, c, d and e.
Give a reason for each answer.

a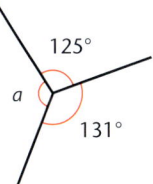

Note
The diagrams are not drawn accurately, so do not try to measure the angles.

b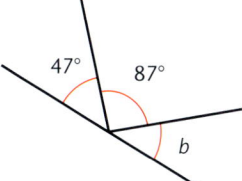

c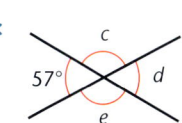

Note
Remember to give a reason for your answer.
State the angle fact that you have used.

Solution

a $a + 125° + 131° = 360°$ The sum of the angles at a point is 360°.
   $a + 256° = 360°$
   $a = 104°$

b $47° + 87° + b = 180°$ The sum of angles on a straight line is 180°.
   $134° + b = 180°$
   $b = 46°$

c $c + 57° = 180°$ The sum of the angles on a straight line is 180°.
   $c = 123°$
   $d = 57°$ Vertically opposite angles are equal.
   $e = 123°$ Vertically opposite angles are equal.

Exercise 37.1

1 Find the size of the angles marked with letters.
   Give reasons for your answers.

   a 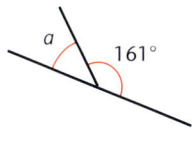   b 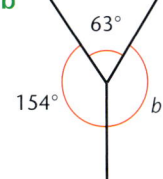   c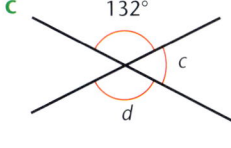

37 ANGLES

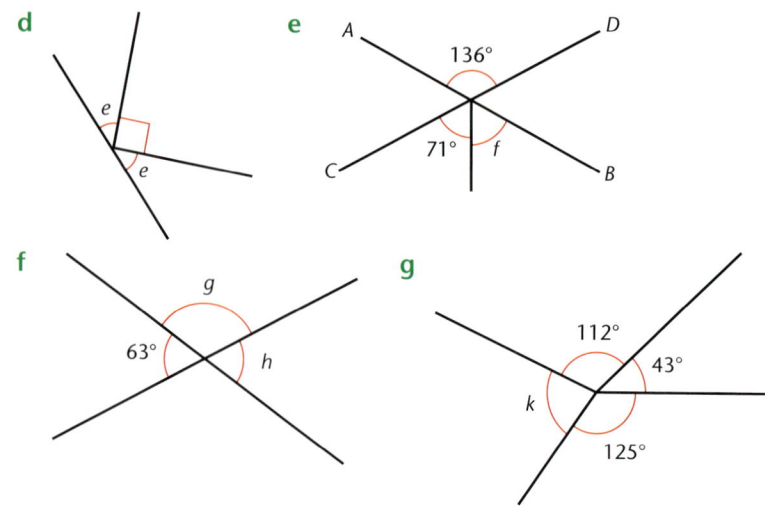

2. Three angles meet at a point. Karim measures them and says the angles are 113°, 123° and 134°.
   Could he be correct?
   Give a reason for your answer.

Angles formed within parallel lines

Alternate angles are equal.

> **Note**
> Thinking of a Z-shape may help you to remember that alternate angles are equal. Alternatively, turn the page round and look at the diagrams upside down and you will see the same shapes!

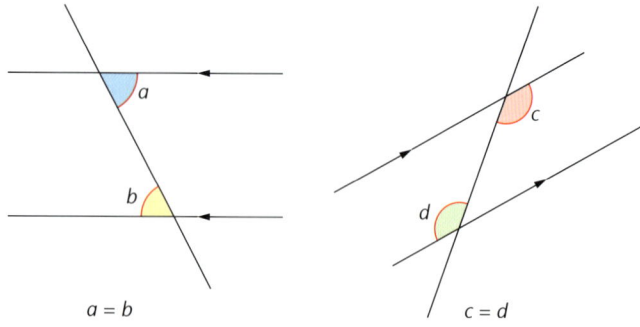

Corresponding angles are equal.

> **Note**
> Thinking of an F-shape or a translation may help you to remember that corresponding angles are equal.

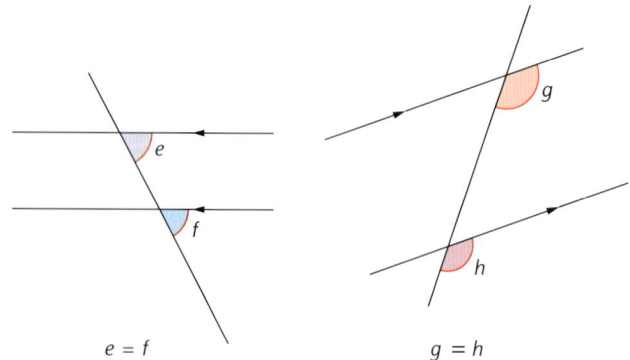

Angles formed within parallel lines

Co-interior angles (sometimes called **allied angles**) are **supplementary**, which means that they add up to 180°.

Note
Remembering a C-shape or using facts about angles on a straight line may help you to remember that co-interior angles add up to 180°.

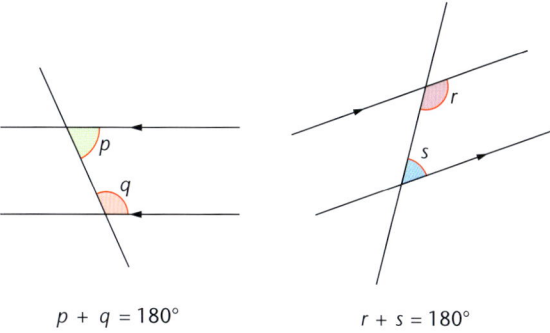

$p + q = 180°$ $r + s = 180°$

Example 37.2

Question
Find the size of each of the lettered angles in these diagrams, giving your reasons.

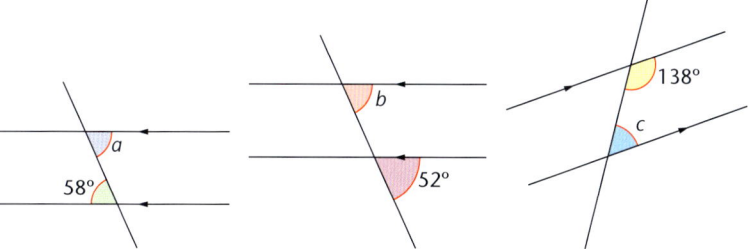

Note
Single letters are often used for angles, as in this chapter so far.

Sometimes, the ends of a line are labelled with letters.

These can be used to name the angles.

In the diagram, angle $ABC = 50°$.

This may also be written $A\bar{B}C = 50°$.

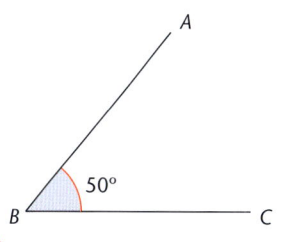

Solution
$a = 58°$ Alternate angles are equal.
$b = 52°$ Corresponding angles are equal.
$c = 42°$ Co-interior angles add up to 180°.

Exercise 37.2

1 Find the value of x, y and z in these diagrams, giving your reasons.

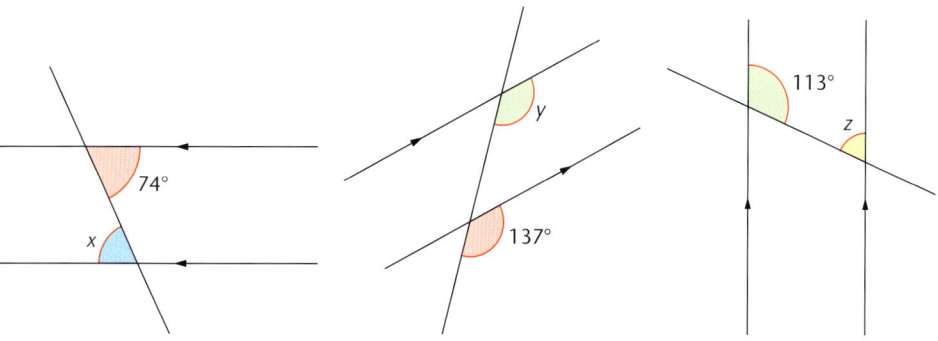

37 ANGLES

2 Find the size of each of the lettered angles in these diagrams.

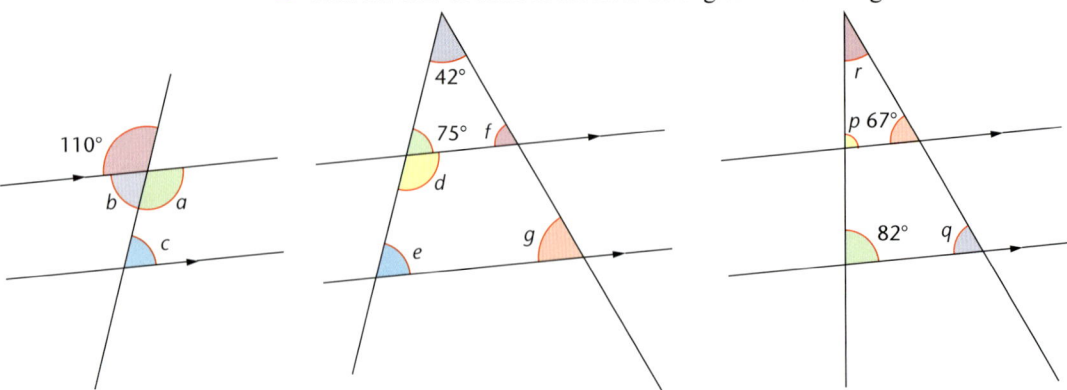

3 In a parallelogram *ABCD*, angle *ABC* = 75°.
   Make a sketch and mark this angle.
   Calculate the other three angles and label them on your sketch.

4 In an isosceles trapezium *ABCD*, angle *BCD* is 127°.
   Make a sketch of the trapezium and mark this angle.
   Calculate the other three angles and label them on your sketch.

5 Find the size of each of the lettered angles in these diagrams and give a reason for your answers.

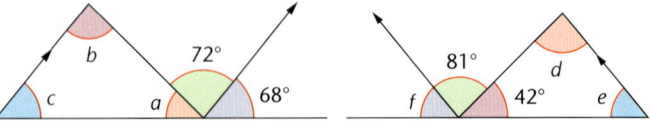

6 Use what you know about triangles to find the sizes of angles *ABE* and *CDE* in this diagram.
   Show why *BE* and *CD* are parallel.

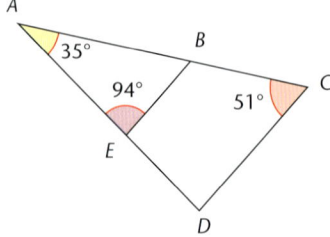

The angles in a triangle

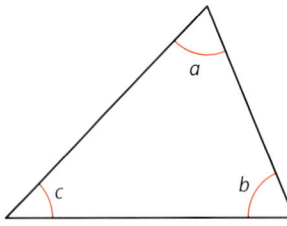

The angles in a triangle add up to 180°.

$a + b + c = 180°$

Here are the three special triangles you have already met.

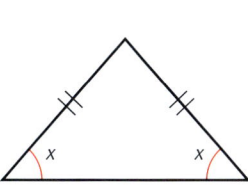
Isosceles triangle

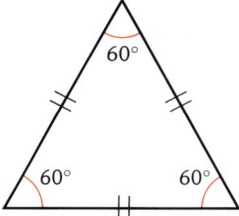
Equilateral triangle

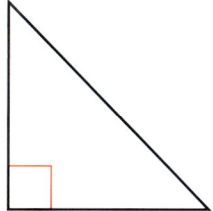
Right-angled triangle

An **isosceles triangle** has two angles the same and two sides the same.

An **equilateral triangle** has all three angles the same (60°) and all three sides the same.

A **right-angled triangle** has one angle of 90°.

A triangle that has all its angles different is called a **scalene triangle**.

These facts can be used to find angles.

Example 37.3

Question

Work out the sizes of the angles in these diagrams.
Sides the same length are marked with two short lines.
Give reasons for your answers.

a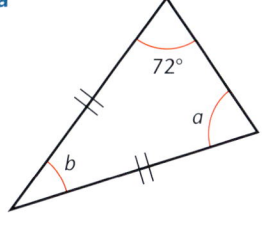

b

c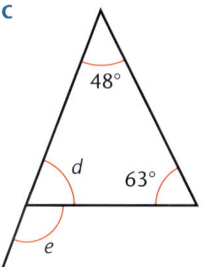

Solution

a $a = 72°$ Equal angles in an isosceles triangle.
   $b + 72° + 72° = 180°$ Angles in a triangle add up to 180°.
   $b + 144° = 180°$
   $b = 36°$

b $c + 63° = 90°$ Other angles in a right-angled triangle add up to 90°, or angles in a triangle add up to 180°.
   $c = 27°$

c $d + 48° + 63° = 180°$ Angles in a triangle add up to 180°.
   $d + 111° = 180°$
   $d = 69°$
   $e + 69° = 180°$ The sum of angles on a straight line is 180°.
   $e = 111°$

Note
Angle $e = 111°$
     $= 63° + 48°$
So the exterior angle of a triangle is equal to the sum of the two opposite interior angles.

37 ANGLES

The angles in a quadrilateral

The angles in a quadrilateral add up to 360°.

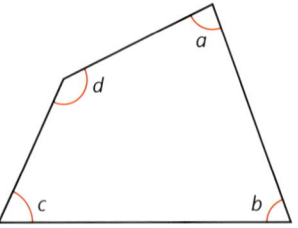

$a + b + c + d = 360°$

Example 37.4

Question

Work out the size of angle x.
Give a reason for your answer.

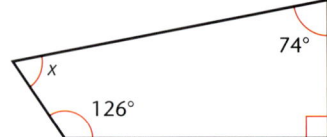

Solution

$x = 360° - (126° + 90° + 74°)$ Angles in a quadrilateral add up to 360°.
$x = 70°$

Exercise 37.3

Calculate the size of the angles marked with letters in these shapes.

Give reasons for your answers.

Note
Sides the same length are marked with two short lines.

Unknown angles marked with the same letter in a diagram are equal in size.

1. (triangle with angles a, 47°, 63°)

2. (triangle with angles 69°, b)

3.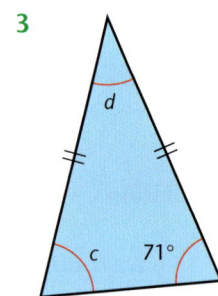

4. (triangle with angles e, f, 36°)

5.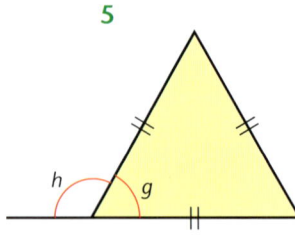

6.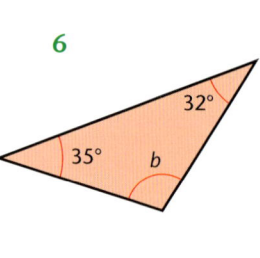

The angles in a quadrilateral

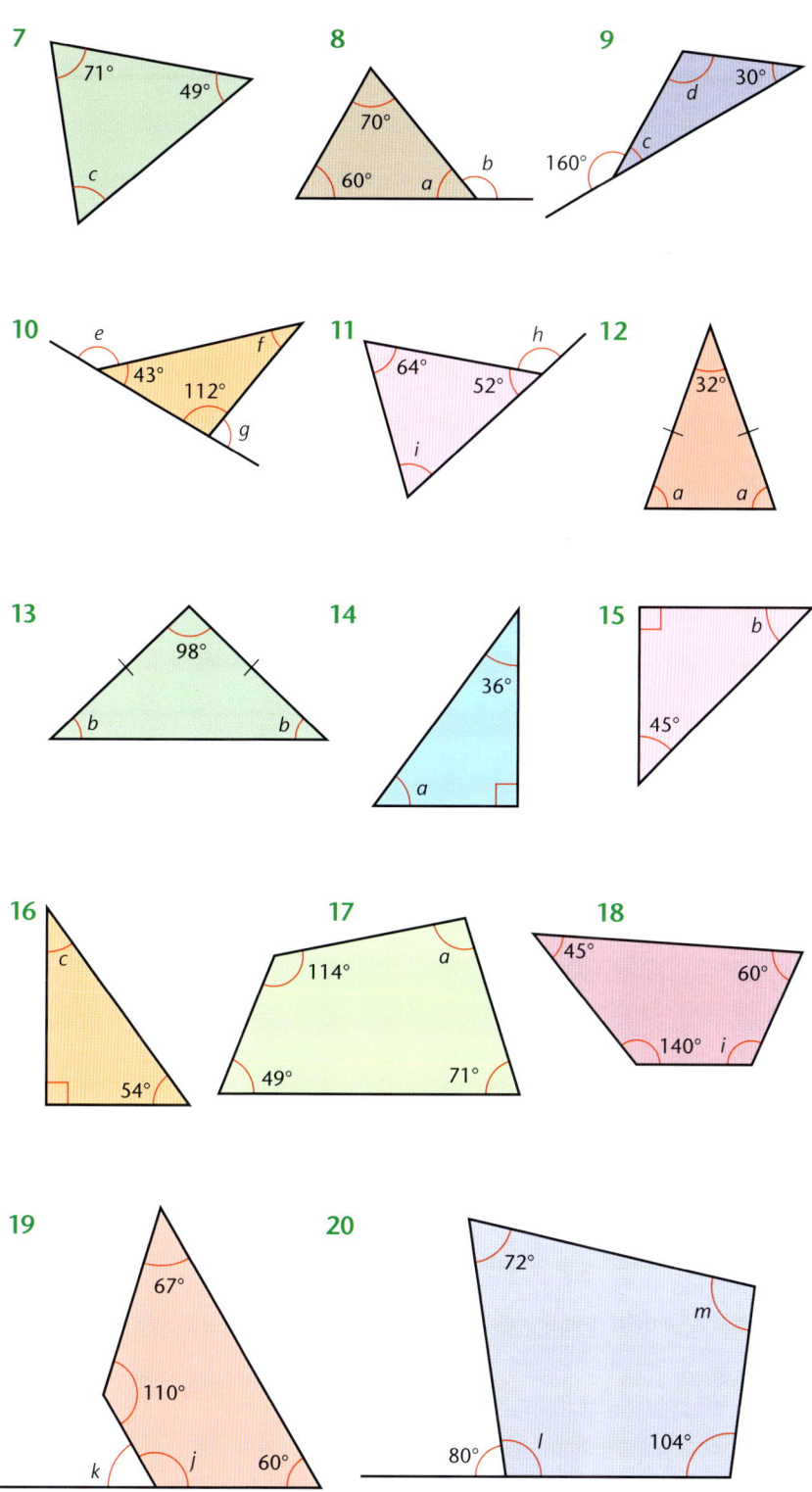

37 ANGLES

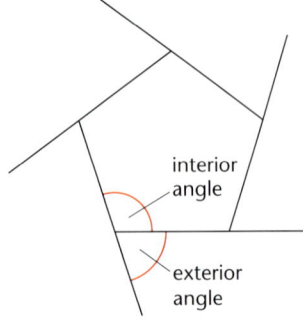

The angles in a polygon

The sum of the exterior angles of any convex polygon is 360°.

At each vertex, the interior and exterior angles make a straight line.

For any convex polygon, at any vertex: interior angle + exterior angle = 180°.

Example 37.5

Question

The two unlabelled exterior angles of this pentagon are equal.
Find their size.

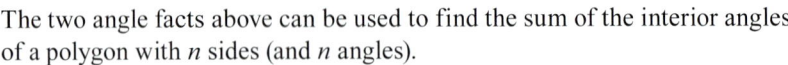

Solution

First, find the sum of the angles that are given.
72° + 80° + 86° = 238°
The sum of all the exterior angles is 360°.
The sum of the remaining two angles = 360° − 238° = 122°.
So each angle is 122 ÷ 2 = 61°.

The two angle facts above can be used to find the sum of the interior angles of a polygon with n sides (and n angles).

At each vertex of a convex polygon, the interior angle + the exterior angle = 180°.

Sum of (interior + exterior angles) for the polygon = 180° × n

But the sum of the exterior angles is 360°.

For an n-sided convex polygon,
the sum of the interior angles = $(180n - 360°) = 180° (n - 2)$

Another way you can find the sum of the interior angles of any polygon is to divide it into triangles.

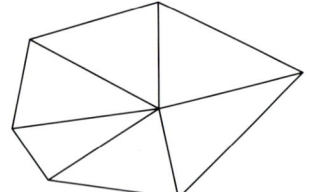

The sum of the angles in all the triangles = 180° × n

The sum of the angles at the centre = 360° and this must be subtracted.

Example 37.6

Question

Find the value of x in this hexagon.

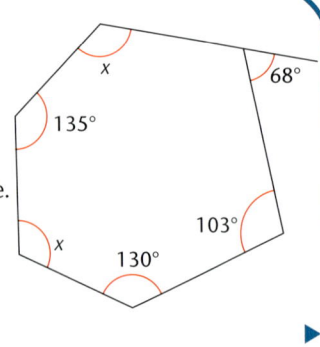

Solution

First, find the size of the missing interior angle.
Interior angle + exterior angle = 180°
Interior angle = 180° − 68°
= 112°

Next, find the sum of the interior angles.
Sum of interior angles = 112° + 103° + 130° + 135° + 2x
= 480° + 2x
The sum of the interior angles of a hexagon = 180°(n − 2)
= 180° × 4
= 720°

Therefore
480° + 2x = 720°
2x = 720° − 480°
2x = 240°
x = 120°

Example 37.7

Question
Find the interior angle of a regular hexagon.

Solution
For a regular hexagon, the exterior angle = 360° ÷ 6 = 60°.
So the interior angle = 180° − 60° = 120°.

Example 37.8

Question
Find the number of sides of a regular polygon with an interior angle of 144°.

Solution
If the interior angle = 144°, the exterior angle = 180° − 144° = 36°.
The sum of the exterior angles = 360°.
Therefore the number of sides = 360° ÷ 36° = 10.

Exercise 37.4

1 Four of the exterior angles of a pentagon are 70°, 59°, 83° and 90°.
   a Find the size of the other exterior angle.
   b Find the size of each interior angle.
2 Five of the exterior angles of a hexagon are 54°, 48°, 65°, 35° and 80°.
   a Find the size of the other exterior angle.
   b Find the size of each interior angle.
3 Four of the exterior angles of a hexagon are 67°, 43°, 91° and 37°.
   a Find the size of the other exterior angles, given that they are equal.
   b Find the size of each interior angle.

4 A regular polygon has nine sides.
  Find the size of each of its exterior and interior angles.
5 Find the size of the interior angle of
  a a regular dodecagon (12 sides)
  b a regular 20-sided polygon.
6 A regular polygon has an exterior angle of 24°.
  How many sides does it have?
7 Six of the angles of a heptagon are 122°, 141°, 137°, 103°, 164° and 126°.
  Calculate the size of the remaining angle.
8 A regular polygon has 15 sides.
  Find the size of each of its exterior and interior angles.
9 A regular polygon has an exterior angle of 30°.
  How many sides does it have?
10 A polygon has 11 sides. Ten of its interior angles add up to 1490°.
  Find the size of the remaining angle.

Key points

- Angles around a point add up to 360°.
- Angles on a straight line add up to 180°.
- Where two lines intersect, opposite angles are equal.
- Alternate angles in parallel lines are equal and form a Z-shape.
- Corresponding angles in parallel lines are equal and form an F-shape.
- Co-interior angles in parallel lines add up to 180° and form a C-shape.
- Angles in a triangle add up to 180°.
- Angles in a quadrilateral add up to 360°.
- The exterior angles of a polygon add up to 360°.
- At any vertex of a convex polygon, the interior angle and the exterior angle add up to 180°.
- Angles in an n-sided convex polygon add up to $180° \times (n - 2)$.

38 CIRCLE THEOREMS

> **BY THE END OF THIS CHAPTER YOU WILL BE ABLE TO:**
> - calculate unknown angles and give explanations using the following geometrical properties of circles:
>   - angle in a semi-circle = 90°
>   - angle between tangent and radius = 90°
>   - angle at the centre is twice the angle at the circumference
>   - angles in the same segment are equal
>   - opposite angles of a cyclic quadrilateral sum to 180° (supplementary)
>   - alternate segment theorem
> - use the following symmetry properties of circles:
>   - equal chords are equidistant from the centre
>   - the perpendicular bisector of a chord passes through the centre
>   - tangents from an external point are equal in length.

> **CHECK YOU CAN:**
> - distinguish between acute, obtuse, reflex and right angles
> - distinguish between equilateral, isosceles, scalene and right-angled triangles
> - understand the meaning of the term *congruent*
> - understand and use geometrical terms.

Symmetry properties of circles

The perpendicular from the centre to a chord

AB is a chord to the circle, centre O.

X is the midpoint of AB.

OA and OB are radii, so triangle OAB is isosceles.

OX is a line of symmetry for triangle OAB.

Therefore OX is perpendicular to AB.

This shows that

> the perpendicular from the centre to a chord bisects the chord.

Equal chords

AB and CD are chords to the circle, centre O.

Chords AB and CD are of equal length.

X is the midpoint of AB and Y is the midpoint of CD.

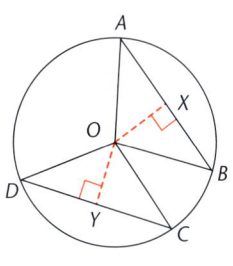

OA, OB, OC and OD are radii so are equal, so triangles OAB and OCD are congruent (SSS).

Therefore $OX = OY$.

This shows that

> equal chords are equidistant from the centre of the circle.

Photocopying is prohibited

38 CIRCLE THEOREMS

Tangents from an external point

T is a point outside a circle, centre O.

TX and TY are the tangents from T to the circle.

OX and OY are radii so are equal.

OT is a line of symmetry and bisects $X\hat{O}Y$, so $X\hat{O}T = Y\hat{O}T$.

OT is common to triangles OXT and OYT.

So triangles OXT and OYT are congruent (SAS).

Therefore $XT = YT$.

This shows that

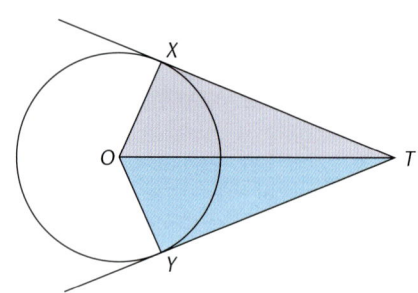

> tangents to a circle from an external point are equal in length.

These facts can be used to solve geometrical problems.

Example 38.1

Question

X is the midpoint of chord AB.
Given that $O\hat{A}B = 42°$, find

a $O\hat{B}A$ b $A\hat{O}B$ c $A\hat{O}X$.

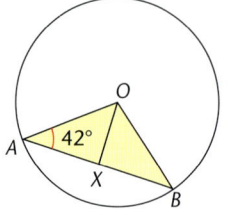

Solution

a $O\hat{B}A = 42°$ Triangles AOX and BOX are congruent.
b $A\hat{O}B = 96°$ Angles in triangle AOB add up to 180°.
c $A\hat{O}X = 48°$ By symmetry, $A\hat{O}X = B\hat{O}X$.

Exercise 38.1

1 AT and BT are tangents to the circle, centre O.
   What is the mathematical name of quadrilateral $OATB$?
   Explain how you know.

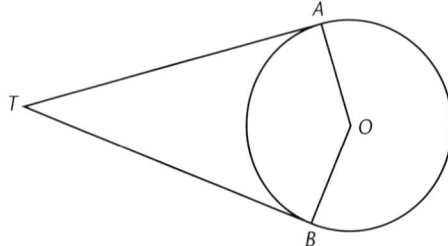

2 P, Q, R and S are points on the circle, centre O.
   X is the midpoint of PQ, Y is the midpoint of RS and OX = OY.
   Find three triangles that are congruent to triangle OXP.
   Explain how you know.

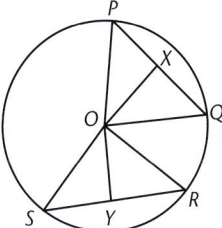

3 A, B and C are points on the circle, centre O.
   M is the midpoint of BC and $C\hat{A}M = 28°$.
   Work out $M\hat{C}A$.
   Give a reason for your answer.

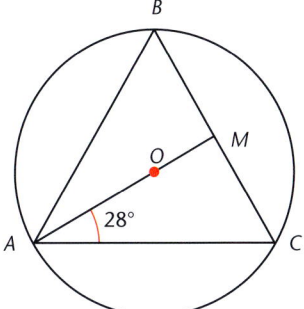

Angles in a circle

The angle subtended by an arc at the centre is twice the angle subtended at the circumference

BC is a chord of the circle whose centre is O.

Arc BC subtends angle BOC at the centre of the circle and angle BAC at the circumference and AD is a straight line.

Here is a simple proof that angle COB = twice angle CAB.

Let angle CAO = x and angle BAO = y.

Then angle CAB = x + y.

Triangles OAB and OAC are isosceles.   OA, OB and OC are radii of the circle.

Angle ACO = x and angle ABO = y   Base angles of isosceles triangles.

Angle DOC = angle OAC + angle OCA = 2x   The exterior angle of a triangle
Angle DOB = angle OAB + angle OBA = 2y   equals the sum of the opposite interior angles.

> **Note**
> An angle is formed or **subtended** at a particular point by an arc when straight lines from its ends are joined at that point.

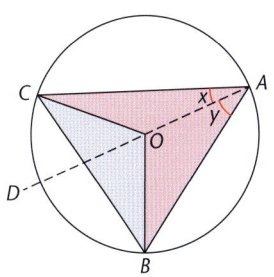

Photocopying is prohibited   325

38 CIRCLE THEOREMS

> **Note**
>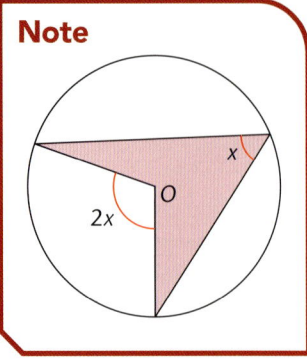

Angle COB = angle DOC + angle DOB

$\qquad = 2x + 2y$

$\qquad = 2(x + y)$

$\qquad = 2 \times$ angle CAB

Example 38.2

Question

In the diagram, O is the centre of the circle.
Calculate the size of angle a.

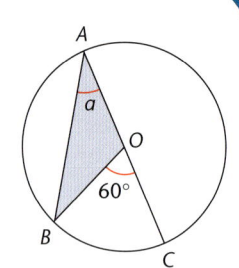

Solution

Angle $a = 30°$ — Angle at the centre = twice the angle at the circumference.

The angle subtended at the circumference in a semi-circle is a right angle

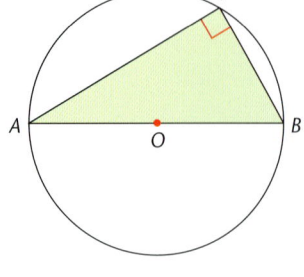

This is a special case of the angle subtended by an arc at the centre is twice the angle subtended at the circumference.

Angle $AOB = 2 \times$ angle APB — The angle at the centre is twice the angle at the circumference.

Angle $AOB = 180°$ — AB is a diameter, that is, a straight line.

Angle $APB = 90°$ — Half of angle AOB.

So the angle in a semi-circle is 90°.

This theorem is often used in conjunction with the angle sum of a triangle, as is shown in the next example.

Example 38.3

Question

Work out the size of angle x.
Give a reason for each step of your work.

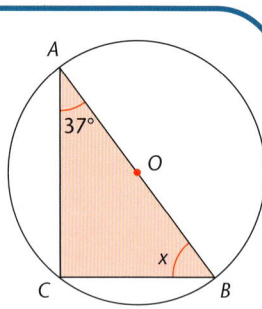

Solution

Angle $ACB = 90°$ — The angle in a semi-circle is a right angle.

$x = 180 - (90 + 37)$ — The angle sum of a triangle is 180°.

$x = 53°$

---

326   Photocopying is prohibited

Angles in a circle

Exercise 38.2

In the diagrams, O is the centre of the circle.

Find the size of each of the lettered angles.

Write down the reasons for each step of your calculations.

Note
When calculating an angle, never assume something is true, for example, 'because it looks like it' on the diagram. You must always give a reason for each step of your calculation.

1. 70°, a

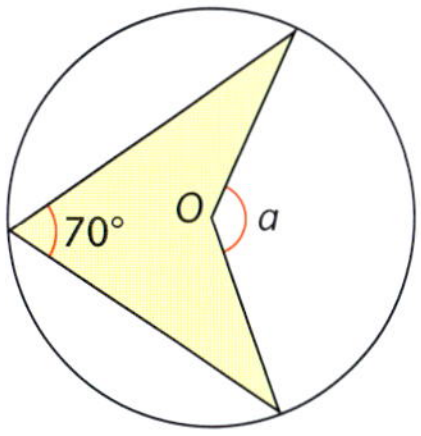

2. b, 90°, c

3. e, d, 40°

4. 120°, f, g

5. 20°, 90°, h

6. 68°, i

7. j, k, 90°

8. 84°, l

9. m, $2m$

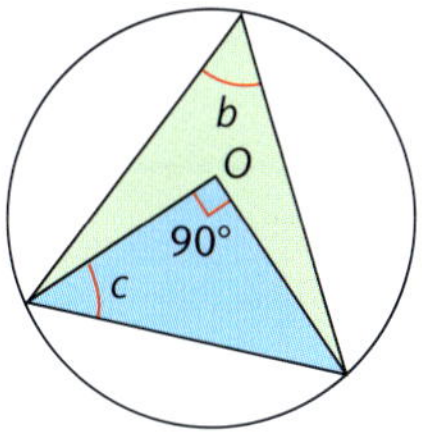

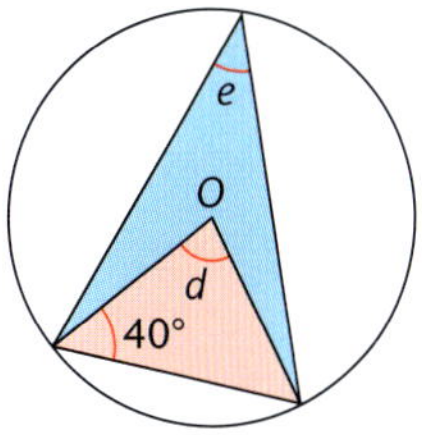

10. n, 240°

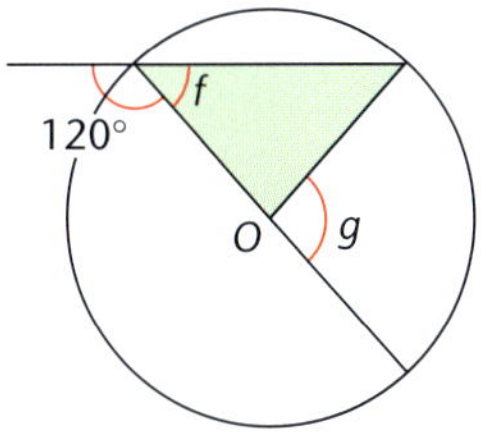

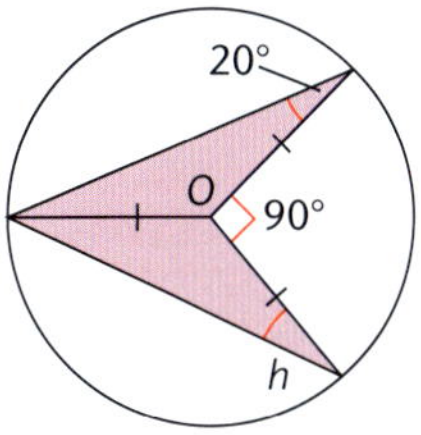

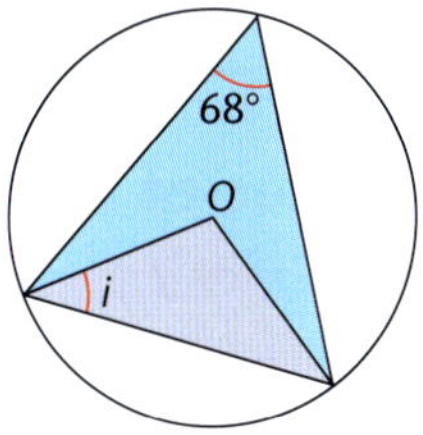

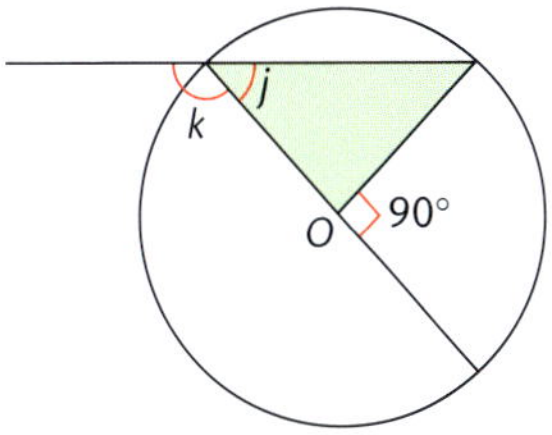

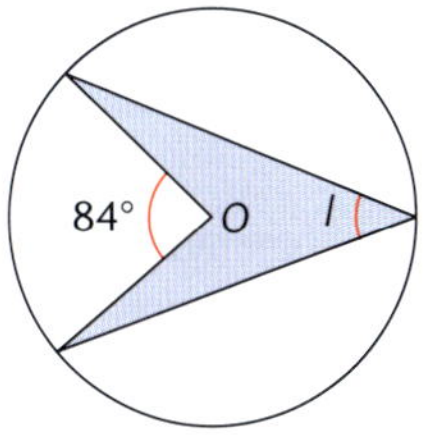

327

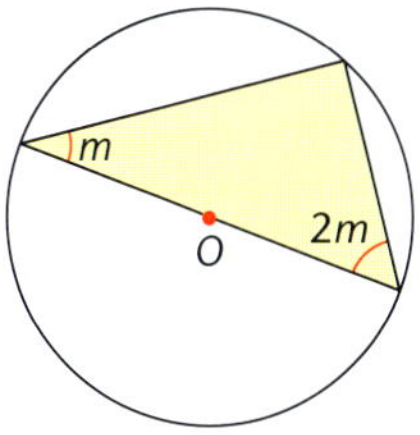

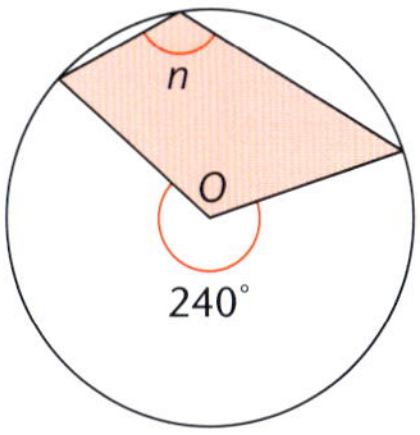

38 CIRCLE THEOREMS

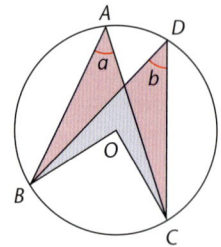

Angles in the same segment of a circle are equal

In the diagram, $a = b$, as both are equal to half the angle subtended at the centre.

To use this property, you have to identify angles subtended by the same arc since these are in the same segment. This is shown in the next example.

Example 38.4

Question

Find the sizes of angles a, b and c.
Give a reason for each step of your work.

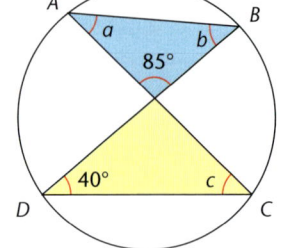

Solution

a = angle BDC = 40°   Angles in the same segment are equal.
b = 180 − (40 + 85)
  = 55°   The angle sum of a triangle is 180°.
$c = b = 55°$ Angles in the same segment are equal.

Opposite angles of a cyclic quadrilateral are supplementary

A **cyclic quadrilateral** has all four vertices on a circle.

This property is also derived from the fact that the angle at the centre is twice the angle at the circumference.

In the diagram

angle $ABC = \frac{1}{2}x$ The angle at the centre is twice the angle at the circumference.

angle $ADC = \frac{1}{2}y$

$x + y = 360°$ The angles around a point add up to 360°.

angle ABC + angle $ADC = \frac{1}{2}(x + y)$

angle ABC + angle $ADC = 180°$

So the opposite angles of a cyclic quadrilateral are supplementary.

Note

Two angles are *supplementary* if they add up to 180°.

Angles in a circle

Example 38.5

Question

Find the sizes of angles c and d.

Give a reason for each step of your work.

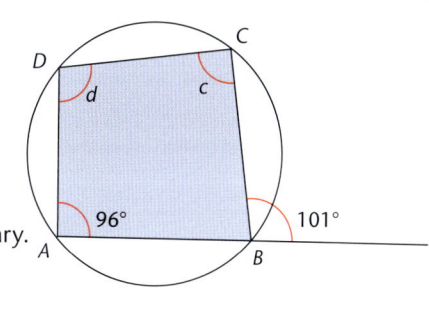

Solution

$c = 180 - 96 = 84°$ — Opposite angles of a cyclic quadrilateral are supplementary.

Angle $ABC = 180 - 101 = 79°$ — Angles on a straight line add up to $180°$.

$d = 180 - 79 = 101°$ — Opposite angles of a cyclic quadrilateral are supplementary.

Exercise 38.3

In the diagrams, O is the centre of the circle.

Find the size of each of the lettered angles.

Write down the reasons for each step of your calculations.

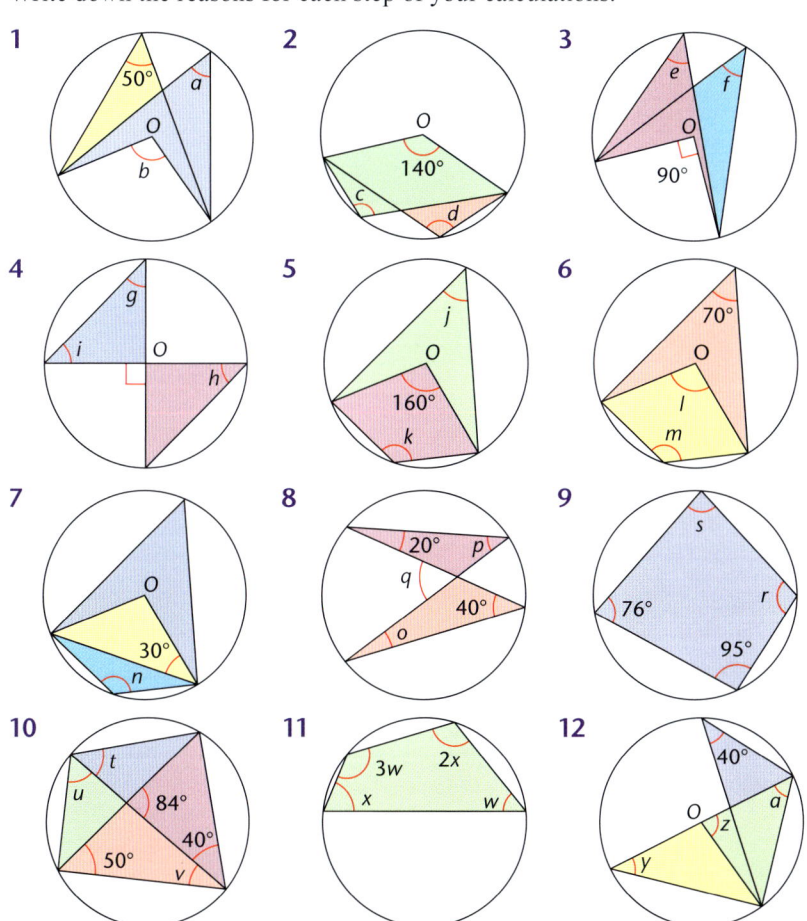

38 CIRCLE THEOREMS

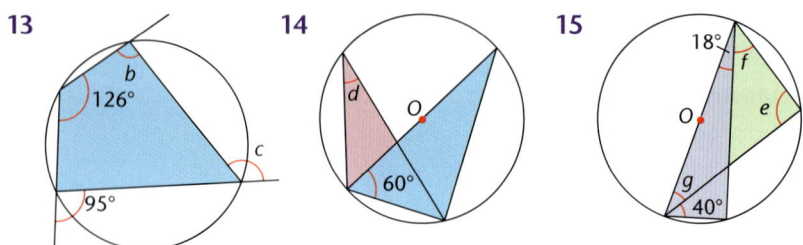

You need to be able to identify when to use the circle theorems you have learned.

Use this exercise to practise.

Exercise 38.4

Calculate the sizes of the angles marked with letters.

O is the centre of each circle.

Give the reasons for each step of your working.

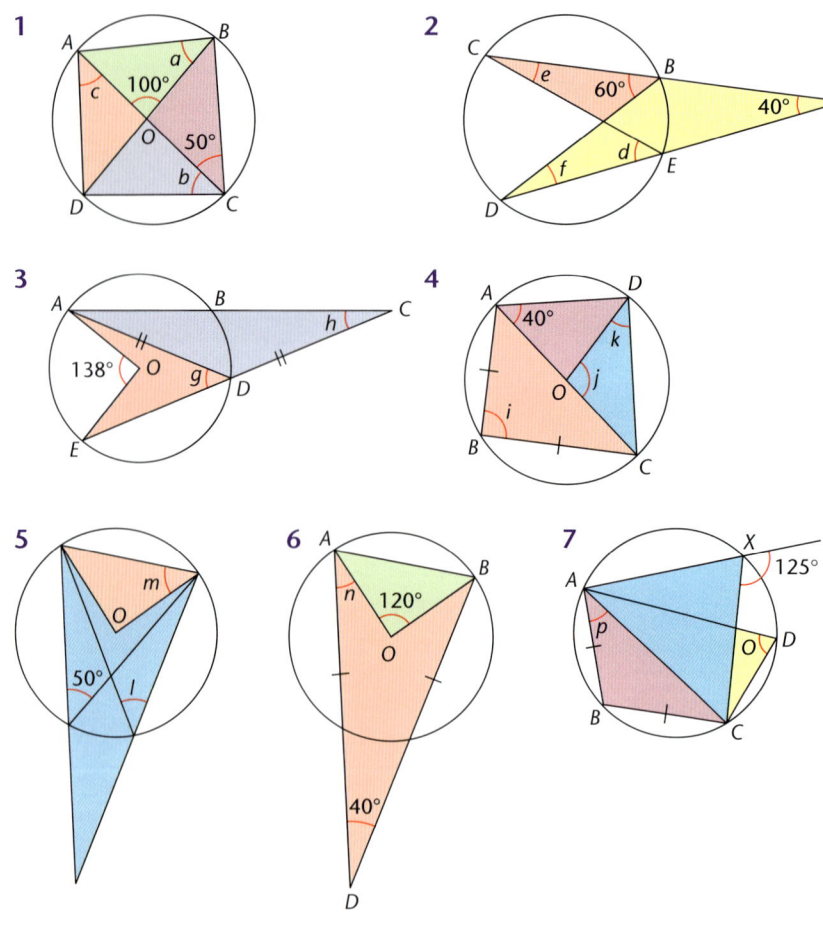

8 9

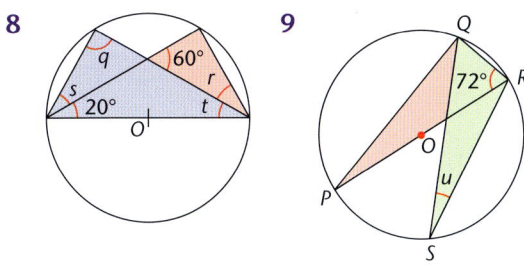

The tangent to a circle is at right angles to the radius at the point of contact

Earlier in this chapter, you saw that the perpendicular bisector of a chord passes through the centre of the circle.

So the line from the centre to the midpoint of the chord is at right angles to the chord.

If you move the chord further from the centre it will eventually become a tangent and *ON* will be a radius.

So the tangent to a circle is at right angles to the radius at the point of contact.

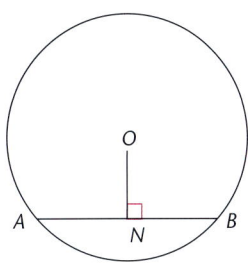 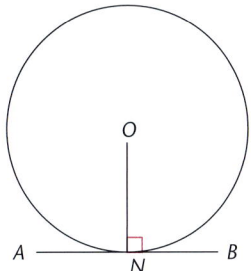

Exercise 38.5

In the diagrams, *O* is the centre of the circle and *X* and *Y* are the points of contact of the tangents to each circle.

Find the size of each of the lettered angles.

Write down the reasons for each step of your calculations.

1 2

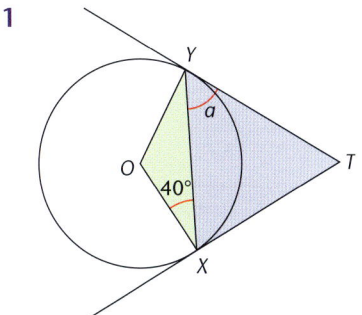 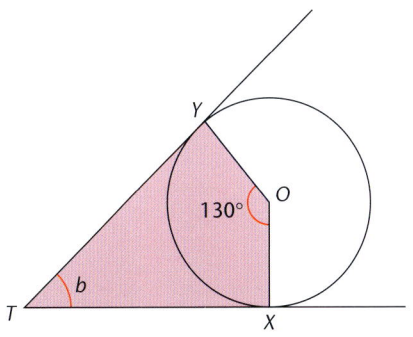

38 CIRCLE THEOREMS

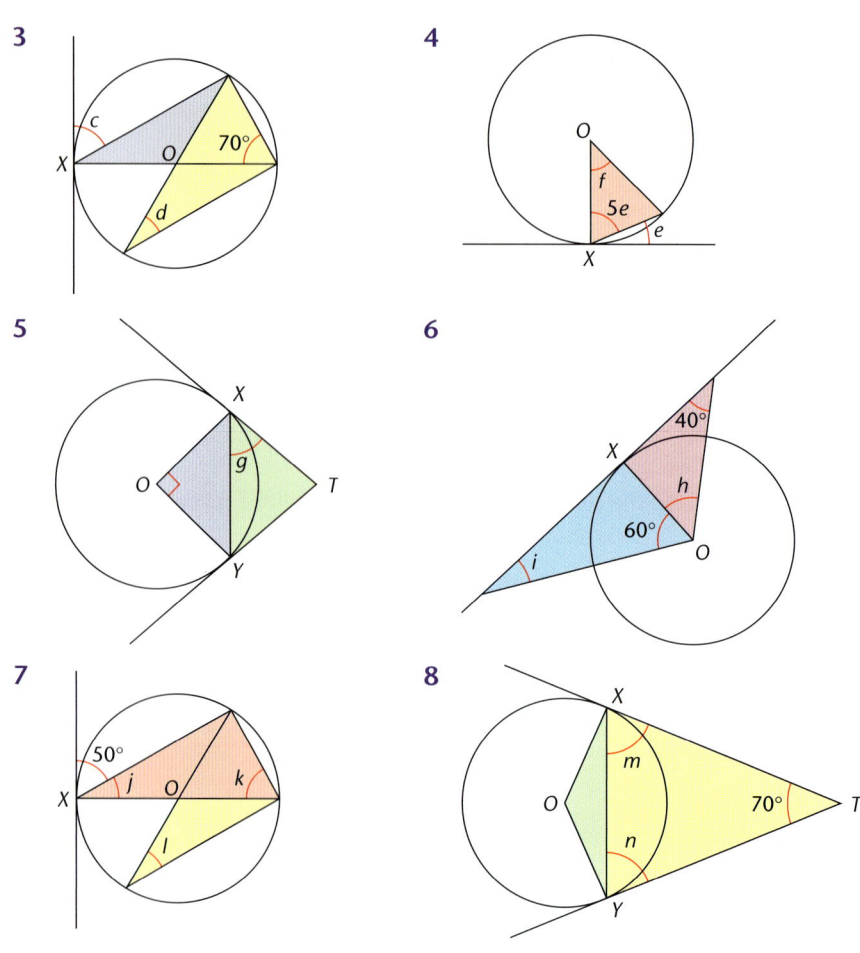

The alternate segment theorem

The angle between a tangent and a chord is equal to any angle made by that chord in the alternate segment of the circle.

$a = b$

and

$c = d$

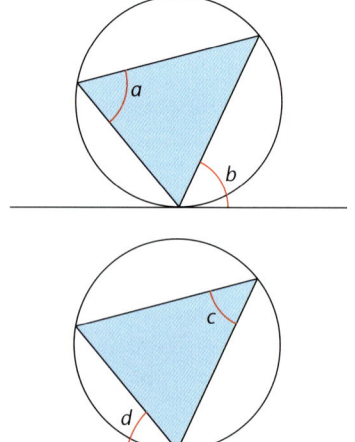

The alternate segment theorem

This can be proved as follows. O is the centre of the circle.

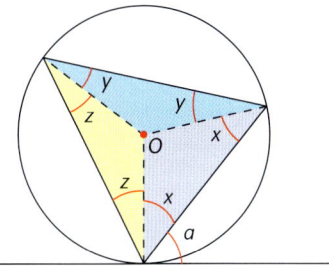

The dotted radii make three isosceles triangles with the base angles equal in each triangle.

$a = 90 - x$ The tangent is perpendicular to the radius.

$2x + 2y + 2z = 180$ Angles in a triangle add up to 180°.

So that $x + y + z = 90$

And $y + z = 90 - x$

Therefore $a = y + z$

So, the angle between the tangent and the chord, a, equals the angle in the alternate segment, $y + z$.

Example 38.6

Question
Find the size of angle x and angle y.

Solution
$x = 180 - 2 \times 24 = 132°$ Angles at the base of an isosceles triangle are equal.

$y = 132°$ Alternate segment theorem.

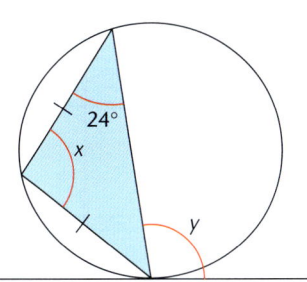

Exercise 38.6

In the following questions, find the sizes of the angles marked with letters. Give reasons for your answers.

1

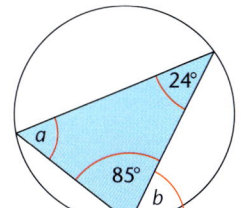

2

38 CIRCLE THEOREMS

3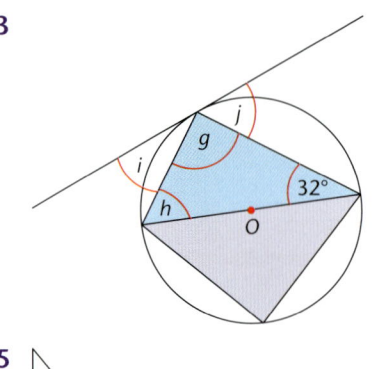

4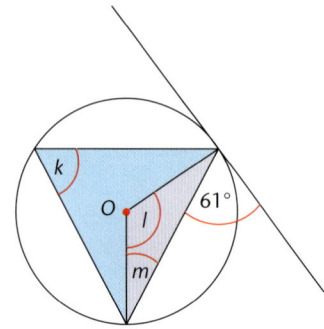

5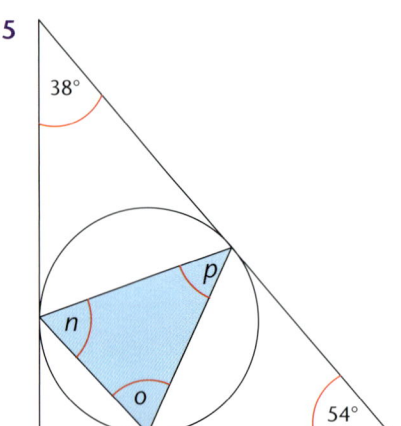

Key points
- The perpendicular from the centre of a circle to a chord bisects the chord.
- Equal chords are equidistant from the centre of the circle.
- Tangents to a circle from an external point are equal in length.
- The angle at the centre of the circle is twice the angle at the circumference.
- The angle in a semi-circle is 90°.
- Angles in the same segment are equal.
- Opposite angles in a cyclic quadrilateral add up to 180°.
- The tangent to a circle is perpendicular to the radius.
- The angle between a tangent and a chord is equal to the angle in the alternate segment.

39 UNITS OF MEASURE

Basic units of length, mass and capacity

These are the connections between the metric units of length.

1 kilometre = 1000 metres	1 km = 1000 m
1 metre = 1000 millimetres	1 m = 1000 mm
1 metre = 100 centimetres	1 m = 100 cm
1 centimetre = 10 millimetres	1 cm = 10 mm

These are the connections between the metric units of mass.

| 1 kilogram = 1000 grams | 1 kg = 1000 g |

When a volume is filled with liquid or gas, it is called the **capacity**.

These are the connections between the metric units of capacity.

| 1 litre = 1000 millilitres | 1 l = 1000 ml |

A millilitre is exactly the same as a cm^3, but is used for liquids rather than cm^3.

BY THE END OF THIS CHAPTER YOU WILL BE ABLE TO:
- use metric units of mass, length, area, volume and capacity in practical situations and convert quantities into larger or smaller units.

CHECK YOU CAN:
- multiply and divide by powers of 10
- add and subtract decimals
- find the volume of a cuboid.

Example 39.1

Question

Misha is baking some cakes.
One recipe needs 1.6 kg of flour, another needs $\frac{1}{2}$ kg of flour.
a How much flour does he need altogether?
b He has a new 3 kg bag of flour.
   How much will be left after he has made the cakes?

Solution

a When you have to add fractions and/or decimal parts of a kilogram, it is usually easier to change all the weights to grams.
   Total = 1.6 kg + $\frac{1}{2}$ kg = 1600 g + 500 g
        = 2100 g or 2.1 kg
b Amount of flour left = 3 kg − 2.1 kg = 3000 g − 2100 g
                       = 0.9 kg or 900 g

Photocopying is prohibited

39 UNITS OF MEASURE

Exercise 39.1

1. Which metric unit would you use to measure each of these?
   - a The width of a book.
   - b The height of a room.
   - c The length of a classroom.
   - d The length of a finger.

2. Change each of these lengths to millimetres.
   - a 4.52 cm
   - b 2 cm
   - c 4.5 cm
   - d 9.35 cm
   - e 2.101 m
   - f 3 m
   - g 2.239 m
   - h 9.1 m

3. Change each of these lengths to centimetres.
   - a 52 m
   - b 5 m
   - c 2.32 m
   - d 18.16 m
   - e 660 mm
   - f 70 mm
   - g 310 mm
   - h 46 mm

4. Change each of these lengths to metres.
   - a 2146.3 mm
   - b 5142 mm
   - c 570 cm
   - d 1146 mm

5. Write each set of lengths in order of size, smallest first.
   - a 2.42 m   1600 mm   284 cm   9 m   31 cm
   - b 423 cm   6100 mm   804 cm   3.2 m   105 mm

6. Marta runs for 3.5 km, walks for 800 m, then runs another 2.4 km. How far has she gone altogether?

7. Change each of these masses to grams.
   - a 0.012 kg
   - b 7 kg
   - c 1.13 kg
   - d 2.14 kg

8. Change each of these masses to kilograms.
   - a 6600 g
   - b 8000 g
   - c 6300 g
   - d 5126 g

9. Write each set of masses in order of size, lightest first.
   - a 4000 g   52 000 g   9.4 kg   874 g   1.7 kg
   - b 4123 g   2104 g   3.4 kg   0.174 kg   2.79 kg

10. Nasrin uses 750 g of sugar from a $1\frac{1}{2}$ kg bag. How much is left?

11. Change these capacities to millilitres.
    - a 52 cl
    - b 7 litres
    - c 1.52 litres
    - d 0.16 litres

12. Change these capacities to litres.
    - a 9503 cl
    - c 2000 ml
    - c 2341 ml

13. A bottle holds 1 litre of lemonade. Abi drinks 350 ml. How much is left in the bottle?

Area and volume measures

> **Note**
> Remember that volume and capacity units are related.
> 1 litre = 1000 cm³
> 1 ml = 1 cm³

You can use the basic relationships between metric units of length to work out the relationships between metric units of area and volume.

For example:

$1\,cm^2 = 1\,cm \times 1\,cm = 10\,mm \times 10\,mm = 100\,mm^2$

$1\,cm^3 = 1\,cm \times 1\,cm \times 1\,cm = 10\,mm \times 10\,mm \times 10\,mm = 1000\,mm^3$

$1\,m^2 = 1\,m \times 1\,m = 100\,cm \times 100\,cm = 10\,000\,cm^2$

$1\,m^3 = 1\,m \times 1\,m \times 1\,m = 100\,cm \times 100\,cm \times 100\,cm = 1\,000\,000\,cm^3$

Area and volume measures

Example 39.2

Question

Change these units.

a $5\,m^3$ to cm^3 b $5600\,cm^2$ to m^2

Solution

a $5\,m^3 = 5 \times 1\,000\,000\,cm^3$ Convert $1\,m^3$ to cm^3 and multiply by 5.
      $= 5\,000\,000\,cm^3$

b $5600\,cm^2 = 5600 \div 10\,000\,m^2$ To convert from m^2 to cm^2 you multiply,
           $= 0.56\,m^2$ so to convert from cm^2 to m^2 you divide.

Note

Make sure you have done the right thing by checking that your answer makes sense. If you had multiplied by 10 000, you would have got $56\,000\,000\,m^2$, which is obviously a much larger area than $5600\,cm^2$.

Exercise 39.2

1 Change these units.
   a $3\,m^2$ to cm^2 b $2.3\,cm^2$ to mm^2 c $9.52\,m^2$ to cm^2 d $0.014\,cm^2$ to mm^2

2 Change these units.
   a $90\,000\,mm^2$ to cm^2 b $8140\,mm^2$ to cm^2 c $7\,200\,000\,cm^2$ to m^2 d $94\,000\,cm^2$ to m^2

3 Change these units.
   a $3.2\,m^3$ to cm^3 b $42\,cm^3$ to m^3 c $5000\,cm^3$ to m^3 d $6.42\,m^3$ to cm^3

4 Change these units.
   a 2.61 litres to cm^3 b 9500 ml to litres   c 2.4 litres to ml   d 910 ml to litres

5 What is wrong with this statement?
   'The trench I have just dug is 5 m long, 2 m wide and 50 cm deep. To fill it in, I would need $500\,m^3$ of concrete.'

6 The carton shown on the right holds 1 litre of juice. How high is it?

7 How many litres are there in 1 cubic metre?

8 1 hectare = $10\,000\,m^2$.
   How many hectares are there in $1\,km^2$?

9 A sugar cube has sides of 15 mm.
   Find how many will fit in a box measuring 11 cm by 11 cm by 5 cm.

10 Two soccer pitches have an area of 1 hectare.
   Each pitch is 100 m long.
   How wide is each pitch?

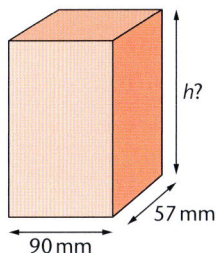

Key points

- Everyday units of length are millimetres (mm), centimetres (cm), metres (m) and kilometres (km).
- There are 10 millimetres in a centimetre, 100 centimetres in a metre and 1000 metres in a kilometre.
- Everyday units of mass are kilograms (kg) and grams (g).
- There are 1000 grams in a kilogram.
- Everyday units of capacity are millilitres (ml) and litres (l).
- There are 1000 millilitres in a litre.
- A millilitre has the same volume as $1\,cm^3$.
- $1\,km^2 = 1\,000\,000\,m^2$. $1\,m^2 = 10\,000\,cm^2$. $1\,cm^2 = 100\,mm^2$.
- 1 litre = $1000\,cm^3$.   1 ml = $1\,cm^3$.
- $1\,m^3 = 1\,000\,000\,cm^3 = 1000$ litres.

Photocopying is prohibited

40 MENSURATION

CHECK YOU CAN:

- use metric units for length, area, volume and capacity
- understand the geometric terms used with shapes
- rearrange formulas and solve equations
- round numbers to a given number of decimal places or significant figures.

For calculating the surface area of a pyramid and a cone, you will also need to be able to:

- use Pythagoras' theorem.

For calculating the volume and surface area of more complex compound shapes, you will also need to be able to:

- use Pythagoras' theorem
- calculate lengths in similar figures
- find the area of a non-right-angled triangle using the formula $\frac{1}{2} bc \sin A$.

BY THE END OF THIS CHAPTER YOU WILL BE ABLE TO:

- use metric units of mass, length, area, volume and capacity in practical situations and convert quantities into larger or smaller units
- carry out calculations involving the perimeter and area of a rectangle, triangle, parallelogram and trapezium
- carry out calculations involving the circumference and area of a circle
- carry out calculations involving arc length and sector area as fractions of the circumference and area of a circle
- carry out calculations and solve problems involving the surface area and volume of a: cuboid, prism, cylinder, sphere, pyramid and cone
- carry out calculations and solve problems involving perimeters and areas of: compound shapes, parts of shapes
- carry out calculations and solve problems involving surface areas and volumes of: compound solids, parts of solids.

The perimeter of a 2-D shape

The **perimeter** of a shape is the distance all the way around the edge of the shape.

Since the perimeter of a shape is a length, you must use units such as centimetres (cm), metres (m) or kilometres (km).

Example 40.1

Question
Find the perimeter of this shape.

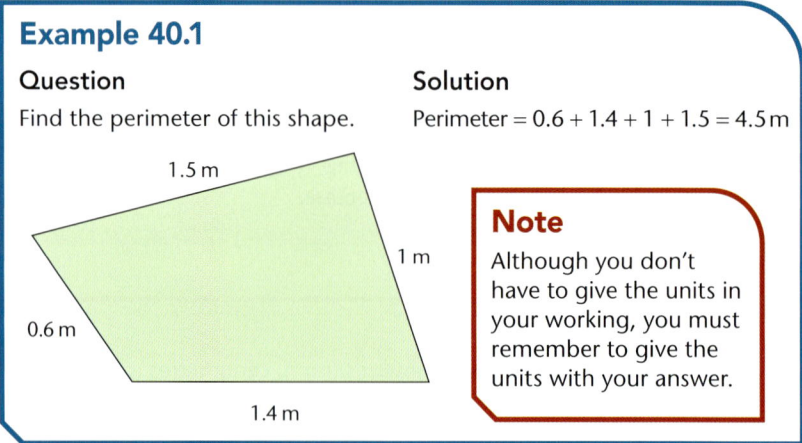

Solution
Perimeter = 0.6 + 1.4 + 1 + 1.5 = 4.5 m

Note
Although you don't have to give the units in your working, you must remember to give the units with your answer.

Sometimes not all of the lengths of the shape are given in the diagram. Before trying to find the perimeter, work out all the lengths.

The perimeter of a 2-D shape

Example 40.2

Question

Find the perimeter of this shape made from rectangles.

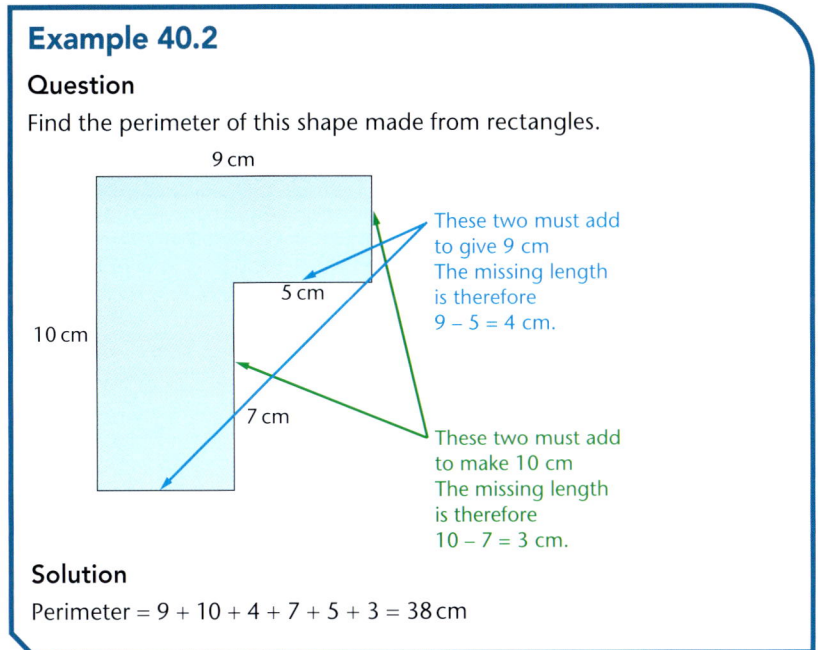

Solution

Perimeter = 9 + 10 + 4 + 7 + 5 + 3 = 38 cm

Exercise 40.1

1 Find the perimeter of each of these rectangles.

   a   **b**

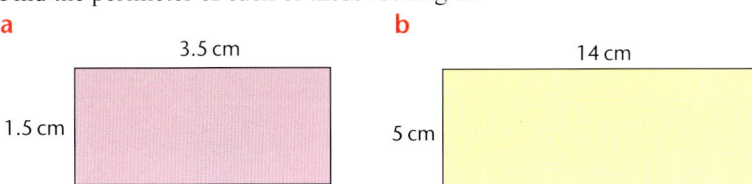

2 Copy each of these diagrams, where the shapes are made from rectangles.
   Find any missing lengths and mark them on your diagram.
   Find the perimeter of each shape.

   a   **b**

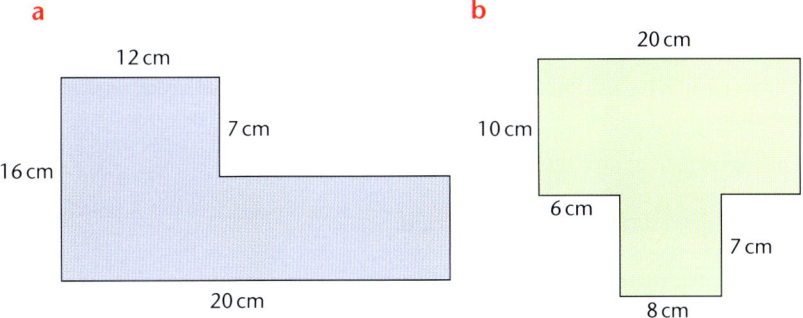

40 MENSURATION

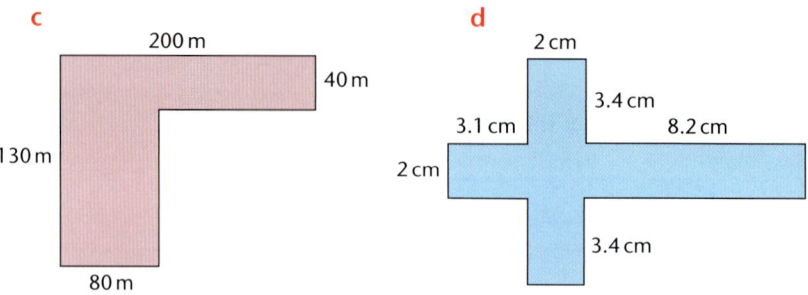

3 Measure each of these shapes accurately.
Work out the perimeter of each shape.

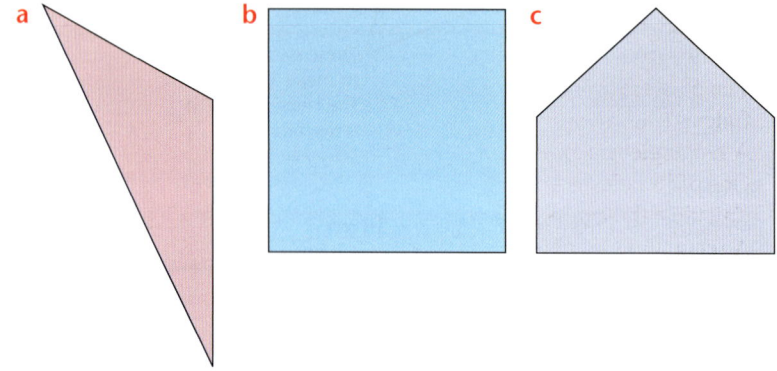

4 The perimeter of a rectangle is 26 cm.
Two sides are 8 cm long.
How long are the other sides?

5 A square has a perimeter of 120 cm.
How long is each side?

6 A rectangle has a perimeter of 60 cm.
The lengths of its sides are in whole centimetres.
What are the possible sizes of the rectangle?

The area of a rectangle

The **area** of a two-dimensional shape is the amount of flat space inside the shape.

For rectangles of any size,

area of a rectangle = length × width

> **Note**
> Whenever you are giving an answer for an area, make sure you include the units.
>
> If the lengths are in centimetres, the area will be in cm².
>
> If the lengths are in metres, the area will be in m².

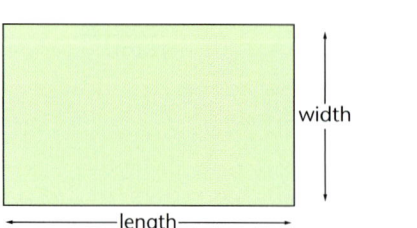

Exercise 40.2

1. A rectangle measures 4.7 cm by 3.6 cm.
   Find its area.
2. A square has sides 2.6 m long.
   Find its area.
3. A rectangle measures 3.62 cm by 4.15 cm.
   Find its area.
4. A rectangular pond measures 4.5 m by 8 m.
   Find its area.
5. An airport is built on a rectangular piece of land 1.8 km long and 1.3 km wide.
   a Calculate the area of the land.
   b Find the length of fencing needed to go round the perimeter of the airport.
6. The perimeter of a square is 28 cm.
   Calculate its area.
7. A rectangle has an area of 240 cm².
   One of its sides is 16 cm long.
   Calculate the length of the other side of the rectangle.
8. A rectangular lawn measures 24 m by 18.5 m.
   a Work out the area of the lawn.
   Lawn weedkiller is spread on the lawn.
   50 g of weedkiller is needed for every square metre.
   b How many kilograms of weedkiller are needed to treat the whole lawn?
   c Weedkiller is sold in 2.5 kg boxes.
   How many boxes are needed?

The area of a triangle

Triangle PQR has base length b and perpendicular height h.

Rectangle $XYQR$ has the same base length and height.

The blue area is the same size as the pink area.

Area of the rectangle is bh, so the area of the triangle is $\frac{1}{2}bh$.

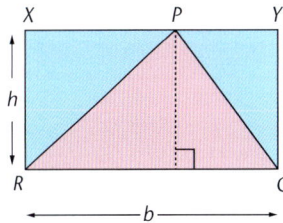

> Area of a triangle $= \frac{1}{2} \times$ base $\times$ perpendicular height or $A = \frac{1}{2}bh$

Example 40.3
Question
a Find the area of this triangle.

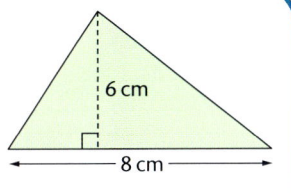

40 MENSURATION

> **Note**
>
> Always use the perpendicular height of the triangle, never the slant height.
>
> In this triangle, area $= \frac{1}{2} \times 6 \times 3 = 9 \text{ cm}^2$.
>
>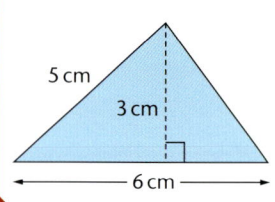

b The area of this triangle is 20 cm^2.
Find the perpendicular height of the triangle.

Solution

a Area $= \frac{1}{2} bh$

$\phantom{\text{Area}} = \frac{1}{2} \times 8 \times 6$

$\phantom{\text{Area}} = 24 \text{ cm}^2$

b Area $= \frac{1}{2} bh$

$20 = \frac{1}{2} \times 8 \times h$

$20 = 4h$

$h = 5$

So the height is 5 cm.

Remember that the units of area are always square units, such as square centimetres or square metres, written cm^2 or m^2.

When using the formula, you can use any of the sides of the triangle as the base, provided you use the perpendicular height that goes with it.

Exercise 40.3

1 Find the area of each of these triangles.

a

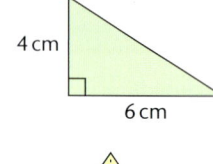

b

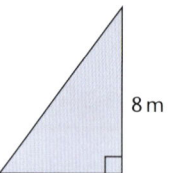

c

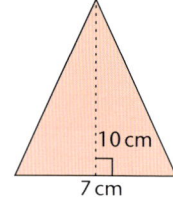

d

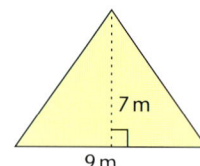

e

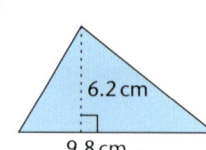

f

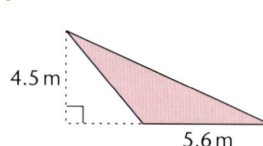

g

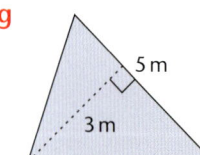

h

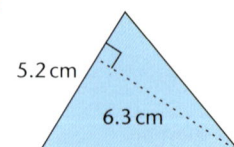

i

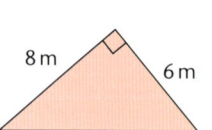

The area of a parallelogram

2 Find the area of each of these triangles.

a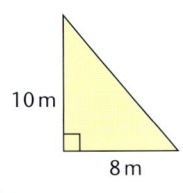
b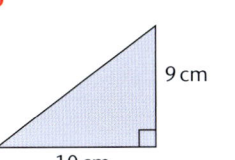
c 9 m / 11 m (perpendicular height shown)

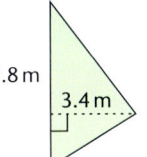

d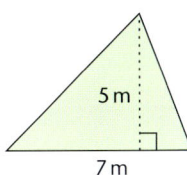
e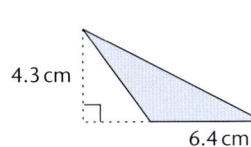
f 5.8 m / 3.4 m

g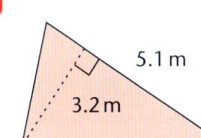
h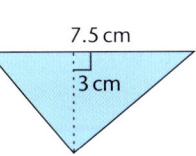
i 8.3 m / 4.6 m

3 In triangle ABC, $AB = 6$ cm, $BC = 8$ cm and $AC = 10$ cm. Angle $ABC = 90°$.
  a Find the area of the triangle.
  b Find the perpendicular height BD.

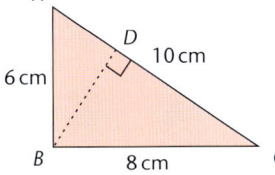

The area of a parallelogram

A parallelogram may be cut up and rearranged to form a rectangle or two congruent triangles.

Area of a rectangle = base × height   Area of each triangle = $\frac{1}{2}$ × base × height

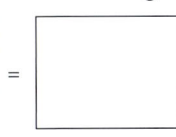   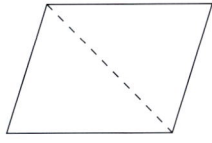

Both these ways of splitting a parallelogram show how to find its area.

Area of a parallelogram = base × height

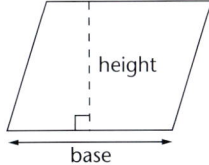

Note

Make sure you use the perpendicular height and not the sloping edge when finding the area of a parallelogram.

Example 40.4

Question

Find the area of this parallelogram.

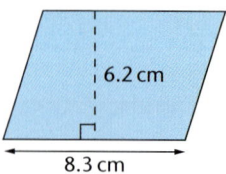

Solution

Area of a parallelogram = base × height
= 8.3 × 6.2
= 51.46 cm²
= 51.5 cm² to 1 decimal place

Note

Always give your final answer to a suitable degree of accuracy, but don't use rounded answers in your working.

Exercise 40.4

1 Find the area of each of these parallelograms.

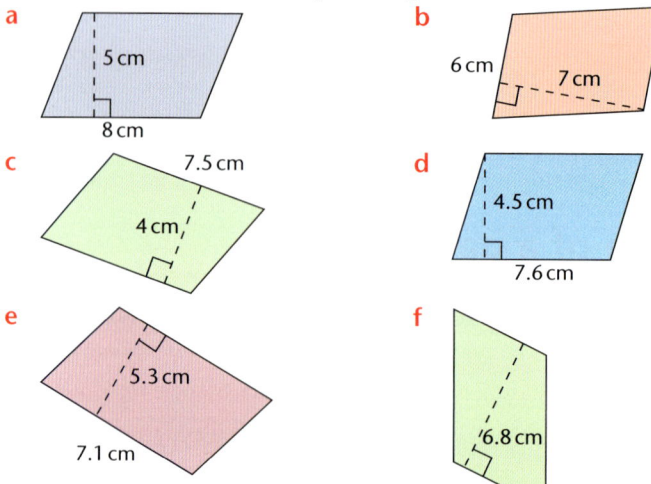

2 Find the area of each of these parallelograms. The lengths are in centimetres.

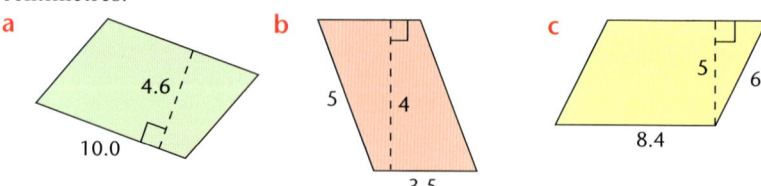

The area of a trapezium

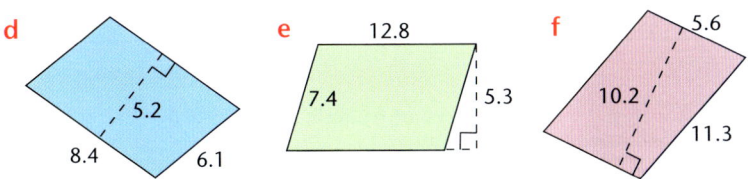

3 Find the values of *a*, *b* and *c*. The lengths are in centimetres.

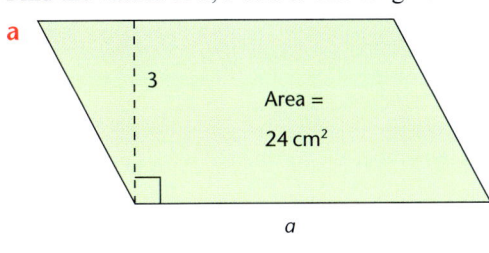

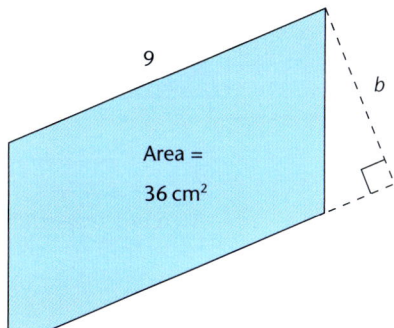

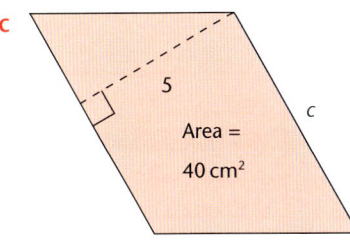

4 Find the values of *x*, *y* and *z*. The lengths are in centimetres.

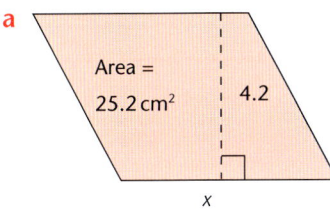

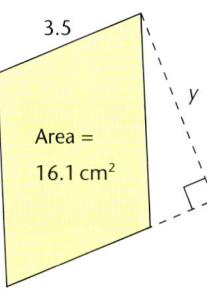

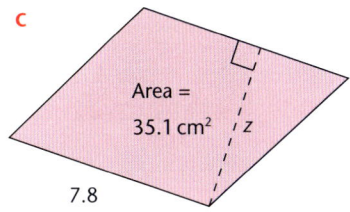

The area of a trapezium

A trapezium has one pair of opposite sides parallel.

It can be split into two triangles.

40 MENSURATION

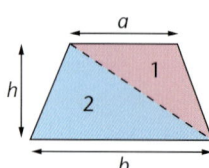

Area of triangle 1 = $\frac{1}{2} \times a \times h$

Area of triangle 2 = $\frac{1}{2} \times b \times h$

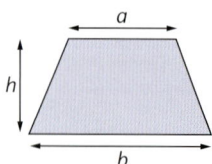

Area of trapezium = $\frac{1}{2} \times a \times h + \frac{1}{2} \times b \times h$

$= \frac{1}{2} \times (a + b) \times h$

$= \frac{1}{2}(a + b)h$

You can remember the formula in words or algebraically.

Area of a trapezium = half the sum of the parallel sides × the height
$= \frac{1}{2}(a + b)h$

Example 40.5

Question

Calculate the area of this trapezium.

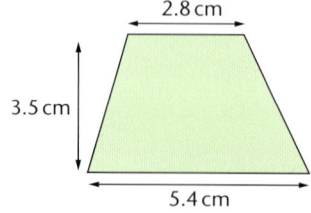

Solution

Area of a trapezium = $\frac{1}{2}(a + b)h$

$= \frac{1}{2} \times (2.8 + 5.4) \times 3.5$

$= 14.35 \text{ cm}^2$

$= 14.4 \text{ cm}^2$ to 1 decimal place

Note

Use the brackets function on your calculator.

Without a calculator, remember to work out the brackets first.

Exercise 40.5

Note

The plural of *trapezium* is *trapezia*.

1 Find the area of each of these trapezia.

a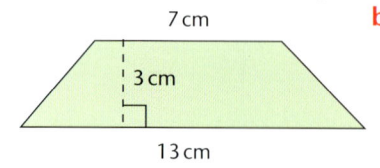

b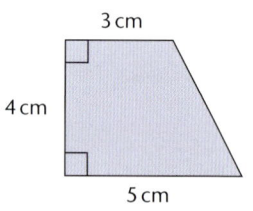

c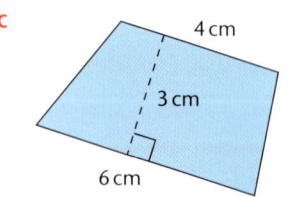

d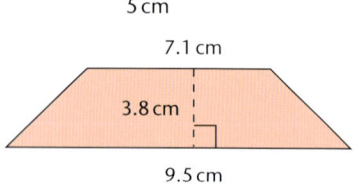

The area of a trapezium

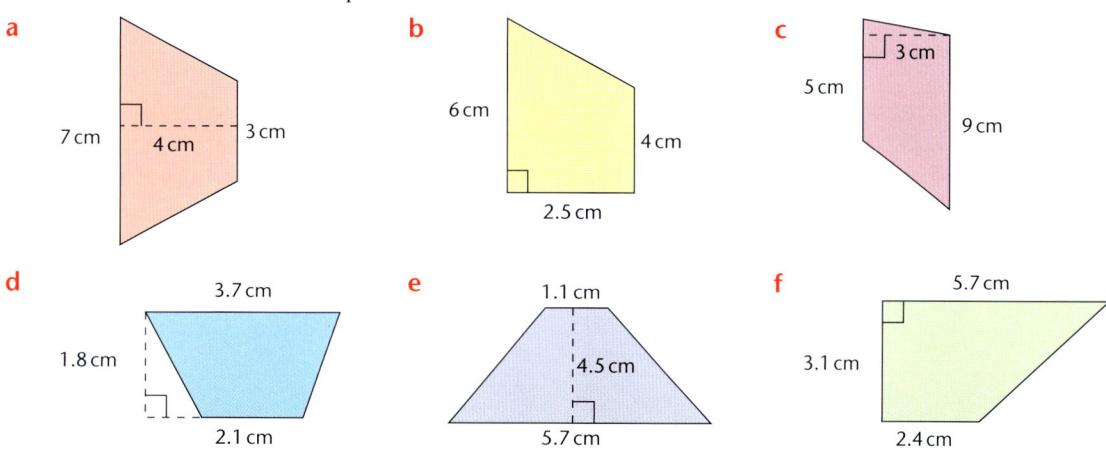

2 Find the area of each of these trapezia.

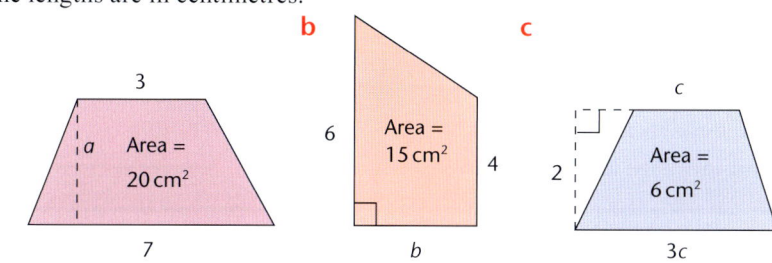

3 Find the values of a, b and c in these trapezia. The lengths are in centimetres.

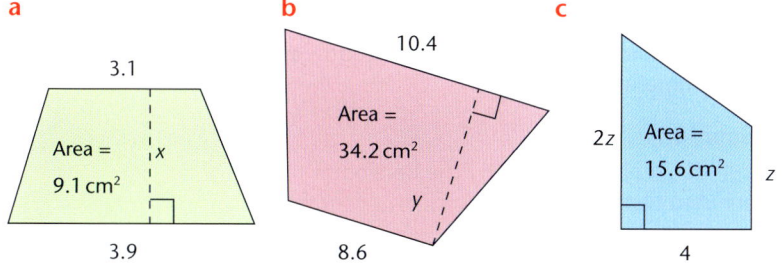

4 Find the values of x, y and z in these trapezia. The lengths are in centimetres.

a
3.1
Area = 9.1 cm² x
3.9

b
10.4
Area = 34.2 cm²
y
8.6

c
$2z$ Area = 15.6 cm² z
4

5 A trapezium has height 4 cm and area 28 cm².
One of its parallel sides is 5 cm long.
How long is the other parallel side?

6 A trapezium has height 6.6 cm and area 42.9 cm².
One of its parallel sides is 5 cm long.
How long is the other parallel side?

40 MENSURATION

The area of shapes made from rectangles and triangles

A way to find the area of many **compound shapes** is to split them up into rectangles and triangles.

Example 40.6

Question

Work out the area of each these shapes. All lengths are in centimetres.

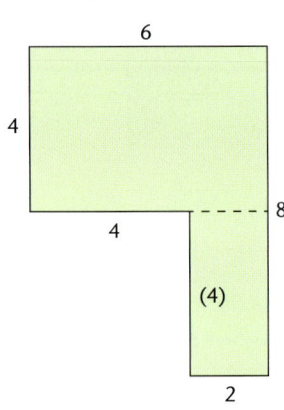

a b

(The lengths marked in brackets were not all given in the question, but have been worked out. The dashed lines have been added as part of the answer.)

Solution

a The shape has been split into two rectangles by a horizontal dotted line. (It could have been split in a different way, by a vertical line.)

Area = 6 × 4 + 2 × 4 = 24 + 8 = 32 cm²

b The shape has been split into three rectangles.

Area = 8 × 2 + 7 × 3 + 5 × 3 = 16 + 21 + 15 = 52 cm²

Note

A common error is to split the shape correctly, but then multiply the wrong numbers to get the area.

Example 40.7

Question

Work out the area of this shape. All lengths are in centimetres.

Solution

The shape has been split by the horizontal line into two triangles.

Area = $\frac{1}{2} \times 4 \times 3 + \frac{1}{2} \times 6 \times 3$

= 6 + 9

= 15 cm²

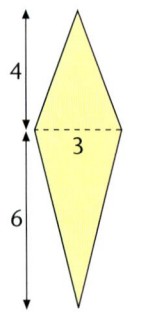

The area of shapes made from rectangles and triangles

Exercise 40.6

1 Work out the area of each of these shapes. All lengths are in centimetres.

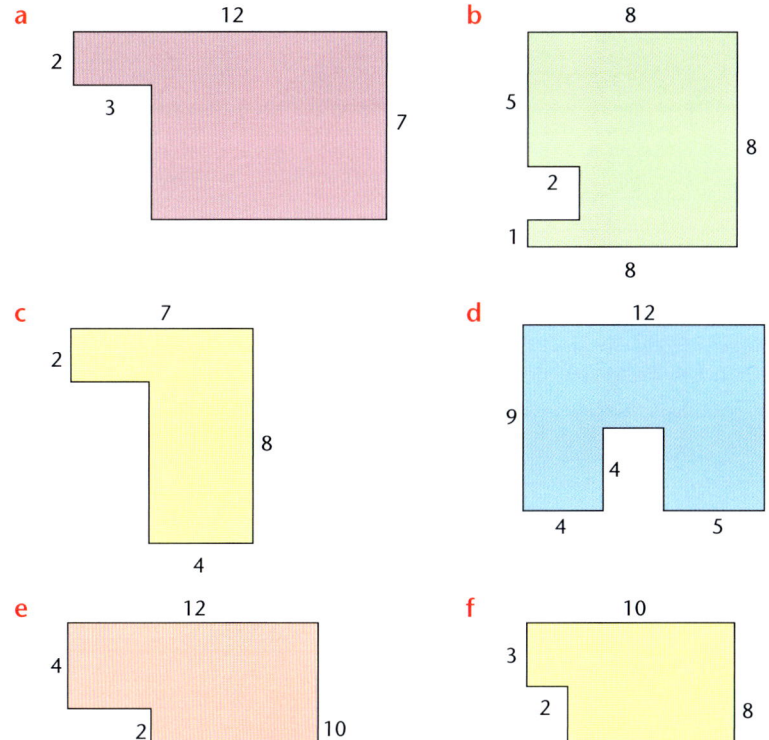

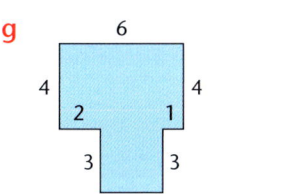

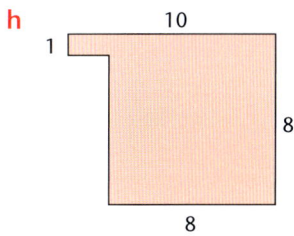

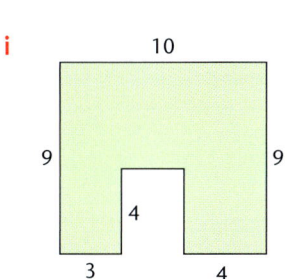

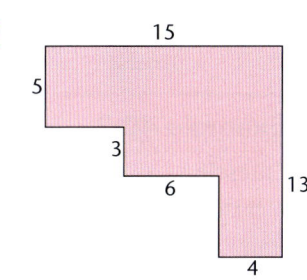

2 Work out the area of each of these shapes. All lengths are in centimetres.

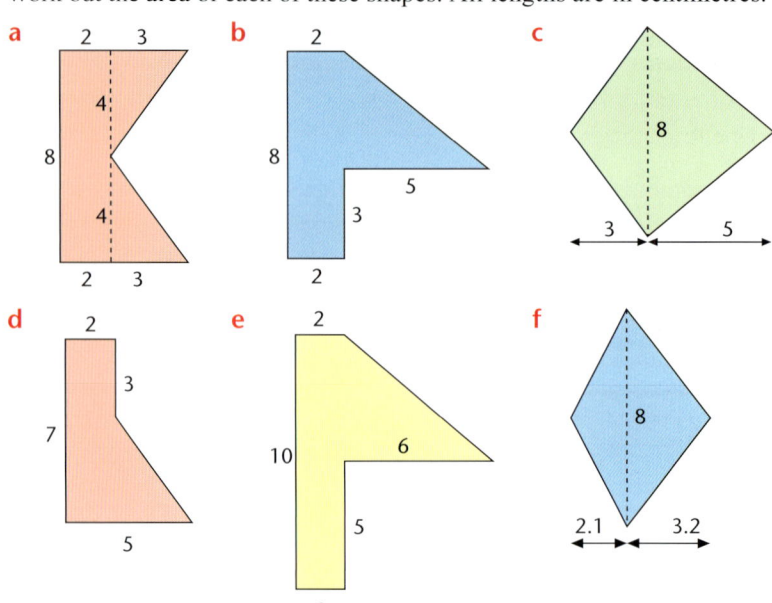

The circumference of a circle

The fact that the **circumference** of a circle is approximately three times the **diameter** has been known for thousands of years. Accurate calculations have found this number to hundreds of decimal places. The number is a never-ending decimal, and is denoted by the Greek letter π.

π is an irrational number. Your calculator has the number which π represents stored in its memory.

Note
If you are using your calculator, always use the π key rather than an approximation.

The formula for the circumference of a circle is

$C = \pi d$

If you know the radius of the circle instead of the diameter, use the fact that the diameter is double the radius, $d = 2r$.

An alternative formula for the circumference of a circle is

$C = 2\pi r$

Example 40.8

Question
A circle has a radius of 8 cm.
Find its circumference.

Solution
Circumference = $2\pi r$
= $2 \times \pi \times 8$
= 50.265...
= 50.3 cm (to 1 d.p.)

The circumference of a circle

Example 40.9

Question
A circle has a circumference of 20 m. Find its diameter.

Solution
Circumference = πd
$20 = \pi d$
$d = 20 \div \pi$
$= 6.366...$
$= 6.37$ m (to 2 d.p.)

In Example 40.10, the answer is given in terms of π.

Example 40.10

Question
Calculate the circumference of a circle with radius 6.5 cm.
Give your answer in terms of π.

Solution
Circumference = $2\pi r$
$= 2 \times \pi \times 6.5$
$= 13\pi$ cm

Exercise 40.7

1. Calculate the circumference of circles with the following diameters, giving your answers correct to 1 decimal place.
   - a 12 cm
   - b 9 cm
   - c 20 m
   - d 16.3 cm
   - e 15.2 m
   - f 25 m
   - g 0.3 cm
   - h 17 m
   - i 5.07 m
   - j 6.5 cm

2. Find the circumference of circles with these radii, giving your answers correct to 1 decimal place.
   - a 5 cm
   - b 7 cm
   - c 16 m
   - d 18.1 m
   - e 5.3 m
   - f 28 cm
   - g 3.2 cm
   - h 60 m
   - i 1.9 m
   - j 73 cm

3. Find the circumference of these circles, leaving your answer in terms of π.
   - a radius 3 cm
   - b diameter 10 cm
   - c radius 3.5 cm
   - d diameter 8 cm

4. The centre circle on a soccer pitch has a radius of 9.15 metres. Calculate the circumference of the circle.

5. The radius of the Earth at the equator is 6378 km. Calculate the circumference of the Earth at the equator.

6. The diagram shows a wastepaper bin in the shape of a cylinder.

   Calculate the circumference of the rim.

40 MENSURATION

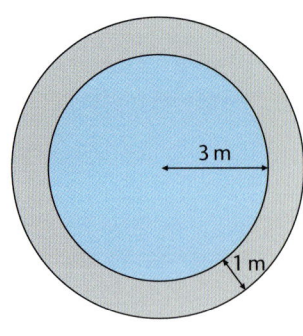

7 Calculate the diameter of the circles with these circumferences, giving your answers correct to 1 decimal place.
  a 75 cm   b 18 cm   c 50 cm

8 A circular racetrack is 300 metres in circumference.
Calculate the diameter of the racetrack.

9 Vivek has a circular pond of radius 3 m.
He wants to surround the pond with a gravel path 1 metre wide.
An edging strip is needed along each side of the gravel path.
What length of edging strip is needed altogether?

The area of a circle

The formula to calculate the area, A, of a circle of radius r is

$$A = \pi r^2$$

Example 40.11

Question
The radius of a circle is 4.3 m.
Calculate the area of the circle.

Solution
$A = \pi r^2$
$= \pi \times 4.3^2$
$= 58.1 \, m^2$ (to 1 d.p.)

Note
Make sure you can use the π key and the square key on your calculator.

Example 40.12

Question
The diameter of a circle is 18.4 cm.
Calculate the area of the circle.

Solution
$r = 18.4 \div 2$
$= 9.2 \, cm$
$A = \pi r^2$
$= \pi \times 9.2^2$
$= 266 \, cm^2$ (to the nearest cm^2)

Note
One of the most common errors is to mix up diameter and radius.
Every time you do a calculation, make sure you have used the right one.

Exercise 40.8

In all of these questions, make sure you state the units of your answer.

1 Find the area of circles with these radii.
  a 4 cm   b 16 m   c 11.3 m   d 13.6 m   e 8.9 cm

2 Find the area of circles with these radii. Give your answers in terms of π.
  a 5 m   b 4 cm   c 12 cm   d 7 m

3 Find the area of each of these circles.
  a 3 cm   b 16 m   c 5.3 cm

Arc length and sector area

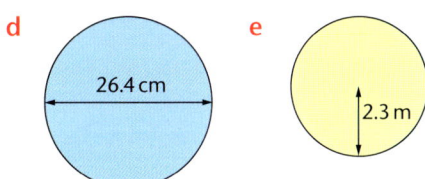
d 26.4 cm
e 2.3 m

4 The radius of a circular fish pond is 1.5 m.
   Find the area of the surface of the water.
5 A circular mouse mat has a radius of 9 cm.
   Find the area of the mouse mat.
6 The diameter of a circular table is 0.8 m.
   Find the area of the table.
7 According to *Guinness World Records*, the largest pizza ever made was 37.4 m in diameter.
   What was the area of the pizza?
8 To make a table mat, a circle of radius 12 cm is cut from a square of side 24 cm, as shown in the diagram.

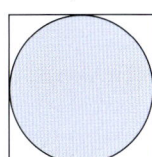

   Calculate the area of the material that is wasted.
9 A square has a side of 3.5 cm and a circle has a radius of 2 cm.
   Which has the bigger area?
   Show your calculations.
10 Cala is making a circular lawn with a radius of 15 m.
   The packets of grass seed say 'sufficient to cover 50 m²'.
   How many packets will she need?
11 Use your calculator to find the area of a circle with radius 6.8 cm.
   Without using your calculator, do an approximate calculation to check your answer.

Note
Remember r^2 must give units such as cm² or m² and so πr^2 must be the area formula.

Arc length and sector area

A **sector** is a fraction of a circle.

It is $\frac{\theta}{360}$ of the circle, where $\theta°$ is the sector angle at the centre of the circle.

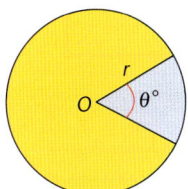
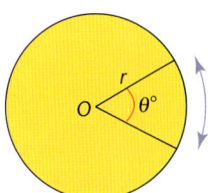

Arc length $= \frac{\theta}{360} \times$ circumference

$\qquad = \frac{\theta}{360} \times 2\pi r$

Sector area $= \frac{\theta}{360} \times$ area of circle

$\qquad = \frac{\theta}{360} \times \pi r^2$

Photocopying is prohibited

40 MENSURATION

Example 40.13

Question

Calculate the arc length and area of this sector.

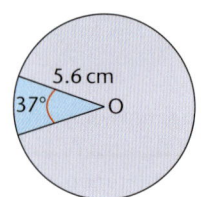

Solution

Arc length $= \frac{\theta}{360} \times 2\pi r$

$= \frac{37}{360} \times 2\pi \times 5.6$

$= 3.62\,\text{cm}$ to 3 significant figures

Sector area $= \frac{\theta}{360} \times \pi r^2$

$= \frac{37}{360} \times \pi \times 5.6^2$

$= 10.1\,\text{cm}^2$ to 3 significant figures

Example 40.14

Question

Calculate the sector angle of a sector with arc length 6.2 cm in a circle with radius 7.5 cm.

Solution

Arc length $= \frac{\theta}{360} \times 2\pi r$

$6.2 = \frac{\theta}{360} \times 2\pi \times 7.5$

$\theta = \frac{6.2 \times 360}{2\pi \times 7.5}$

$= 47.4°$ to 3 significant figures.

Example 40.15

Question

A sector makes an angle of 54° at the centre of a circle.

The area of the sector is 15 cm².

Calculate the radius of the circle.

Solution

Sector area $= \frac{\theta}{360} \times \pi r^2$

$15 = \frac{54}{360} \times \pi r^2$

$r^2 = \frac{15 \times 360}{54 \times \pi}$

$r^2 = 31.830\ldots$

$r = \sqrt{31.83\ldots}$

$= 5.64\,\text{cm}$ to 3 significant figures.

Note

You can rearrange the formula before you substitute, if you prefer.

Arc length and sector area

Exercise 40.9

1 Calculate the arc length of each of these sectors.
Give your answers to 3 significant figures.

a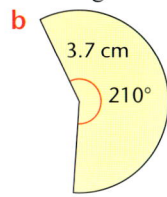
b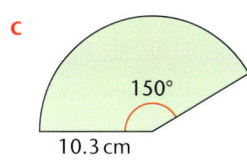
c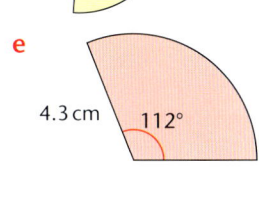
d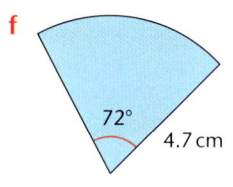
e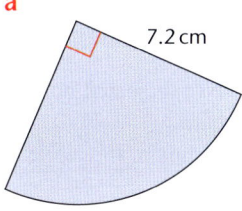
f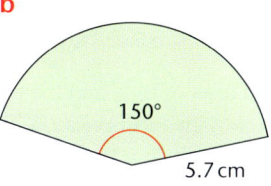

2 Calculate the area of each of the sectors in question **1**.
Give your answers to 3 significant figures.

3 Calculate the perimeter of each of these sectors.
Give your answers to 3 significant figures.

a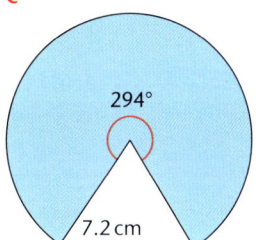
b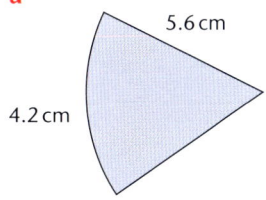
c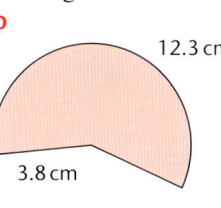

4 Calculate the sector angle in each of these sectors.
Give your answers to the nearest degree.

a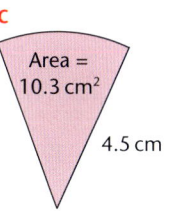
b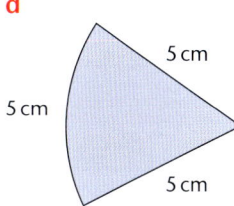
c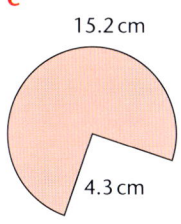
d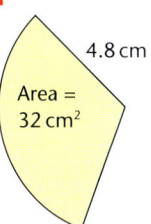
e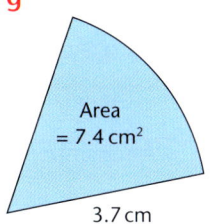
f
g

5 Calculate the radius of each of these sectors.

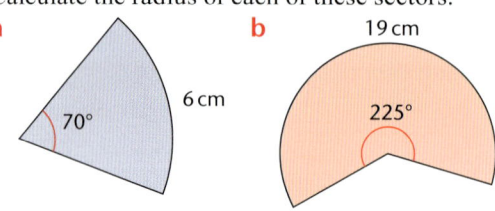

6 Calculate the radius of each of these sectors.
  a Sector area = 19.7 cm², sector angle = 52°.
  b Sector area = 2.7 cm², sector angle = 136°.
  c Sector area = 6.2 m², sector angle = 218°.

7 A shape consists of two separate sectors of a circle, each with angle 35° and radius 32 mm.
They are to be painted blue with a thin black border.

Calculate the total area of blue and the total length of the black border. Give your answers to an appropriate degree of accuracy.

8 A sector of a circle has its arc length equal to the radius of the circle. The sector angle is defined as 1 radian.
What is this angle in degrees?

The volume of a prism

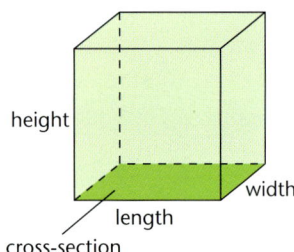

A **prism** is a three-dimensional shape that has the same cross-section throughout its length.

A **cuboid** is a prism with a rectangular cross-section.

> Volume of a cuboid = length × width × height

You can also think of this as

> Volume of a cuboid = area of cross-section × height

The general formula for the volume of a prism is

> Volume of a prism = area of cross-section × length

Another important prism is the **cylinder**.

The cross-section of a cylinder is a circle, which has area πr^2.

> Volume of a cylinder = $\pi r^2 h$

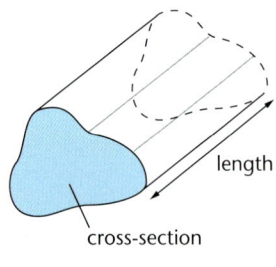

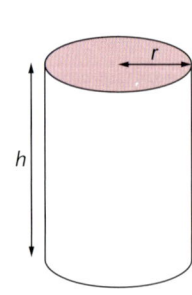

The volume of a prism

Example 40.16

Question

Calculate the volume of a cylinder with base diameter 15 cm and height 10 cm.

Solution

Radius of base = $\frac{15}{2}$ = 7.5 cm.

Volume of a cylinder = $\pi r^2 h$

$= \pi \times 7.5^2 \times 10$

$= 1767 \text{ cm}^3$, to the nearest whole number.

Example 40.17

Question

A chocolate box is a prism with a trapezium as cross-section, as shown. Calculate the volume of the prism.

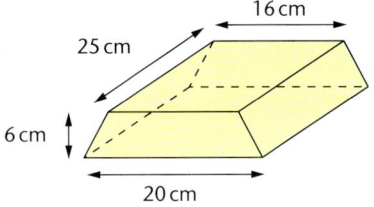

Solution

Area of a trapezium = $\frac{1}{2}(a + b)h$

$= \frac{1}{2}(20 + 16) \times 6$

$= 108 \text{ cm}^2$

Volume of a prism = area of cross-section × length

$= 108 \times 25$

$= 2700 \text{ cm}^3$

Example 40.18

Question

A cylinder has volume 100 cm³ and is 4.2 cm high.
Find the radius of its base.
Give your answer to the nearest millimetre.

Solution

Volume of a cylinder = $\pi r^2 h$

$100 = \pi \times r^2 \times 4.2$

$r^2 = \dfrac{100}{\pi \times 4.2}$

$= 7.578...$

$r = \sqrt{7.578}.$

$= 2.752...$

$= 2.8 \text{ cm}$, to the nearest millimetre.

Exercise 40.10

1. A cuboid has edges 3 cm, 5 cm and 2 cm long.
   Work out the volume.

2. The edges of a cube are 2 cm long.
   Work out the volume.

40 MENSURATION

3 A rectangular cupboard is 2 m by 1 m by 1.5 m.
Find its volume.

4 A cuboid has a base 3 cm by 4 cm and its volume is 48 cm³.
What is the height of the cuboid?

5 Calculate the volume of a cylinder with base radius 5.6 cm and height 8.5 cm.

6 A cylinder is 12 cm long and has radius 2.4 cm.
Find its volume.

7 A cylinder has diameter 8 cm and height 8 cm.
Calculate its volume.

8 Calculate the volume of a prism 15 cm long with each of these cross-sections.

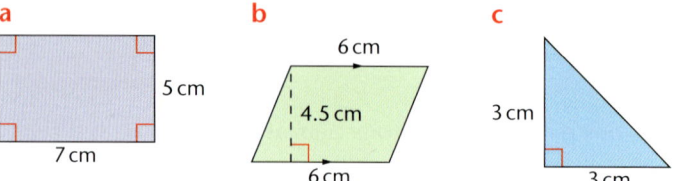

9 Calculate the volume of a prism 12 cm long with each of these cross-sections.

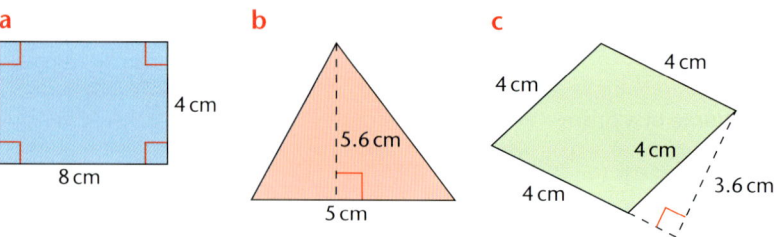

10 A chocolate bar is in the shape of a triangular prism.
Calculate its volume.

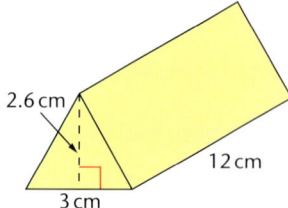

11 A cylindrical water tank is 4.2 m high and has radius 3.6 m.
Find its volume.

12 A gift box is a prism with a triangular base.
Calculate its volume.

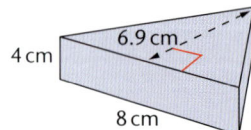

13 The area of cross-section of a prism is 75 cm².
Its volume is 1200 cm³.
Calculate its length.

14 The volume of a cylinder is 800 cm³.
   Its radius is 5.3 cm.
   Calculate its height.
15 The volume of a cylindrical tank is 600 m³.
   Its height is 4.6 m.
   Calculate the radius of its base.
16 A vase is a prism with a trapezium as its base.
   The internal measurements are as shown.
   How much water can the vase hold?
   Give your answer in litres.
   (Remember that 1 litre = 1000 cm³.)

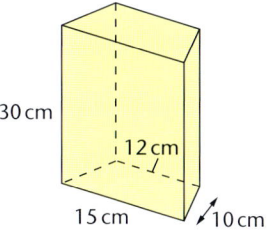

The surface area of a prism

To find the total surface area of a prism, you add up the surface area of all the individual surfaces.

Example 40.19

Question
Find the total surface area of this prism.

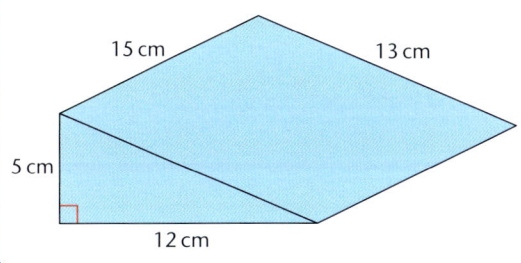

Solution
Area of end = $\frac{1}{2} \times 12 \times 5$ = 30 cm²
Area of other end   = 30 cm²
Area of base = 12 × 15 = 180 cm²
Area of top   = 13 × 15 = 195 cm²
Area of back = 5 × 15   = 75 cm²
Total surface area   = 510 cm²

For a cylinder there are three surfaces.

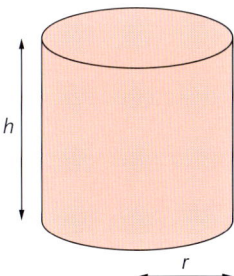

The two ends are circles and each have area πr^2.

If the cylinder were made out of paper, the curved surface would open out to a rectangle.

Photocopying is prohibited

40 MENSURATION

The length of the rectangle is the circumference of the cylinder, so is $2\pi r$.

The curved surface area is therefore $2\pi r \times h = 2\pi rh$.

Total surface area of a cylinder = $2\pi rh + 2\pi r^2$

Example 40.20

Question
Calculate the total surface area of a cylinder with base diameter 15 cm and height 10 cm.

Solution
Radius of base = $\frac{15}{2}$ = 7.5 cm

Area of two ends = $2 \times \pi r^2 = 2 \times \pi \times 7.5^2 = 353.429…$ cm²

Curved surface area = $2\pi rh = 2 \times \pi \times 7.5 \times 10 = 471.238…$ cm²

Total surface area = 353.429… + 471.238… = 825 cm², to the nearest whole number.

Exercise 40.11

Note
You may find it helpful to make a rough sketch of the net of the prism before calculating its surface area. This should stop you missing out a face.

1 Find the surface area of these cuboids.

   a

   b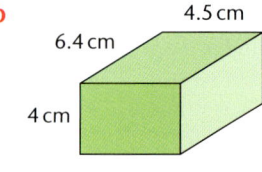

2 Find the surface area of a cube of side 20 cm.
3 The diagram shows the net of a shoe box. Calculate its surface area.

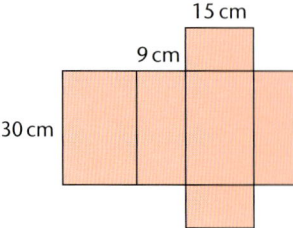

4 A fish tank is 80 cm long, 45 cm wide and 40 cm high.
   It does not have a lid.
   Calculate the surface area of the glass used to make the fish tank.
5 Calculate the surface area of this triangular prism.

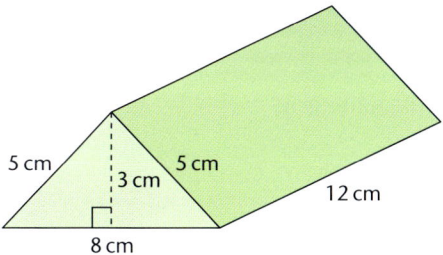

6 Find the total surface area of these shapes from Exercise 40.10.
   a The cylinder in question 5
   b The cylinder in question 6
   c The cylinder in question 7
   d The prism in question 8a
   e The prism in question 9a
7 Find the total surface area of each of these prisms.
   a
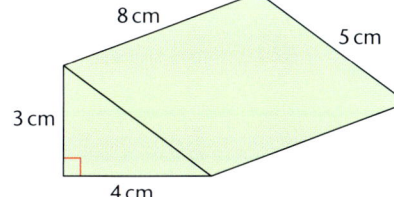
   b
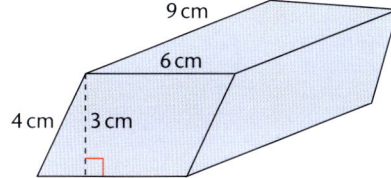

The volume of a pyramid, a cone and a sphere

Not all three-dimensional shapes are prisms.

Some shapes have a cross-section which, though similar, decreases to a point.

Such shapes include the pyramid and the cone.

The volume of a pyramid or cone is given by

Volume = $\frac{1}{3}$ × area of base × height

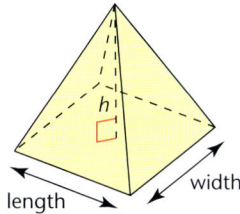

40 MENSURATION

Because a cone has a circular base, for a cone with base radius r and height h,

$$\text{Volume} = \tfrac{1}{3}\pi r^2 h$$

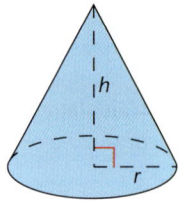

You also need to know about a different type of three-dimensional shape – the sphere.

For a sphere of radius r,

$$\text{Volume} = \tfrac{4}{3}\pi r^3$$

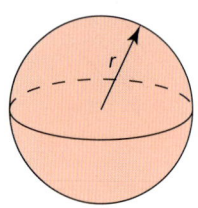

You do not need to learn these formulas but you do need to be able to use them.

Example 40.21

Question
Find the volume of this cone.

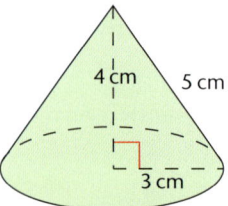

Solution
$$\begin{aligned}\text{Volume} &= \tfrac{1}{3}\pi r^2 h \\ &= \tfrac{1}{3} \times \pi \times 3^2 \times 4 \\ &= 37.7 \text{ cm}^3 \text{ to 3 significant figures.}\end{aligned}$$

Note
A hemisphere is half a sphere.

Example 40.22

Question
A bowl is in the shape of a hemisphere of diameter 25 cm.
How much water can the bowl hold?
Give your answer in litres.

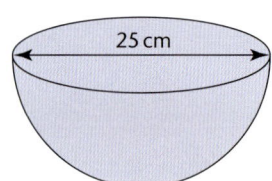

Solution
Volume of hemisphere = $\tfrac{1}{2}$ × volume of sphere
$$\begin{aligned} &= \tfrac{1}{2} \times \tfrac{4}{3}\pi r^3 \\ &= \tfrac{2}{3}\pi r^3 \\ &= \tfrac{2}{3} \times \pi \times 12.5^3 \\ &= 4090.6\ldots \text{ cm}^3 \qquad 1 \text{ litre} = 1000 \text{ cm}^3 \\ &= 4.09 \text{ litres to 3 significant figures.}\end{aligned}$$

The volume of a pyramid, a cone and a sphere

Exercise 40.12

Use the formulas given on pages 361 and 362 to answer these questions.

1 Calculate the volume of each of these pyramids. Their bases are squares or rectangles.

a
6 cm, 3 cm, 3 cm

b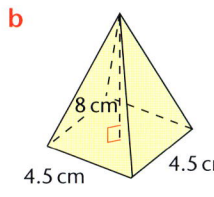
8 cm, 4.5 cm, 4.5 cm

c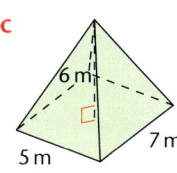
6 m, 5 m, 7 m

d
6 cm, 5 cm, 5 cm

e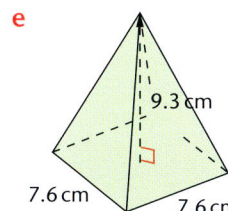
9.3 cm, 7.6 cm, 7.6 cm

f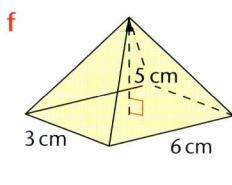
5 cm, 3 cm, 6 cm

2 Calculate the volume of each of these cones.

a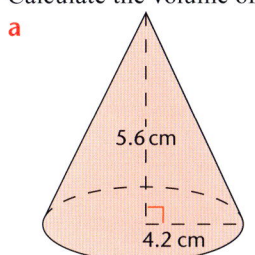
5.6 cm, 4.2 cm

b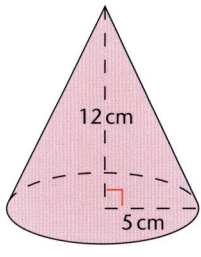
12 cm, 5 cm

c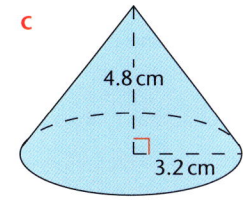
4.8 cm, 3.2 cm

d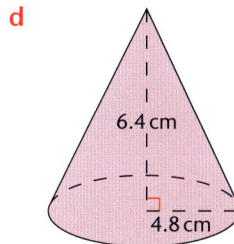
6.4 cm, 4.8 cm

e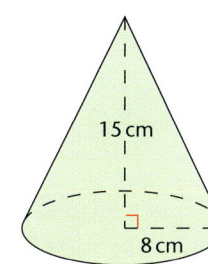
15 cm, 8 cm

f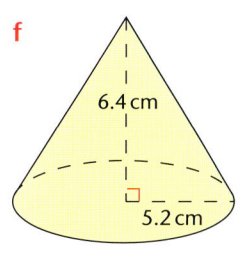
6.4 cm, 5.2 cm

3 Find the volume of a sphere of these radii.
   a 5 cm   b 6.2 cm   c 2 mm
   d 3 cm   e 4.7 cm   f 7.8 mm

4 A pyramid has a square base with sides of 8 cm.
Its volume is 256 cm^3.
Find its height.

5 Find the radius of the base of a cone with these dimensions.
   a Volume 114 cm^3, height 8.2 cm
   b Volume 52.9 cm^3, height 5.4 cm
   c Volume 500 cm^3, height 12.5 cm

6 Find the volume of a sphere with these dimensions.
   a Radius 5.1 cm   b Radius 8.2 cm   c Diameter 20 cm

40 MENSURATION

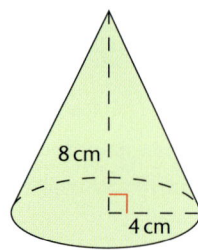

7 Find the radius of a sphere of these volumes.
   a 1200 cm³   b 8000 cm³
8 How many ball bearings of radius 0.3 cm can be made from 10 cm³ of metal when it is melted?
9 A solid cone and a solid cylinder both have base radius 6 cm.
   The height of the cylinder is 4 cm.
   The cone and the cylinder both have the same volume.
   Find the height of the cone.
10 A sphere has the same volume as this cone.
   Calculate the radius of the sphere.

The surface area of a pyramid, a cone and a sphere

As with the surface area of a prism, the surface area of a pyramid, cone or sphere is the total area of all its surfaces.

All the surfaces of a pyramid that have a base in the shape of a polygon are flat.

Apart from the base, the faces are triangular.

To find the total surface area, you find the area of each face and add them together.

A cone has a base in the shape of a circle and a curved surface.

The curved surface of a cone can be opened out to form a sector of a circle of radius l, where l is the slant height of the cone.

The arc length of the sector is the circumference of the base of the cone.

> **Note**
> Make sure you distinguish between the perpendicular height, h, and the slant height, l, of a cone – and don't read l as 1!

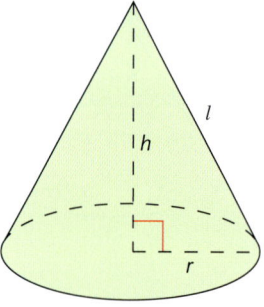 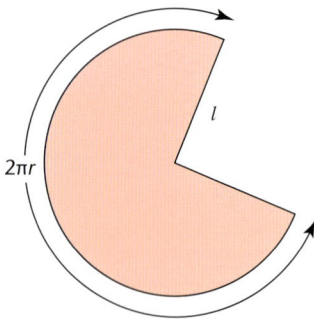

Curved surface area of a cone = $\pi r l$

All of the surface of a sphere is curved.

The formula for the surface area of a sphere of radius r is

surface area of a sphere = $4\pi r^2$

You do not need to learn these formulas but you do need to be able to use them.

The surface area of a pyramid, a cone and a sphere

Example 40.23

Question
This cone has a solid base.
Find the total surface area of the cone.

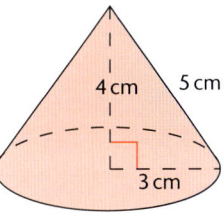

Solution
Curved surface area = $\pi r l$
$= \pi \times 3 \times 5 = 15\pi$
Area of base = πr^2
$= \pi \times 3^2 = 9\pi$
Total surface area = $15\pi + 9\pi$
$= 24\pi$
$= 75.4 \, cm^2$ to 3 significant figures.

Note
Leaving the two areas in terms of π means that you are using exact values and avoids rounding errors.

Example 40.24

Question
Calculate the curved surface area of this cone.

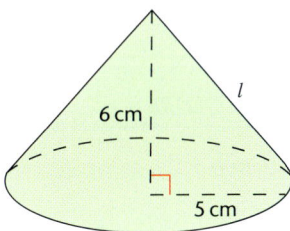

Solution
First the slant height l must be found.
Using Pythagoras' theorem,
$l^2 = 5^2 + 6^2$
$= 61$
$l = \sqrt{61}$
Curved surface area = $\pi r l$
$= \pi \times 5 \times \sqrt{61}$
$= 123 \, cm^2$ to 3 significant figures

Note
Leaving the value for l in surd form means that you are using an exact value and avoids rounding errors.

Example 40.25

Question
Calculate the surface area of this pyramid.

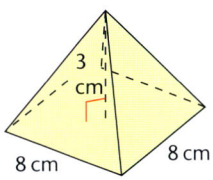

Solution
This is the section through midpoints of opposite sides of the base.
It is an isosceles triangle.
The height splits it into two right-angled triangles.
The hypotenuse of one of the right-angled triangles is the height of one of the triangular faces of the pyramid.
Using Pythagoras' theorem,
height$^2 = 4^2 + 3^2 = 25$
height = 5

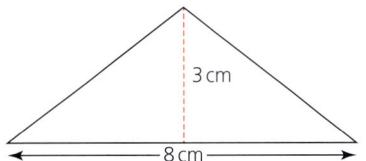

Area of one triangular face = $\frac{1}{2} \times 8 \times 5 = 20\,\text{cm}^2$.
Total surface area of the pyramid = 4 × area of one triangular face + the area of the square base
$$= 4 \times 20 + 8 \times 8$$
$$= 144\,\text{cm}^2$$

Exercise 40.13

Use the formulas given on page 364 to answer these questions.

Some of the questions in this exercise also ask for the volume of the shape.

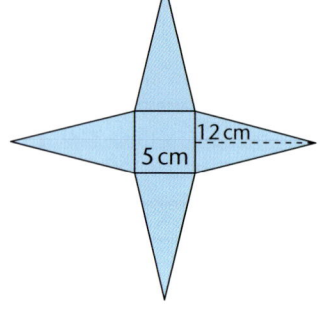

1. This is a sketch of the net for a square-based pyramid.
   Calculate the surface area of the pyramid.
2. Calculate the curved surface area of each of these cones.

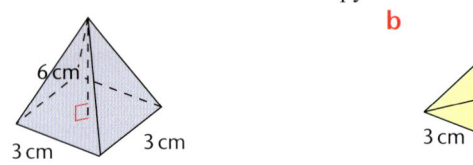

3. Calculate the surface area of these pyramids.

4. Calculate the surface area of a sphere of each of these radii.
   a 5 cm   b 6.2 cm   c 2 mm
   d 3 cm   e 4.7 cm   f 7.8 mm
5. A sphere has a surface area of 47.6 cm².
   Calculate its radius.
6. A solid cone has a base radius of 4.5 cm and a height of 6.3 cm.
   Calculate its total surface area.
7. Calculate the total surface area of a solid cone with base radius 7.1 cm and slant height 9.7 cm.
8. Calculate the total surface area of a solid hemisphere of radius 5.2 cm.
9. Calculate the slant height of a cone of base radius 5 cm and curved surface area 120 cm².
10. The flat surface of a hemisphere has an area of 85 cm².
    Calculate the curved surface area.
11. A sphere has a surface area of 157.6 cm².
    Calculate its radius.
12. Calculate the base area of a cone with slant height 8.2 cm and curved surface area 126 cm².
13. A cone has slant height 7 cm and curved surface area 84 cm².
    Calculate the total surface area of the cone.

14 The radius of Jupiter is 7.14×10^4 km.
Assume that Jupiter is a sphere.
Find the surface area of Jupiter.
Give your answer in standard form, correct to 2 significant figures.

15 A sphere containing liquid has a capacity of 1 litre.
Calculate the surface area of the sphere.

16 A spherical ball has a curved surface area of 120 cm².
Calculate its volume.

17 A cone has a base of radius 4.2 cm and a slant height of 7.8 cm.
Calculate
  a its surface area   b its height   c its volume.

18 All the edges of this square-based pyramid are 5 cm.
Calculate
  a its surface area
  b its height
  c its volume.

19 A cone has height 6 cm and volume 70 cm³.
Calculate
  a its base radius   b its curved surface area.

20 A cone has a base of radius 5 cm and a perpendicular height of 12 cm.
Calculate its curved surface area.

The area and volume of compound shapes

You have already met the areas of compound shapes made from rectangles and triangles.

In this section we look at examples of compound shapes made from other shapes you have met in this chapter.

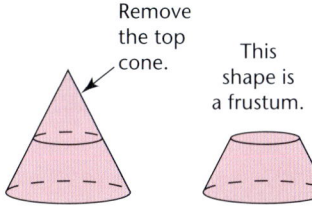

Remove the top cone.

This shape is a frustum.

The circle on the top of the frustum is in a plane parallel to the base.

Volume of a frustum = volume of whole cone − volume of missing cone

Example 40.26

Question

Find the volume of the frustum remaining when a cone of height 8 cm is removed from a cone of height 12 cm and base radius 6 cm.

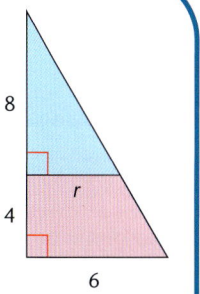

Solution

First, use similar triangles to find the base radius, $r$cm, of the cone which has been removed.

$\frac{r}{8} = \frac{6}{12}$

$r = 4$

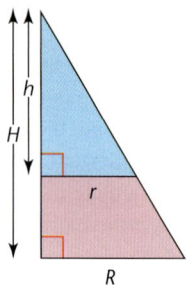

Then find the volume of the frustum.

Volume of frustum = volume of whole cone − volume of missing cone

$= \frac{1}{3}\pi R^2 H - \frac{1}{3}\pi r^2 h$

$= \frac{1}{3} \times \pi \times 6^2 \times 12 - \frac{1}{3} \times \pi \times 4^2 \times 8$

$= 318 \, cm^2$ to 3 significant figures.

Example 40.27

Question

Calculate the curved surface area of the frustum in Example 40.26.

Solution

Slant height of complete cone = $\sqrt{6^2 + 12^2} = \sqrt{180}$

Curved surface area of complete cone = $\pi \times 6 \times \sqrt{180}$

Slant height of removed cone = $\sqrt{4^2 + 8^2} = \sqrt{80}$

Curved surface area of removed cone = $\pi \times 4 \times \sqrt{80}$

Curved surface area of frustum = curved surface area of whole cone − curved surface area of missing cone

$= \pi \times 6 \times \sqrt{180} - \pi \times 4 \times \sqrt{80}$

140 cm² to 3 significant figures.

Example 40.28

Question

Calculate the area of the purple minor segment.

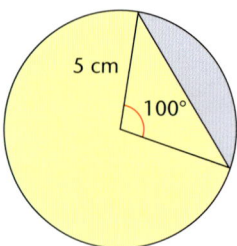

Solution

Area of minor segment = area of sector − area of triangle

Area of sector = $\frac{\theta}{360}\pi r^2$

$= \frac{100}{360} \times \pi \times 5^2$

$= 21.816... \, cm^2$

Area of triangle = $\frac{1}{2}ab \sin C$

$= \frac{1}{2} \times 5^2 \times \sin 100°$

$= 12.310... \, cm^2$

Area of minor segment = 21.816... − 12.310...

$= 9.51 \, cm^2$ to 3 significant figures.

Note

Write down more figures than you need in the working and round the final answer.

Using the calculator memory means you do not have to re-key the figures.

The area and volume of compound shapes

> **Note**
> When dealing with problems where you first have to work out how to solve them, follow these steps.
> - Read the question carefully and plan.
>   What do I know?
>   What do I have to find?
>   What methods can I apply?
> - When you have finished, ask, 'Have I answered the question?'
>   There may be one last step you have forgotten to do.

Exercise 40.14

You may need to refer to the formulas given throughout this chapter.

1. A tower of a castle has walls in the form of a cylinder and a roof in the shape of a cone.
   The diameter of the tower is 5.6 m and the height of the wall is 12 m.
   The roof has a height of 6.2 m.
   Calculate the volume of the tower.

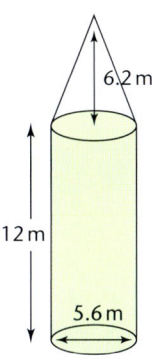

2. Calculate
   a. the length of the chord AB
   b. the perimeter of the pink minor segment.

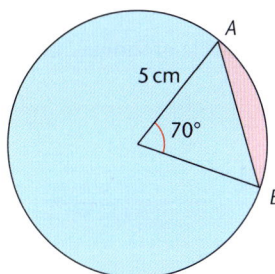

3. A cone of height 15 cm and base radius 9 cm has a cone of height 5 cm removed from its top as shown.
   a. What is the radius of the base of the top cone?
   b. Calculate the volume of the remaining frustum of the cone.
   c. Calculate the surface area of this frustum.

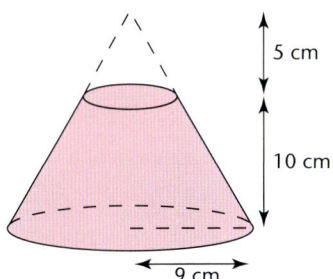

40 MENSURATION

4 Show that the volume of this frustum is $\frac{7}{3}\pi r^2 h$.

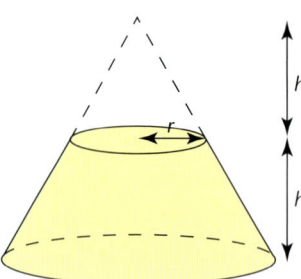

5 A piece of cheese is a prism whose cross-section is the sector of a circle with measurements as shown.
Calculate the volume of the piece of cheese.

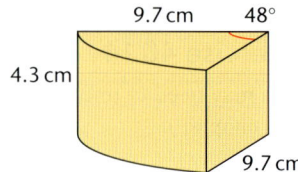

6 Calculate the area of the purple segment.

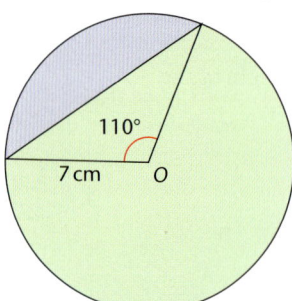

7 Calculate the perimeter of this segment of a circle of radius 5 cm.

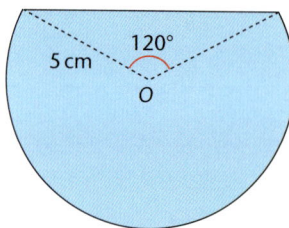

8 The top of a flowerpot is a circle of radius 10 cm.
Its base is a circle of radius 8 cm.
The height of the flowerpot is 10 cm.
  a Show that the flowerpot is a frustum of an inverted cone of complete height 50 cm and base radius 10 cm.
  b Calculate how many litres of soil the flowerpot can contain.

9 The cross-section of a railway tunnel is a major segment of a circle with diameter 7.6 m.
The height of the tunnel is 6.5 m.
The tunnel is 5.3 km long.
Calculate the volume of soil removed when the tunnel was constructed.

The area and volume of compound shapes

Key points

- The perimeter of a shape is the distance all the way around the edge of the shape.
- Area of a rectangle = length × width.
- Area of a triangle = $\frac{1}{2}$ × base × perpendicular height.
- Area of a parallelogram = base × perpendicular height.
- Area of a trapezium = half the sum of the parallel sides × the height.
- To find the area and perimeter of many compound shapes, split them up into simpler basic shapes.
- The circumference of a circle is πd or $2\pi r$, where d is the diameter and r is the radius.
- The area of a circle of radius r is πr^2.
- A sector is a fraction of a circle. Where $\theta°$ is the angle at the centre of the circle, this means that
  Arc length = $\frac{\theta}{360} \times 2\pi r$ and Sector area = $\frac{\theta}{360} \times \pi r^2$
- Area of minor segment = area of minor sector − area of triangle.
- Area of major segment = area of major sector + area of triangle, or area of circle − area of minor segment.
- Volume of a prism = area of cross-section × length.
- Volume of a cylinder of radius r and height $h = \pi r^2 h$.
- Curved surface area of a cylinder = $2\pi rh$.
- To find the total surface area of a solid, add the areas of all the individual surfaces.
- Volume of a sphere of radius $r = \frac{4}{3}\pi r^3$.
- Surface area of a sphere = $4\pi r^2$.
- Volume of a pyramid = $\frac{1}{3}$ × area of base × height.
- Volume of a cone of base radius r and height $h = \frac{1}{3}\pi r^2 h$.
- Curved surface area of a cone = πrl, where l is the slant height.
- Volume of a frustum of a cone = Volume of cone − volume of missing cone.

Note

You should know how to use the formulas for the volume of a sphere, pyramid and cone. You should also know how to use the formulas for the curved surface area of a cone and the surface area of a sphere. You do not need to memorise the formulas but you should be able to use them correctly.

Photocopying is prohibited

REVIEW EXERCISE 7

Ch 32 1 Complete these statements.
    a A cube has .... faces. [1]
    b A cylinder has .... vertices. [1]
    c A triangular prism has .... edges. [1]

Ch 32
Ch 33
Ch 34

2 The scale diagram shows two islands at *A* and *B*.

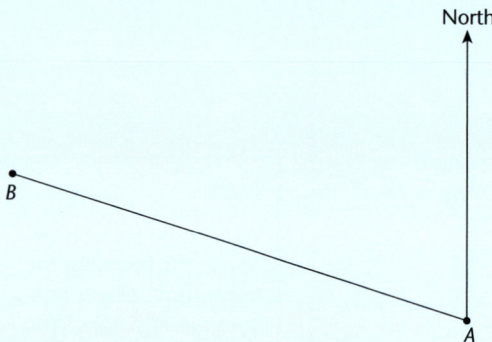

Scale: 2 cm to 1 km
    a Write the scale 2 cm to 1 km in the form 1 : *n*. [1]
    b By measurement, find the bearing of *B* from *A*. [1]
    c An island at *C* is on the northern side of *AB*.
       It is 3 km from *A* and 2.5 km from *B*.
       Use a ruler and a pair of compasses to construct triangle *ABC*. [2]

Cambridge O Level Mathematics Syllabus D (4024) Paper 12 Q13, November 2018

Ch 38 3 In the diagram *B*, *C* and *D* are on the circumference of the circle with centre *O*.
    AC is a tangent to the circle at *C*.

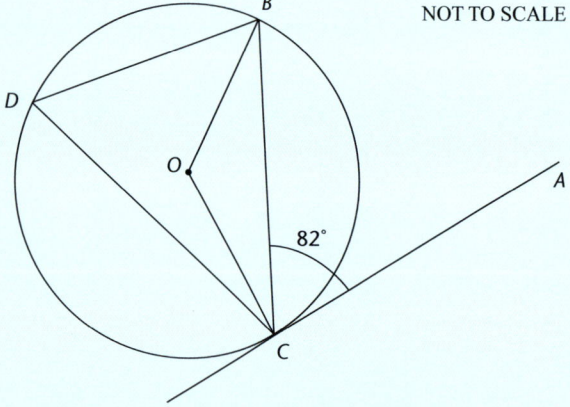

NOT TO SCALE

    a Work out the size of angle *BDC* and give a reason for your answer. [2]
    b Work out the size of angle *BOC* and give a reason for your answer. [2]

4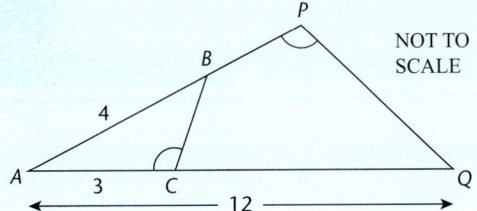

In the diagram, ABP and ACQ are straight lines.

Angle ACB = angle APQ

a Show that triangle ABC is similar to triangle AQP.
   Give a reason for each statement you make. [2]
b AB = 4 cm, AC = 3 cm and AQ = 12 cm.
   Calculate AP. [2]
c The area of triangle ABC is x cm².
   Find an expression, in terms of x, for the area of quadrilateral BPQC. [1]

Cambridge O Level Mathematics Syllabus D (4024) Paper 11 Q25, November 2020

5 a Change 36 m³ into cm³. [1]
  b Work out the size of an interior angle of a regular decagon. [2]

6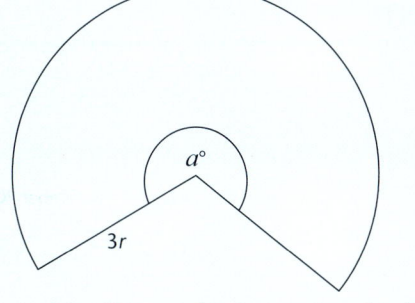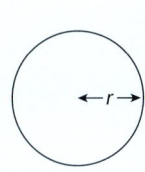

The diagram shows a sector of a circle with radius 3r cm and angle a° and a circle with radius r cm.
The ratio of the area of the sector to the area of the circle with radius r cm is 8 : 1.

a Find the value of a. [3]
b Find an expression, in terms of π and r, for the perimeter of the sector. [2]

Cambridge O Level Mathematics Syllabus D (4024) Paper 12 Q24, June 2016

41 PYTHAGORAS' THEOREM AND TRIGONOMETRY

CHECK YOU CAN:

- use a ruler and a protractor
- find the area of a triangle
- use your calculator to find squares and square roots
- solve simple equations.

BY THE END OF THIS CHAPTER YOU WILL BE ABLE TO:

- know and use Pythagoras' theorem
- know and use the sine, cosine and tangent ratios for acute angles in calculations involving sides and angles of a right-angled triangle
- solve problems in two dimensions using Pythagoras' theorem and trigonometry
- know that the perpendicular distance from a point to a line is the shortest distance to the line
- carry out calculations involving angles of elevation and depression
- use the sine and cosine rules in calculations involving lengths and angles for any triangle
- use the formula
  Area of triangle = $\frac{1}{2}ab \sin C$
- carry out calculations and solve problems in three dimensions using Pythagoras' theorem and trigonometry, including calculating the angle between a line and a plane.

Pythagoras' theorem

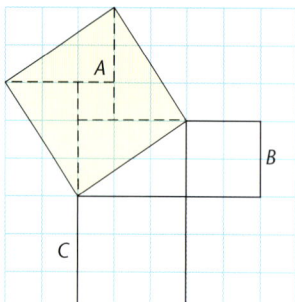

Look at square A.

Area of square A = areas of the four triangles + small square

$$= 4\left(\frac{1}{2} \times 2 \times 3\right) + 1$$

$$= 13$$

Area of square B + area of square C = 4 + 9 = 13

This is an example of the rule linking the areas of squares around a right-angled triangle, known as Pythagoras' theorem.

The largest square will always be on the longest side of the triangle – this is called the **hypotenuse** of the right-angled triangle.

Pythagoras' theorem

Pythagoras' theorem can be stated like this.

The area of the square on the hypotenuse = the sum of the areas of the squares on the other two sides.

Exercise 41.1

Calculate the missing area in each of these diagrams.

1
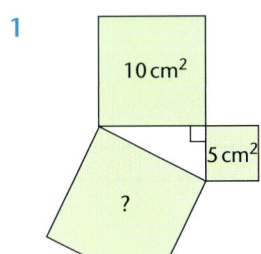

2
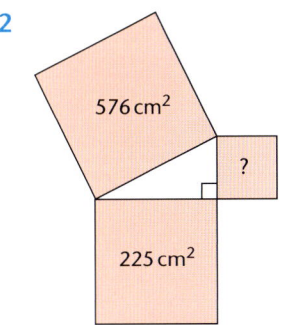

3
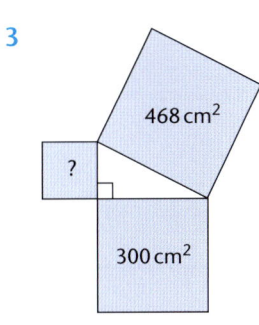

4
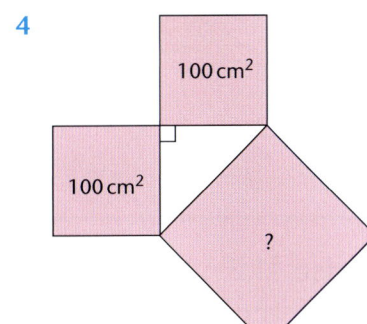

Using Pythagoras' theorem

If you know the lengths of two sides of a right-angled triangle you can use Pythagoras' theorem to find the length of the third side.

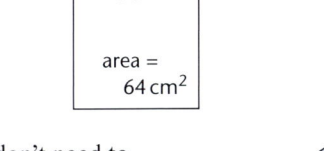

The unknown area = 64 + 36 = 100 cm².

This means that the sides of the unknown square have a length of $\sqrt{100}$ = 10 cm.

When using Pythagoras' theorem, you don't need to draw the squares – you can simply use the rule.

$a^2 = b^2 + c^2$

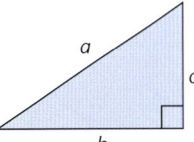

41 PYTHAGORAS' THEOREM AND TRIGONOMETRY

Example 41.1

Question
Find the length a in the diagram.

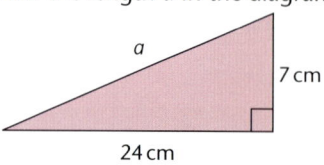

Solution
$a^2 = 7^2 + 24^2$
$a^2 = 49 + 576$
$a^2 = 625$
$a = \sqrt{625}$
$a = 25$ cm

Example 41.2

Question
Find the length c in the diagram.

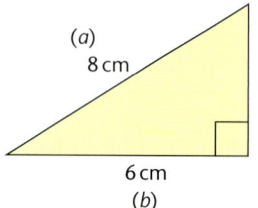

Solution
$a^2 = b^2 + c^2$
$8^2 = 6^2 + c^2$
$64 = 36 + c^2$
$c^2 = 64 - 36$
$c = \sqrt{28}$
$c = 5.29$ cm (to 2 decimal places)

Exercise 41.2

For each of the triangles in this exercise, find the length of the side marked with a letter.

Give your answers either exactly, or correct to 2 decimal places.

1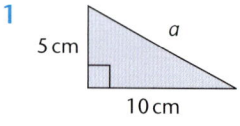

2

3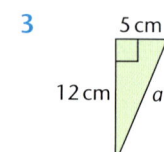

4

5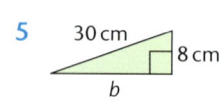

6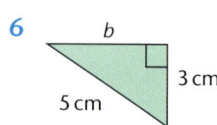

7

8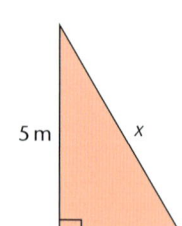

9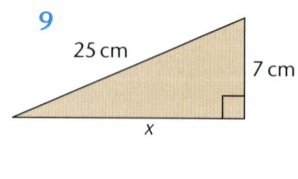

Pythagoras' theorem

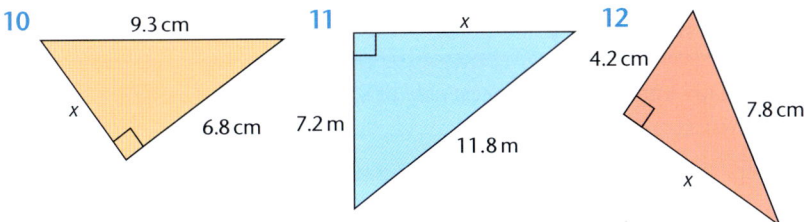

Using Pythagoras' theorem to solve problems

You can use Pythagoras' theorem to solve problems.

It is a good idea to draw a sketch if a diagram isn't given.

Try to draw it roughly to scale and mark on it any lengths you know.

Example 41.3

Question

Tao is standing 115 m from a vertical tower.
The tower is 20 m tall.
Work out the distance from Tao directly to the top of the tower.

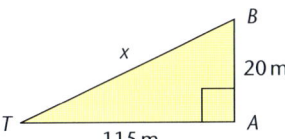

Solution

$x^2 = 115^2 + 20^2$
$ = 13\,625$
$x = \sqrt{13\,625}$
$ = 116.7\,m$ (to 1 decimal place)

Tao is 116.7 m (to 1 decimal place) from the top of the tower.

Exercise 41.3

1. A rectangular field is 225 m long and 110 m wide.
   Find the length of the diagonal path across it.
2. A rectangular field is 25 m long.
   A footpath 38.0 m long crosses the field diagonally.
   Find the width of the field.
3. A ladder is 7 m long.
   It is resting against a wall, with the top of the ladder 5 m above the ground.
   How far from the wall is the base of the ladder?
4. Haleef is making a kite for his sister.
   This is his diagram of the kite.
   The kite is 30 cm wide.
   Haleef needs to buy some cane to make the struts AC and DB.
   What length of cane does he need to buy?

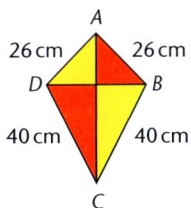

41 PYTHAGORAS' THEOREM AND TRIGONOMETRY

Trigonometry

> **Note**
> Label the sides in the order hypotenuse, opposite, adjacent.
> To identify the opposite side, go straight out from the middle of the angle. The side you hit is the opposite.
> You can shorten the labels to 'H', 'O' and 'A'.
> θ is the Greek letter 'theta'.

You already know that the longest side of a right-angled triangle is called the **hypotenuse**.

The side opposite the angle you are using (θ) is called the **opposite**.

The remaining side is called the **adjacent**.

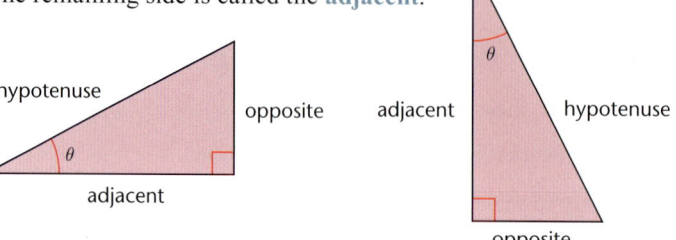

For a given angle $\theta°$, all right-angled triangles with an angle $\theta°$ will be similar (angles in each triangle of 90°, $\theta°$ and (90 – $\theta°$)). It follows that the ratios of the sides will be constant for that value of θ.

The ratio $\frac{\text{Opposite}}{\text{Hypotenuse}}$ is called the **sine** of the angle.

This is often shortened to 'sin'.

$$\sin\theta = \frac{\text{Opposite}}{\text{Hypotenuse}}$$

> **Note**
> Notice that the ratio of the lengths is written as a fraction, $\frac{\text{Opposite}}{\text{Hypotenuse}}$, rather than a ratio, Opposite : Hypotenuse.

The ratio $\frac{\text{Adjacent}}{\text{Hypotenuse}}$ is called the **cosine** of the angle.

This is often shortened to 'cos'.

$$\cos\theta = \frac{\text{Adjacent}}{\text{Hypotenuse}}$$

The ratio $\frac{\text{Opposite}}{\text{Adjacent}}$ is called the **tangent** of the angle.

This is often shortened to 'tan'.

$$\tan\theta = \frac{\text{Opposite}}{\text{Adjacent}}$$

> **Note**
> You need to learn the three ratios.
> $\sin\theta = \frac{O}{H}$
> $\cos\theta = \frac{A}{H}$
> $\tan\theta = \frac{O}{A}$
> There are various ways of remembering these, but one of the most popular is to learn the 'word' 'SOHCAHTOA'.

Trigonometry

Using the ratios 1

When you need to solve a problem using one of the ratios, you should follow these steps.

- Draw a clearly labelled diagram.
- Label the sides H, O and A.
- Decide which ratio you need to use.
- Solve the equation.

In one type of problem you will encounter, you are required to find the numerator (top) of the fraction. This is demonstrated in the following examples.

Example 41.4

Question

Find the length marked *x*.

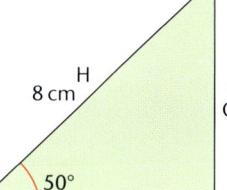

Solution

Since you know the hypotenuse (H) and want to find the opposite (O), you use the sine ratio.

$\sin 50° = \frac{O}{H}$

$\sin 50° = \frac{x}{8}$

$8 × \sin 50° = x$ Multiply both sides by 8.

$x = 6.128\,35… = 6.13$ cm correct to 3 significant figures.

Note
Press these keys on your calculator to find *x*.

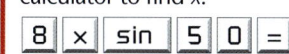

Note
Make sure that your calculator is set to degrees.
This is the default setting but, if you see 'rad' or 'R' or 'grad' or 'G' in the window, change the setting using the key labelled 'mode' or 'DRG' or 'set up'.

Example 41.5

Question

In triangle *ABC*, *BC* = 12 cm, angle *B* = 90° and angle *C* = 35°.
Find the length *AB*.

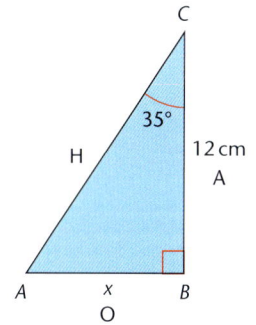

Solution

Draw the triangle and label the sides.
Since you know the adjacent (A) and want to find the opposite (O), you use the tangent ratio.

$\tan 35° = \frac{O}{A}$

$\tan 35° = \frac{x}{12}$

$12 × \tan 35° = x$ Multiply both sides by 12.

$x = 8.402\,49… = 8.40$ cm correct to 3 significant figures.

Note
Press these keys on your calculator to find *x*.

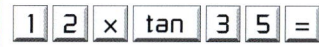

Photocopying is prohibited

41 PYTHAGORAS' THEOREM AND TRIGONOMETRY

Exercise 41.4

1 In these diagrams find the lengths marked a, b, c, d, e, f, g and h.

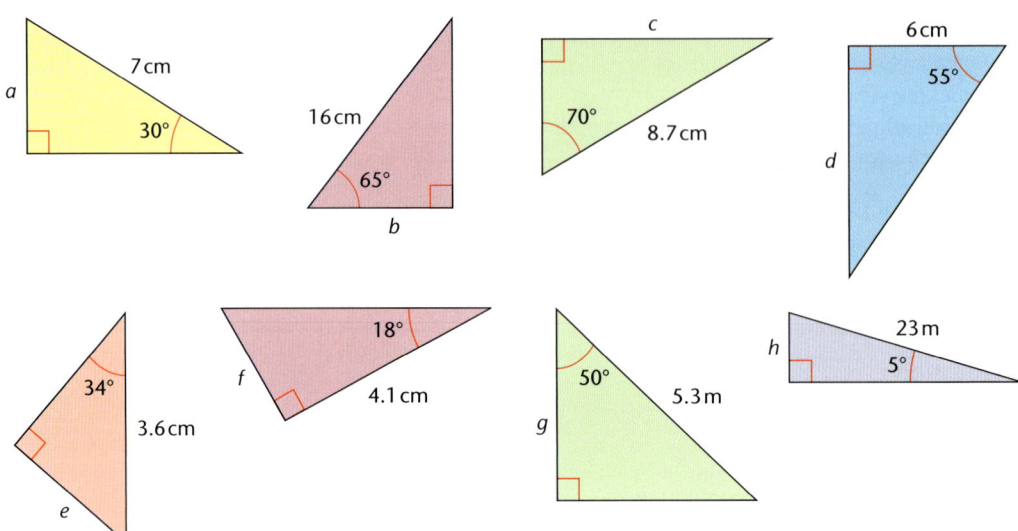

2 The ladder shown here is 6 metres long.
The angle between the ladder and the ground is 70°.
How far from the wall is the foot of the ladder?

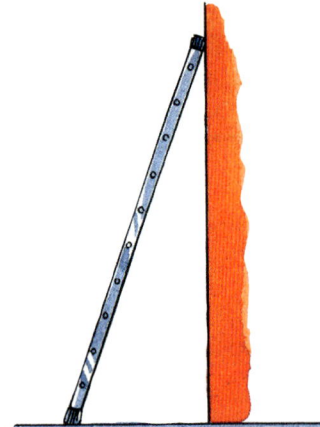

3 a Find the height, h, of the triangle.
b Use the height you found in part a to find the area of the triangle.

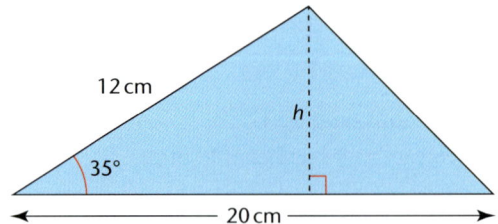

Trigonometry

4 A ship sails on a bearing of 070° for 250 km.
   a Find how far north the ship has travelled.
   b Find how far east the ship has travelled.

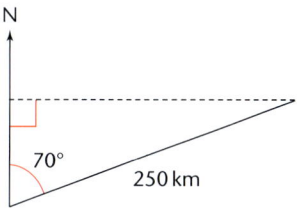

> **Note**
> You can read more about bearings in Chapter 34.

5 The diagram shows a crane.
   The length of the crane's arm is 20 metres.
   The crane can operate with the arm anywhere between 15° and 80° to the vertical.
   Calculate the minimum and maximum values of x, the distance from the crane at which a load can be lowered.

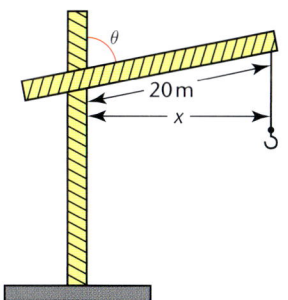

Using the ratios 2

In the second type of problem you will encounter, you are required to find the denominator (bottom) of the fraction. This is demonstrated in the following example.

Example 41.6

Question
Find the length marked x.

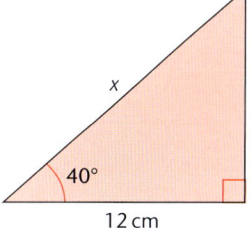 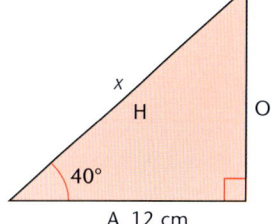

Solution
Since you know A and want to find H, you use the cosine ratio.

$\cos 40° = \frac{A}{H}$

$\cos 40° = \frac{12}{x}$

$x = \frac{12}{\cos 40°}$

$x = 15.66488\ldots = 15.7$ cm correct to 3 significant figures.

> **Note**
> Always look to see whether the length you are trying to find should be longer or shorter than the one you are given.
>
> If your answer is obviously wrong, you have probably multiplied instead of divided.

Photocopying is prohibited

Exercise 41.5

1. In these diagrams find the lengths marked a, b, c, d, e, f, g and h.

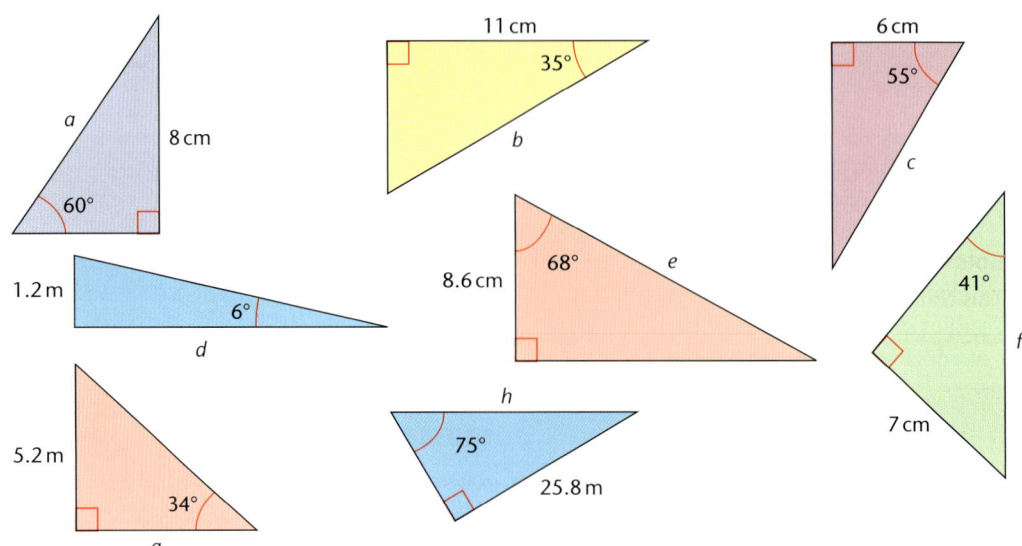

2. The bearing of A from B is $040°$.
   A is 8 kilometres east of B.
   Calculate how far A is north of B.

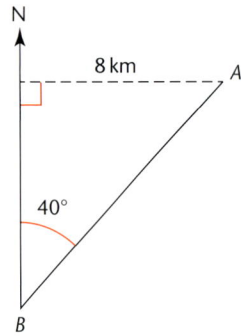

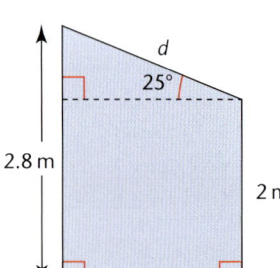

3. The diagram shows a shed.
   a. Find the length d.
   b. The length of the shed is 2.5 m.
      Find the area of the roof.

4. A ship sailed on a bearing of $140°$.
   It was then 90 km south of its original position.
   a. Draw a diagram to show the ship's journey.
   b. How far east of its original position is it?

5. Mrs Khan wants to buy a ladder.
   Her house is 5.3 metres high and she needs to reach the top.
   The ladders are in two sections, each section being the same length.
   When extended, there must be an overlap of 1.5 metres between the two sections.
   The safe operating angle is $76°$.
   Calculate the length of each of the sections of ladder she needs to buy.

Trigonometry

Using the ratios 3

In the third type of problem you will encounter, you are given the value of two sides and are required to find the angle. This is demonstrated in the following examples.

Example 41.7

Question

Find the angle θ.

Solution

This time, look at the two sides you know.
Since they are O and H, you use the sine ratio.

$\sin\theta = \frac{O}{H}$ $\sin\theta = \frac{5}{8}$

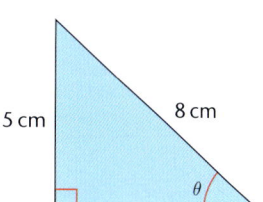

Work out $5 \div 8 = 0.625$ on your calculator and leave this number in the display.
Now use the $\sin^{-1}$ function (the inverse of sine).
With 0.625 still in the display, press SHIFT sin = , or the equivalent on your calculator.
You should see 38.682 18... .

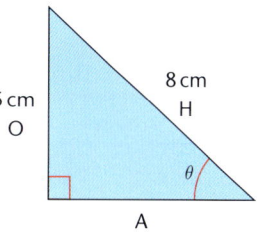

So $\theta = 38.7°$ correct to 3 significant figures, or 39° correct to the nearest degree.

Example 41.8

Question

Find the angle θ.

Solution

Since this is an isosceles triangle, not a right-angled triangle, you need to use the fact that the line of symmetry splits an isosceles triangle into two equal right-angled triangles.

The sides you know are A and H so you use the cosine ratio.

$\cos\theta = \frac{A}{H} = \frac{6}{8}$

$\theta = \cos^{-1}\frac{6}{8}$

$\theta = 41.4°$ correct to 3 significant figures.

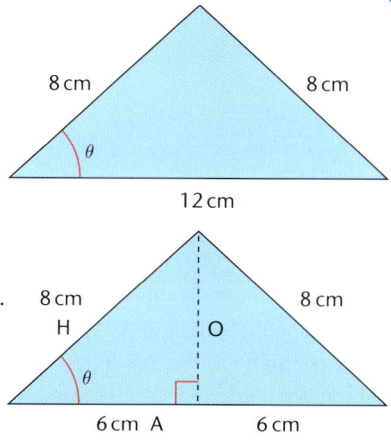

Note

Example 41.8 shows how to deal with isosceles triangles.

You use the line of symmetry to split the triangle into two equal right-angled triangles.

This works only with isosceles triangles, because they have a line of symmetry.

Exercise 41.6

1 In these diagrams find the angles marked *a, b, c, d, e, f, g* and *h*.

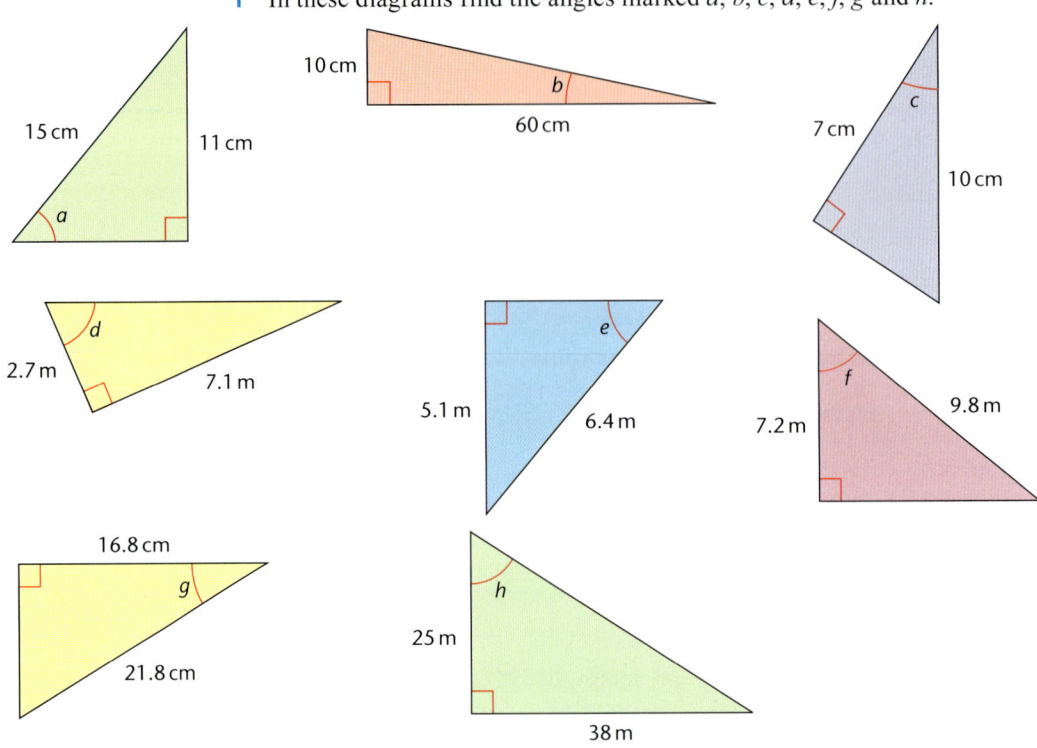

2 The diagram represents a ladder leaning against a wall.
Find the angle the ladder makes with the horizontal.

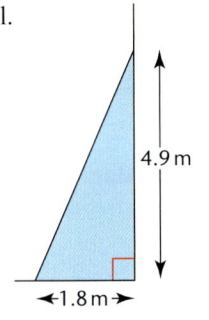

3 In the picture, the kite is 15 metres above the girl.
The string is 25 metres long.
Find the angle the string makes with the horizontal.

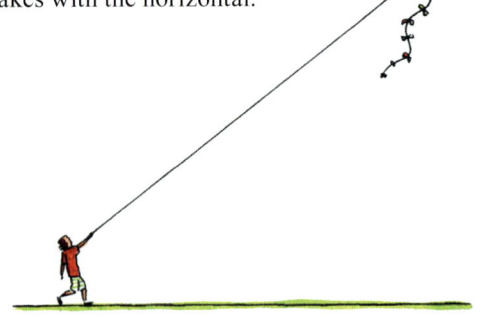

Trigonometry

4 The diagram represents a pair of step ladders standing on a horizontal floor.
   Find the angle, θ, between the two parts of the step ladder.

5 An aircraft flies 180 km due east from A to B.
   It then flies 115 km due south from B to C.
   a Draw a diagram to show the positions of A, B and C.
   b Calculate the bearing of C from A.

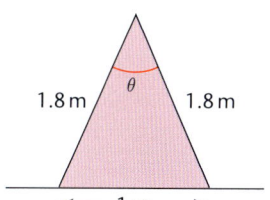

6 A television mast is 54 metres high and stands on horizontal ground.
   Six guy wires keep the mast upright.
   Three of these are attached to the top and a point on the ground.
   These three make an angle of 16.5° with the vertical.
   a Calculate the total length of these three wires.
   The other three wires are attached $\frac{2}{3}$ of the way up the mast.
   They are attached to the same points on the ground as the previous three.
   b Calculate the angle these make with the vertical.

7 A ship sails for 150 km on a bearing of 115° from A to B.
   It then sails for 250 km on a bearing of 230° from B to C.
   Calculate the distance AC and the bearing of C from A.

8 The diagrams show a bridge in the shut and open positions.

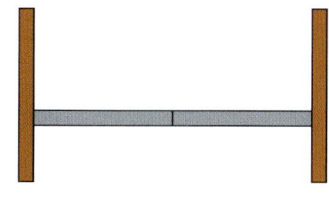

When opened, the bridge sections raise by 50°.
The distance between the supports is 38 metres.
Calculate the distance d.

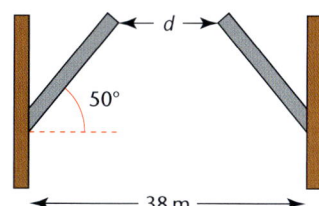

Angle of elevation and angle of depression

The angle that a line to an object makes above a horizontal plane is called the angle of elevation.

The angle of elevation of the top of this tree, A, from B, is θ.

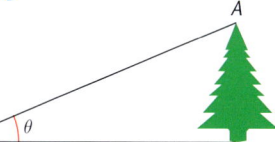

The angle a line makes to an object below a horizontal plane is called the angle of depression.

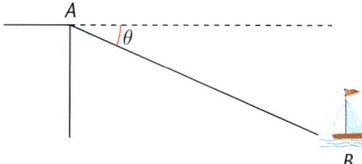

The angle of depression of the boat, B, from A, is θ.

> **Note**
> Since alternate angles are equal, the angle of elevation of A from B equals the angle of depression of B from A.

Example 41.9

Question

The height of this tree is 24 m.
A is 68 m from the foot of the tree.

Find the angle of elevation of the top of the tree from A.

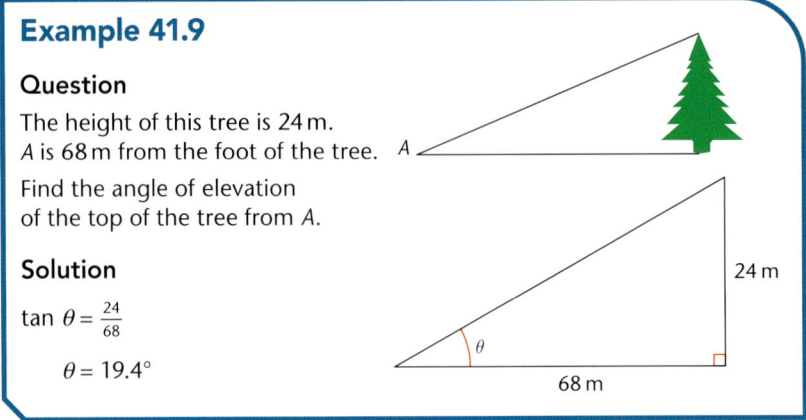

Solution

$\tan \theta = \frac{24}{68}$

$\theta = 19.4°$

Exercise 41.7

1. Find the angle of elevation of A from B.

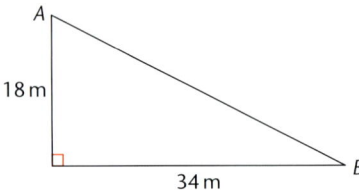

2. The angle of elevation of the top of a tree from A is 25°.
   A is 38 m from the base of the tree.
   Calculate the height of the tree.

3. AB is a vertical cliff 32 m high.
   The boat is on the sea, 75 metres from the bottom of the cliff.
   Calculate the angle of depression of the boat from the top of the cliff, A.

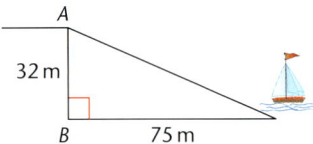

4. The diagram shows a tall building.
   B is a point on the ground 48 m from the foot of the building.
   The angle of depression of B from a point A, on the top of the building, is 28°.
   Calculate the height of the building.

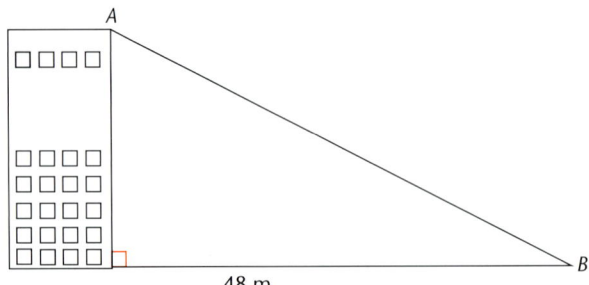

5 Adil is standing on horizontal ground looking at the top of a building 18 m high.
   The angle of elevation of the top of the building from Adil's eyes is 24°.
   Adil's eyes are 1.8 m above the ground.
   Find how far away from the building Adil is standing.

The sine and cosine functions for obtuse angles

So far, you have only dealt with the sine and cosine functions for right-angled triangles, so the angles have all been acute.

However, your calculator will give you the sine and cosine of any angle.

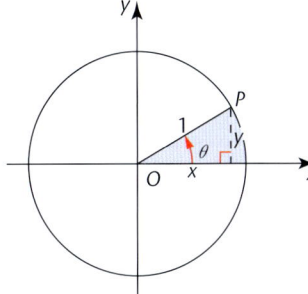

In this diagram you can see that for an acute angle

$\cos\theta = \frac{x}{1}$ so $x = \cos\theta$

$\sin\theta = \frac{y}{1}$ so $y = \sin\theta$

So P has coordinates $(\cos\theta, \sin\theta)$.

For other angles, the trigonometric functions are defined in a similar way, where the angle is measured anticlockwise from the x-axis.

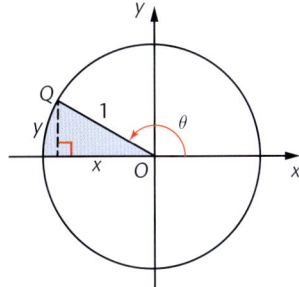

By symmetry, Q has coordinates $(-\cos\theta, \sin\theta)$.

So

$\sin\theta = \sin(180 - \theta)$

$\cos\theta = -\cos(180° - \theta)$

By symmetry, $\sin\theta = y = \sin(180° - \theta)$
and $\cos\theta = x = -\cos(180° - \theta)$.

So for an obtuse angle θ, $\sin\theta$ is equal to $\sin(180° - \theta)$
and $\cos\theta$ takes the same value as $\cos(180° - \theta)$ but is negative.

41 PYTHAGORAS' THEOREM AND TRIGONOMETRY

Example 41.10

Question

You are given that sin 40° = 0.6428 and cos 40° = 0.7660.
Without using your calculator, write down the values of
a sin 140°
b cos 140°.

Solution

a sin 140° = sin (180° − 40°) = sin 40° = 0.6428
b cos 140° = cos (180° − 40°) = −cos 40° = −0.7660
Check these values by finding sin 140° and cos 140° directly on your calculator.

Example 41.11

Question

Solve the equation sin x = 0.8 for 0° ⩽ x ⩽ 180°.

Solution

Using a calculator gives x = 53.13°, so this is one solution.
But sin x is also positive for values of x between 90° and 180°.
So another solution is 180° − 53.13° = 126.87°.
So x = 53.13° or 126.87°

Exercise 41.8

1 You are given that sin 60° = 0.8660 and cos 60° = 0.5.
   Without using your calculator, write down the values of
   a sin 120°
   b cos 120°.
2 Solve the equation sin x = 0.3 for 0° ⩽ x ⩽ 180°.
3 Solve the equation cos x = 0.7 for 0° ⩽ x ⩽ 180°.
4 In the diagram, BCD is a straight line.
   BC = 5 cm, AB = 12 cm, AC = 13 cm and $A\hat{B}C$ = 90°.
   Find
   a tan $B\hat{A}C$
   b cos $A\hat{C}D$.

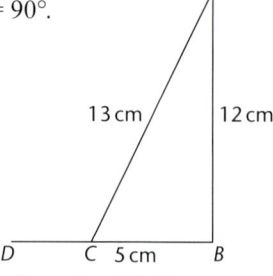

5 A is the point (1, 2), B is the point (4, 6) and C is the point (−5, 2).
   a Calculate the length of AB.
   b Find the value of sin $C\hat{A}B$.
   c Find the value of cos $C\hat{A}B$.

6 P is the point (−6, 8).
   Find the value of
   a sin θ   b cos θ.

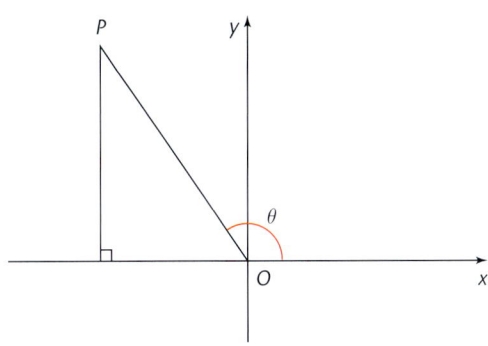

Non-right-angled triangles

All the trigonometry that you have learned so far has been based on finding lengths and angles in right-angled triangles. However, many triangles are not right-angled. You need a method to find lengths and angles in these other triangles.

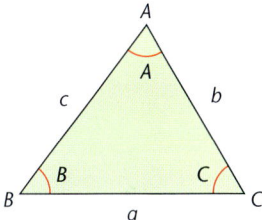

For simplicity, we shall use a single letter to represent each side and each angle of the triangle, a capital letter for an angle and a lowercase letter for a side. The side opposite an angle takes the same, lowercase letter, as shown in the diagram above.

There are two rules for dealing with non-right-angled triangles; they are called the **sine rule** and the **cosine rule**.

Learn the formulas so that you can use them easily but you do not need to memorise them.

The sine rule

The sine rule is derived from the trigonometry in right-angled triangles that you have already met.

Use triangle BCP to find an expression for h in terms of a and C.

$\frac{h}{a} = \sin C$ so $h = a \sin C$

Use triangle ABP to find an expression for h in terms of c and A.

$\frac{h}{c} = \sin A$ so $h = c \sin A$

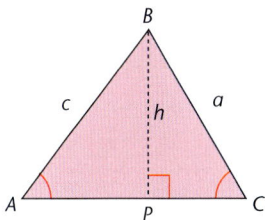

Equating the expressions for h gives

$h = a \sin C = c \sin A$

Dividing both sides by (sin A sin C) gives

$$\frac{a}{\sin A} = \frac{c}{\sin C}$$

41 PYTHAGORAS' THEOREM AND TRIGONOMETRY

> **Note**
> You do not need to be able to prove the sine rule.

Similarly, by drawing a perpendicular from point C to side AB we can show that

$$\frac{a}{\sin A} = \frac{b}{\sin B}$$

So the full sine rule is

$$\frac{a}{\sin A} = \frac{b}{\sin B} = \frac{c}{\sin C}$$

or

$$\frac{\sin A}{a} = \frac{\sin B}{b} = \frac{\sin C}{c}$$

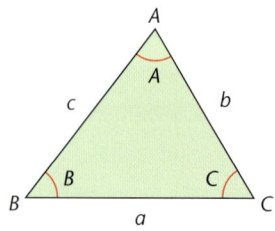

To use the sine rule you must know one side and the angle opposite it. You also need to know one other angle or one other side.

Example 41.12

Question

Find these.

a Length b **b** Length c

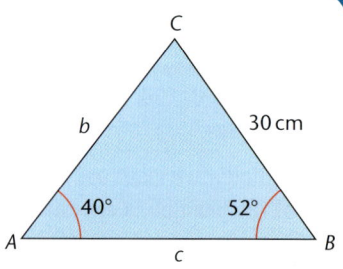

Solution

a Since you are finding a length, use the formula with lengths on top.

Choose pairs of angles and opposite sides where three of the four values are known and substitute into the formula.

$$\frac{b}{\sin B} = \frac{a}{\sin A}$$

$$\frac{b}{\sin 52°} = \frac{30}{\sin 40°}$$

$$b = \frac{30}{\sin 40°} \times \sin 52°$$

$$b = 36.8 \text{ cm to 1 decimal place}$$

> **Note**
> Although you could use the pair b and B instead of a and A, you should, whenever possible, use values that are given rather than values that have been calculated.

b Before you can find c, you need to find angle C.

$$C = 180 - (40 + 52)$$
$$C = 88°$$

$$\frac{c}{\sin C} = \frac{a}{\sin A}$$

$$\frac{c}{\sin 88°} = \frac{30}{\sin 40°}$$

$$c = \frac{30}{\sin 40°} \times \sin 88°$$

$$c = 46.6 \text{ cm to 1 decimal place}$$

Non-right-angled triangles

Example 41.13

Question

Find C.

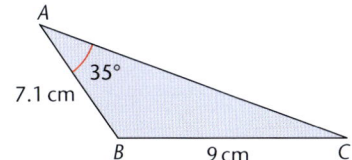

Solution

Since you are finding an angle, use the formula with angles on top.

$$\frac{\sin C}{c} = \frac{\sin A}{a}$$

$$\frac{\sin C}{7.1} = \frac{\sin 35°}{9}$$

$$\sin C = \frac{\sin 35°}{9} \times 7.1$$

$$\sin C = 0.4524...$$

$$C = \sin^{-1}(0.4524...)$$

$$C = 26.9° \text{ to 1 decimal place}$$

Exercise 41.9

1 Find c, A and a.

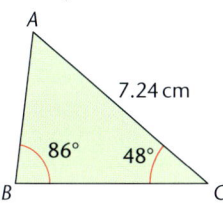

2 Find p, R and r.

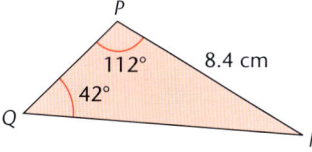

3 Find B, C and c.

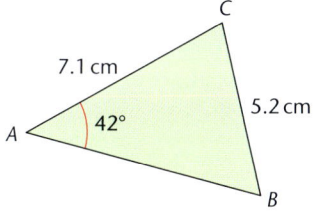

4 Find M, N and n.

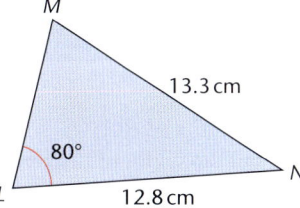

5 Find P, R and r.

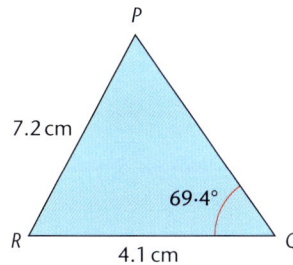

6 Find Y, Z and z.

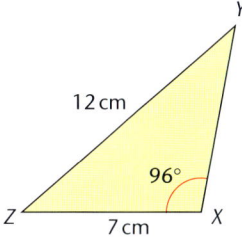

41 PYTHAGORAS' THEOREM AND TRIGONOMETRY

7 Find y, Z and z.

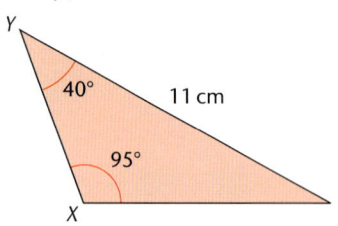

8 Find s, T and t.

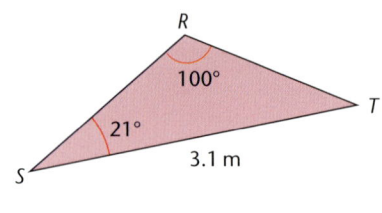

9 In triangle ABC, $c = 23$ cm, $b = 19.4$ cm and angle C is $54°$.
Find the length of BC.

10 Calculate the largest angle of the triangle ABC given that $A = 35°$, $a = 8.9$ cm and $c = 12$ cm.

11 Two points, A and B, are 30 m apart on horizontal ground.
The angles of elevation of the top, T, of a vertical tower, TC, from A and B are $27°$ and $40°$ respectively.
  a Find AT and BT.
  b Find the height of the tower, TC.

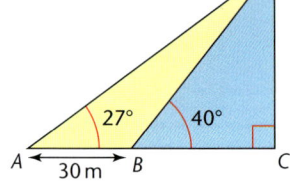

> **Note**
> The shortest distance between a point and a line is the perpendicular distance to the line.

12 A river has two parallel banks.
Points A and C are on one side of the river, 45 m apart.
The angles from these points to a tree, B, on the opposite bank are $68°$ and $34°$, as shown in the diagram.

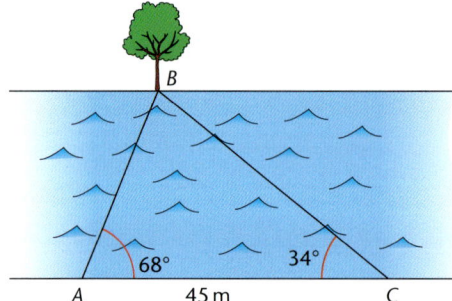

  a Find AB and BC.
  b Find the width of the river.

13 Town B is 45 km due east of town A.
Town C is on a bearing of $057°$ from town A and a bearing of $341°$ from town B.
Find how far town C is from towns A and B.

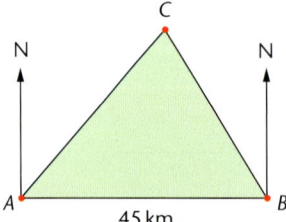

Non-right-angled triangles

The ambiguous case of the sine rule

Construct accurately triangle ABC, given that $AC = 7$ cm, $CB = 5$ cm and angle A is 30°.

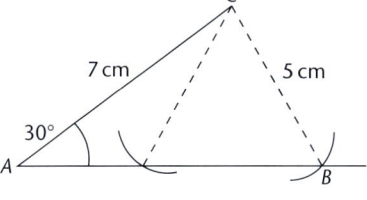

There are two possible triangles that you can draw with this information.

Using the sine rule to calculate angle B, with $a = 5$ cm, $b = 7$ cm and $A = 30°$.

$$\frac{\sin B}{7} = \frac{\sin 30}{5}$$

$$\sin B = 7 \times \frac{\sin 30}{5}$$

$$= 0.7$$

$$B = 44.4 \text{ or } (180 - 44.4)$$

$$= 44.4° \text{ or } 135.6°$$

In this triangle, the obtuse angle also gives a possible triangle, since 135.6° + 30° is less than 180. Then $C = 14.4°$. Using $B = 44.4°$ gives $C = 105.6°$.

In triangles you have met in the past, this did not happen.

The ambiguous case can only happen when you are finding an angle opposite the longest side you are given. This is because the largest angle in a triangle is opposite the longest side. The smallest angle in a triangle is opposite the shortest side.

Use these facts with the sine rule to help you check whether you need to consider the ambiguous case.

Exercise 41.10

1. In triangle ABC, $C = 42°$, $a = 8.2$ cm and $c = 6.1$ cm.
   Use the sine rule to show that there are two possible triangles with this information.
   Find the values of the other angles and side in each case.
2. a. Show, using the sine rule, that it is not possible to draw triangle ABC when $B = 50°$, $c = 10$ cm and $b = 7$ cm.
   b. Calculate the smallest value of b that does give a possible triangle with these values of B and c.
3. In triangle ABC, $A = 25°$, $a = 5.1$ cm and $c = 9.7$ cm.
   Use the sine rule to show that there are two possible triangles with this information.
   Find the values of the other angles and side in each case.

Note
Sketch a diagram to help you. Remember how to find the shortest distance from a point to a line.

41 PYTHAGORAS' THEOREM AND TRIGONOMETRY

The cosine rule

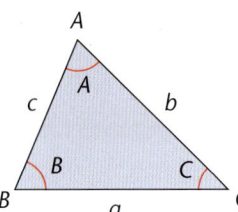

To derive the cosine rule, you combine right-angled trigonometry with Pythagoras' theorem.

Use Pythagoras' theorem for triangle *BPA*.

$h^2 + x^2 = c^2$ so $h^2 = c^2 - x^2$.

Use Pythagoras' theorem for triangle *BPC*.

$h^2 + (b-x)^2 = a^2$ so $h^2 = a^2 - (b-x)^2$.

Equate the two expressions for h^2 $h^2 = a^2 - (b-x)^2 = c^2 - x^2$

Expand the bracket and simplify $a^2 - (b^2 - 2bx + x^2) = c^2 - x^2$

$$a^2 - b^2 + 2bx - x^2 = c^2 - x^2$$

$$a^2 - b^2 + 2bx = c^2$$

Rearrange $a^2 = c^2 + b^2 - 2bx$

But $\frac{x}{c} = \cos A$ so $x = c \cos A$ giving $a^2 = c^2 + b^2 - 2bc \cos A$

This is the cosine rule.

> **Note**
> You do not need to be able to prove the cosine rule.

Changing the letters and rearranging gives the six formulas for the cosine rule.

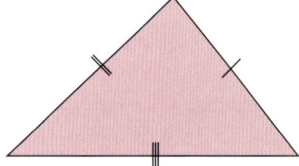

$a^2 = c^2 + b^2 - 2bc \cos A$

$b^2 = c^2 + a^2 - 2ca \cos B$

$c^2 = a^2 + b^2 - 2ab \cos C$

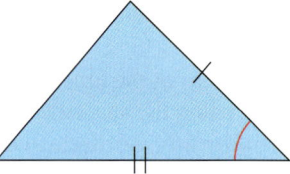

$\cos A = \dfrac{b^2 + c^2 - a^2}{2bc}$

$\cos B = \dfrac{c^2 + a^2 - b^2}{2ca}$

$\cos C = \dfrac{a^2 + b^2 - c^2}{2ab}$

Use the cosine rule

- when you know all the sides
- when you know two sides and the included angle.

Example 41.14

Question

Find *c*.

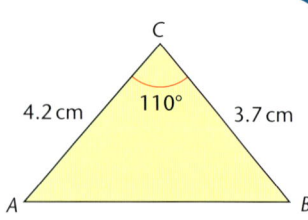

Solution

$c^2 = a^2 + b^2 - 2ab \cos C$

$c^2 = 3.7^2 + 4.2^2 - (2 \times 3.7 \times 4.2 \times \cos 110°)$

$c^2 = 41.959...$

$c = 6.48\,\text{cm}$ to 2 decimal places

Non-right-angled triangles

Example 41.15

Question

Find R.

Solution

$\cos R = \dfrac{p^2 + q^2 - r^2}{2pq}$

$\cos R = \dfrac{6^2 + 8^2 - 5^2}{2 \times 6 \times 8}$

$\cos R = 0.781\,25$

$\quad R = \cos^{-1}(0.781\,25)$

$\quad R = 38.6°$ to 1 decimal place

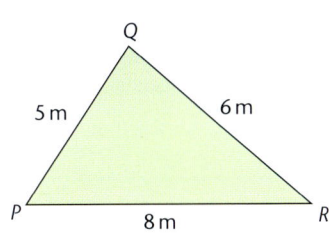

Exercise 41.11

1 Find a.

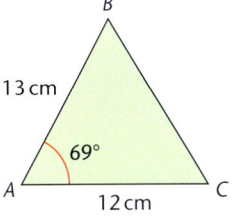

2 Find c.

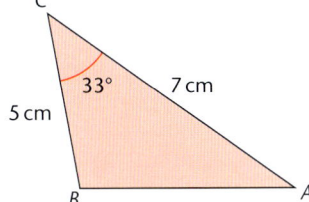

3 Find C.

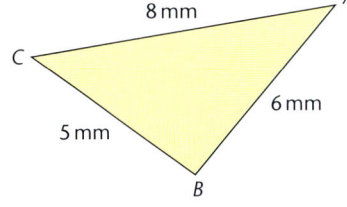

4 Find B.

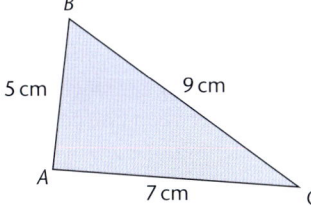

5 Find a.

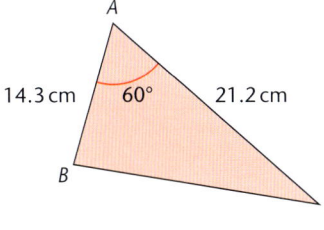

6 Find B.

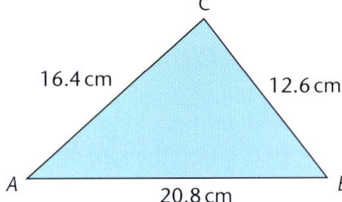

7 Find B.

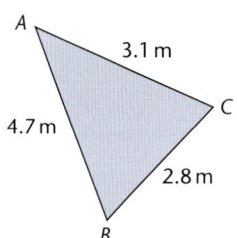

8 Three towns, A, B and C, are shown in the diagram. Find $A\hat{C}B$.

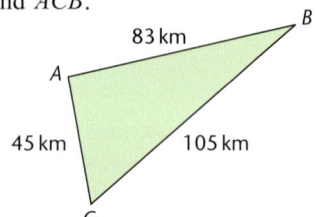

9 Find the smallest angle in this triangle.

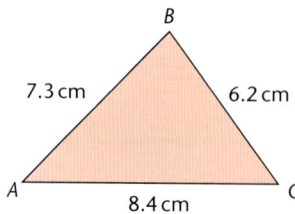

10 A cross-country runner runs 4 km due north and then 6.7 km in a south-east direction. How far is she from her starting point?

11 From a boat C, A is 9 km away on a bearing of 058° and B is 12 km away on a bearing of 110°. Calculate the distance AB.

12 The diagram shows the cross-section of a roof. The house is 12 m wide. Calculate the length of the roof, x, and the angle of the slope, y.

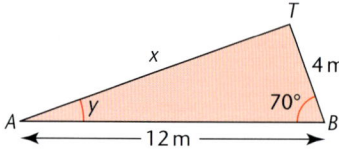

13 A parallelogram has sides of length 5.1 cm and 2.5 cm. The angle between one pair of adjacent sides is 70°. Find the length of each of the diagonals of the parallelogram.

14 Three points, A, B and C, form an equilateral triangle on horizontal ground with sides 10 m. Vertical posts PA, QB and RC are placed in the ground. $PA = 4$ m, $QB = 10$ m and $RC = 6$ m.
   a i Find PQ.
      ii Find PR.
      iii Find QR.
   b Hence find the size of
      i $R\hat{P}Q$
      ii $P\hat{R}Q$.

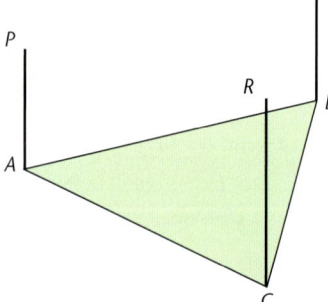

Note

Always draw a sketch if you are not given a diagram.

Label the sides and angles with the information given.

Non-right-angled triangles

The general formula for the area of any triangle

To find the area of a triangle when you do not know the perpendicular height, you can derive a formula from the trigonometry of right-angled triangles.

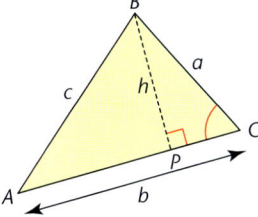

Area of triangle $ABC = \frac{1}{2} \times b \times h$

But $\frac{h}{a} = \sin C$ so $h = a \sin C$

Substituting this expression in the formula for the area of a triangle gives

Area of triangle $ABC = \frac{1}{2} ab \sin C$

> Area of triangle $ABC = \frac{1}{2} ab \sin C$
> $= \frac{1}{2} bc \sin A$
> $= \frac{1}{2} ca \sin B$

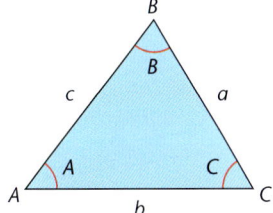

Example 41.16

Question
Find the area of the triangle shown.

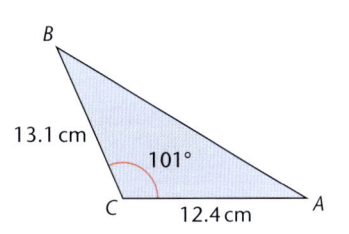

Solution
Area $= \frac{1}{2} ab \sin C$
$= \frac{1}{2} \times 13.1 \times 12.4 \times \sin 101°$
$= 79.7 \text{ cm}^2$ to 1 decimal place

Exercise 41.12

1 Find the area of each of these triangles.

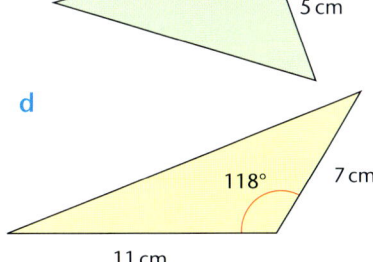

a 8.6 cm, 60°, 2.4 cm
b 8 cm, 100°, 5 cm
c 8.1 cm, 42.4°, 7.5 cm
d 11 cm, 118°, 7 cm
e 6.5 m, 81°, 67°, 7 m

41 PYTHAGORAS' THEOREM AND TRIGONOMETRY

> **Note**
> Remember to draw a diagram and label it with the information you know.

2 In triangle ABC, $a = 10$ cm, $c = 6$ cm and $B = 150°$. Find the area of the triangle.

3 In triangle ABC, $a = 4$ cm, $c = 7$ cm and its area is 13.4 cm^2. Find the size of angle B.

4 The area of triangle PQR is 273 cm^2. Given that $PQ = 12.8$ cm and $P\hat{Q}R = 107°$, find QR.

5 Calculate the area of parallelogram $ABCD$ in which $AB = 6$ cm, $BC = 9$ cm and $A\hat{B}C = 41.4°$.

6 Calculate the area of this triangle.

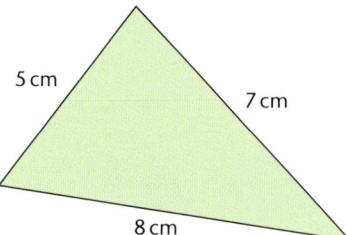

7 $ABCD$ is a field with dimensions as shown in the diagram. Calculate the area of the field.

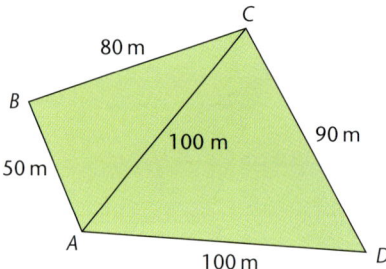

Finding lengths and angles in three dimensions

You can find lengths and angles of three-dimensional objects by identifying right-angled triangles within the object and using Pythagoras' theorem or trigonometry.

Finding lengths and angles in three dimensions

Example 41.17

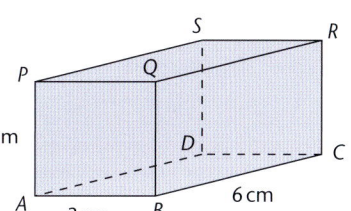

Question
A rectangular box measures 4 cm by 3 cm by 6 cm.
Calculate these.

a AC **b** AR **c** Angle RBC **d** Angle ARC

Solution

a Identify the right-angled triangle required and draw a sketch.

$AC^2 = AB^2 + BC^2$
$AC^2 = 3^2 + 6^2$
$AC^2 = 45$
$AC = 6.71$ cm to 2 decimal places

Note
Don't use 6.71^2 as this has been rounded.
You need AC^2 and this is 45.

b $AR^2 = AC^2 + RC^2$
$AR^2 = 45 + 16$
$AR^2 = 61$
$AR = 7.81$ cm to 2 decimal places

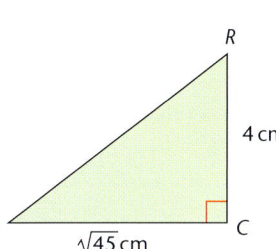

Note
When there is a choice of formulas to use, use the one which contains as many given values as possible. You may have calculated a value incorrectly.

c $\tan x = \frac{4}{6}$

$x = \tan^{-1} \frac{4}{6}$

$x = 33.7°$ to 1 decimal place

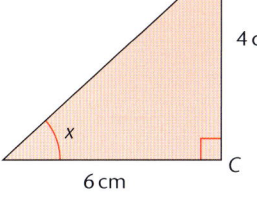

Note
You can't avoid using a calculated value here.

Using 6.71 will give an answer correct to 3 significant figures but it is good practice to use the more accurate value, $\sqrt{45}$, and round at the end.

d $\tan x = \frac{\sqrt{45}}{4}$

$x = \tan^{-1} \frac{\sqrt{45}}{4}$

$x = 59.2°$ to 1 decimal place

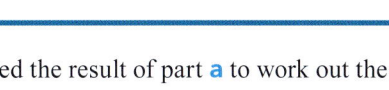

In Example 41.17, notice how you used the result of part **a** to work out the result of part **b**.

Photocopying is prohibited

41 PYTHAGORAS' THEOREM AND TRIGONOMETRY

You can use this method to derive a general formula for the length of the diagonal of a cuboid measuring a by b by c.

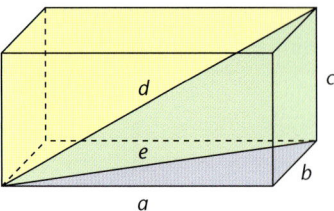

The diagonal of the rectangular base is labelled e.

$$e^2 = a^2 + b^2$$

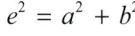

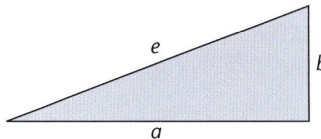

The diagonal of the cuboid is labelled d.

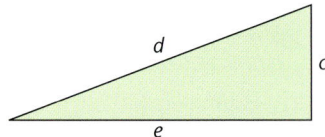

$d^2 = e^2 + c^2$

But $e^2 = a^2 + b^2$

so

$d^2 = a^2 + b^2 + c^2$

$d = \sqrt{a^2 + b^2 + c^2}$

Example 41.18

Question
Find the length of the diagonal of this cuboid.

Solution
The length of the diagonal of a cuboid $= \sqrt{a^2 + b^2 + c^2}$.
In this cuboid $a = 5\,\text{cm}$, $b = 2\,\text{cm}$ and $c = 3\,\text{cm}$.
Length of diagonal $= \sqrt{5^2 + 2^2 + 3^2}$
$= \sqrt{38}$
$= 6.2\,\text{cm}$ to 1 decimal place

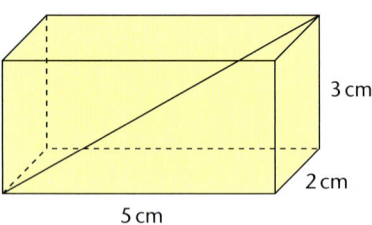

Finding lengths and angles in three dimensions

Example 41.19

Question

A tree, TC, is 20 m north of point A.

The angle of elevation of the top of the tree, T, from A, is 35°.

Point B is 30 m east of point A.

A, B and C are on horizontal ground.

Calculate

a the height of the tree, TC
b the length BC
c the angle of elevation of T from B.

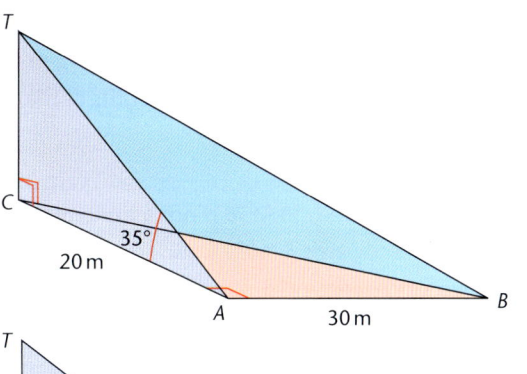

Solution

First, draw a diagram.

a $\tan 35° = \frac{TC}{20}$

   $TC = 20 \tan 35°$

   $TC = 14.0$ m to 1 decimal place

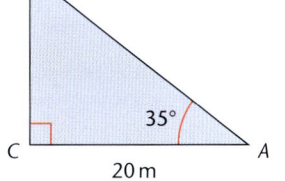

b $BC^2 = 20^2 + 30^2$

   $BC^2 = 1300$

   $BC = 36.1$ m to 1 decimal place

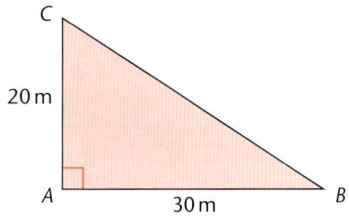

c $\tan x = \frac{14.0...}{36.1...}$

   $x = \tan^{-1}\left(\frac{14.0...}{36.1...}\right)$

   $x = 21.2°$ to 1 decimal place

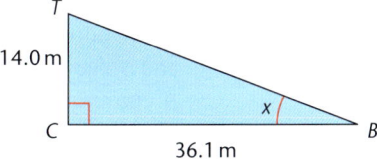

Example 41.20

Question

A mast, MG, is 50 m high.

It is supported by two ropes, AM and BM, as shown.

ABG is horizontal.

Other measurements are shown on the diagram.

Is the mast vertical?

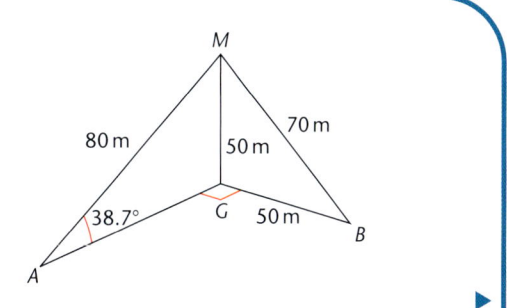

41 PYTHAGORAS' THEOREM AND TRIGONOMETRY

Solution

If $M\hat{G}A$ is a right angle then $\sin 38.7° = \frac{50}{80}$.

$\sin^{-1}\left(\frac{50}{80}\right) = 38.68...$

So $M\hat{G}A$ is a right angle.

If $M\hat{G}B$ is a right angle then the side lengths of triangle MGB will fit Pythagoras' theorem.

$50^2 + 50^2 = 5000$

$\sqrt{5000} = 70.7...$

MB is shorter than this, so $M\hat{G}B$ is not a right angle.

The mast leans towards B ($M\hat{G}B < 90°$).

Exercise 41.13

1 $ABCDEFGH$ is a cuboid with dimensions as shown.
   Calculate these.
   a $G\hat{D}C$
   b EG
   c EC
   d $G\hat{E}C$

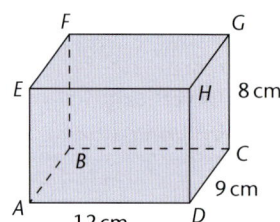

2 A wedge has a rectangular base, $ABCD$, on horizontal ground.
   The rectangular face, $BCEF$, is vertical.
   Calculate these.
   a FC
   b $D\hat{F}C$
   c FD
   d $E\hat{D}F$

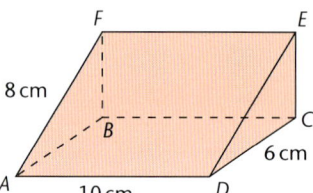

3 Three points, A, B and C, are on horizontal ground, with B due west of C and A due south of C.
   A chimney, CT, at C, is 50 m high.
   The angle of elevation of the top, T, of the chimney is 26° from A and 38° from B.
   Calculate these.
   a How far A and B are from C.
   b The distance between A and B.
   c The bearing of B from A.

4 The diagram shows a square-based pyramid with V vertically above the centre, X, of the square $ABCD$.
   $AB = 8$ cm and $AV = 14$ cm.
   Calculate these.
   a $C\hat{A}B$ b $V\hat{C}B$
   c AC d AX
   e VX f $V\hat{A}X$

Finding lengths and angles in three dimensions

5 *ABCDPQ* is a triangular prism with *ABPQ* horizontal and *ADP* vertical. Calculate these.
   a *DP*
   b *AC*
   c *CÂQ*

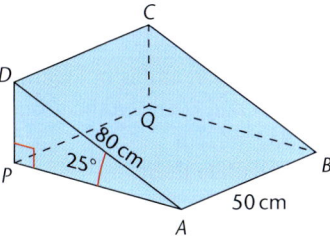

6 A pyramid has a rectangular base, *ABCD*, with *AB* = 15 cm and *BC* = 8 cm. The vertex, *V*, of the pyramid is directly above the centre, *X*, of *ABCD* with *VX* = 10 cm.
   a Calculate these.
      i *AC*   ii *AV*   iii *AV̂B*
   b *M* is the midpoint of *AB*. Calculate these.
      i *VM*   ii *VM̂X*

7 Triangle *ABC* is horizontal. *X* is vertically above *A*. *AC* = *AX* = 15 cm. *AĈB* = 27° and *BÂC* = 90°. Calculate these.
   a *XC*
   b *BC*
   c *BX*

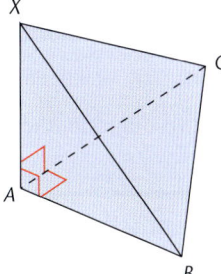

8 A field, *ABCD*, is a quadrilateral with opposite sides equal. *AD* = *BC* = 80 m. *DC* = *AB* = 35 m. A vertical post, *CE*, is at one corner of the field. The angles of elevation of the top of the post are 7.8° from *A* and 8.5° from *B*. Is the field a rectangle? Show how you decide.

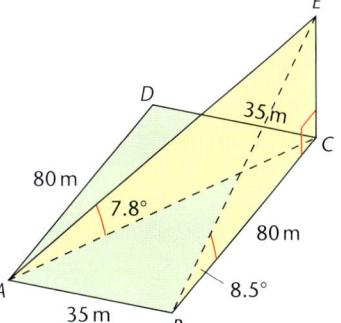

9 The pyramid *OABCD* has a horizontal rectangular base *ABCD* as shown. *O* is vertically above *A*. Calculate these.
   a The length of *OC*
   b *OĈA*

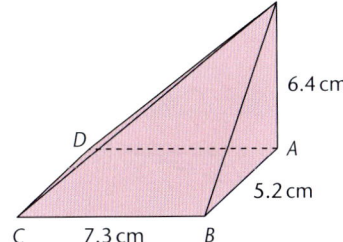

41 PYTHAGORAS' THEOREM AND TRIGONOMETRY

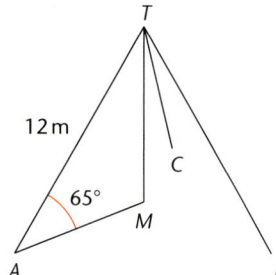

10 A vertical mast, *MT*, has its base, *M*, on horizontal ground.
It is supported by a wire, *AT*, which makes an angle of 65° with the horizontal and is of length 12 m, and two more wires, *BT* and *CT*, where *A*, *B* and *C* are on the ground.
BM = 4.2 m.
Calculate these.
   a The height of the mast.
   b The angle which *BT* makes with the ground.
   c The length of wire *BT*.

The angle between a line and a plane

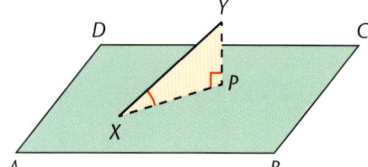

To find the angle between a line and a plane, you need to identify the correct angle.

The shortest distance between a point and a plane is the perpendicular distance to it. Use this fact to identify a right-angled triangle.

The hypotenuse will be the given line. One side will be a line in the plane.

The angle you need is the angle between these lines.

In this diagram, the angle between the line *XY* and the plane *ABCD* is the angle *YXP*.

Example 41.21

Question

For this triangular prism, sketch the triangle and label the angle between the line and the plane given.

   a *DY* and *ABCD*   **c** *AY* and *BCYX*
   b *AY* and *ABCD*   **d** *BY* and *ABX*

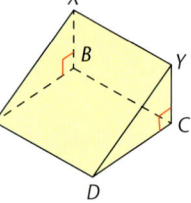

Solution

a Angle *YDC*

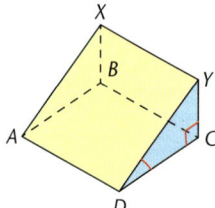

b Angle *YAC*

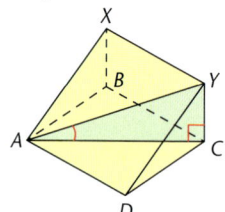

c Angle *AYB*

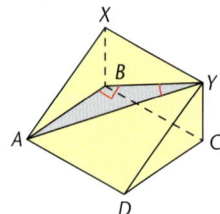

d Angle *YBX*

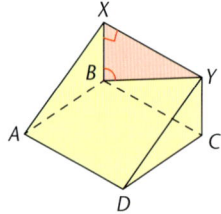

The angle between a line and a plane

Example 41.22

Question

ABCDVWXY is a cuboid.

Calculate the angle between the following lines and planes.

a DW and ABCD
b DW and ABVW

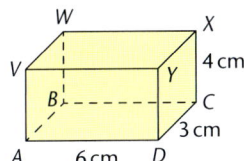

Solution

a Angle BDW is the required angle.

First, find length BD.

$BD^2 = 6^2 + 3^2$
$BD^2 = 45$
$BD = \sqrt{45}$ cm
$\tan x = \frac{4}{\sqrt{45}}$
$x = \tan^{-1}\frac{4}{\sqrt{45}}$

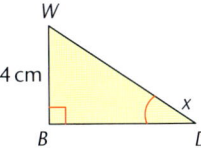

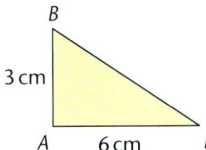

$x = 30.8°$ to 1 decimal place

b Angle DWA is the required angle.

First, find length WA.

$WA^2 = 4^2 + 3^2$
$WA^2 = 25$
$WA = 5$ cm
$\tan x = \frac{6}{5}$
$x = \tan^{-1}\frac{6}{5}$

$x = 50.2°$ to 1 decimal place

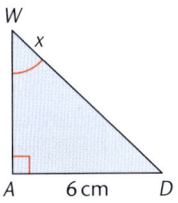

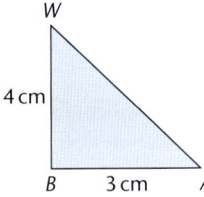

Exercise 41.14

1 The diagram shows the cuboid ABCDEFGH. For this cuboid, sketch the triangle and label the angle between these lines and planes.

   a EB and ABCD
   b EB and ADHE
   c AG and ABCD
   d AG and CDHG

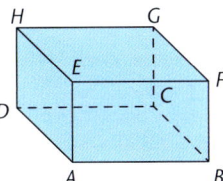

41 PYTHAGORAS' THEOREM AND TRIGONOMETRY

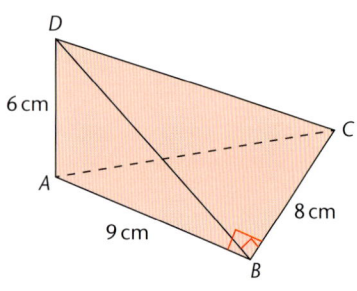

2 Given that, for the cuboid in question 1, $AB = 8\,cm$, $BC = 6\,cm$ and $GC = 5\,cm$, calculate the angles between the lines and the planes listed in that question.

3 The diagram shows the tetrahedron $ABCD$.
   ABC is a right-angled triangle on a horizontal plane.
   D is vertically above A and angle $DBC = 90°$.
   Calculate these.
   a The angle between BD and ABC.
   b The angle between DC and ABD.

4 In the pyramid, $ABCD$ is a rectangle and V is vertically above the centre of the rectangle.
   M is the midpoint of AD.

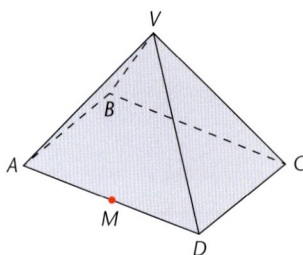

Sketch the triangle and label the angle between these lines and planes.
   a VC and $ABCD$
   b VM and $ABCD$

5 Given that, for the pyramid in question 4, $AD = 8\,cm$, $DC = 6\,cm$ and that $VA = VB = VC = VD = 12\,cm$, calculate the angles between the lines and the planes listed above.

6 A cuboid has a base of sides $5.6\,cm$ and $8.2\,cm$.
   Its height is $4.3\,cm$.
   Calculate the angle between the diagonal of the cuboid and
   a the base
   b a $5.6\,cm$ by $4.3\,cm$ face.

7 A pyramid is $8\,cm$ high and has a square base of side $6\,cm$.
   Its sloping edges are all of equal length.
   Calculate the angle between a sloping edge and the base.

8 $ABCDEF$ is a triangular wedge.
   The base $ADFE$ is a horizontal rectangle.
   C is vertically above B.

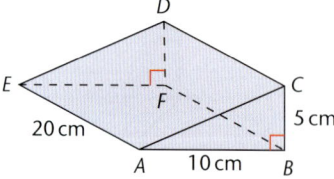

   a Calculate the length AD.
   b Calculate the angle that AD makes with
      i the base
      ii the face ABC.

The angle between a line and a plane

Key points
- The angle that a line to an object makes above a horizontal plane is called the angle of elevation.
- The angle that a line to an object makes below a horizontal plane is called the angle of depression.
- The shortest distance between a point and a line is the perpendicular distance.
- In right-angled triangles
  - Use Pythagoras' theorem when you know two sides and need to find the third.
  $a^2 = b^2 + c^2$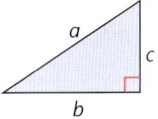
  - Use trigonometry when you need to find an angle or you need to find a side when you know one side and an angle.

  $\sin \theta = \dfrac{\text{Opposite}}{\text{Hypotenuse}} \quad \cos \theta = \dfrac{\text{Adjacent}}{\text{Hypotenuse}} \quad \tan \theta = \dfrac{\text{Opposite}}{\text{Adjacent}}$

  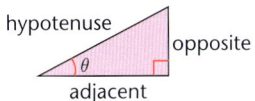

- In triangles without right angles
  - For obtuse angles, $\sin \theta = \sin(180 - \theta)$ and $\cos \theta = -\cos(180 - \theta)$.
  - The sine rule is

  $\dfrac{a}{\sin A} = \dfrac{b}{\sin B} = \dfrac{c}{\sin C}$ or $\dfrac{\sin A}{a} = \dfrac{\sin B}{b} = \dfrac{\sin C}{c}$

  Use the sine rule to find lengths and angles when you know
    - any two angles and a side
    - two sides and an angle opposite one of these.
  - The cosine rule is
  $c^2 = a^2 + b^2 - 2ab \cos C$ or $\cos C = \dfrac{a^2 + b^2 - c^2}{2ab}$
  Use the cosine rule
    - to find the third side when you know two sides and the angle between them
    - to find an angle when you know all three sides.
  - The area of a triangle is given by Area $= \dfrac{1}{2} ab \sin C$.
- In three dimensions
  - The length of the diagonal of a cuboid with dimensions a, b and c is $\sqrt{a^2 + b^2 + c^2}$.
  - To solve problems in three dimensions, identify a right-angled triangle containing the length or angle you need to find. Draw a separate sketch of this triangle. Then, use Pythagoras's theorem and/or trigonometry, as needed.

Note
You do not need to memorise the formulas for the sine and cosine rule and the area of a triangle but it is a good idea to learn them so that you can use them easily.

42 TRANSFORMATIONS

BY THE END OF THIS CHAPTER YOU WILL BE ABLE TO:

Recognise, describe and draw the following transformations:
- reflection of a shape in a straight line
- rotation of a shape about a centre through multiples of 90°
- enlargement of a shape from a centre by a scale factor
- translation of a shape by a vector $\begin{pmatrix} x \\ y \end{pmatrix}$.

CHECK YOU CAN:
- plot points in all four quadrants
- draw and recognise simple straight lines on a graph such as $x = k$, $y = k$ and $y = \pm x$
- use a protractor.

The language of transformations

When a shape is transformed, the original shape is called the **object** and the new shape is called the **image**.

A transformation **maps** the object on to the image.

If the vertices of an object are labelled A, B, C, ..., the corresponding vertices of the image can be labelled A', B', C' ... .

Anything that stays the same when a transformation is performed is **invariant**.

Reflection

Reflections are carried out in **mirror lines**.

In a reflection, corresponding points are the same distance from the mirror line but on the opposite side.

The object and the image are **congruent**, but the image is reversed.

Points on the mirror lines are invariant points under a reflection.

In this diagram, the shape has been reflected in the line PQ.

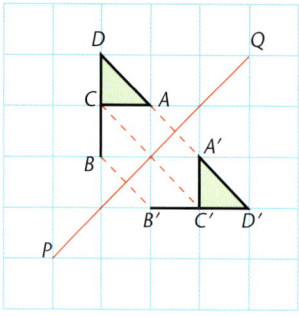

Instead of counting squares you could trace the shape and the mirror line, then turn the tracing paper over and line up the mirror line and your tracing of it.

Example 42.1

Question

Reflect the trapezium in the line $y = 1$.

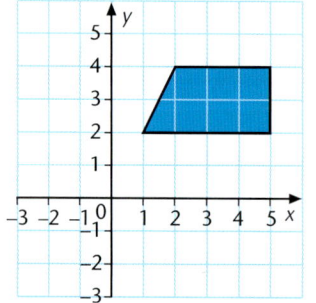

Solution

The diagram shows the reflection.

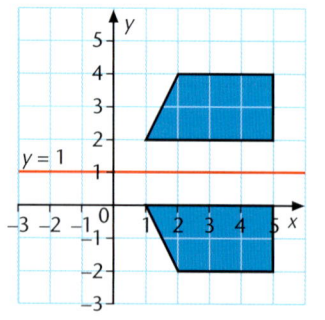

408   *Photocopying is prohibited*

Reflection

Example 42.2

Question

Reflect this shape in the line $y = -x$.

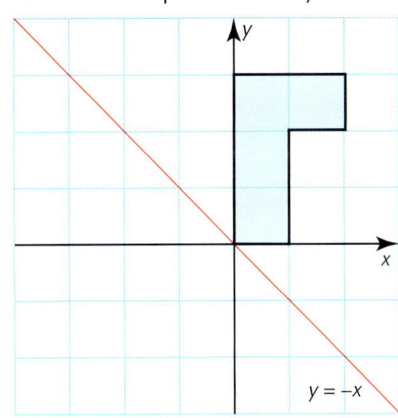

Solution

The diagram shows the reflection.

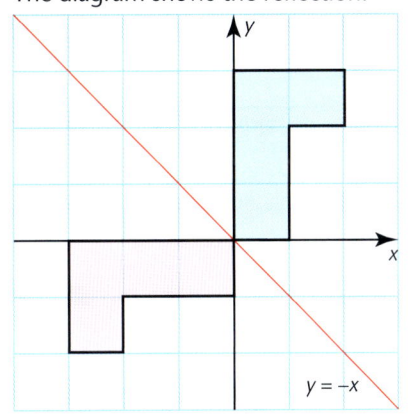

Note

When you have drawn a reflection in a sloping line, check it by turning the page so the mirror line is vertical. Then you can easily see if it has been reflected correctly.

Exercise 42.1

In each of questions **1** to **6**, copy the diagram on to squared paper first.

1 Reflect the parallelogram in the y-axis.

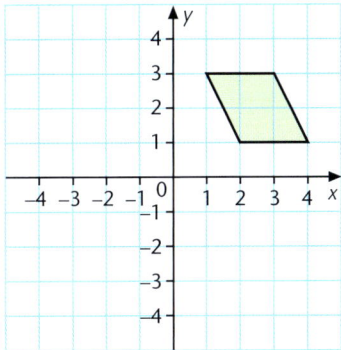

Note

You can count the squares or use tracing paper to help you draw the image.

2 Reflect the triangle in the x-axis.

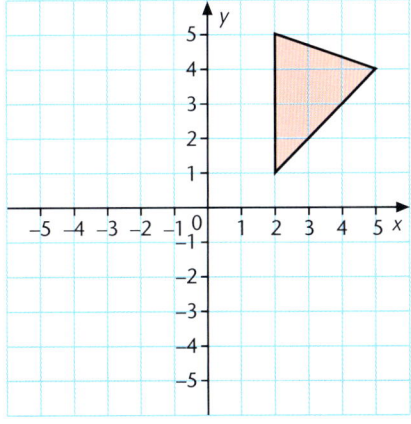

Note

When drawing transformations, you should use a ruler for all straight edges.

3 Reflect the trapezium in the x-axis.

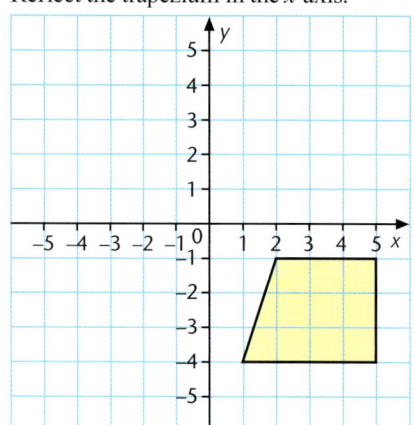

4 Reflect the rectangle in the line $x = -1$.

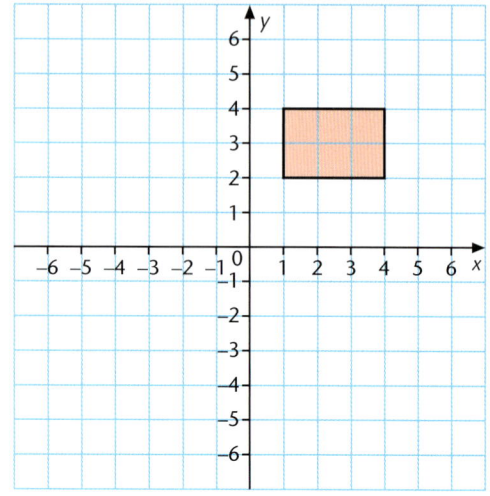

5 Reflect the triangle in the line $y = -2$.

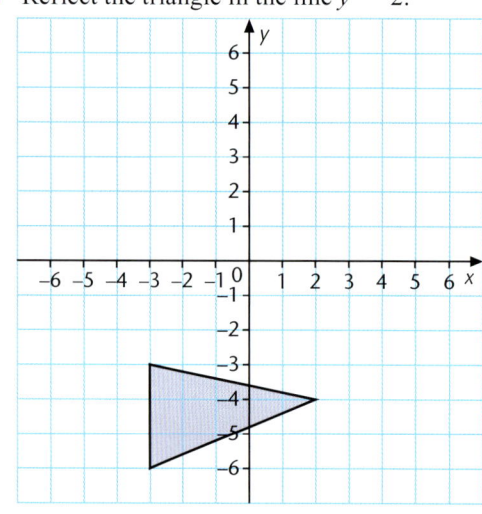

6 Reflect the quadrilateral in the line $y = 2$.

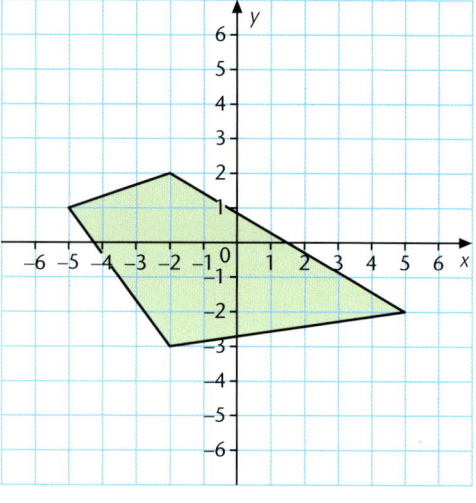

7 Draw a pair of axes and label them −4 to 4 for x and y.
  a Draw a triangle with vertices at (1, 0), (1, −2) and (2, −2). Label it A.
  b Reflect triangle A in the line $y = 1$. Label it B.
  c Reflect triangle B in the line $y = x$. Label it C.

8 Draw a pair of axes and label them −4 to 4 for x and y.
  a Draw a triangle with vertices at (1, 1), (2, 3) and (3, 3). Label it A.
  b Reflect triangle A in the line $y = 2$. Label it B.
  c Reflect triangle A in the line $y = -x$. Label it C.

Rotation

Rotation

In a **rotation** you turn an object about a point.

The point is called the **centre of rotation**.

When an object is rotated, the object and its image are congruent.

Unlike reflections, when an object is rotated, the object and image are the same way round.

In this diagram, shape P has been rotated 90° anticlockwise about O.

To rotate a shape you need to know three things

- the angle of rotation
- the direction of rotation
- the centre of rotation.

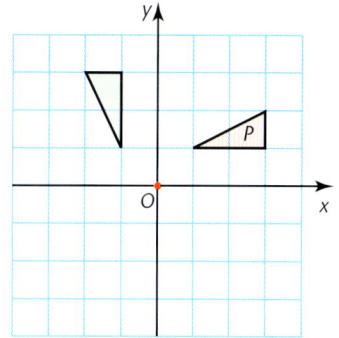

Example 42.3

Question

Rotate this shape 270° anticlockwise about O.

Solution

You can think of 270° anticlockwise as being the same as 90° clockwise.

You can draw the rotation by counting squares.

A is 2 squares above O so its image, A', is 2 squares to the right.

B is 3 squares above O so its image, B', is 3 squares to the right.

C is 2 squares above and 2 to the right of O so its image, C', is 2 to the right and 2 below O.

D is 3 squares above and 2 to the right of O so its image, D', is 3 to the right and 2 below O.

Alternatively, you could do the rotation using tracing paper.

Trace the shape.

Hold the centre of rotation still with the point of a pin or a pencil and rotate the paper through a quarter-turn clockwise.

Use another pin or the point of your compasses to prick through the corners.

Join the pin holes to form the image.

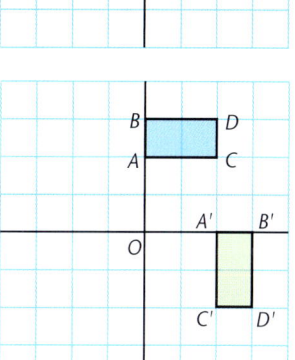

42 TRANSFORMATIONS

Exercise 42.2

1 Make three copies of this diagram and answer each part of the question on a separate diagram.

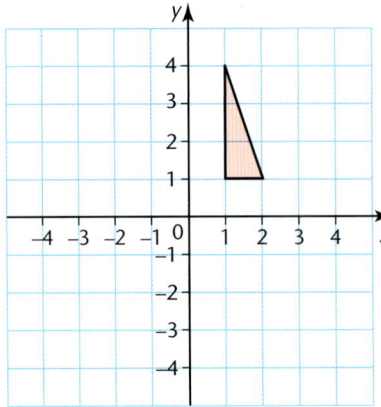

a Rotate the shape through 180° about the origin.
b Rotate the shape through 90° clockwise about the origin.
c Rotate the shape through 90° anticlockwise about the origin.

2 Make three copies of this diagram and answer each part of the question on a separate diagram.

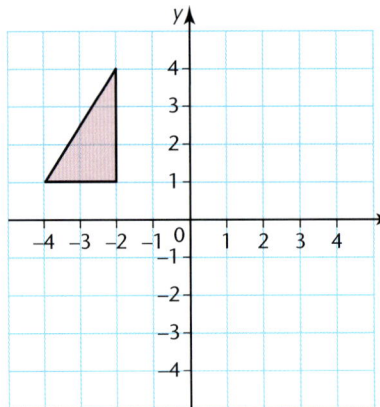

a Rotate the shape through 90° anticlockwise about the origin.
b Rotate the shape through 90° clockwise about the origin.
c Rotate the shape through 180° about the origin.

In each of questions 3 to 5, copy the diagram on to squared paper first.

3 Rotate the shape through 180° about the point (4, 3).

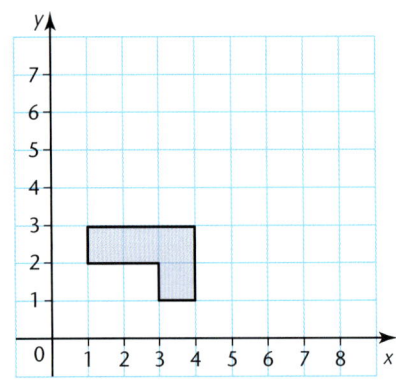

4 Rotate the shape through 180° about the point (5, 3).

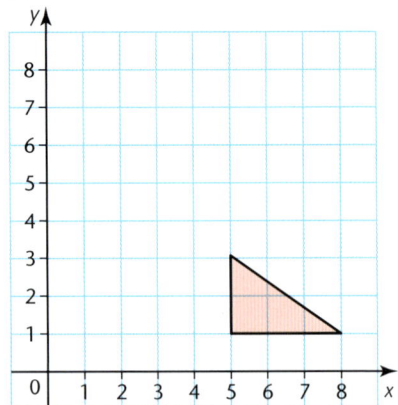

5 Rotate the shape through 90° clockwise about the point (−1, −2).

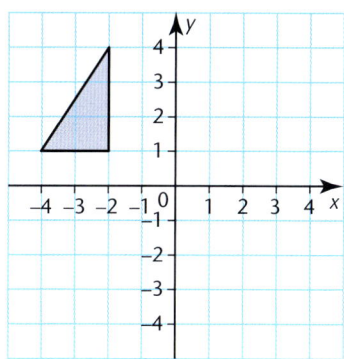

Translation

In a **translation**, every point on the object moves the same distance in the same direction as every other point.

The object and the image are **congruent**.

In this diagram, triangle A maps on to triangle B.

Each vertex of the triangle moves 6 squares to the right and 4 squares down.

The translation is given by the vector $\begin{pmatrix} 6 \\ -4 \end{pmatrix}$.

Note
Take care with the counting.
Choose a point on both the object and the image and count the units from one to the other.

Note
You will learn about vectors in Chapter 43.

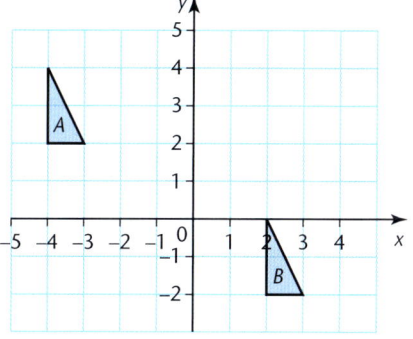

42 TRANSFORMATIONS

Exercise 42.3

In each of questions **1** to **3**, copy the diagram on to squared paper first.

1

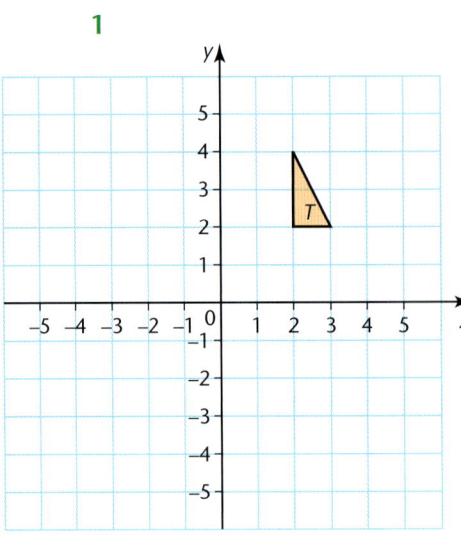

a Translate triangle T by $\begin{pmatrix}2\\1\end{pmatrix}$. Label the image A.

b Translate triangle T by $\begin{pmatrix}1\\-3\end{pmatrix}$. Label the image B.

c Translate triangle T by $\begin{pmatrix}-4\\1\end{pmatrix}$. Label the image C.

d Translate triangle T by $\begin{pmatrix}-6\\-6\end{pmatrix}$. Label the image D.

e Translate triangle T by $\begin{pmatrix}-2\\-5\end{pmatrix}$. Label the image E.

2

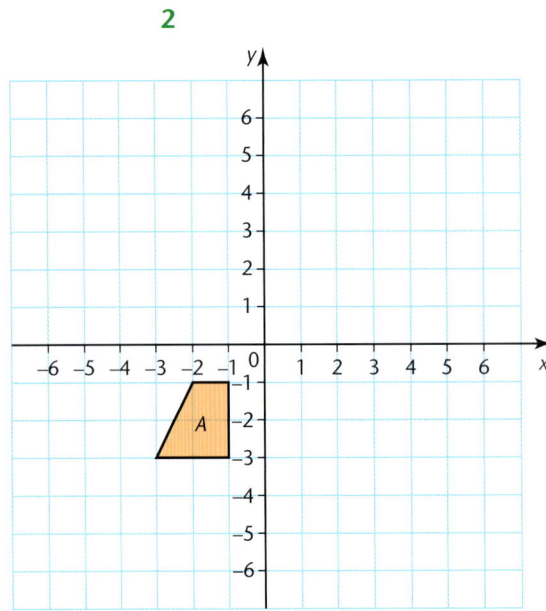

a Translate shape A by $\begin{pmatrix}-3\\6\end{pmatrix}$. Label the image B.

b Translate shape B by $\begin{pmatrix}7\\-8\end{pmatrix}$. Label the image C.

c Translate shape C by $\begin{pmatrix}1\\5\end{pmatrix}$. Label the image D.

d Translate shape D by $\begin{pmatrix}-5\\-3\end{pmatrix}$. What do you notice?

Enlargement

3 a Translate shape A by $\begin{pmatrix} -2 \\ -6 \end{pmatrix}$.
      Label the image B.

   b Translate shape B by $\begin{pmatrix} 9 \\ 5 \end{pmatrix}$.
      Label the image C.

   c Translate shape C by $\begin{pmatrix} 2 \\ -4 \end{pmatrix}$.
      Label the image D.

   d What must you translate D by to get back to A?

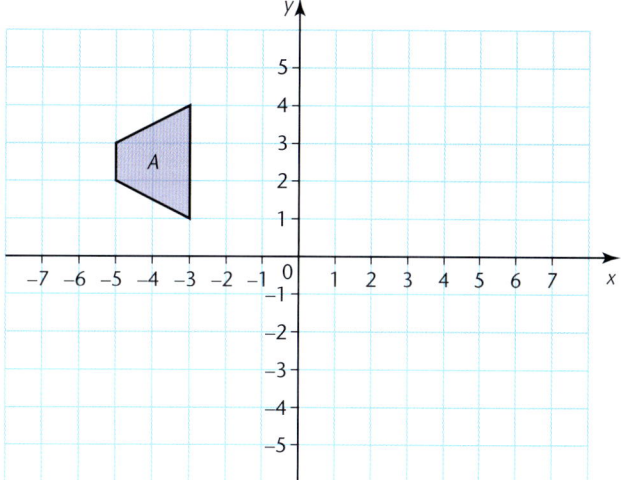

Enlargement

In an **enlargement** of a shape, the image is the same shape as the object, but it is larger or smaller.

The object and image are not congruent, because they are different sizes, but they are **similar**.

An enlargement is drawn from a given point, the **centre of enlargement**.

For a '2 times enlargement', each point on the image must be twice as far from the centre as the corresponding point on the object.

If the centre is on the object, that point will not move and the enlargement will overlap the original shape.

The number used to multiply the lengths for the enlargement is the **scale factor**.

Example 42.4

Question

Enlarge this shape by scale factor 3 with the origin as the centre of enlargement.

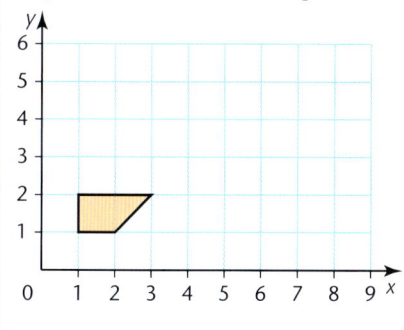

Solution

Scale factor 3 means this is a '3 times enlargement', so each point in the image is three times as far from the origin as the corresponding point in the original shape.

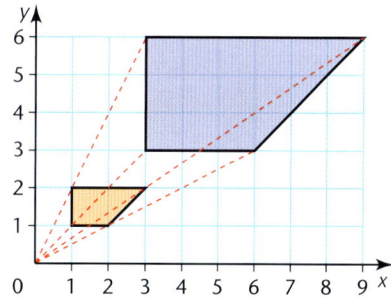

Photocopying is prohibited

42 TRANSFORMATIONS

Exercise 42.4

In each of questions **1** to **3**, copy the diagram on to squared paper first.

1 Enlarge the shape by scale factor 2, centre the origin.

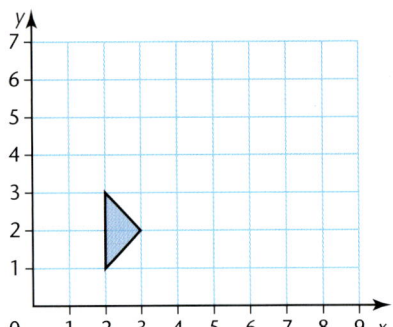

2 Enlarge the shape by scale factor 3, centre the origin.

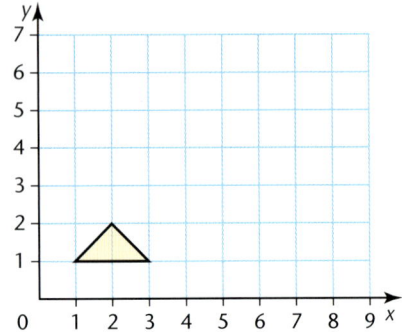

3 Enlarge the shape by scale factor 2, centre the point (2, 1).

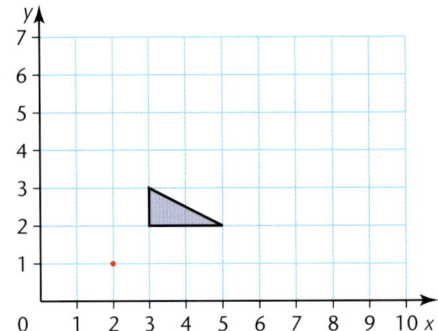

Enlargement with a fractional scale factor

The technique for drawing an enlargement with a fractional scale factor is the same as for a positive integral scale factor. You can see how this is done in the following example.

Example 42.5

Question

Draw an enlargement of triangle *ABC* with centre *O* and scale factor $\frac{1}{2}$.

Enlargement

Solution

Draw lines from O to A, B and C.

By counting squares, mark A' on OA so that A' is the midpoint of OA.

Similarly mark B' and C' at the midpoints of OB and OC.

Join A'B'C'.

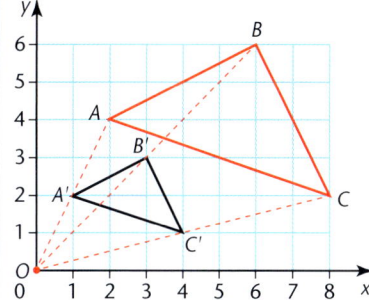

A'B'C' is the required enlargement with scale factor $\frac{1}{2}$.

Check that the lengths of A'B', B'C' and A'C' are $\frac{1}{2}$ the lengths of the corresponding lengths on the original triangle.

An enlargement with scale factor $\frac{1}{2}$ is the inverse of an enlargement with scale factor 2.

The transformation is still called an enlargement even though the image is smaller than the original.

Exercise 42.5

In each of questions **1** to **4**, copy the diagram on to squared paper first.

1 Enlarge the rectangle with scale factor $\frac{1}{2}$ and centre the origin.

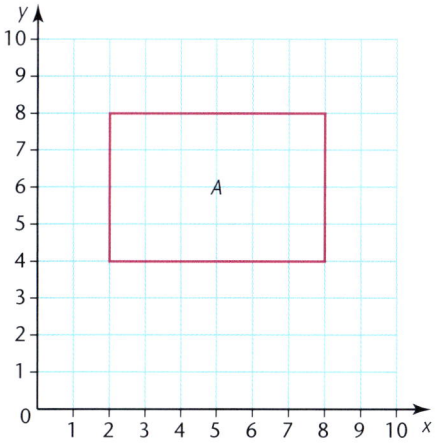

2 Enlarge the triangle with scale factor $\frac{1}{3}$ and centre (1, 6).

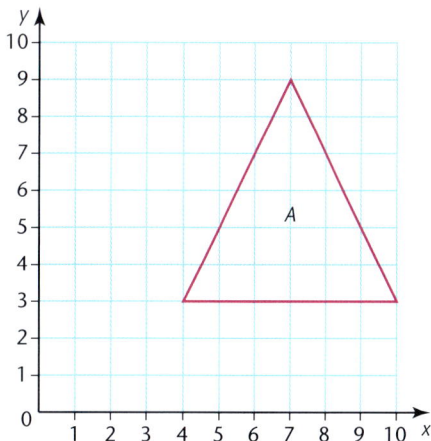

42 TRANSFORMATIONS

3 Enlarge the shape A with scale factor $1\frac{1}{2}$ and centre (0, 2).

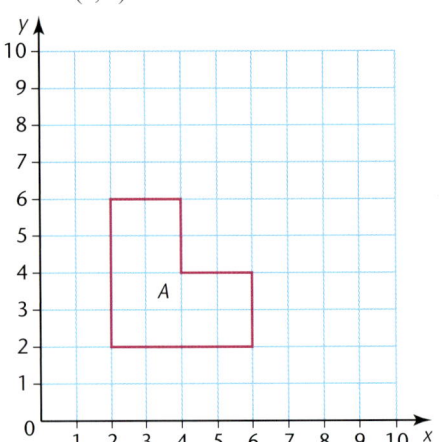

4 Enlarge the shape A with scale factor $\frac{1}{2}$ and centre (5, 4).

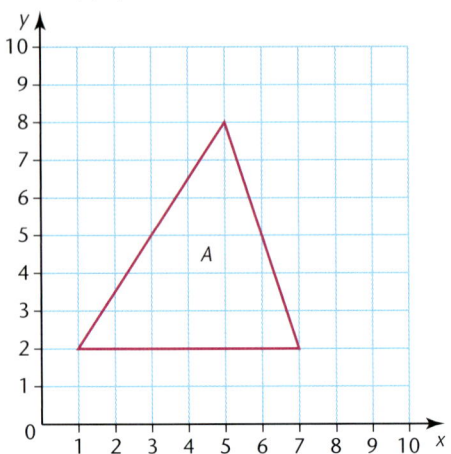

Enlargement with a negative scale factor

When the scale factor of an enlargement is negative, the image is on the opposite side of the centre of enlargement from the object and the image is inverted.

Example 42.6

Question

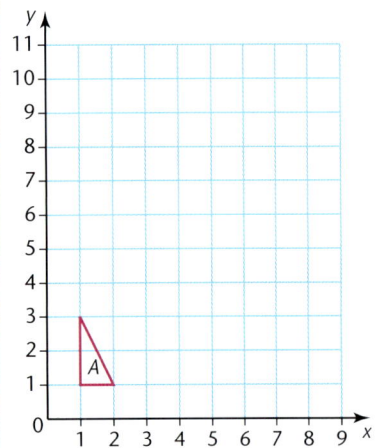

Enlarge triangle A with scale factor −2 and centre (3, 4).

Solution

Plot the centre of enlargement, O, at (3, 4).

Draw the line AO and extend it. Measure the length OA and mark the point A' so that OA' = 2 × OA.

Note that A' is on the opposite side of O to A.

Or, by counting squares, O is 2 to the right and one above A, so A' is 4 to the right and 2 above O.

Do the same for BO to give point B' and CO to give point C'.

A'B'C' is the required image of ABC.

Check that the sides of triangle A'B'C' are twice as long as the corresponding sides of triangle ABC.

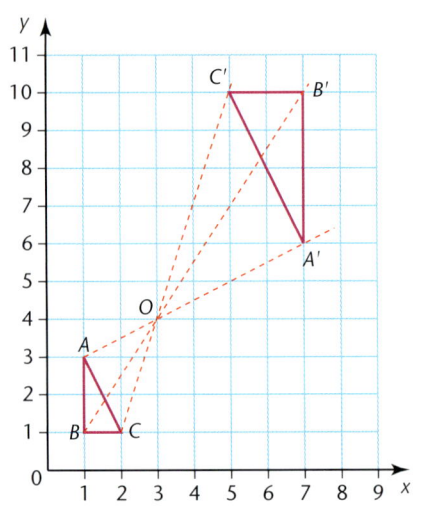

Enlargement

Exercise 42.6

In each of questions **1** to **4**, copy the diagram on to squared paper first.

1 Enlarge rectangle A with scale factor -2 and centre $(3, 4)$.

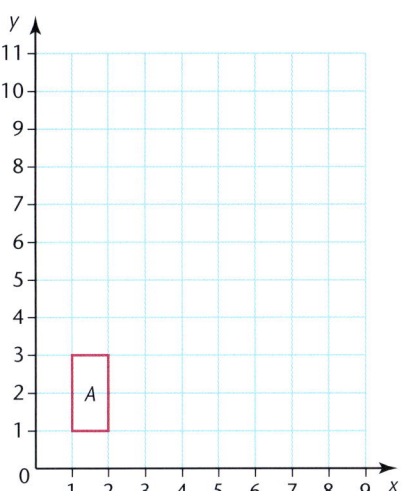

> **Note**
> It is easy to confuse negative and fractional scale factors.
>
> Remember that fractional scale factors make the image smaller than the object.
>
> With negative scale factors, the image is inverted and on the opposite side of the centre of enlargement to the object.

2 Enlarge triangle A with scale factor -3 and centre $(3, 5)$.

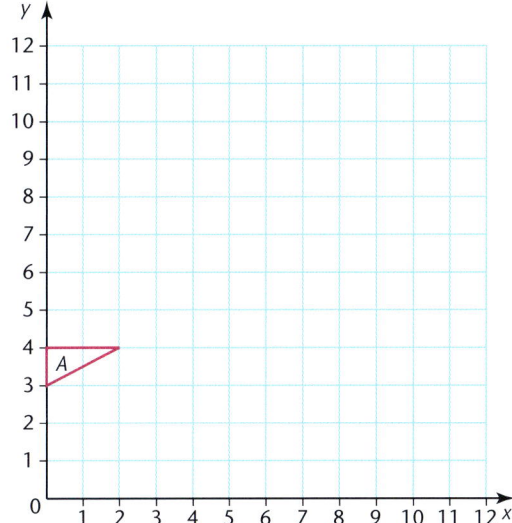

3 Enlarge shape A with scale factor $-\frac{1}{2}$ and centre (7, 6).

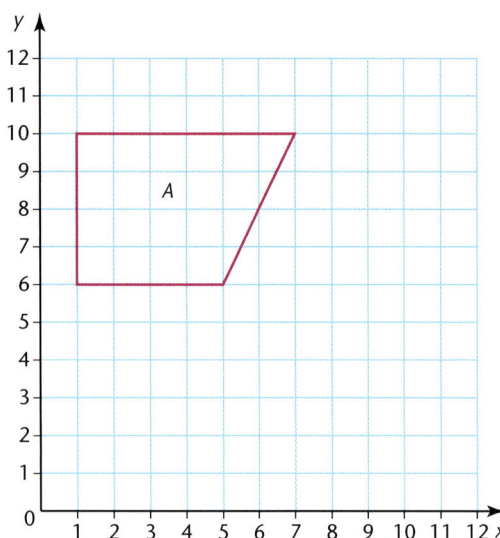

4 Enlarge triangle A with scale factor $-\frac{1}{3}$ and centre (4, 5).

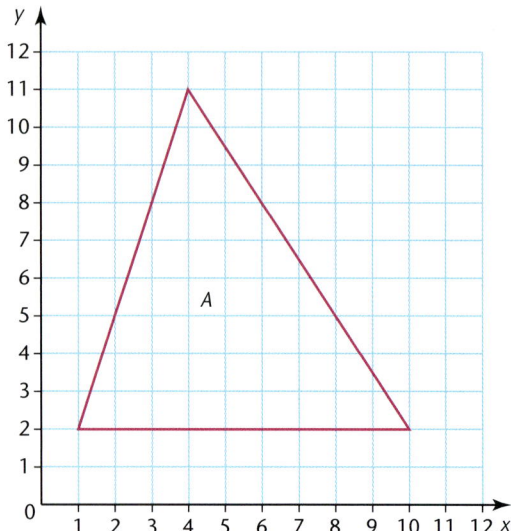

Recognising and describing transformations

You should know how to describe fully a transformation if given the object and the image.

You must first state which type of transformation it is.

Trace the object.

If you have to turn it over to fit on the image, the transformation is a reflection.

If you have to turn the tracing round but not turn it over, the transformation is a rotation.

If you just slide the tracing without turning over or turning round, the transformation is a translation.

If the object is not the same size as the image, the transformation is an enlargement.

You then need to complete the description.

Recognising and describing transformations

The extra information required is shown in this table.

Transformation	Extra information required
Reflection	The equation of the mirror line or a statement that it is the *x*-axis or the *y*-axis
Rotation	Angle
	Direction
	Centre
Translation	Column vector
Enlargement	Scale factor
	Centre

Note
When describing fully a *single* transformation, do not give a combination of transformations.

Reflection

Example 42.7

Question
Describe fully the single transformation that maps triangle *ABC* on to triangle *A'B'C'*.

Solution
You can check that the transformation is a reflection by using tracing paper.

To find the mirror line, put a ruler between two corresponding points (*B* and *B'*) and mark a point halfway between them, at (3, 3).

Repeat this for two other corresponding points (*C* and *C'*). The midpoint is (4, 4).

Join the two midpoints to find the mirror line. The mirror line is $y = x$.

The transformation is a reflection in the line $y = x$.

Again, the result can be checked using tracing paper.

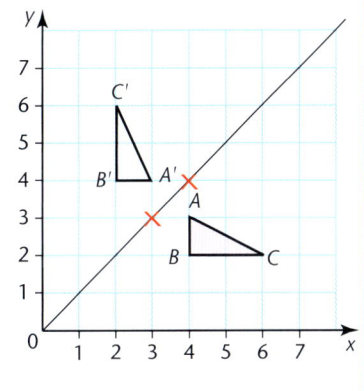

Rotation

To find the angle of rotation, find a pair of sides that correspond in the object and the image. Measure the angle between them. You may need to extend both of the sides to do this.

If the centre of rotation is not on the object, its position may not be obvious. However, you can usually find it, either by counting squares or using tracing paper.

Always remember to state whether the rotation is clockwise or anticlockwise.

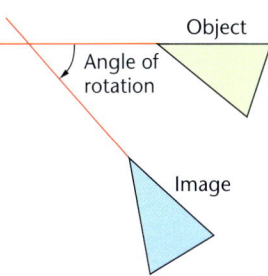

42 TRANSFORMATIONS

Example 42.8

Question

Describe fully the single transformation that maps flag A on to flag B.

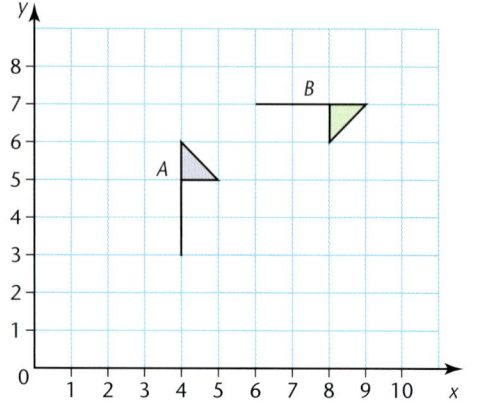

Solution

You have to turn the shape round but you don't turn it over, so the transformation is a rotation.

The angle is 90° clockwise.

You may need to make a few trials, using tracing paper and a compass point centred on different points, to find that the centre of rotation is (7, 4).

So the transformation that maps flag A on to flag B is a rotation of 90° clockwise, with centre of rotation (7, 4).

Translation

Example 42.9

Question

Describe fully the single transformation that maps shape ABC on to shape A'B'C'.

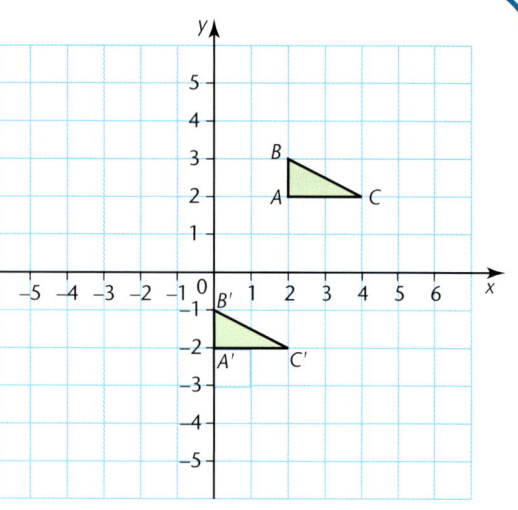

Solution

The shape stays the same way up and is the same size so the transformation is a translation.

A moves from (2, 2) to (0, −2). This is a movement of 2 to the left and 4 down.

The transformation is a translation by the vector $\begin{pmatrix} -2 \\ -4 \end{pmatrix}$.

Recognising and describing transformations

Enlargement

Example 42.10

Question

Describe fully the transformation that maps

a   A on to B   b   B on to A.

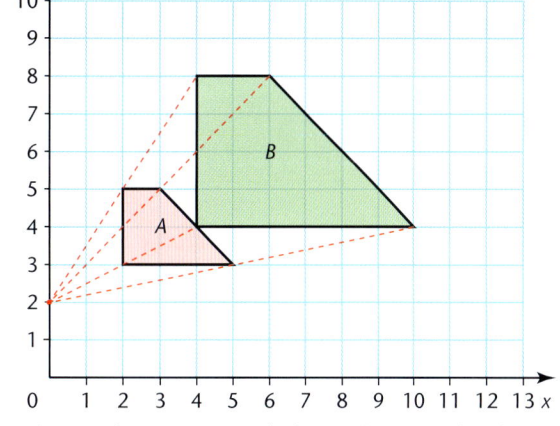

Solution

By measuring the sides you can see that shape B is twice as big as shape A.

To find the centre of enlargement, you join corresponding vertices of the two shapes and extend them until they cross.

The point where they cross is the centre of enlargement.

The centre of enlargement is (0, 2).

a   The transformation that maps shape A on to shape B is an enlargement, scale factor 2, centre (0, 2).

b   The transformation that maps shape B on to shape A is an enlargement, scale factor $\frac{1}{2}$, centre (0, 2).

Exercise 42.7

1 Describe fully the transformation that maps A on to B.

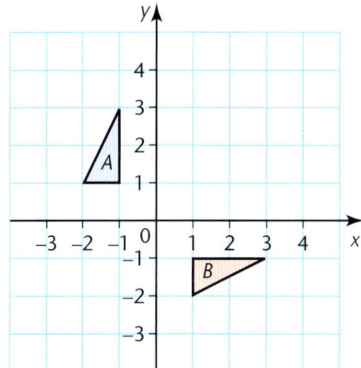

2 Describe fully the transformation that maps A on to B.

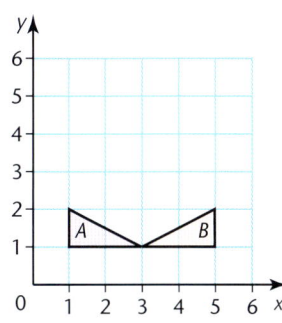

3 Describe fully the transformation that maps A on to B.

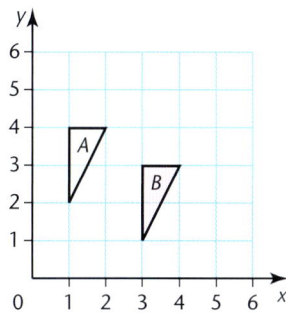

42 TRANSFORMATIONS

4 Describe fully the single transformation that maps
  a triangle *A* on to triangle *B*
  b triangle *A* on to triangle *C*
  c triangle *A* on to triangle *D*
  d triangle *B* on to triangle *D*.

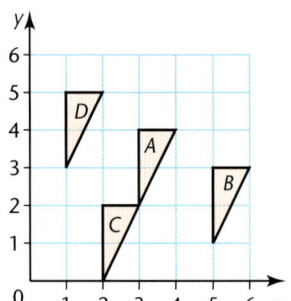

5 Describe fully the single transformation that maps
  a flag *A* on to flag *B*
  b flag *A* on to flag *C*
  c flag *B* on to flag *D*.

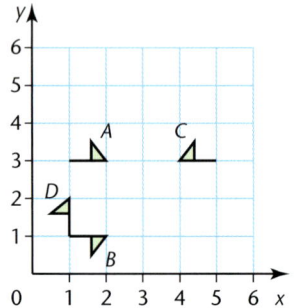

6 Describe fully the single transformation that maps
  a triangle *A* on to triangle *B*
  b triangle *A* on to triangle *C*
  c triangle *C* on to triangle *D*
  d triangle *A* on to triangle *E*
  e triangle *A* on to triangle *F*
  f triangle *G* on to triangle *A*.

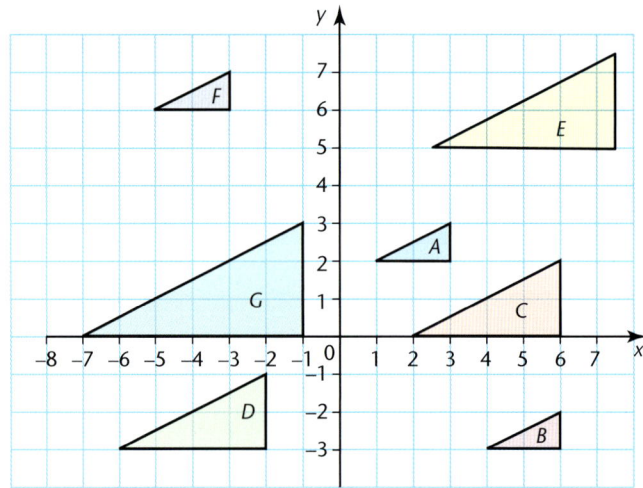

7 Describe fully the single transformation that maps shape *ABCD* on to shape *A'B'C'D'*.

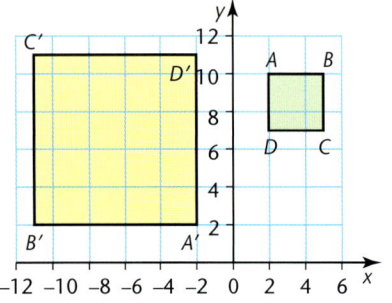

8 Describe fully the single transformation that maps shape *ABC* on to shape *A'B'C'*.

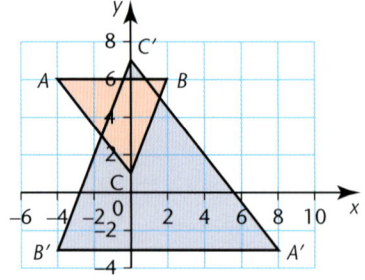

9 Describe fully the single transformation that maps shape $ABCD$ on to shape $A'B'C'D'$.

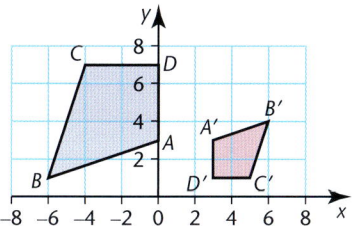

Combining transformations

If a transformation is followed by another transformation, the result is sometimes equivalent to a third transformation.

Note

It is very easy to assume PQ means P followed by Q.

It does not. It means Q followed by P.

This notation is like that used for functions.

Example 42.11

Question

Describe fully the single transformation equivalent to a rotation of 90° clockwise about the origin, followed by a reflection in the y-axis.

Solution

Choose a simple shape like triangle A to start with.

Rotating triangle A 90° clockwise about the origin gives triangle B.

Reflecting triangle B in the y-axis gives triangle C.

You need to find the single transformation that will map triangle A on to triangle C.

If you trace triangle A, you will need to turn the tracing over to fit it on to triangle C, so the transformation must be a reflection.

The mirror line is the $y = -x$ line.

So the single transformation is a reflection in the line $y = -x$.

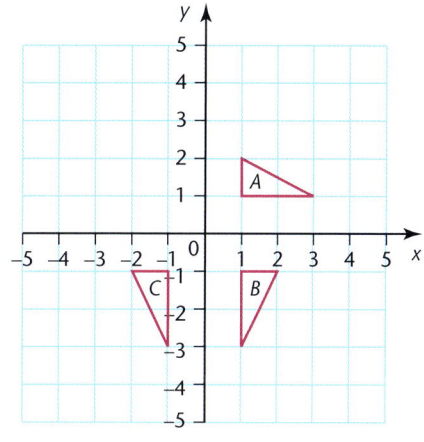

Note

Make sure you carry out the transformations in the order stated in the question. It usually makes a difference.

When describing a single transformation, do not give a combination of transformations.

42 TRANSFORMATIONS

Example 42.12

Question

E is the transformation: an enlargement, scale factor 3, centre (3, 3).

T is the transformation: a translation by the vector $\begin{pmatrix} 2 \\ 2 \end{pmatrix}$.

Describe fully the single transformation that is equivalent to E followed by T.

Solution

Again, start with a simple object like triangle A.
Enlarge triangle A with scale factor 3, centre (3, 3).
This maps triangle A on to triangle B.

Then translate triangle B by the vector $\begin{pmatrix} 2 \\ 2 \end{pmatrix}$.

This maps triangle B on to triangle C.
Clearly the equivalent transformation to E followed by T is still an enlargement with scale factor 3. It just remains to find the centre of enlargement.
Draw lines through the corresponding vertices of triangle A and triangle C. These intersect at (2, 2).
So the single transformation is an enlargement, scale factor 3, with centre (2, 2).

Exercise 42.8

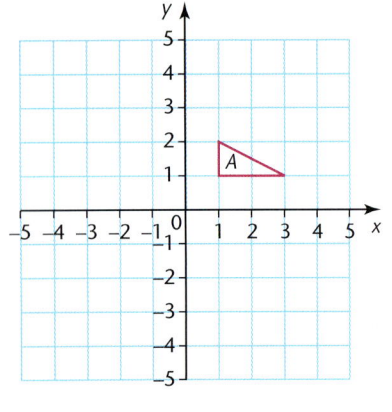

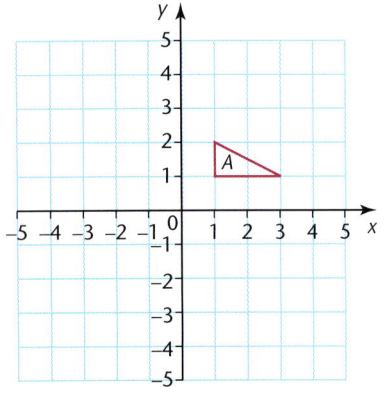

1 **a** Rotate triangle A 180° about the origin.
Label the image B.
b Reflect triangle B in the x-axis.
Label the image C.
c Describe fully the single transformation equivalent to a rotation of 180° about the origin, followed by a reflection in the x-axis.

2 **a** Translate triangle A by the vector $\begin{pmatrix} -2 \\ 4 \end{pmatrix}$.
Label the image B.
b Rotate triangle B 90° clockwise about the point (−2, 3).
Label the image C.
c Describe fully the single transformation equivalent to a translation

by the vector $\begin{pmatrix} -2 \\ 4 \end{pmatrix}$ followed by a rotation, 90° clockwise,

about the point (−2, 3).

In questions **3** to **7**, carry out the transformations on a simple shape of your choice.

3 Find the single transformation equivalent to a reflection in the y-axis, followed by a rotation of 90° clockwise about the origin.

4 Describe fully the single transformation equivalent to a reflection in the line x = 3, followed by a reflection in the line y = −2.

5 Describe fully the single transformation equivalent to a reflection in $y = 3$, followed by a reflection in $y = -1$.
6 R is a rotation of 90° anticlockwise about the point (2, 3).
M is a reflection in the line $x = 2$.
Describe fully the single transformation equivalent to R followed by M.
7 E is an enlargement with scale factor 2, centre the origin.
R is a rotation of 180° about (0, 3).
Describe fully the single transformation equivalent to E followed by R.

Key points
- When a shape is reflected, rotated or translated, it stays exactly the same shape. The shape and its image are congruent.
- In a translation, every point on the shape moves the same distance, in the same direction.
- In a rotation, all the points move through the same angle about the same centre.
- In a reflection, each point on the shape and the corresponding point on its image are the same distance from the mirror line, but on opposite sides.
- When a shape is enlarged by, for example, a scale factor of 2, from a given centre, each point on the enlargement will be 2 times as far away from the centre as the corresponding point on the original shape.
- A shape and its enlargement are similar.
- To find the scale factor of an enlargement, divide the length of a side of the image by the length of the corresponding side of the object.
- To find the centre of enlargement, join the corresponding corners of the two shapes with straight lines using a ruler and extend the lines until they cross. The point where they cross is the centre of enlargement.
- When the scale factor is between 0 and 1, such as $\frac{1}{2}$, the image is smaller than the object. However, the transformation is still called an enlargement.
- When the scale factor is negative, the image is on the opposite side of the centre of enlargement from the object, and the image is inverted.
- When describing a transformation, first give the name of the transformation. Then, give the extra information required.
- When combining transformations, make sure you carry out the operations in the correct order.

43 VECTORS

BY THE END OF THIS CHAPTER YOU WILL BE ABLE TO:

- describe a translation using a vector represented by $\begin{pmatrix} x \\ y \end{pmatrix}$, $\overrightarrow{AB}$ or **a**
- add and subtract vectors
- multiply a vector by a scalar
- calculate the magnitude of a vector $\begin{pmatrix} x \\ y \end{pmatrix}$ as $\sqrt{x^2 + y^2}$
- represent vectors by directed line segments
- use position vectors
- use the sum and difference of two or more vectors to express given vectors in terms of two coplanar vectors
- use vectors to reason and to solve geometric problems.

CHECK YOU CAN:

- add, subtract, multiply and divide numerical fractions
- simplify and factorise algebraic expressions
- rearrange formulae.

Vectors and translations

A transformation that moves points in a given direction for a given distance is called a **translation**.

The distance and direction that a shape moves in a translation can be written as a **column vector**.

The top number tells you how far the shape moves across, or in the x-direction.

The bottom number tells you how far the shape moves up or down, or in the y-direction.

A positive top number is a move to the right.

A negative top number is a move to the left.

A positive bottom number is a move up.

A negative bottom number is a move down.

A translation of 3 to the right and 2 down is written as $\begin{pmatrix} 3 \\ -2 \end{pmatrix}$.

Example 43.1

Question

Translate the triangle ABC by $\begin{pmatrix} -3 \\ 4 \end{pmatrix}$.

Solution

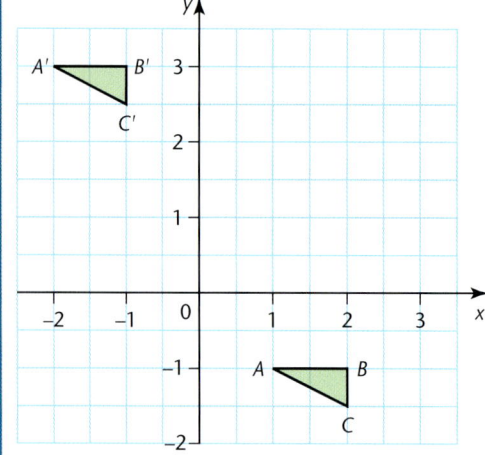

$\begin{pmatrix} -3 \\ 4 \end{pmatrix}$ means move 3 units left and 4 units up.

Point A moves from $(1, -1)$ to $(-2, 3)$.

Point B moves from $(2, -1)$ to $(-1, 3)$.

Point C moves from $(2, -1.5)$ to $(-1, 2.5)$.

428 Photocopying is prohibited

Vectors and translations

Example 43.2

Question
Describe fully the single transformation that maps shape A on to shape B.

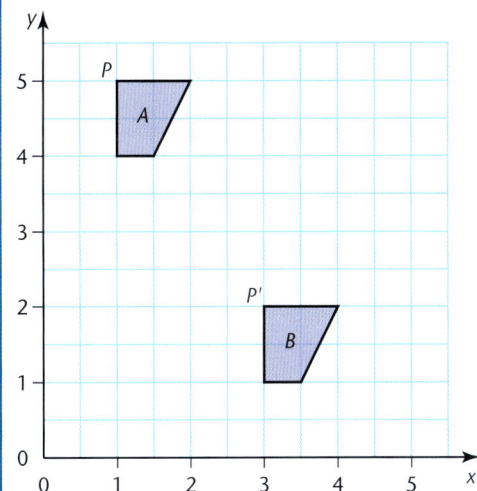

Note
Try not to confuse the words *transformation* and *translation*.

Transformation is the general name for all changes made to shapes.

Translation is the particular transformation where all points of an object move the same distance in the same direction.

Solution
It is clearly a translation as the shape stays the same way up and the same size.

To find the movement, choose one point on the original shape and the image and count the units moved.

For example, P moves from (1, 5) to (3, 2). This is a movement of 2 to the right and 3 down.

The transformation is a translation of $\begin{pmatrix} 2 \\ -3 \end{pmatrix}$.

Note
You must state that the transformation is a translation and give the column vector.

Exercise 43.1

1 Copy the diagram.
   a Translate triangle T 2 units to the right and 1 up. Label the image A.
   b Translate triangle T 1 unit to the right and 3 down. Label the image B.
   c Translate triangle T by $\begin{pmatrix} -4 \\ 1 \end{pmatrix}$. Label the image C.
   d Translate triangle T by $\begin{pmatrix} -6 \\ -6 \end{pmatrix}$. Label the image D.
   e Translate triangle T by $\begin{pmatrix} -2 \\ -5 \end{pmatrix}$. Label the image E.

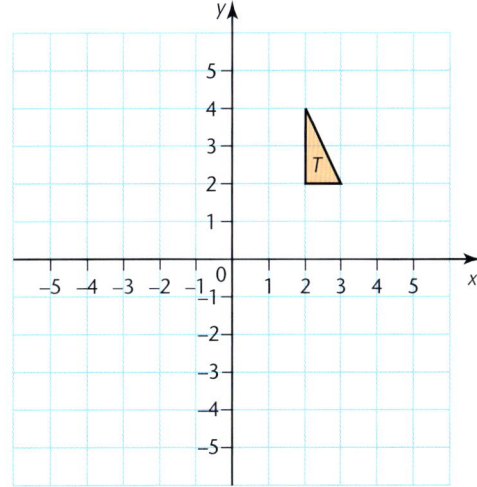

Photocopying is prohibited

43 VECTORS

2 Look at the diagram.

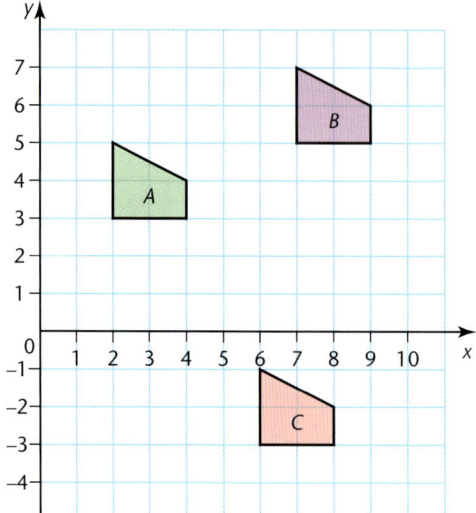

Describe the translation that maps
- **a** shape A on to shape B
- **b** shape A on to shape C
- **c** shape C on to shape B
- **d** shape C on to shape A.

Vector notation

You have already met directed numbers; these are numbers which go in either a positive or a negative direction. Vectors, however, can go in any direction. They have two parts, a **magnitude** (length) and a **direction**. They are very useful for the study of motion.

Vectors are usually written in one of three ways.

- The most common notation for a vector is a column vector, for example $\binom{3}{2}$, meaning 3 in the positive x-direction and 2 in the positive y-direction.

  These are similar to coordinates, except written in a column, and they can start anywhere (not just at the origin).

- $\overrightarrow{AB}$ meaning the vector starts at A and finishes at B.
- **a**. This is the algebraic form of a vector.

Column vectors

If a vector is drawn on a grid then it can be described by a column vector $\binom{x}{y}$, where x is the length across to the right and y is the length upwards.

In the diagram, $\mathbf{a} = \binom{2}{1}$, $\mathbf{b} = \binom{3}{-2}$ and $\mathbf{c} = \binom{-3}{-1}$.

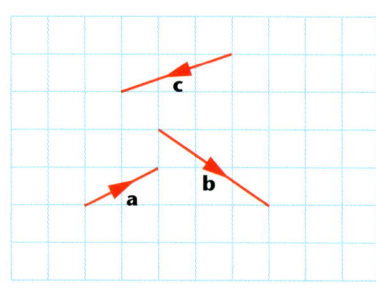

Vector notation

Example 43.3

Question
Write down these column vectors.

a The vector that maps point A to point B.
b The vector $\overrightarrow{CD}$.
c The vector $\overrightarrow{DC}$.

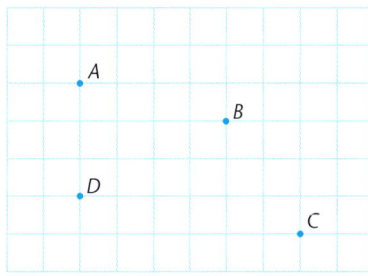

Note
Students often make an error of 1 when working out the values for the vector.
Take care with the counting.

Solution

a $\overrightarrow{AB} = \begin{pmatrix} 4 \\ -1 \end{pmatrix}$
b $\overrightarrow{CD} = \begin{pmatrix} -6 \\ 1 \end{pmatrix}$
c $\overrightarrow{DC} = \begin{pmatrix} 6 \\ -1 \end{pmatrix}$

Example 43.4

Question
Find the column vector that maps (−1, 6) on to (5, 2).

Solution
The x-coordinate has changed from −1 to 5 so the increase is 6.
The y-coordinate has changed from 6 to 2 so the decrease is 4.

The vector is $\begin{pmatrix} 6 \\ -4 \end{pmatrix}$.

Note
As you can see, it is not necessary to plot the points to do this question. However, you may prefer to do so.

General vectors

A vector has both length and direction but can be in any position. The vector going from A to B can be labelled $\overrightarrow{AB}$ or it can be given a letter **a**, in bold type.

When you write a vector, you must indicate that the quantity is a vector in some definite way. You can use an arrow, for example, $\overrightarrow{AB}$, to show the vector that goes from A to B or you can use underlining, for example, a̲, to show a vector that in print is bold, **a**.

43 VECTORS

All four lines drawn below are of equal length and go in the same direction and they can all be called **a**.

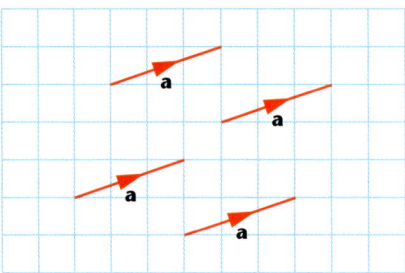

Look at the diagram below.
$\overrightarrow{AB}$ = **b**.

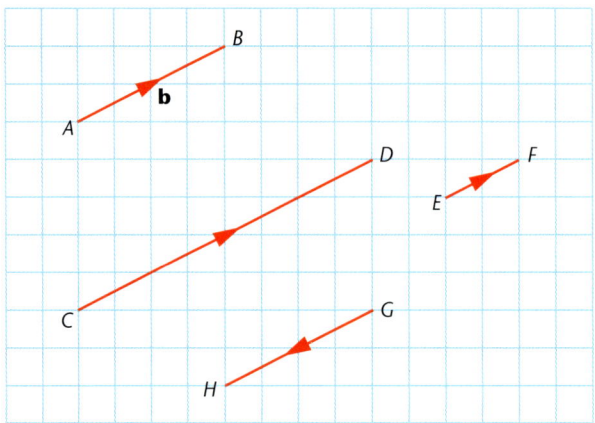

The line CD is parallel to AB and twice as long so $\overrightarrow{CD}$ = 2**b**.

EF is parallel to AB and half the length so $\overrightarrow{EF} = \frac{1}{2}$**b**.

GH is parallel and equal in length to AB but in the opposite direction so $\overrightarrow{GH}$ = −**b**.

Example 43.5

Question
For the diagram, write down the vectors $\overrightarrow{CD}$, $\overrightarrow{EF}$, $\overrightarrow{GH}$ and $\overrightarrow{PQ}$ in terms of **a**.

Solution
$\overrightarrow{CD}$ = **a**, $\overrightarrow{EF}$ = 3**a**, $\overrightarrow{GH} = -\frac{1}{2}$**a**, $\overrightarrow{PQ} = \frac{3}{2}$**a**

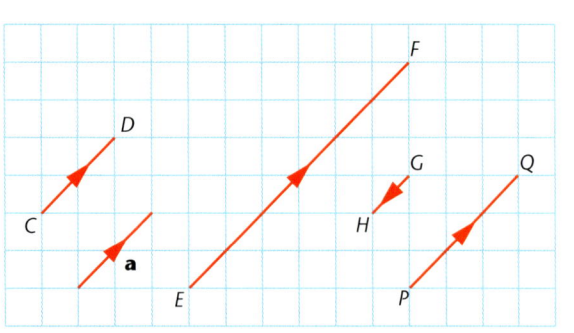

Vector notation

Example 43.6

Question

ABCD is a rectangle and E, F, G, H are the midpoints of the sides. $\overrightarrow{AB} = \mathbf{a}$ and $\overrightarrow{AD} = \mathbf{b}$.

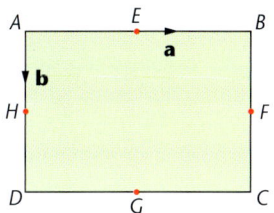

Write the vectors $\overrightarrow{BC}, \overrightarrow{CD}, \overrightarrow{AE}, \overrightarrow{AH}, \overrightarrow{EG}, \overrightarrow{CF}$ and $\overrightarrow{FH}$ in terms of **a** or **b**.

Solution

$\overrightarrow{BC} = \mathbf{b}, \quad \overrightarrow{CD} = -\mathbf{a}, \quad \overrightarrow{AE} = \tfrac{1}{2}\mathbf{a}, \quad \overrightarrow{AH} = \tfrac{1}{2}\mathbf{b}, \quad \overrightarrow{EG} = \mathbf{b}, \quad \overrightarrow{CF} = -\tfrac{1}{2}\mathbf{b}, \quad \overrightarrow{FH} = -\mathbf{a}$

Exercise 43.2

1 Write down the column vectors for $\overrightarrow{AB}, \overrightarrow{CD}, \overrightarrow{CB}, \overrightarrow{AD}$ and $\overrightarrow{CA}$.

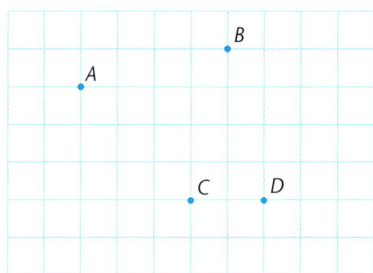

2 Write down the column vectors for $\overrightarrow{EF}, \overrightarrow{GH}, \overrightarrow{EH}, \overrightarrow{GF}$ and $\overrightarrow{FH}$.

3 Find the column vector that describes these translations.
   a (1, 2) to (1, 4)
   b (2, 3) to (−2, 3)
   c (1, 0) to (−1, 3)
   d (4, 2) to (5, 9)
   e (−3, 2) to (5, −4)
   f (6, 1) to (0, 5)

43 VECTORS

4 Copy and complete this table.

	Original point	Vector	New point
a	(1, 2)	$\begin{pmatrix} 3 \\ 2 \end{pmatrix}$	
b	(2, 3)	$\begin{pmatrix} 4 \\ 1 \end{pmatrix}$	
c	(1, 0)	$\begin{pmatrix} -3 \\ 2 \end{pmatrix}$	
d	(4, 2)	$\begin{pmatrix} 0 \\ -3 \end{pmatrix}$	
e	(−3, 2)	$\begin{pmatrix} -5 \\ -2 \end{pmatrix}$	
f	(6, 1)	$\begin{pmatrix} -6 \\ -1 \end{pmatrix}$	

5 For the diagram below, write down the vectors $\overrightarrow{AB}$, $\overrightarrow{CD}$, $\overrightarrow{EF}$, $\overrightarrow{GH}$, $\overrightarrow{PQ}$ and $\overrightarrow{RS}$ in terms of **a**.

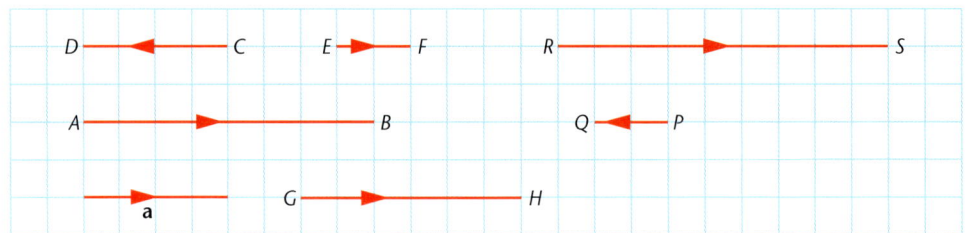

6 For the diagram below, write down the vectors $\overrightarrow{AB}$, $\overrightarrow{CD}$, $\overrightarrow{EF}$, $\overrightarrow{GH}$, $\overrightarrow{PQ}$ and $\overrightarrow{RS}$ in terms of **a** or **b**.

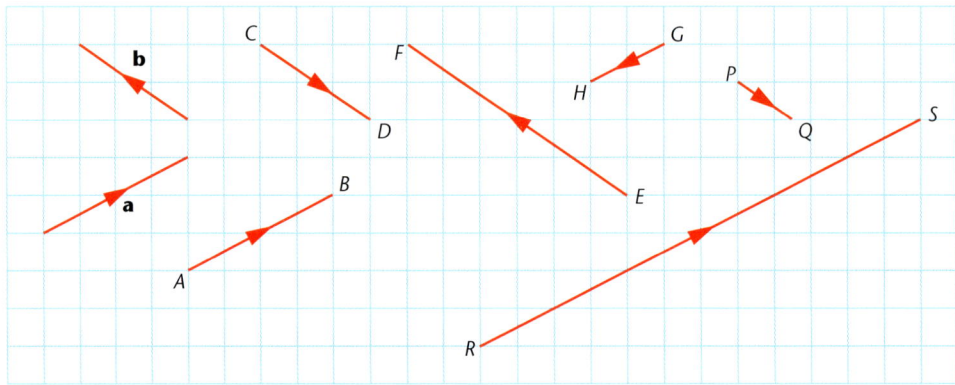

Addition and subtraction of column vectors

7 *ABCD* is a parallelogram. *E, F, G, H* are the midpoints of the sides.

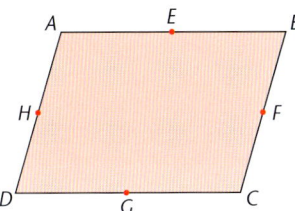

$\vec{AE} = \mathbf{a}$ and $\vec{AH} = \mathbf{b}$.
Write down the vectors for $\vec{AB}, \vec{CD}, \vec{EB}, \vec{GD}, \vec{HF}$ and $\vec{FC}$ in terms of **a** or **b**.

8 *ABCD* is a square. *E, F, G, H* are the midpoints of sides *AB, BC, CD, DA*, respectively.
$\vec{AB} = \mathbf{a}$ and $\vec{AD} = \mathbf{b}$.
Write down the vectors $\vec{BC}, \vec{CD}, \vec{EB}, \vec{GD}, \vec{HF}$ and $\vec{FC}$ in terms of **a** or **b**.

Multiplying a vector by a scalar

A quantity that has magnitude but not direction is called a **scalar**.

Multiplying a vector by a scalar produces a vector in the same direction but longer by a factor equal to the scalar.

If you know that $\vec{AB} = k\vec{CD}$, you can conclude that

- $\vec{AB}$ is parallel to $\vec{CD}$
- $\vec{AB}$ is *k* times the length of $\vec{CD}$.

If you know that $\vec{AB} = k\vec{AC}$ and there is a common point *A*, you can conclude that

- *A*, *B* and *C* are in a straight line
- $\vec{AB}$ is *k* times the length of $\vec{AC}$.

Addition and subtraction of column vectors

In the diagram, $\mathbf{a} = \begin{pmatrix} 3 \\ 1 \end{pmatrix}$ and $\mathbf{b} = \begin{pmatrix} 2 \\ -3 \end{pmatrix}$.

You can see that

$\mathbf{a} + \mathbf{b} = \begin{pmatrix} 3 \\ 1 \end{pmatrix} + \begin{pmatrix} 2 \\ -3 \end{pmatrix} = \begin{pmatrix} 5 \\ -2 \end{pmatrix}$

and $\mathbf{a} - \mathbf{b} = \mathbf{a} + (-\mathbf{b}) = \begin{pmatrix} 3 \\ 1 \end{pmatrix} - \begin{pmatrix} 2 \\ -3 \end{pmatrix} = \begin{pmatrix} 1 \\ 4 \end{pmatrix}$.

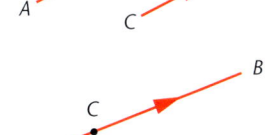
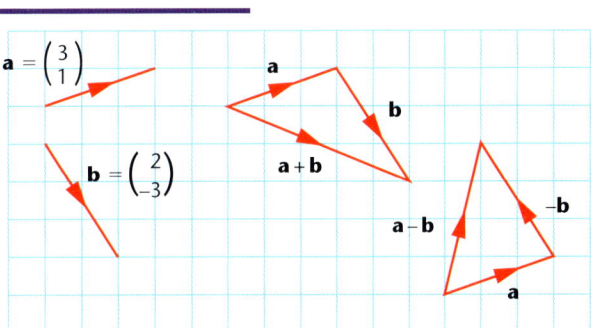

43 VECTORS

So, to add or subtract column vectors, you add or subtract the components.

$$\begin{pmatrix} a \\ b \end{pmatrix} + \begin{pmatrix} c \\ d \end{pmatrix} = \begin{pmatrix} a+c \\ b+d \end{pmatrix} \text{ and } \begin{pmatrix} a \\ b \end{pmatrix} - \begin{pmatrix} c \\ d \end{pmatrix} = \begin{pmatrix} a-c \\ b-d \end{pmatrix}$$

Example 43.7

Question

Given $\mathbf{a} = \begin{pmatrix} 3 \\ 1 \end{pmatrix}$, $\mathbf{b} = \begin{pmatrix} 1 \\ 3 \end{pmatrix}$ and $\mathbf{c} = \begin{pmatrix} -2 \\ 1 \end{pmatrix}$, work out these.

a $2\mathbf{a}$
b $\mathbf{a} + 2\mathbf{b}$
c $\mathbf{a} - \mathbf{b} + \mathbf{c}$
d $2\mathbf{a} + \mathbf{b} - \mathbf{c}$
e $\frac{1}{2}\mathbf{a}$

Solution

a $2\mathbf{a} = 2 \times \begin{pmatrix} 3 \\ 1 \end{pmatrix}$
$= \begin{pmatrix} 6 \\ 2 \end{pmatrix}$

b $\mathbf{a} + 2\mathbf{b} = \begin{pmatrix} 3 \\ 1 \end{pmatrix} + 2 \times \begin{pmatrix} 1 \\ 3 \end{pmatrix}$
$= \begin{pmatrix} 3 \\ 1 \end{pmatrix} + \begin{pmatrix} 2 \\ 6 \end{pmatrix}$
$= \begin{pmatrix} 5 \\ 7 \end{pmatrix}$

c $\mathbf{a} - \mathbf{b} + \mathbf{c} = \begin{pmatrix} 3 \\ 1 \end{pmatrix} - \begin{pmatrix} 1 \\ 3 \end{pmatrix} + \begin{pmatrix} -2 \\ 1 \end{pmatrix}$
$= \begin{pmatrix} 0 \\ -1 \end{pmatrix}$

d $2\mathbf{a} + \mathbf{b} - \mathbf{c} = 2 \times \begin{pmatrix} 3 \\ 1 \end{pmatrix} + \begin{pmatrix} 1 \\ 3 \end{pmatrix} - \begin{pmatrix} -2 \\ 1 \end{pmatrix}$
$= \begin{pmatrix} 6 \\ 2 \end{pmatrix} + \begin{pmatrix} 1 \\ 3 \end{pmatrix} - \begin{pmatrix} -2 \\ 1 \end{pmatrix}$
$= \begin{pmatrix} 9 \\ 4 \end{pmatrix}$

e $\frac{1}{2}\mathbf{a} = \frac{1}{2} \times \begin{pmatrix} 3 \\ 1 \end{pmatrix}$
$= \begin{pmatrix} 1.5 \\ 0.5 \end{pmatrix}$

Note

When adding and subtracting column vectors, be very careful with the signs as most errors are made in that way.

Exercise 43.3

1 Work out these.

a $2 \times \begin{pmatrix} 2 \\ 3 \end{pmatrix}$
b $\begin{pmatrix} 6 \\ 2 \end{pmatrix} + \begin{pmatrix} 3 \\ 1 \end{pmatrix}$
c $\frac{1}{2}\begin{pmatrix} 4 \\ 6 \end{pmatrix}$
d $\begin{pmatrix} 3 \\ 1 \end{pmatrix} - \begin{pmatrix} 2 \\ 1 \end{pmatrix}$
e $\begin{pmatrix} 3 \\ 4 \end{pmatrix} + 2 \times \begin{pmatrix} 1 \\ 4 \end{pmatrix}$

2 Work out these.

a $2 \times \begin{pmatrix} -3 \\ 0 \end{pmatrix}$
b $\begin{pmatrix} 3 \\ 1 \end{pmatrix} - \begin{pmatrix} 4 \\ 3 \end{pmatrix}$
c $\frac{1}{2}\begin{pmatrix} 1 \\ -3 \end{pmatrix}$
d $\begin{pmatrix} 2 \\ -1 \end{pmatrix} + 2 \times \begin{pmatrix} 2 \\ 1 \end{pmatrix}$
e $\frac{1}{2}\begin{pmatrix} 1 \\ 4 \end{pmatrix} - \frac{1}{4}\begin{pmatrix} 2 \\ 4 \end{pmatrix}$

3 Work out these.

a $3 \times \begin{pmatrix} 1 \\ 4 \end{pmatrix}$
b $\begin{pmatrix} 3 \\ 4 \end{pmatrix} + \begin{pmatrix} 5 \\ 8 \end{pmatrix}$
c $\frac{1}{2}\begin{pmatrix} 8 \\ 10 \end{pmatrix}$
d $2 \times \begin{pmatrix} 5 \\ 4 \end{pmatrix} - \begin{pmatrix} 3 \\ 4 \end{pmatrix}$
e $2 \times \begin{pmatrix} 1 \\ 4 \end{pmatrix} + 5 \times \begin{pmatrix} 1 \\ 2 \end{pmatrix}$

4 Work out these.

a $2 \times \begin{pmatrix} -1 \\ 0 \end{pmatrix}$
b $\begin{pmatrix} 1 \\ 6 \end{pmatrix} - \begin{pmatrix} 7 \\ 3 \end{pmatrix}$
c $\frac{1}{2}\begin{pmatrix} -2 \\ 4 \end{pmatrix}$
d $\begin{pmatrix} 1 \\ -4 \end{pmatrix} - 2 \times \begin{pmatrix} 2 \\ 3 \end{pmatrix}$
e $\frac{1}{2}\begin{pmatrix} 2 \\ 6 \end{pmatrix} - \frac{1}{2}\begin{pmatrix} 3 \\ -5 \end{pmatrix}$

Position vectors

The vector $\overrightarrow{OA}$ is the **position vector** of A in relation to O.

Similarly $\overrightarrow{OB}$ is the position vector of B in relation to O.

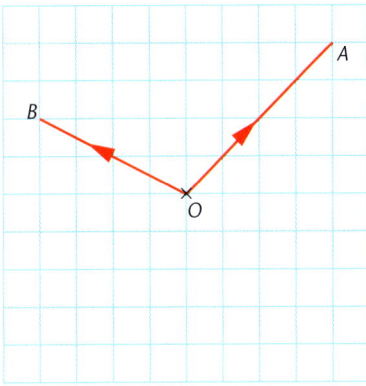

Exercise 43.4

1 Find, as column vectors, the position vectors relative to O of P, Q and R.

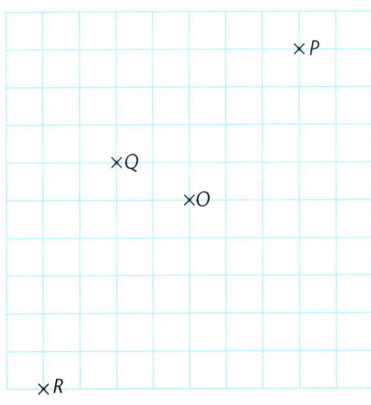

2 A is the point $(-2, 1)$, B is the point $(4, 3)$ and C is the point $(7, 4)$.

  a What are the position vectors $\overrightarrow{OA}$, $\overrightarrow{OB}$ and $\overrightarrow{OC}$?
  b Work out these column vectors.
   i $\overrightarrow{AB}$
   ii $\overrightarrow{BC}$
  c What can you say about A, B and C?

43 VECTORS

3 *A* is the point (2, 1), *B* is the point (4, 4), *C* is the point (7, 4) and *D* is the point (3, −2).
  a What are the position vectors $\overrightarrow{OA}$, $\overrightarrow{OB}$ and $\overrightarrow{OC}$?
  b Work out these column vectors.
    i $\overrightarrow{AB}$ ii $\overrightarrow{CD}$
  c What can you say about *AB* and *CD*?

The magnitude of a vector

The magnitude of a vector is its length.

The magnitude of the vector $\overrightarrow{AB}$ is written as $|\overrightarrow{AB}|$.

The magnitude of the vector **a** is written as $|\mathbf{a}|$.

Pythagoras' theorem can be used to calculate the magnitude of a vector.

> The **magnitude** of the column vector $\begin{pmatrix} x \\ y \end{pmatrix}$ is $\sqrt{x^2 + y^2}$.

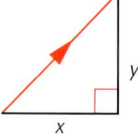

Example 43.8

Question

$\overrightarrow{CD} = \begin{pmatrix} 5 \\ -2 \end{pmatrix}$

Find $|\overrightarrow{CD}|$.

Solution

$|\overrightarrow{CD}| = \sqrt{5^2 + (-2)^2}$

$= \sqrt{25 + 4}$

$= \sqrt{29}$

= 5.39 to 3 significant figures.

Exercise 43.5

1 Given that $\mathbf{a} = \begin{pmatrix} 6 \\ 3 \end{pmatrix}$, work out these.
  a 2**a**   b −**a**   c 4**a**
  d $\frac{1}{2}\mathbf{a}$ e $-\frac{1}{3}\mathbf{a}$ f $|\mathbf{a}|$

2 Given that $\mathbf{a} = \begin{pmatrix} 1 \\ 3 \end{pmatrix}$ and $\mathbf{b} = \begin{pmatrix} 3 \\ 4 \end{pmatrix}$, work out these.
  a 3**a**   b **a** + **b**   c **b** − **a**
  d 2**a** + **b**   e 3**a** − 2**b**   f $|\mathbf{a}|$
  g $|\mathbf{b}|$

3 Given that $\mathbf{a} = \begin{pmatrix} 2 \\ 3 \end{pmatrix}$, $\mathbf{b} = \begin{pmatrix} -3 \\ 4 \end{pmatrix}$ and $\mathbf{c} = \begin{pmatrix} -1 \\ -3 \end{pmatrix}$, work out these.
  a 3**c**   b 4**c** − 2**b**   c **a** − **b** + **c**
  d 2**a** + 3**b** + 2**c**   e $\frac{1}{2}\mathbf{a} - \mathbf{b} - \frac{1}{2}\mathbf{c}$ f $|\mathbf{b}|$
  g $|\mathbf{c}|$

4 Given that $\mathbf{p} = \begin{pmatrix} 5 \\ 8 \end{pmatrix}$, work out these.

  a $4\mathbf{p}$ **b** $-2\mathbf{p}$ **c** $\frac{1}{2}\mathbf{p}$

  d $9\mathbf{p}$ **e** $0.4\mathbf{p}$ **f** $|\mathbf{p}|$

5 Given that $\mathbf{p} = \begin{pmatrix} 4 \\ 1 \end{pmatrix}$ and $\mathbf{q} = \begin{pmatrix} 5 \\ 3 \end{pmatrix}$, work out these.

  a $2\mathbf{p}$ **b** $\mathbf{p} + \mathbf{q}$ **c** $\mathbf{q} - \mathbf{p}$

  d $2\mathbf{p} + \mathbf{q}$ **e** $3\mathbf{q} - 2\mathbf{p}$ **f** $|\mathbf{p} - \mathbf{q}|$

6 Given that $\mathbf{a} = \begin{pmatrix} -2 \\ 4 \end{pmatrix}$, $\mathbf{b} = \begin{pmatrix} 3 \\ 5 \end{pmatrix}$ and $\mathbf{c} = \begin{pmatrix} -2 \\ -3 \end{pmatrix}$, work out these.

  a $3\mathbf{c}$ **b** $3\mathbf{c} + 2\mathbf{b}$ **c** $\mathbf{a} - \mathbf{b} + \mathbf{c}$

  d $\mathbf{a} + 4\mathbf{b} - 2\mathbf{c}$ **e** $\frac{1}{2}\mathbf{a} + \mathbf{b} - \frac{1}{2}\mathbf{c}$ **f** $|\mathbf{a} - \mathbf{c}|$

Vector geometry

In the diagram, $\overrightarrow{OA} = \mathbf{a}$, $\overrightarrow{OB} = \mathbf{b}$ and $OACB$ is a parallelogram.

AC is parallel and equal to OB, so $AC = \mathbf{b}$.

$\overrightarrow{OC} = \overrightarrow{OA} + \overrightarrow{AC} = \mathbf{a} + \mathbf{b}$.

You can also see in the diagram that $\overrightarrow{OC} = \overrightarrow{OB} + \overrightarrow{BC} = \mathbf{b} + \mathbf{a}$.

This shows that $\mathbf{a} + \mathbf{b} = \mathbf{b} + \mathbf{a}$.

The vectors can be added in either order.

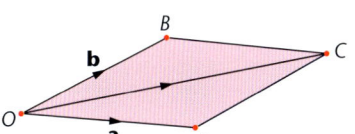

To subtract vectors you use the fact that $\mathbf{p} - \mathbf{q} = \mathbf{p} + (-\mathbf{q})$.

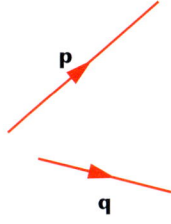
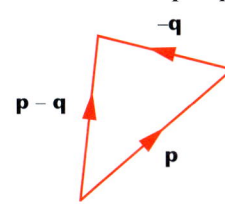

There are different routes you can use to get from A to E on this diagram.

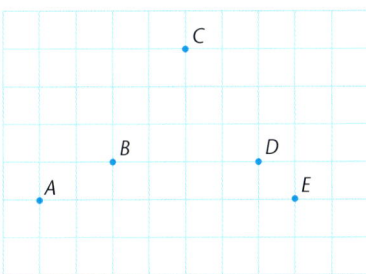

For instance $\overrightarrow{AE} = \overrightarrow{AB} + \overrightarrow{BC} + \overrightarrow{CE} = \begin{pmatrix} 2 \\ 1 \end{pmatrix} + \begin{pmatrix} 2 \\ 3 \end{pmatrix} + \begin{pmatrix} 3 \\ -4 \end{pmatrix} = \begin{pmatrix} 7 \\ 0 \end{pmatrix}$.

Whichever route you use, adding the column vectors gives the same result,
$\vec{AE} = \begin{pmatrix} 7 \\ 0 \end{pmatrix}$.

The fact that the result is the same is a very important rule which, together with your knowledge of how to multiply a vector by a scalar, is used to find vectors in geometrical diagrams.

Example 43.9

Question

In the triangle ABC, $\vec{AB} = \mathbf{p}$, $\vec{AC} = \mathbf{q}$ and D is the midpoint of BC.

Write these vectors in terms of $\mathbf{p}$ and $\mathbf{q}$.

a $\vec{BC}$ b $\vec{BD}$ c $\vec{AD}$

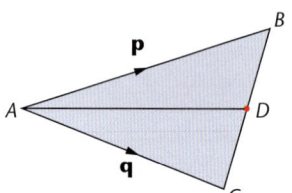

Solution

a $\vec{BC} = \vec{BA} + \vec{AC}$
$= -\mathbf{p} + \mathbf{q}$
$= \mathbf{q} - \mathbf{p}$

b $\vec{BD} = \tfrac{1}{2}\vec{BC}$
$= \tfrac{1}{2}(\mathbf{q} - \mathbf{p})$

c $\vec{AD} = \vec{AB} + \vec{BD}$
$= \mathbf{p} + \tfrac{1}{2}(\mathbf{q} - \mathbf{p})$
$= \mathbf{p} + \tfrac{1}{2}\mathbf{q} - \tfrac{1}{2}\mathbf{p}$
$= \tfrac{1}{2}\mathbf{p} + \tfrac{1}{2}\mathbf{q}$
$= \tfrac{1}{2}(\mathbf{p} + \mathbf{q})$

Example 43.10

Question

In this diagram, $OC = 2 \times OA$ and $OD = 2 \times OB$.
$\vec{OA} = \mathbf{a}$ and $\vec{OB} = \mathbf{b}$.

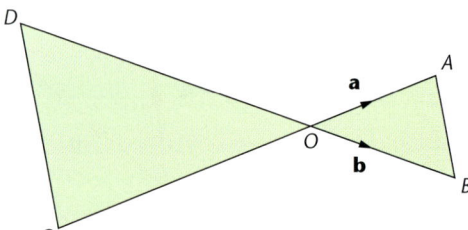

a Write these vectors in terms of $\mathbf{a}$ and $\mathbf{b}$.
  i $\vec{OC}$ ii $\vec{OD}$ iii $\vec{AB}$ iv $\vec{DC}$

b What does this show about the lines AB and DC?

Solution

a i OC is on the same line as OA, in the opposite direction and twice as long.
$\vec{OC} = -2 \times \vec{OA} = -2\mathbf{a}$

ii By the same reasoning as above, $\vec{OD} = -2 \times \vec{OB} = -2\mathbf{b}$.

iii $\vec{AB} = \vec{AO} + \vec{OB} = -\mathbf{a} + \mathbf{b} = \mathbf{b} - \mathbf{a}$

iv $\vec{DC} = \vec{DO} + \vec{OC} = 2\mathbf{b} - 2\mathbf{a} = 2(\mathbf{b} - \mathbf{a})$

b The vector for DC is twice the vector for AB.
So AB and DC are parallel and DC is twice as long as AB.

Vector geometry

Exercise 43.6

1. In the triangle on the right, $\overrightarrow{AB} = \mathbf{a}$ and $\overrightarrow{AC} = 2\mathbf{b}$.
   Find the vector $\overrightarrow{BC}$ in terms of $\mathbf{a}$ and $\mathbf{b}$.

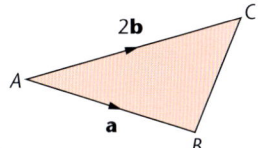

2. In triangle PQR on the right, $\overrightarrow{PQ} = 2\mathbf{a}$ and $\overrightarrow{RQ} = 3\mathbf{b}$.
   Work out the vector $\overrightarrow{PR}$.

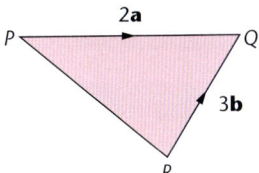

3. In the diagram below, $\overrightarrow{OA} = 2\mathbf{a}$, $\overrightarrow{OB} = \mathbf{a} - \mathbf{b}$, and $\overrightarrow{OC} = 2\mathbf{b} - 3\mathbf{a}$.

   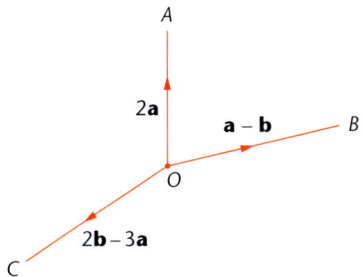

   Write these vectors in terms of $\mathbf{a}$ and/or $\mathbf{b}$, as simply as possible.

   a $\overrightarrow{AB}$ b $\overrightarrow{BC}$ c $\overrightarrow{AC}$

4. $ABCD$ is a parallelogram. $\overrightarrow{AB} = \mathbf{a}$ and $\overrightarrow{AD} = \mathbf{b}$.

   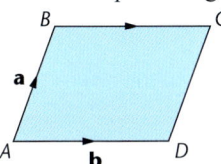

   Work out the vectors $\overrightarrow{BC}$, $\overrightarrow{CD}$, $\overrightarrow{BD}$ and $\overrightarrow{AC}$ in terms of $\mathbf{a}$ and/or $\mathbf{b}$.

5. In the triangle OAB, C is a point on AB so that $AC = 2 \times CB$.
   $\overrightarrow{OA} = \mathbf{a}$ and $\overrightarrow{OB} = \mathbf{b}$.
   Work out the vectors $\overrightarrow{AB}$, $\overrightarrow{CB}$ and $\overrightarrow{OC}$ in terms of $\mathbf{a}$ and/or $\mathbf{b}$.

   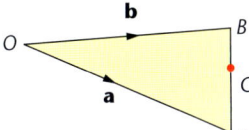

6. $ABCD$ is a parallelogram.
   E is the midpoint of the DC line.
   $\overrightarrow{AB} = \mathbf{a}$ and $\overrightarrow{AD} = \mathbf{b}$.
   Write down the vector $\overrightarrow{EB}$ in terms of $\mathbf{a}$ and/or $\mathbf{b}$.

   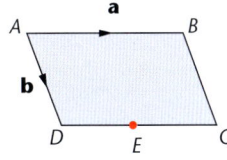

7. OAB is a triangle with E a point on OA so that $OE = 2 \times EA$.
   $\overrightarrow{OA} = \mathbf{a}$ and $\overrightarrow{OB} = \mathbf{b}$.
   Work out the vector $\overrightarrow{EB}$ in terms of $\mathbf{a}$ and/or $\mathbf{b}$.

   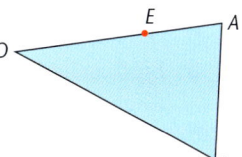

43 VECTORS

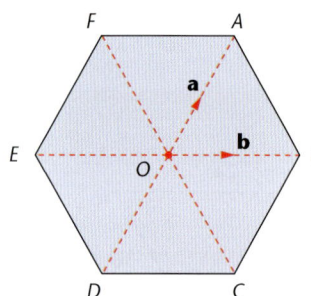

8 *ABCDEF* is a regular hexagon.
  O is the centre of the hexagon.
  $\overrightarrow{OA} = \mathbf{a}$ and $\overrightarrow{OB} = \mathbf{b}$.
  Find each of these vectors in terms of **a** and/or **b**.
  a $\overrightarrow{FA}$ **b** $\overrightarrow{BD}$ **c** $\overrightarrow{AB}$ **d** $\overrightarrow{AC}$

9 In the diagram below, *P* is one-third of the way along *AB*.
  The position vectors of *A* and *B* in relation to *O* are **a** and **b**, respectively.

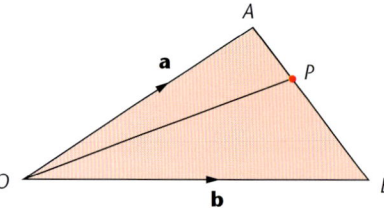

Find each of these vectors, as simply as possible, in terms of **a** and **b**.
  a $\overrightarrow{AB}$ **b** $\overrightarrow{AP}$ **c** $\overrightarrow{OP}$

10 Triangle *AEF* is a 3 times enlargement of triangle *ABC*.
  $\overrightarrow{AB} = \mathbf{a}$ and $\overrightarrow{AC} = \mathbf{b}$.
  a Write down the vectors $\overrightarrow{AE}, \overrightarrow{AF}, \overrightarrow{BC}$
  and $\overrightarrow{EF}$ in terms of **a** and/or **b**.
  b What do the vectors show about *BC* and *EF*?

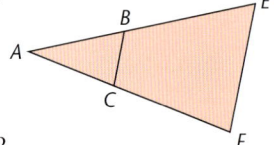

11 In the triangle *OCD*, $AC = 3 \times OA$ and $BD = 3 \times OB$.
  $\overrightarrow{OA} = \mathbf{a}$ and $\overrightarrow{OB} = \mathbf{b}$.
  a Use vectors to show that *AB* is parallel to *CD*.
  b What is the ratio of the lengths of *AB* and *CD*?

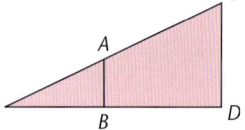

12 In the diagram, $\overrightarrow{OA} = \overrightarrow{AD} = \overrightarrow{CB} = \overrightarrow{BE} = \mathbf{a}$ and $\overrightarrow{OC} = \overrightarrow{AB} = \overrightarrow{DE} = \mathbf{c}$.
  F is one third of the way along *AC*.

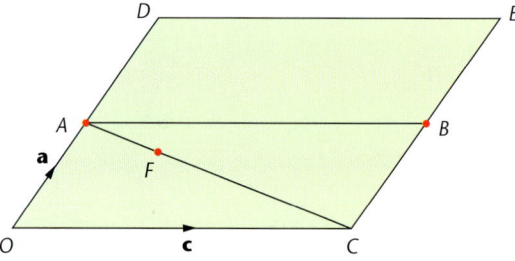

  a Find each of these vectors in terms of **a** and **c**.
  Simplify your answers where possible.
  i $\overrightarrow{OE}$ **ii** $\overrightarrow{AC}$ **iii** $\overrightarrow{OF}$

  b What two facts can you conclude about the points *O*, *F* and *E*?

13 $OABC$ is a quadrilateral. $\vec{OA} = \mathbf{a}$
$\vec{OC} = \vec{AB} = \mathbf{c}$

D is the midpoint of AC.

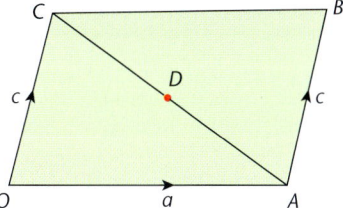

a Find these vectors in terms of **a** and **c**.
  i $\vec{AC}$ ii $\vec{AD}$ iii $\vec{OD}$ iv $\vec{OB}$

b What do these results show about the points O, D and B?

Key points

- A translation moves every point on an object the same distance, in the same direction.
- You can use a column vector to describe a translation.
  Translate by $\begin{pmatrix} -4 \\ 2 \end{pmatrix}$ means move 4 units left and 2 units up.
- A vector has magnitude and direction but can start at any point.
- Multiplying a vector by a scalar (ordinary number) changes its length but not its direction.
- If you know that $\vec{AB} = k \times \vec{CD}$, then you know that $\vec{AB}$ and $\vec{CD}$ are parallel and that the length AB is k times the length CD.
- If you know that $\vec{AB} = k \times \vec{AC}$, then you know that A, B and C are in a straight line (collinear) and that the length of AB is k times the length of AC.
- To multiply a column vector by a scalar, multiply each component by the scalar.
- To add or subtract column vectors, add or subtract the components.
- The vector $\vec{OA}$ is the position vector of A in relation to O.
- The magnitude of a vector is its length.
- The magnitude of the vector $\vec{AB}$ is written as $|\vec{AB}|$.
- The magnitude of the vector **a** is written as $|\mathbf{a}|$.
- Pythagoras' theorem can be used to calculate the magnitude of a vector.
  The magnitude of the column vector $\begin{pmatrix} x \\ y \end{pmatrix}$ is $\sqrt{x^2 + y^2}$.

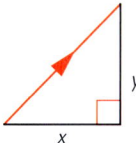

- In vector geometry, $\mathbf{a} + \mathbf{b} = \mathbf{b} + \mathbf{a}$. The vectors can be added in either order.

REVIEW EXERCISE 8

Ch 41 **1** Ammaar walks from home, 3.2 km due east, and then he walks 4.7 km due south.
How far away from home is Ammaar now? [2]

Ch 41 **2**

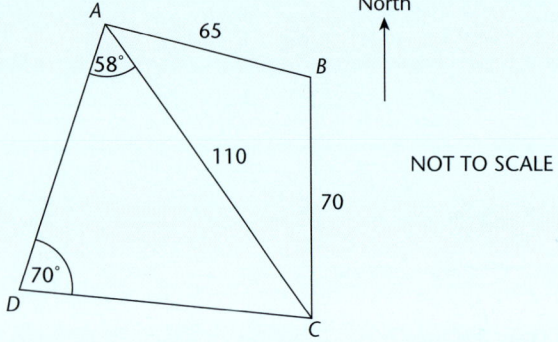

NOT TO SCALE

ABCD is a level playing field.
AB = 65 m, BC = 70 m and CA = 110 m.
Angle CDA = 70°, Angle DAC = 58° and C is due South of B.
  a Calculate the bearing of A from C. [4]
  b Calculate AD. [3]

Cambridge O Level Mathematics Syllabus D (4024) Paper 22 Q9a & b, November 2016

Ch 43 **3**

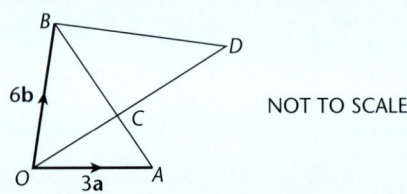

NOT TO SCALE

ACB and OCD are straight lines.
AC : CB = 1 : 2.
$\overrightarrow{OA}$ = 3**a** and $\overrightarrow{OB}$ = 6**b**.

  a Express $\overrightarrow{AB}$ in terms of **a** and **b**. [1]
  b Express $\overrightarrow{AC}$ in terms of **a** and **b**. [1]
  c $\overrightarrow{BD}$ = 5**a** − **b**.
     Showing your working clearly, find OC : CD. [4]

Cambridge O Level Mathematics Syllabus D (4024) Paper 22 Q10a, November 2016

4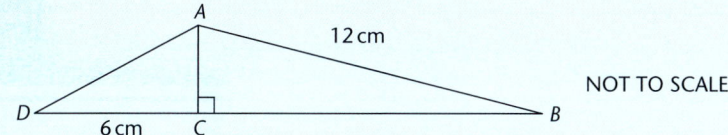

In the diagram, *BCD* is a straight line and angle *ACB* = 90°.
AB = 12 cm, *DC* = 6 cm and the angle of elevation of *A* from *B* is 32°.
Work out angle *ADC*. [4]

5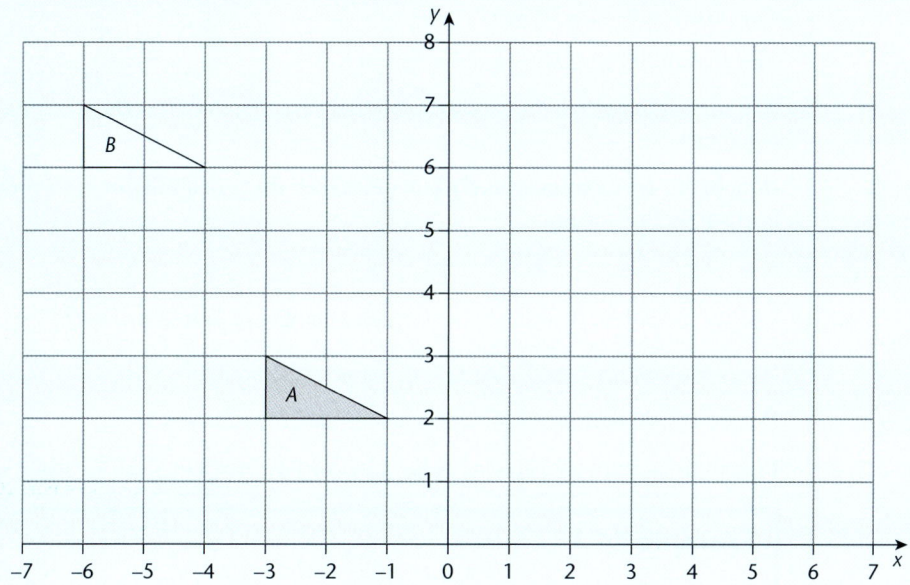

Triangle *A* and triangle *B* are drawn on the grid.
  a Describe fully the **single** transformation that maps triangle *A* onto triangle *B*. [2]
  b Triangle *A* is mapped onto triangle *C* by an enlargement with centre (0, 3) and scale factor −2.
On a copy of the grid, draw triangle *C*. [2]

Adapted from Cambridge O Level Mathematics Syllabus D (4024) Paper 12 Q16, June 2019

6 The position vector of point *A* is $\begin{pmatrix} -5 \\ 2 \end{pmatrix}$ and the position vector of point *B* is $\begin{pmatrix} 7 \\ -4 \end{pmatrix}$.
  a Work out $\overrightarrow{AB}$. [1]
  b Work out $3\overrightarrow{AB}$. [1]
  c Work out $|\overrightarrow{AB}|$. [2]

44 PROBABILITY

BY THE END OF THIS CHAPTER YOU WILL BE ABLE TO:

- understand and use the probability scale from 0 to 1
- understand and use probability notation
- calculate the probability of a single event
- understand that the probability of an event not occurring = 1 − the probability of the event occurring
- understand relative frequency as an estimate of probability
- calculate expected frequencies
- calculate the probability of combined events using, where appropriate: sample space diagrams, Venn diagrams and tree diagrams.

CHECK YOU CAN:

- calculate with fractions
- use the probability scale from 0 to 1.

The probability of a single event

When you toss a coin, there are two possible **outcomes**, heads or tails.

If the coin is fair, these outcomes are equally likely.

The probability of getting a head is $\frac{1}{2}$ or 0.5.

The probability of getting a tail is $\frac{1}{2}$ or 0.5.

The sum of these two probabilities is 1, because there are no other possible outcomes.

If there is a set of equally likely outcomes then the probability of each **event** is given by this formula.

$$\text{Probability of an event} = \frac{\text{number of outcomes which give this event}}{\text{total number of outcomes}}$$

If it is **impossible** that the event will happen, the probability is 0.

If it is **certain** that the event will happen, the probability is 1.

Probabilities can be written as either fractions or decimals.

Example 44.1

Question

A bag contains six red counters, three blue counters and one green counter.

A counter is taken at random from the bag.

Find the probability that the counter is

a red b green c black.

Note
When you write a probability as a fraction, write your final answer as a fraction in its simplest form.

Solution

a There are 6 + 3 + 1 = 10 counters in the bag, so there are ten equally likely outcomes.

Six of the counters are red, so six outcomes give the required event.
The probability that the counter is red is $\frac{6}{10}$ which can be simplified to $\frac{3}{5}$.

b One of the counters is green, so only one outcome gives the required event.
The probability that the counter is green is $\frac{1}{10}$. This fraction is in its simplest form.

c There are no black counters in the bag, so this event is impossible.
The probability that the counter is black is 0.

The probability of a single event

Exercise 44.1

In this exercise, give your answers as fractions in their simplest form.

1. A fair spinner numbered 1 to 6 is spun.
   Find the probability of getting
   a a 6
   b an even number
   c a number less than 5.

2. There are six blue balls and four red balls in a bag.
   A ball is taken at random from the bag.
   Find the probability that the ball is
   a red
   b blue.

3. There are seven blue balls, three red balls and ten yellow balls in a bag.
   A ball is taken at random from the bag.
   Find the probability that the ball is
   a blue
   b red
   c yellow.

4. A letter is chosen at random from the word REPRESENT.
   Find the probability that it is
   a an E
   b an R.

5. The table shows information about a group of 40 children.

	Boy	Girl
Left-handed	6	8
Right-handed	11	15

   One of the children is selected at random.
   Find the probability that the child is
   a a left-handed boy
   b a right-handed girl
   c a girl.

6. The bar chart shows the colours of cars parked in a car park.

   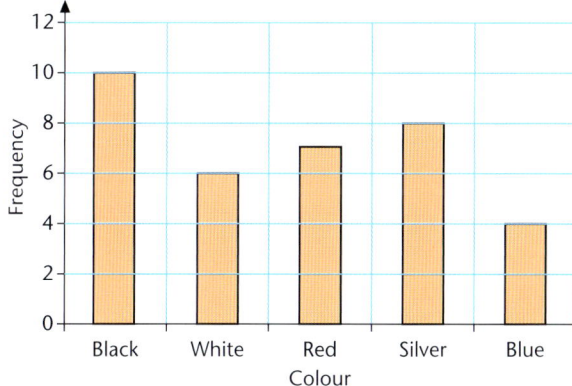

   One of the cars is selected at random.
   Find the probability that it is
   a black
   b silver.

44 PROBABILITY

The probability of an event not occurring

There are ten counters in a bag.

Seven of the counters are black.

A counter is taken at random from the bag.

The probability that the counter is black is $\frac{7}{10}$.

You can write this probability as P(black) = $\frac{7}{10}$.

To find the probability that the counter is not black, you need to think about how many outcomes there are where the counter is not black.

As there are ten counters in the bag, there are three counters that are not black.

The probability that the counter is not black is $\frac{3}{10}$.

You can write this as P(not black) = $\frac{3}{10}$

Note that P(black) + P(not black) = $\frac{7}{10} + \frac{3}{10} = 1$.

This is because there are no other possible outcomes for this experiment.

In probability notation, if A is an event, then you can write A' for the event not occurring. A' is the event 'not A'.

> The probability of an event not occurring = 1 − probability of the event occurring.
>
> So P(A') = 1 − P(A).

This rule can also be used when there are more than two possible events and you know the probability of all but one of them occurring.

Example 44.2

Question

a The probability that Nayim's bus is late is $\frac{1}{8}$.
   What is the probability that Nayim's bus is not late?

b Salma can travel to school by bus, car or bike.
   The probability that she travels by bus is 0.5.
   The probability that she travels by car is 0.2.
   What is the probability that she travels by bike?

Solution

a P(not late) = 1 − P(late)
            = 1 − $\frac{1}{8}$
            = $\frac{7}{8}$

b P(bike) = 1 − P(bus) − P(car)   The only possibilities are bus,
         = 1 − 0.5 − 0.2   car and bike, so these three
         = 0.3   probabilities sum to 1.

Exercise 44.2

1. The probability that Husna will do the washing up tonight is $\frac{3}{8}$.
   What is the probability that she will not do the washing up?
2. The probability that Chris will get a motorcycle for his sixteenth birthday is 0.001.
   What is the probability that he will not get a motorcycle?
3. In a game of tennis you can only win or lose. A draw is not possible.
   Qasim plays Roshan at tennis.
   The probability that Qasim wins is 0.7.
   What is the probability that Roshan wins?
4. James has ten T-shirts.
   Four of them have pictures on.
   James takes one of the T-shirts at random.
   What is the probability that it does not have a picture?
5. A bag contains red, white and blue balls.
   I pick one ball out of the bag at random.
   The probability that I pick a red ball is $\frac{1}{12}$.
   The probability that I pick a white ball is $\frac{7}{12}$.
   What is the probability that I pick a blue ball?
6. A shop has black, grey and blue dresses on a rail.
   Mina picks one at random.
   The probability that she picks a grey dress is 0.2.
   The probability that she picks a black dress is 0.1.
   What is the probability that she picks a blue dress?
7. When she goes to the gym, Yasmin runs, cycles or rows first.
   The probability that she runs first is $\frac{7}{20}$.
   The probability that she cycles first is $\frac{9}{20}$.
   What is the probability that she rows first?
8. The table shows the probability of getting some of the scores when a biased spinner numbered 1 to 6 is spun.

Score	1	2	3	4	5	6
Probability	0.27	0.16	0.14		0.22	0.1

   What is the probability of getting 4?

> **Note**
> A spinner is **biased** if the outcomes are *not* all equally likely. It is **fair** if all outcomes are equally likely.

Estimating from a population

If you know the probability of an event happening, then you can use this probability to make an estimate of the number of times this event will occur.

Example 44.3

Question

Saira goes to work by bus each morning.

The probability that the bus is late is 0.15.

Estimate the number of times out of 40 mornings that the bus is expected to be late.

Solution

Number of times late = 0.15 × 40 = 6

The bus is expected to be late on approximately six out of the 40 mornings.

Note

This does not mean that the bus will definitely be late six times.

It is an estimate based on the probability that has been given.

Exercise 44.3

1. A fair coin is tossed 250 times.
   Estimate the number of times it will land on heads.
2. A fair spinner numbered 1 to 6 is spun 300 times.
   Estimate the number of times it will land on 6.
3. The probability that a soccer team will win a match is 0.6.
   How many of their next 30 matches are they expected to win?
4. A factory makes plates.
   The probability that a plate is faulty is $\frac{1}{20}$.
   In a batch of 1000 plates, how many would you expect to be faulty?
5. A biased spinner is numbered from 1 to 5.
   The probability that it lands on 3 is 0.4.
   It is spun 50 times.
   Estimate the number of times it will land on 3.

Relative frequency and probability

It is not always possible to find probabilities by looking at equally likely outcomes.

For example, supposing you have to work out the probability of scoring a 6 with a spinner that is numbered 1 to 6 that may be biased, the probability of a driver having a car accident or the probability that a person will visit a certain shop.

For this type of event, you need to set up an experiment, carry out a survey or use past results.

Take the example of scoring a 6 with a spinner numbered 1 to 6 that may be biased.

For a fair (unbiased) spinner, the probability of getting a 6 is $\frac{1}{6} = 0.166\ldots$
= 0.17 approximately.

Relative frequency and probability

You can record the number of 6s you get when you spin the spinner a number of times, and use the results to decide whether the spinner may be biased.

The proportion of times a 6 occurs is known as the **relative frequency**.

$$\text{Relative frequency} = \frac{\text{number of times an outcome occurs}}{\text{total number of trials}}$$

The table shows the results of an experiment.

Number of trials	Number of 6s	Relative frequency
10	4	$\frac{4}{10} = 0.4$
50	18	$\frac{18}{50} = 0.36$
100	35	$\frac{35}{100} = 0.35$
500	180	$\frac{180}{500} = 0.36$

The relative frequency gives an estimate of the probability of scoring a 6.

The relative frequency changes depending on the number of trials.

The greater the number of trials, the better the estimate of the probability will be.

The results of this experiment suggest that the spinner is biased.

An estimate for the probability of scoring a 6 with this spinner is 0.36.

Example 44.4

Question

Farid carries out a survey on the colours of the cars passing his school. The table shows his results.

Colour	Black	Red	Blue	White	Green	Other	Total
Number of cars	51	85	64	55	71	90	416

Use his results to estimate the probability that the next car that passes will be

a red   b not red.

Solution

Relative frequency can be used as an estimate of probability.

a Relative frequency of red cars $= \frac{\text{number of red cars}}{\text{total number of cars}} = \frac{85}{416}$

Estimate of probability $= \frac{85}{416} = 0.204$ to 3 decimal places.

b P(not red) = 1 − P(red)

Estimate of P(not red) $= 1 - \frac{85}{416} = \frac{331}{416} = 0.796$ to 3 decimal places.

Note

The relative frequency can be written as either a fraction or a decimal.

44 PROBABILITY

Exercise 44.4

1. Use the figures from Farid's survey in Example 44.4 to estimate the probability that the next car will be
   a. blue
   b. black or white.
   Give your answers correct to 3 decimal places.

2. Solomon has a spinner in the shape of a pentagon, with sides labelled 1, 2, 3, 4, 5. Solomon spins the spinner 500 times.
   His results are shown in the table.

Number on spinner	Frequency
1	102
2	103
3	98
4	96
5	101

   a. What is the relative frequency of scoring
      i. 2
      ii. 4?
   b. Do the results suggest that Solomon's spinner is fair? Explain your answer.

3. In an experiment, a drawing pin is thrown.
   It can land either point up or point down.
   It lands point up 87 times in 210 throws.
   Use these figures to estimate the probability that, the next time it is thrown, it will land
   a. point up
   b. point down.
   Give your answers correct to 2 decimal places.

4. The table shows the results of a survey on the type of detergent households use to do their washing.

Type of detergent	Number of households
Liquid	120
Powder	233
Tablets	85

   Use these figures to estimate, correct to 2 decimal places, the probability that the next household surveyed will use
   a. liquid
   b. tablets.

5. Dina surveys 80 students from her school to find out how they travel to school.
   Of these students, 28 travel by bus.
   a. Estimate the probability that a student chosen at random travels to school by bus.
   b. There are 1200 students in the school.
      Estimate the number of students in the school who travel by bus.

6 Pavan tested a four-sided spinner.
   He spun it 120 times and it landed on red 24 times.
   a Estimate the probability that the spinner lands on red.
   b How many times would he expect it to land on red if he spun it 300 times?

The probability of combined events

To find the probability of more than one event happening, you need to make sure that you consider all of the possible outcomes.

One way of finding all of the outcomes is to use a **sample space diagram**.

If two fair coins are tossed, the first coin can show either a head or a tail and the second coin can show either a head or a tail.

The sample space diagram below shows all of the possible outcomes.

	1st coin	
	H	T
H (2nd coin)	x	x
T	x	x

This shows that there are four possible outcomes: HH, HT, TH, TT.

These are all equally likely.

The probability of each event can be found from the sample space diagram.

P(2 heads) = $\frac{1}{4}$

P(1 head, 1 tail) = $\frac{2}{4} = \frac{1}{2}$

P(2 tails) = $\frac{1}{4}$

When the events are numerical – such as when two fair spinners numbered 1 to 6 are spun – it can be helpful to show the numerical outcomes on the sample space diagram rather than using a cross.

The sample space diagram here shows the outcomes for spinning two fair spinners numbered 1 to 6 when the two scores are added together.

This shows that there are 36 possible outcomes.

Each outcome is equally likely.

The diagram can be used to find the probability of different events.

There are five different outcomes that give a score of 8, so P(8) = $\frac{5}{36}$.

There are two different outcomes that give a score of 11, so P(11) = $\frac{2}{36}$.

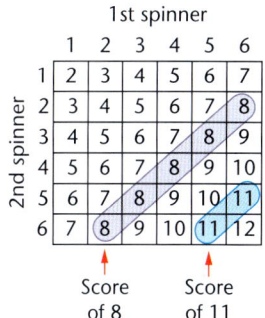

Suppose that you want to find the probability of scoring 8 **or** 11.

From the sample space diagram, you can see that there are a total of seven outcomes that give a score of 8 or 11, so P(8 **or** 11) = $\frac{7}{36}$.

You can also see that P(8) + P(11) = $\frac{5}{36} + \frac{2}{36} = \frac{7}{36}$.

So P(8 **or** 11) = P(8) + P(11).

When two events cannot happen at the same time they are known as **mutually exclusive events**.

The events 'scoring 8' and 'scoring 11' are mutually exclusive because they cannot both happen at the same time.

> If events A and B are mutually exclusive, then P(A **or** B) = P(A) + P(B).

1st spinner

2nd spinner	1	2	3	4	5	6
1	2	3	4	5	6	7
2	3	4	5	6	7	8
3	4	5	6	7	8	9
4	5	6	7	8	9	10
5	6	7	8	9	10	11
6	7	8	9	10	11	12

Score of 8 — 4 on 1st spinner

This rule does not apply if the events can happen at the same time, for example 'scoring 8' and 'scoring a 4 on the first spinner'.

$P(8) = \frac{5}{36}$

$P(4 \text{ on first spinner}) = \frac{6}{36}$

From the sample space diagram, you can see that there are 10 outcomes that give a score of 8 **or** score a 4 on the first spinner, so

P(8 **or** 4 on first spinner) = $\frac{10}{36}$.

P(8 **or** 4 on first spinner) ≠ P(8) + P(4 on first spinner).

This is because scoring a double 4 fits both events.

These two events are not mutually exclusive, so the probabilities cannot be added.

Example 44.5

Question

Ravi spins two fair spinners numbered 1 to 6 and adds the scores.

Find the probability that he scores

a 12

b less than 6

c 10 or 11.

Solution

The outcomes can be seen in the sample space diagram shown above.

a There is only one way to score 12, so $P(12) = \frac{1}{36}$.

b It can be seen from the diagram that there are 10 outcomes with a score of less than 6.

P(less than 6) = $\frac{10}{36} = \frac{5}{18}$ Write the fraction in its simplest form.

c It can be seen from the diagram that there are five outcomes with a score of 10 or 11.

P(10 or 11) = $\frac{5}{36}$

Note that $P(10) = \frac{3}{36}$ and $P(11) = \frac{2}{36}$.

Scoring 10 and 11 are mutually exclusive,
so P(10 or 11) = P(10) + P(11)

$= \frac{3}{36} + \frac{2}{36} = \frac{5}{36}$.

Exercise 44.5

1. **a** Copy and complete this table to show the total score when a fair spinner numbered 1 to 6 is spun and a fair spinner numbered from 1 to 4 is spun.
   b What is the probability of getting a total of 9?
   c What is the probability of getting a total of less than 6?
   d What is the probability of getting a total that is an even number?

   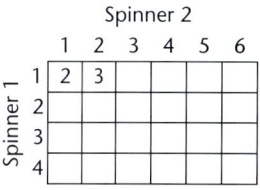

2. Fatima spins a fair spinner that is numbered 1 to 6 and tosses a coin.
   a Draw a sample space diagram to show all the possible outcomes.
   b Find the probability that Fatima scores
     i a head and a 6   **ii** a tail and an odd number.

3. A fair spinner is numbered from 1 to 4.
   The spinner is spun twice and the score is the *product* of the two numbers shown.
   a Draw a sample space diagram to show all of the possible outcomes.
   b Find the probability that the score is
     i 4   **ii** more than 8   **iii** an odd number.

4. A fair spinner is numbered 2, 4, 6, 8, 10.
   The spinner is spun twice and the score is the sum of the two numbers shown.
   a Draw a sample space diagram to show all the possible outcomes.
   b What is the probability that the score is greater than 10?
   c What is the probability that the score is 8 or 14?

5. The Venn diagram shows the number of students studying history (H) and the number of students studying geography (G) in a class.

   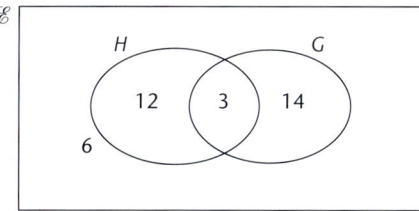

   One student is chosen at random.
   Find the probability that this student
   a studies history
   b studies both history and geography
   c doesn't study history or geography.

   > **Note**
   > Venn diagrams can also be used to represent combined events. This question could be written using set notation. Part **b** asks for $P(H \cap G)$. Part **c** asks for $1 - P(H \cup G)$.
   >
   > For more detail about Venn diagrams and set notation, see Chapter 2.

6. The Venn diagram shows the elements of the universal set and its subsets, A and B.

   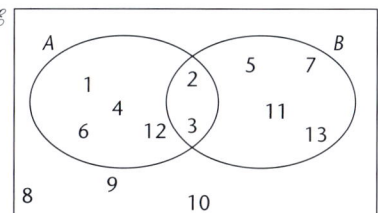

   One element is chosen at random. Find
   a $P(A)$,   **b** $P(A \cap B)$,   **c** $P(B')$.

44 PROBABILITY

Independent events

When you toss a coin twice, the outcome of the second toss is not affected by the outcome of the first toss.

Two events are said to be **independent events** when the outcome of the second event is not affected by the outcome of the first.

> If events A and B are independent, $P(A \text{ and } B) = P(A) \times P(B)$.

The probability of getting a head when you toss a fair coin is $\frac{1}{2}$.

You have seen from the sample space diagram that the probability of getting two heads when you toss a coin twice is $\frac{1}{4}$.

$P(\text{head and head}) = \frac{1}{2} \times \frac{1}{2} = \frac{1}{4}$

The result for independent events can be extended to more than two events. The probability of all of the events happening is found by multiplying together the probabilities of each individual event.

Example 44.6

Question

a The probability that Imran's bus to school is late is 0.2.
  Assuming that the events are independent, find the probability that the bus is late
  i two mornings in a row
  ii three mornings in a row.
b There are four green balls and one blue ball in a bag.
  Anna selects a ball, notes its colour and replaces it.
  She then selects another ball.
  What is the probability that the first ball is green and the second ball is blue?

Solution

a The events are independent, so the probabilities are multiplied.
  i P(late two mornings) = P(late) × P(late) = 0.2 × 0.2 = 0.04
  ii P(late three mornings) = P(late) × P(late) × P(late) = 0.2 × 0.2 × 0.2 = 0.008
b The first ball is replaced in the bag before the second is taken, so the events are independent.
  $P(\text{green}) = \frac{4}{5}$
  $P(\text{blue}) = \frac{1}{5}$
  P(first green **and** second blue) = P(green) × P(blue) = $\frac{4}{5} \times \frac{1}{5} = \frac{4}{25}$

Note
The larger the number of repeats of the event, the smaller the probability that it occurs.

Exercise 44.6

1. There are five green balls, three red balls and two yellow balls in a bag.
   Jai chooses a ball at random, notes its colour and puts it back in the bag.
   He then does this a second time.
   Find the probability that both Jai's choices are red.
2. The probability that I cycle to work is 0.4.
   The probability that I take the bus to work is 0.6.
   Assuming the events are independent, what is the probability that I cycle on Monday and take the bus on Tuesday?
3. a What is the probability that I get a multiple of 3 when I spin a fair spinner numbered from 1 to 6?
   b If I spin the spinner twice, what is the probability that both spins give a multiple of 3?
4. There is an equal likelihood that someone is born on any day of the week. What is the probability that Gary and Rushna were both born on a Monday?
5. The probability that Hira wins the 100-metre race is 0.4.
   The probability that Rabia wins the 400-metre race is 0.3.
   What is the probability that
   a both girls win their race
   b neither of them wins their race?
6. Salma spins this five-sided spinner three times.
   What is the probability that all Salma's spins land on 1?

Tree diagrams for combined events

You have seen that sample space diagrams are one way of showing all of the outcomes for two combined events.

A tree diagram can be used to show the outcomes, and their probabilities, for two or more events.

The tree diagram for tossing two coins looks like this.

First coin	Second coin	Outcome	Probability
H (½)	H (½)	HH	$\frac{1}{2} \times \frac{1}{2} = \frac{1}{4}$
	T (½)	HT	$\frac{1}{2} \times \frac{1}{2} = \frac{1}{4}$
T (½)	H (½)	TH	$\frac{1}{2} \times \frac{1}{2} = \frac{1}{4}$
	T (½)	TT	$\frac{1}{2} \times \frac{1}{2} = \frac{1}{4}$

To find the probability of one outcome, *multiply* the probabilities on the branches.

To find the probability of more than one outcome, *add* the probabilities of each outcome.

44 PROBABILITY

Notice that on each pair of branches the probabilities add up to 1.

Also the probabilities of all the four possible outcomes add up to 1.

If there are more than two combined events, further sets of branches can be added to the tree.

Example 44.7

Question

There are six red balls and four black balls in a bag.
Gita selects a ball, notes its colour and replaces it.
She then selects another ball.
What is the probability that Gita selects
a two red balls
b one of each colour?

Solution

A tree diagram can be drawn to show this information.

First ball	Second ball	Outcome	Probability
0.6 R	0.6 R	RR	$0.6 \times 0.6 = 0.36$
	0.4 B	RB	$0.6 \times 0.4 = 0.24$
0.4 B	0.6 R	BR	$0.4 \times 0.6 = 0.24$
	0.4 B	BB	$0.4 \times 0.4 = 0.16$

a $P(RR) = 0.6 \times 0.6 = 0.36$

b For one of each colour, the outcome is either RB or BR.
   P(one of each colour) = P(RB) + P(BR)
   $= (0.6 \times 0.4) + (0.4 \times 0.6)$
   $= 0.24 + 0.24 = 0.48$

Note
Some questions may not ask you to draw a tree diagram. However, drawing a tree diagram will help you to make sure that you have considered all of the possible outcomes.
You may be able to answer some probability questions without a tree diagram.

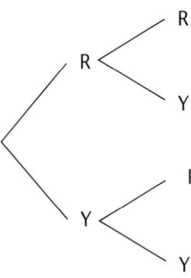

 Exercise 44.7

1 There are seven red balls and three yellow balls in a bag.
   Lee chooses a ball at random, notes its colour and replaces it.
   He then chooses another.
   Copy and complete the tree diagram to show Lee's choices.
   What is the probability that Lee chooses
   a two red balls
   b a red ball and then a yellow ball
   c a yellow ball and then a red ball
   d a red ball and a yellow ball in either order?

2 On any day, the probability that Sarah's bus is late is 0.2.
   a Copy the tree diagram and complete it for two days.

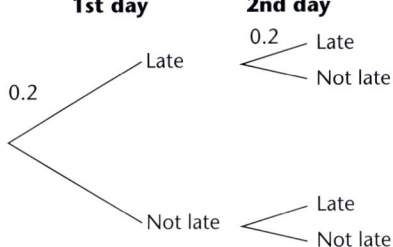

   b Calculate the probability that Sarah's bus is
      i late on both days   ii late on just one of the two days.
3 In an experiment, a drawing pin falls point up 300 times in 500 throws.
   a Write down, as a fraction in its simplest form, the probability of the
     pin landing point up.
   b Draw a tree diagram to show the result of two throws, and the pin
     landing point up or point down.
   c Find the probability that the pin lands point up on
      i both throws   ii just one of the two throws.
4 Extend the tree diagram you drew for question 3 to show the results of
  three throws.
  Find the probability that the pin lands point up on
   a all three throws   b just one of the three throws.
5 There are ten red balls, three blue balls and seven yellow balls in a bag.
  Waseem chooses a ball at random, notes its colour and replaces it.
  He then chooses another.
   a Draw a tree diagram to show the results of Waseem's choices.
   b Calculate the probability that Waseem chooses
      i   two blue balls
      ii two balls of the same colour
      iii two balls of different colours. (Look for the quick way of doing it.)
6 The probability that a train arrives on time is 0.7.
  Find the probability that on three consecutive days
   a the train arrives on time
   b the train is late on the first two days
   c the train is on time for two out of the three days.

The probability of dependent events

In the last section, all of the events were independent.

The outcome of the second event was not affected by the outcome of the first.

In some situations, the outcome of the second event *is* affected by the
first event.

In this situation, the probability of the second event depends on what
happened in the first event.

In this case the events are called **dependent events**.

44 PROBABILITY

You can solve problems involving dependent events using tree diagrams.

When events are dependent, the probabilities on the second sets of branches are not the same as the probabilities on the first set.

Example 44.8

Question
There are seven blue balls and three red balls in a bag.
A ball is selected at random and **not** replaced.
A second ball is then selected.
Find the probability that
a two blue balls are chosen
b two balls of the same colour are chosen.

Solution
A tree diagram can be used to find the probabilities of the outcomes.
The probabilities for the second ball depend on the colour of the first ball.
If the first ball was blue there are now six blue balls and three red balls left in the bag.
The probabilities are therefore $\frac{6}{9}$ and $\frac{3}{9}$, respectively.
If the first ball was red, there are now seven blue balls and two red balls left in the bag.
The probabilities are therefore $\frac{7}{9}$ and $\frac{2}{9}$, respectively.
The tree diagram looks like this.

First ball	Second ball	Outcome	Probability
$\frac{7}{10}$ B	$\frac{6}{9}$ B	BB	$\frac{7}{10} \times \frac{6}{9} = \frac{42}{90}$
	$\frac{3}{9}$ R	BR	$\frac{7}{10} \times \frac{3}{9} = \frac{21}{90}$
$\frac{3}{10}$ R	$\frac{7}{9}$ B	RB	$\frac{3}{10} \times \frac{7}{9} = \frac{21}{90}$
	$\frac{2}{9}$ R	RR	$\frac{3}{10} \times \frac{2}{9} = \frac{6}{90}$

> **Note**
> You should give your final answer as a fraction in its simplest form, but it is usually better not to simplify the fractions in the tree diagram. This is because you may need to add the probabilities and you will need them with a common denominator.

a $P(BB) = \frac{7}{10} \times \frac{6}{9}$

$= \frac{42}{90}$

$= \frac{7}{15}$

b $P(BB \text{ or } RR) = \frac{7}{10} \times \frac{6}{9} + \frac{3}{10} \times \frac{2}{9}$

$= \frac{42}{90} + \frac{6}{90}$

$= \frac{48}{90}$

$= \frac{8}{15}$

You use the tree diagram in the same way as you would for independent events.
The multiplication and addition rules still apply.
It is the probabilities that are different from the independent case.

Exercise 44.8

1. There are five blue balls in a bag and four white balls.
   If a white ball is selected first, what is the probability that the second ball is blue?

2. The probability that a school will win the first hockey match of the season is 0.6.
   If they win the first match of the season the probability that they win the second is 0.7, otherwise it is 0.4.
   a Copy and complete the tree diagram.
   b Find the probability that they win one of the first two matches only.
   c Find the probability that they win at least one of the first two matches.

   1st match
   — Win
   — Do not win

3. There are eight green balls in a bag and two white balls.
   Two balls are selected at random without replacement.
   a Draw a tree diagram to show the probabilities of the possible outcomes.
   b Find the probability of selecting two white balls.
   c Find the probability of selecting two balls of different colours.

4. Salima walks to school, cycles or goes by bus.
   The probability that she walks is 0.5.
   The probability that she cycles is 0.2.
   If she walks, the probability that she is late is 0.2. If she cycles, it is 0.1 and if she goes by bus, it is 0.4.
   Find the probability that, on any given day, she is on time for school.

5. There are three red pens and five blue pens in a pencil case.
   Basma picks two pens at random.
   Find the probability that Basma picks at least one blue pen.

Key points

- The probability of an event = $\dfrac{\text{number of outcomes that give the event}}{\text{total number of outcomes}}$.
- Probability can have a value from 0 to 1 or from 0% to 100%.
- Probability can be written as a fraction, a decimal or a percentage only.
- Probability of an event not occurring =
  1 − probability of event occurring or $P(A') = 1 - P(A)$.
- Expected frequency = probability of an event occurring × total number of observations of an event.
- Relative frequency is an estimate of probability:
  Relative frequency = $\dfrac{\text{number of times an event occurs}}{\text{total number of trials}}$
- A sample space diagram is a two-way table listing all possible outcomes of two separate events occurring.
- For mutually exclusive events A and B
  $P(A \text{ or } B) = P(A) + P(B)$.
- For independent events A and B
  $P(A \text{ and } B) = P(A) \times P(B)$.
- A probability tree diagram shows the outcomes and probabilities of two or more events.

45 CATEGORICAL, NUMERICAL AND GROUPED DATA

BY THE END OF THIS CHAPTER YOU WILL BE ABLE TO:
- classify and tabulate statistical data
- read, interpret and draw inferences from tables and statistical diagrams
- compare sets of data using tables, graphs and statistical measures
- appreciate restrictions on drawing conclusions from given data
- calculate the mean, median, mode and range for individual data and distinguish between the purposes for which these are used
- calculate an estimate of the mean for grouped discrete or grouped continuous data
- identify the modal class from a grouped frequency distribution.

CHECK YOU CAN:
- make tally charts
- add, subtract, multiply and divide numbers.

Collecting and grouping data

You may have seen headlines like these in newspapers or on television.

THE BEATLES ARE THE MOST POPULAR BAND

or

TEENAGERS VOTE DIET COKE AS THE BEST SOFT DRINK

You may not agree with either headline, but how is such information found out?

One way is to ask people questions – that is, to carry out a survey using a questionnaire and collect the answers on a data collection sheet. The answers are then analysed to give the results.

Example 45.1

Question

Jamila is asked to find out how many people there are in cars that pass the school gates each day between 8.30 a.m. and 9.00 a.m.

She writes down the number of people in the first 40 cars that pass on the first Monday.

1	2	4	1	1	3	1	2	2	3
4	2	1	1	1	2	2	3	2	1
1	1	2	2	1	3	4	1	1	1
2	1	3	3	1	1	4	2	1	2

Make a frequency table to show her results.

Solution

You make a tally chart in the usual way.

Remember to group the tally marks into fives, which are shown like this, ||||, for easier counting.

A frequency table is a tally chart with an extra column for the totals.

The frequencies are the total of the tallies.

Number of people	Tally	Frequency
1																	18
2												12
3							6
4						4

Surveys

When you are trying to find out some statistical information, it is often impossible to find this out from everyone concerned. Instead, you ask a smaller section of the population. This selection is called a **sample**.

Choosing a sample

The size of your sample is important. If your sample is too small, the results may not be very reliable. In general, the sample size needs to be at least 30. If your sample is too large, the data may take a long time to collect and analyse. You need to decide what represents a reasonable sample size for the hypothesis you are investigating.

You also need to eliminate **bias**. A biased sample is unreliable because it means that some results are more likely than others.

It is often a good idea to use a **random sample**, where every person or piece of data has an equal chance of being selected. You may want, however, to make sure that your sample has certain characteristics.

For example, when investigating the hypothesis 'boys are taller than girls', random sampling within the whole school could mean that all the boys selected happen to be in Year 7 and all the girls in Year 11: this would be a biased sample, as older children tend to be taller. So you may instead want to use random sampling to select five girls and five boys from within each year group.

Example 45.2

Question

Candace is doing a survey about school meals.

She asks every tenth person going into lunch.

Why may this not be a good method of sampling?

Solution

She will not get the opinions of those who dislike school meals and have stopped having them.

Data collection sheets

When you collect large amounts of data, you may need to group it in order to analyse it or to present it clearly. It is usually best to use equal class widths for this.

Tally charts are a good way of obtaining a frequency table, or you can use a spreadsheet or another statistics program to help you. Before you collect your data, make sure you design a suitable data collection sheet or spreadsheet.

Think first how you will show the results, as this may influence the way you collect the data.

You could quite easily collect data from every student in your class on the number of children in their family. One way to collect this data would be to ask each person individually and make a list like the following one.

45 CATEGORICAL, NUMERICAL AND GROUPED DATA

Class 10G

1	2	1	1	2	3	2	1	2	1	1	2	4	2	1
5	2	3	1	1	4	10	3	2	5	1	2	1	1	2

The data for Class 10G is shown in the completed frequency table.

Using the same method to collect data for all the students in your year could get very messy. One way to make the collection of data easier is to use a data collection sheet. Designing a table like the one here can make collecting the data easy and quick.

Number of children	Tally	Frequency
1	ЖЖ II	12
2	ЖЖ	10
3	III	3
4	II	2
5	II	2
6		0
7		0
8		0
9		0
10	I	1

Before you design a data collection sheet it is useful to know what the answers might be, although this is not always possible. For example, what would happen if you were using the tally chart shown above and someone said there were 13 children in their family?

One way to deal with this problem is to have an extra line at the end of the table to record all other responses. The table here shows how this might look.

Number of children	Tally	Frequency
1	ЖЖ II	12
2	ЖЖ	10
3	III	3
4	II	2
5	II	2
More than 5	I	1

Adding the 'More than 5' line allows all possible responses to be recorded and gets rid of some of the lines where there are no responses.

Designing a questionnaire

A **questionnaire** is often a good way of collecting data.

You need to think carefully about what information you need and how you will analyse the answers to each question. This will help you get the data in the form you need.

Here are some points to bear in mind when you design a questionnaire.

- Make the questions short, clear and relevant to your task.
- Only ask one thing at a time.
- Make sure your questions are not 'leading' questions. Leading questions show bias. They 'lead' the person answering them towards a particular answer. For example: 'Because of global warming, don't you think that families should be banned from having more than one car?'
- If you give a choice of answers, make sure there are neither too few nor too many.

Example 45.3

Question

Suggest a sensible way of asking an adult their age.

Solution

Please tick your age group:

☐ 18–25 years ☐ 26–30 years ☐ 31–40 years
☐ 41–50 years ☐ 51–60 years ☐ Over 60 years

This means that the person does not have to tell you their exact age.

Note
When you are using groups, make sure that all the possibilities are covered and that there are no overlaps.

When you have written your questionnaire, test it out on a few people.

This is called doing a **pilot survey**.

Try also to analyse the data from the pilot survey, so that you can check whether it is possible. You may then wish to reword one or two questions, regroup your data or change your method of sampling before you do the proper survey.

Any practical problems encountered in collecting the data would be described in a report of the survey.

Two-way tables

Sometimes the data collected involve two factors.

For example, data can be collected about the colour of a car and where it was made.

You can show both of these factors in a two-way table.

45 CATEGORICAL, NUMERICAL AND GROUPED DATA

Example 45.4

Question

Peter has collected data about cars in a car park.

He has recorded the colour of each car and where it was made.

The two-way table below shows some of his data.

Complete the table.

	Made in Europe	Made in Asia	Made in the USA	Total
Red	15	4	2	
Not red	83			154
Total		73		

Solution

	Made in Europe	Made in Asia	Made in the USA	Total
Red	15	4	2	21
Not red	83	69	2	154
Total	98	73	4	175

All the totals can now be completed by adding across the rows or down the columns.

73 cars are made in Asia so 73 − 4 are not red.

154 cars are not red so 154 − 83 − 69 are not red and are made in the USA.

Note

A useful check is to calculate the grand total (the number in the bottom right corner of the table) twice.

The number you get by adding down the last column should be the same as the number you get by adding across the bottom row.

Exercise 45.1

1 These are the number of letters delivered one morning to 80 houses.

0	2	5	3	1	2	0	3	4	7
2	1	0	0	1	4	6	3	2	1
5	3	2	1	1	2	0	3	2	7
1	2	0	3	5	2	0	1	1	2
2	0	1	2	0	2	1	0	0	3
8	1	6	2	0	1	1	2	3	2
0	1	2	6	2	4	1	0	3	2
2	1	4	0	1	2	3	1	4	0

a Make a frequency table.
b How many houses received three letters?

2 Kamil wants to survey public opinion about the local swimming pool.
   Give one disadvantage of each of the following sampling situations.
   a Selecting people to ring at random from the local phone directory.
   b Asking people who are shopping on Saturday morning.
3 Paul plans to ask 50 people at random how long they spent doing homework yesterday evening.
   Here is the first draft of his data collection sheet.

Time spent	Tally	Frequency
Up to 1 hour		
1–2 hours		
2–3 hours		

   Give two ways in which Paul could improve his collection sheet.
4 State what is wrong with each of these survey questions.
   Then write a better version for each of them.
   a What is your favourite sport: cricket, tennis or athletics?
   b Do you do lots of exercise each week?
   c Don't you think this government should encourage more people to recycle waste?
5 Design a questionnaire to investigate the use of a school library or resource centre.
   You need to know
   • which year group the student is in
   • how often they use the library
   • how many books they usually borrow on each visit.
6 Here is a two-way table showing the results of a car survey.
   a Copy and complete the table.

	Japanese	Not Japanese	Total
Red	35	65	
Not red	72	438	
Total			

   b How many cars were surveyed?
   c How many Japanese cars were in the survey?
   d How many of the Japanese cars were not red?
   e How many red cars were in the survey?
7 A drugs company compared a new type of drug for hay fever with an existing drug.
   The two-way table shows the results of the trial.
   a Copy and complete the table.

	Existing drug	New drug	Total
Symptoms eased	700	550	
No change in symptoms	350	250	
Total			

   b How many people took part in the trial?
   c How many people using the new drug had their symptoms eased?

8 At the indoor athletics championships, the USA, Germany and China won the most medals.

  a Copy and complete the table.

	Gold	Silver	Bronze	Total
USA	31		10	
Germany	18	16		43
China		9	11	42
Total		43		

  b Which country won the most gold medals?
  c Which country won the most bronze medals?

Averages and range

The mode and the median

The **mode** and the **median** are two different types of 'average'.

> The mode is the most common number in a set of numbers.

A set of data may have more than one mode or it may have no mode.

> The median is the middle number of a set of numbers arranged in order.

In a list of n numbers, the middle number is $\frac{1}{2}(n + 1)$.

Example 45.5

Question
Here is a list of the weights of people in an exercise class.
73 kg, 58 kg, 61 kg, 43 kg, 81 kg, 53 kg, 73 kg, 70 kg, 73 kg, 62 kg, 60 kg, 85 kg
The instructor wants to know the average weight for the group.

Solution
One way is to find the most common weight, the mode.
The mode is 73 kg.
Another way is to put the weights in order and find the middle weight, the median.
There are 12 numbers.
$\frac{1}{2}(n + 1) = \frac{1}{2}(12 + 1) = 6\frac{1}{2}$

The middle numbers are the sixth and seventh.
The median is halfway between the two.
It is found by adding the values in the two middle positions and dividing by 2.
43 kg, 53 kg, 58 kg, 60 kg, 61 kg, 62 kg, 70 kg, 73 kg, 73 kg, 73 kg, 81 kg, 85 kg
The median = $\frac{62 + 70}{2}$ = 66 kg.

Exercise 45.2

1. Find the median mark for each of these tests.
   Remember to write the marks in order, smallest to largest, first.
   a 3 5 6 7 9
   b 8 9 4 3 7 3 1 7

2. Twelve people have their handspan measured.
   These are the results.
   225 mm   216 mm   188 mm   212 mm   205 mm   198 mm
   194 mm   180 mm   194 mm   198 mm   200 mm   194 mm
   a How many of the group had a handspan greater than 200 mm?
   b What are the mode and the median?

3. These are the earnings of ten workers in a small company.
   $10 000 $10 000 $10 000 $10 000 $13 000
   $13 000 $15 000 $21 000 $23 000 $70 000
   a Find the mode and the median for the data.
   b Which of these averages do you think gives the better impression of the average pay?
      Explain your answer.

4. Here are the recent scores of two basketball players.
   Harvey   24   0   0   32   0   17   29   19   23
   Nick   12   9   3   16   8   6   9   0   11   13
   Work out the mode and the median for each player.

   Who would you pick for the next match?

   Explain your answer.

Mean and range

The **mean** is another type of average.

$$\text{Mean} = \frac{\text{sum of data values}}{\text{number of data values}}$$

The **range** is a measure of how spread out the data is.

Range = largest data value − smallest data value

Example 45.6

Question

Madea and Parveen play cricket.
These are the number of runs they score in five innings.
Madea   7   8   5   4   7
Parveen   10   10   2   1   6

a Calculate the mean number of runs per innings for each player.
b Who is the more consistent batter?

45 CATEGORICAL, NUMERICAL AND GROUPED DATA

> **Solution**
> a Mean = $\dfrac{\text{total number of runs}}{\text{number of innings}}$
>
> Mean for Madea = 31 ÷ 5 = 6.2
> Mean for Parveen = 29 ÷ 5 = 5.8
>
> b Use the range to determine who has more consistent scores.
> Range = largest data value − smallest data value
> Range for Madea = 8 − 4 = 4
> Range for Parveen = 10 − 1 = 9
> Madea has a smaller range so she is the more consistent batter.

Exercise 45.3

1 Find the mean and range for each of these sets of data.
   a   3   12   4   6   8   5   4
   b   12   1   10   1   9   3   4   9   7   9
   c   15   17   12   29   21   18   31   22
   d   313 550 711 365 165 439 921 264

2 Copy and complete the table for each of these sets of data.
   Data set A: 1, 1, 2, 2, 3, 3, 3, 4, 5, 6, 7
   Data set B: 1, 1, 2, 2, 3, 3, 3, 4, 5, 6, 7, 1, 1, 2, 2, 3, 3, 3, 4, 5, 6, 7
   Data set C: 2, 2, 4, 4, 6, 6, 6, 8, 10, 12, 14

	Data set A	Data set B	Data set C
Range			
Mean			

Write down anything that you notice about these results.

3 a Find the mean of this set of data.
      1, 2, 2, 3, 7
   b Write down the mean and range of each of these sets of data.
      i   1, 2, 2, 3, 7, 1, 2, 2, 3, 7
      ii 10, 20, 20, 30, 70
      iii 110, 120, 120, 130, 170

4 These are the salaries of ten workers in a small company.
   $10 000 $10 000 $10 000 $10 000 $13 000
   $13 000 $15 000 $21 000 $23 000 $70 000
   Find the mean and range for the data.

5 These are marks scored in a test.
   20   16   18   17   16   18   14   13
   18   18   15   18   19   9   12   13
   a Find the mean.
   b Find the range.

6 Here are the weekly wages of five people who work in a shop.
$157 $185 $189 $177 $171

   a **i** Find the range of these wages.
      ii Find the mean of these wages.
   b A new employee starts who earns $249.
      i Find the new range of these wages.
      ii Find the new mean of these wages.

7 A group of 12 people have each measured the length of their stride to the nearest centimetre.
These are the results.
87   78   91   73   84   98   77   82   83   73   89   81
Work out
   a the mean length of stride.
   b the range of these lengths.

8 **a** Nine students scored a mean of 7 marks for their last maths test.
   What was their total mark?
   b A tenth student scored 10.
   What was the mean mark of all ten students?

Which average to use when comparing data

When you are comparing sets of data, you need to compare the size of the data values and how spread out the data values are.

You compare the size of the data values using an average. You have now met three types of average: the mode, the median and the mean. You can use any of the three to compare sets of data, but in some circumstances one is better than the others.

The **mean** has the advantage that it uses every data value but it can be misleading if there are one or two very high or very low data values.

The **median** is not affected by a few high or low values.

The **mode** has the advantage that it is often quick and easy to find, but can be unreliable, particularly with a small sample of numbers.

Also, some sets of data will not have a mode and some will have more than one. However, as you will see in Example 45.8 on pages 472 and 473, in certain circumstances it can be very useful.

The range is a measure of how spread out the data values are.

The greater the range, the more spread out the data values are. A greater range tells you that the data values are less consistent.

The smaller the range, the less spread out the data values are. A smaller range tells you that the data values are more consistent.

You will meet another measure of spread in Chapter 46.

45 CATEGORICAL, NUMERICAL AND GROUPED DATA

> **Note**
> Always state your answer in the context of the question when comparing two sets of data.
>
> So, in Example 45.7, don't just state that the potatoes from the market had a higher mean, state that this means that they are heavier on average.
>
> Also don't just state that the range is smaller for the potatoes from the market, state that this means they are more consistently the same mass.

Example 45.7

Question

These are the masses, in grams, of 12 potatoes in a bag bought from a supermarket.

200 410 300 250 280 290 420 380 310 280 210 320

These are the masses, in grams, of 12 potatoes in a bag bought from a local market.

230 400 350 360 270 390 410 370 360 380 410 380

Compare the masses.

Solution

Arrange the data in order and calculate the averages and range.

Supermarket potatoes:

200 210 250 280 280 290 300 310 320 380 410 420

The mode is 280 g.

The median is halfway between 290 and 300, so it is 295 g.

Mean = 3650 ÷ 12 = 304 g (to the nearest gram)

Range = 420 − 200 = 220 g

Market potatoes:

230 270 350 360 360 370 380 380 390 400 410 410

There are three modes, 360 g, 380 g and 410 g.

The median is halfway between 370 and 380, so it is 375 g.

Mean = 4310 ÷ 12 = 359 g (to the nearest gram)

Range = 410 − 230 = 180 g

The mean shows that the potatoes from the market are heavier on average.

The range shows that the potatoes from the market are more consistently the same mass than the ones from the supermarket.

In Example 45.7, the mode was not useful. Either the median or the mean could be used to compare the masses, but there were no extreme values affecting the mean and the mean uses all the data. Also, the total quantity of the potatoes is important. The bags of potatoes from the market weigh more in total.

Example 45.8

Question

At the local shoe shop 15 pairs of ladies' shoes were sold.

These were the sizes sold. (They do not sell half sizes.)

3, 5, 6, 3, 3, 4, 4, 5, 5, 5, 5, 7, 3, 4 and 4

a Work out the mean, median and mode for these sizes.

b Which measure of average is the most useful in this case, and why?

Solution

a Arrange the data in order and calculate the averages and range.

3, 3, 3, 3, 4, 4, 4, 4, 5, 5, 5, 5, 5, 6, 7

Mean = 66 ÷ 15 = 4.4

The median is the 8th value.

Median = 4

Mode = 5

b The most useful measure in this case is the mode, because it can be used to find out which size of shoe to order most of.

Example 45.9

Question

The label on a matchbox states, 'Average contents 50 matches'.

A survey of the number of matches in ten matchboxes produced the following data.

45   51   48   47   46   47   49   50   52   47

a Find the mode, median, mean and range of the contents.
b What do these results tell you about the statement on the label?

Solution

a Arrange the data in order and calculate the averages and range.

45   46   47   47   47   48   49   50   51   52

Mode = 47

The median is halfway between the 5th and 6th values.

Median = $\frac{47 + 48}{2}$ = 47.5

Mean = $\frac{482}{10}$ = 48.2

Range = 52 − 45 = 7

b The three measures of average – the mode, the median and the mean – are all below 50, so the statement on the label is untrue.

The range shows that the number of matches varies greatly from one box to another.

Exercise 45.4

1 The number of points scored in eight games by two players, Carl and Adam, in a basketball team are shown in the table.

Carl	17	19	21	24	16	18	14	23
Adam	19	20	14	25	26	29	12	13

a Work out the mean and range for each player.
b Compare the number of points scored by the two players.

2 The number of hours of sunshine in June during the last eight years at two holiday resorts, A and B, are given in the table.

| Resort A | 180 | 170 | 109 | 171 | 165 | 190 | 162 | 173 |
| Resort B | 170 | 168 | 120 | 172 | 158 | 178 | 157 | 169 |

   a Work out the mean, median and range for each resort.
   b Compare the amount of sunshine at the two resorts.

3 A shopkeeper records the sizes of the pairs of jeans sold.
   These are the sizes sold.
   34, 28, 36, 30, 32, 30, 28, 32, 30, 36,
   34, 30, 26, 30, 32, 34, 36, 28, 34, 32
   a Work out the mean, median and mode for these figures.
   b Which measure of average is most useful when the shopkeeper wants to order more jeans?

4 At a gymnastics competition, Olga was given these marks by the ten judges.
   8.7, 8.9, 9.3, 9.1, 8.4, 8.1, 8.9, 9.1, 9.2, 9.1
   a Work out the mean, the median and the range for Olga.
   b Janice had a mean of 9.1, a median of 8.8 and a range of 1.6.
      Which is the best average to use to compare the two girls?
      Give a reason for your answer.

5 In an ice-skating competition, Jamilla was given the following marks by the ten judges.
   9.1, 9.0, 8.7, 9.5, 8.8, 8.2, 7.9, 9.2, 8.9, 8.7
   The highest and lowest marks are ignored and the mean of the remaining marks is her score.
   a Work out Jamilla's score.
   b Is this better or worse than calculating the mean mark from all the judges?

6 At Prothero fabric factory, the salaries have a mean of $25 000, a median of $20 000 and a range of $21 000.
   At Jaline fabric factory, the salaries have a mean of $23 250, a median of $20 000 and a range of $10 000.
   Compare the salaries at the two factories.

Working with larger data sets

When working with larger data sets, it is often easier to see any patterns in the data if it is displayed in a frequency table.

Number of goals	Frequency
0	4
1	6
2	4
3	3
4	2
5	0
6	1

For example, this is a list of goals scored in 20 matches.

1 1 3 2 0 0 1 4 0 2
2 0 6 3 4 1 1 3 2 1

This is the same data displayed in a frequency table.

From the table, it is easy to identify the mode. It is the number of goals with the greatest frequency. Here the mode is 1 goal.

The median is the middle value. In this case, it is halfway between the 10th and 11th values.

Looking at the frequency column in the table, the 10th value is 1 and the 11th value is 2.

The median is, therefore, 1.5.

The table can also be used to calculate the mean.

There are: four matches with 0 goals $4 \times 0 = 0$ goals

six matches with 1 goal $6 \times 1 = 6$ goals

four matches with 2 goals $4 \times 2 = 8$ goals

and so on.

To find the total number of goals scored altogether, multiply each number of goals by its frequency and then add the results. You can add an extra column to your table to help you work out the values multiplied by their frequencies.

Number of goals	Frequency	Number of goals × frequency
0	4	0
1	6	6
2	4	8
3	3	9
4	2	8
5	0	0
6	1	6
Totals	20	37

Then, dividing by the total number of matches (20) gives the mean.

Mean = $37 \div 20 = 1.85$ goals

Example 45.10

Question

Work out the mean, mode and range for the number of children in the houses in Berry Road, listed in this table.

Number of children (c)	Frequency (number of houses)	Total number of children (c × frequency)
0	6	0
1	4	4
2	5	10
3	7	21
4	1	4
5	2	10
Totals	25	

Solution

Mean = $49 \div 25 = 1.96$ children

Mode = 3 children

Range = $5 - 0 = 5$ children

45 CATEGORICAL, NUMERICAL AND GROUPED DATA

Exercise 45.5

1. A taxi company did a survey of the number of passengers carried in their taxis on each trip in one shift.
   Calculate the mean number of passengers per taxi.

Number of passengers in a taxi	Frequency
1	84
2	63
3	34
4	15
5	4
Total	200

2. The table shows the number of minutes late that the students in a class arrived for registration one Monday morning.

Number of minutes late	Frequency
1	0
2	1
3	5
4	7
5	11
6	2
7	1
8	0
9	1
10	0

   a What was the modal number of minutes late?
   b What is the mean number of minutes late?

3. A soccer team recorded the number of goals it scored each match one season.

Number of goals	Frequency
0	7
1	10
2	9
3	4

   a How many matches were played?
   b What was the total number of goals scored?
   c What was the mean number of goals scored?
   d What was the median number of goals scored?

4. Bus tickets cost 50c, $1.00, $1.50 or $2.00 depending on the length of the journey.
   The frequency table shows the numbers of tickets sold on one Friday.
   Calculate the mean fare paid on that Friday.

Price of ticket ($)	Number of tickets
0.50	140
1.00	207
1.50	96
2.00	57

5 800 people were asked how many cups of coffee they had bought one week. Calculate the mean, median and mode of the number of cups of coffee bought.

Number of coffees	Frequency
0	20
1	24
2	35
3	26
4	28
5	49
6	97
7	126
8	106
9	54
10	83
11	38
12	67
13	21
14	26

Working with grouped and continuous data

So far in this chapter, most of the data has been whole numbers only. The information is obtained through counting, not measuring.

Data obtained by counting is called **discrete data**.

Data obtained by measuring is called **continuous data**.

This is because you have the whole range of measurement between any two values.

For example, a length L given as 18 cm to the nearest centimetre means $17.5 \leqslant L < 18.5$.

Any length between these values will be recorded as 18 cm.

Often, however, the groups are larger, to make handling the data easier.

Discrete data can also be grouped.

For example, if you are looking at examination results in an examination marked out of 100, it is difficult to draw any conclusions from individual scores, so it is often better to group the marks.

For example, 1 to 10 marks, 11 to 20 marks, 21 to 30 marks, etc.

When you group data, some of the detail is lost.

45 CATEGORICAL, NUMERICAL AND GROUPED DATA

For example, a frequency table may show that one student scored between 1 and 10 marks. You can no longer tell whether that student scored 1 mark, or 10 marks or any of the scores in between.

For this reason, you cannot obtain exact values for the mode, median, mean or range.

The **modal class** may be found when data are given as a table or bar graph. It is the class with the highest frequency.

Similarly, you cannot find the median from grouped data, but you can find the class which contains the median.

You can use the midpoints of the largest and smallest classes to estimate the range.

Similarly, you can use the midpoint of the classes to calculate an estimate of the mean.

Example 45.11

Question

The frequency table shows some information about the heights of a year group in a school.

Height h (cm)	Frequency
$155 < h \leqslant 160$	2
$160 < h \leqslant 165$	6
$165 < h \leqslant 170$	18
$170 < h \leqslant 175$	25
$175 < h \leqslant 180$	9
$180 < h \leqslant 185$	4
$185 < h \leqslant 190$	1

Note

Add two columns to the frequency table to help you work out the mean: one column for the midpoint of each class and one for the midpoint multiplied by the frequency of the class.

a State the modal class.
b State the class which contains the median.
c Estimate the range.
d Calculate an estimate of the mean.

Solution

a The highest frequency is 25 so the modal class is $170\,\text{cm} < h \leqslant 175\,\text{cm}$.
b The total frequency is 65 so the median is the 33rd data value.
Adding the frequencies $2 + 6 = 8$, $8 + 18 = 26$, so 26 are less than 170 cm.
$26 + 25 = 51$ so all of the 27th to 51st heights are in the $170 < h \leqslant 175$ class.
So the 33rd height is in the $170\,\text{cm} < h \leqslant 175\,\text{cm}$ class.
c Using the midpoints of the first and last classes,
estimate of the range = $187.5 - 157.5 = 30$

d

Height h (cm)	Frequency	Midpoint	Midpoint × frequency
$155 < h \leq 160$	2	157.5	315
$160 < h \leq 165$	6	162.5	975
$165 < h \leq 170$	18	167.5	3 015
$170 < h \leq 175$	25	172.5	4 312.5
$175 < h \leq 180$	9	177.5	1 597.5
$180 < h \leq 185$	4	182.5	730
$185 < h \leq 190$	1	187.5	187.5
Total	65		11 132.5

Estimate of the mean = 11 132.5 ÷ 65
= 171.3 cm (to 1 decimal place)

Exercise 45.6

1 The table shows a distribution of times.
   a State the modal class.
   b State the class which contains the median.
   c Estimate the range.
   d Calculate an estimate of the mean.

Time t (seconds)	Frequency
$0 < t \leq 2$	4
$2 < t \leq 4$	6
$4 < t \leq 6$	3
$6 < t \leq 8$	2
$8 < t \leq 10$	7

2 The table shows a distribution of heights.
   a State the modal class.
   b State the class which contains the median.
   c Estimate the range.
   d Calculate an estimate of the mean.

Height h (cm)	Frequency
$50 < h \leq 60$	15
$60 < h \leq 70$	23
$70 < h \leq 80$	38
$80 < h \leq 90$	17
$90 < h \leq 100$	7

3 The table shows a distribution of lengths.
   a State the modal class.
   b State the class which contains the median.
   c Estimate the range.
   d Calculate an estimate of the mean.

Length l (metres)	Frequency
$1.0 < l \leq 1.2$	2
$1.2 < l \leq 1.4$	7
$1.4 < l \leq 1.6$	13
$1.6 < l \leq 1.8$	5
$1.8 < l \leq 2.0$	3

45 CATEGORICAL, NUMERICAL AND GROUPED DATA

4 Calculate an estimate of the mean length for this distribution.

Length y (cm)	Frequency
$10 < y \leq 20$	2
$20 < y \leq 30$	6
$30 < y \leq 40$	9
$40 < y \leq 50$	5
$50 < y \leq 60$	3

5 Calculate an estimate of the mean mass of these tomatoes.

Mass of tomato t (g)	Frequency
$35 < t \leq 40$	7
$40 < t \leq 45$	13
$45 < t \leq 50$	20
$50 < t \leq 55$	16
$55 < t \leq 60$	4

6 Calculate an estimate of the mean of these times.

Time t (seconds)	Frequency
$0 < t \leq 20$	4
$20 < t \leq 40$	9
$40 < t \leq 60$	13
$60 < t \leq 80$	8
$80 < t \leq 100$	6

7 Calculate an estimate of the mean of these heights.

Height h (metres)	Frequency
$0 < h \leq 2$	12
$2 < h \leq 4$	26
$4 < h \leq 6$	34
$6 < h \leq 8$	23
$8 < h \leq 10$	5

8 Calculate an estimate of the mean of these lengths.

Length l (cm)	Frequency
$3.0 < l \leq 3.2$	3
$3.2 < l \leq 3.4$	8
$3.4 < l \leq 3.6$	11
$3.6 < l \leq 3.8$	5
$3.8 < l \leq 4.0$	3

Working with grouped and continuous data

Grouped discrete data

One method for finding the midpoint of a class is to add the end values of the class and divide by 2.

For example, if the class is $60 \leq s < 80$, then the midpoint is $\frac{60 + 80}{2} = 70$.

Strictly, this method is not correct. The class $60 \leq s < 80$ comprises the digits 60, 61, 62, 63, …, 78, 79. The midpoint of these is $\frac{60 + 79}{2} = 69.5$.

However, using midpoints is an approximation and the error in using the simpler calculation is not significant.

Example 45.12

Question

The table shows the scores of 40 students in a history test.

Score (s)	Frequency
$0 < s \leq 20$	2
$20 < s \leq 40$	4
$40 < s \leq 60$	14
$60 < s \leq 80$	16
$80 < s \leq 100$	4
Total	40

Calculate an estimate of the mean score.

Solution

Score (s)	Frequency	Midpoint	Midpoint × frequency
$0 < s \leq 20$	2	10	20
$20 < s \leq 40$	4	30	120
$40 < s \leq 60$	14	50	700
$60 < s \leq 80$	16	70	1120
$80 < s \leq 100$	4	90	360
Total	40		2320

Mean = 2320 ÷ 40 = 58

Note

Don't forget to divide by the *total frequency*, not by the number of groups.

Exercise 45.7

1 The table shows a summary of the marks scored by students in a test.
   a State the modal class.
   b State the class which contains the median.
   c Estimate the range.
   d Calculate an estimate of the mean.

Marks (m)	Frequency
$0 < m \leq 20$	1
$20 < m \leq 40$	3
$40 < m \leq 60$	15
$60 < m \leq 80$	9
$80 < m \leq 100$	2

2 The table summarises the number of words in each sentence on one page of a book.

Number of words	Frequency
1–3	0
4–6	2
7–9	7
10–12	6
13–15	15

a State the modal class.
b State the class which contains the median.
c Estimate the range.
d Calculate an estimate of the mean.

3 The table shows a summary of the attendance at the first 30 games of a soccer club.

Attendance (a)	Frequency
$0 < a \leqslant 4000$	0
$4000 < a \leqslant 8000$	2
$8000 < a \leqslant 12\,000$	7
$12\,000 < a \leqslant 16\,000$	6
$16\,000 < a \leqslant 20\,000$	15

a State the modal class.
b State the class which contains the median.
c Estimate the range.
d Calculate an estimate of the mean.

4 The table shows a summary of the number of points scored by a school basketball player.
Calculate an estimate of the mean number of points scored.

Points (p)	Frequency
$0 < p \leqslant 5$	2
$5 < p \leqslant 10$	3
$10 < p \leqslant 15$	8
$15 < p \leqslant 20$	9
$20 < p \leqslant 25$	12
$25 < p \leqslant 30$	3

Key points
- The mode of a set of numbers is the most common value.
- The median of an ordered list of numbers is the middle value.
- Range = largest value − smallest value.
- The mean of a set of values = $\dfrac{\text{the sum of all values}}{\text{the number of values}}$.
- When comparing two sets of data, compare a measure of average and a measure of spread from each set. This is usually the mean and the range.
- For a discrete frequency table,

  Mean = $\dfrac{\text{total of (value} \times \text{frequency) products}}{\text{total of all frequencies}}$.
- Large sets of data are easier to deal with if they are grouped.
- For a grouped frequency table,

  Mean = $\dfrac{\text{total of (mid-group value} \times \text{frequency) products}}{\text{total of all frequencies}}$.
- The modal class of a grouped frequency distribution is the group with the largest frequency.

46 STATISTICAL DIAGRAMS

Bar charts and pictograms

Bar charts and **pictograms** can both be used to represent data that is divided into categories.

This bar chart shows the number of children in each of the 30 families of Class 11S.

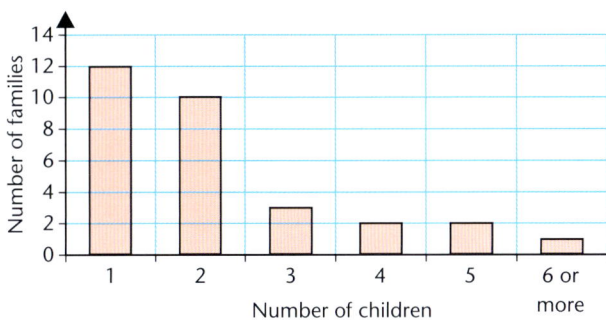

You can find the number of families with different numbers of children from the bar chart.

There are 10 families with 2 children.

There is 1 family with 6 or more children.

Example 46.1

Question

This pictogram shows people's favourite colour for their mp3 player.

 represents 2 people.

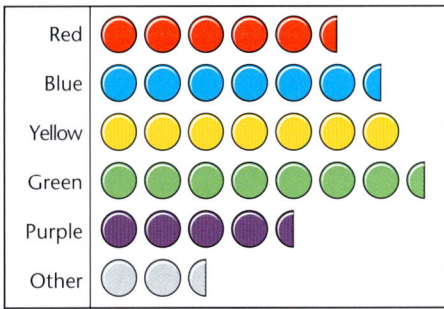

a   What does ◖ represent?
b   How many people chose blue?
c   How many people altogether were asked about their favourite colour of mp3 player?

BY THE END OF THIS CHAPTER YOU WILL BE ABLE TO:

- draw and interpret: bar charts, pie charts, pictograms, simple frequency distributions
- draw and interpret scatter diagrams
- understand what is meant by positive, negative and zero correlation
- draw by eye, interpret and use a straight line of best fit
- draw and interpret cumulative frequency tables and diagrams
- estimate and interpret the median, percentiles, quartiles and interquartile range from cumulative frequency diagrams
- draw and interpret histograms
- calculate with frequency density.

CHECK YOU CAN:

- calculate with fractions and percentages
- draw simple bar charts and pictograms
- measure angles
- understand grouped data
- find the median and the mode.

Photocopying is prohibited

46 STATISTICAL DIAGRAMS

Solution

a ● represents 2 people.

◗ is half a symbol, so it represents 1 person.

b There are $6\frac{1}{2}$ symbols for blue.
Each symbol represents 2 people, so 13 people chose blue.

c There are a total of $33\frac{1}{2}$ symbols in the pictogram, so 67 people were asked altogether.

There are other forms of bar charts, dual (side-by-side) bar charts and composite (stacked) bar charts.

Example 46.2

Question

A shop records the numbers of computers and printers it sells in four weeks.

The information is shown in this dual bar chart.

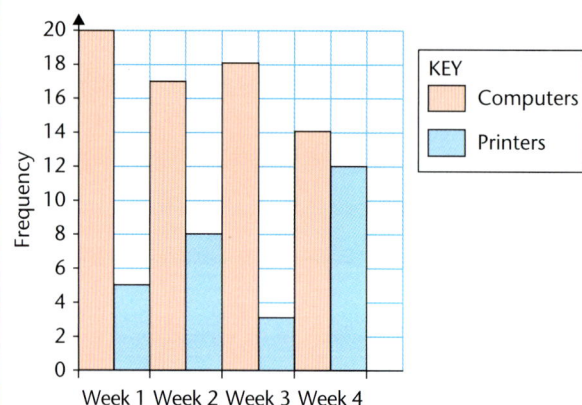

a In which week did the shop sell the most printers?
b Work out how many computers were sold altogether in the four weeks.
c Work out how many more printers were sold in Week 4 than in Week 1.
d In which week did the shop sell the greatest total number of computers and printers?

Solution

a Week 4
b 69
c 7
d Week 4

Bar charts and pictograms

Example 46.3

Question

The composite bar chart shows the numbers of boys and girls in one class absent from school one week.

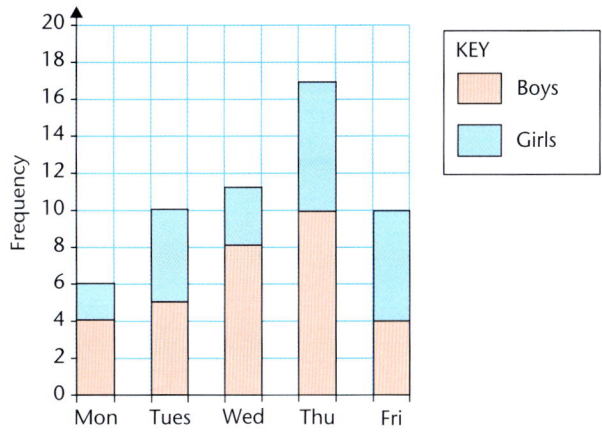

a How many absences were there altogether in the week?
b On which day were more girls than boys absent?
c On which day was the largest number of girls absent?
d How many more boys were absent on Thursday than on Tuesday?

Solution

a 54
b Friday
c Thursday
d 5

Exercise 46.1

1 Zainab drew this pictogram to show the number of books borrowed from the school library one week.

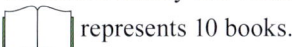 represents 10 books.

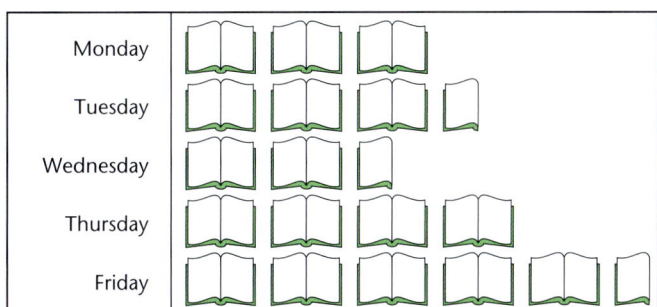

a How many books were borrowed on each of the days?
b Which day was the most popular? Why do you think that might be?

Photocopying is prohibited   485

46 STATISTICAL DIAGRAMS

2 The bar chart shows the eye colour of students in Class 10F.

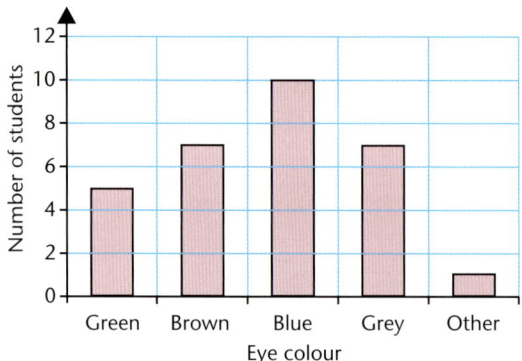

a How many students have brown eyes?
b How many more students have blue eyes than have grey eyes?
c How many students are there in the class?

3 The table shows the numbers of students absent from school during one week.

Day	Frequency
Monday	8
Tuesday	5
Wednesday	7
Thursday	10
Friday	6

Draw a bar chart to represent this information.

4 The table shows the numbers of bikes sold by a shop in four weeks.

Draw a pictogram to represent this information.

Use the symbol ⊛ to represent 8 bikes.

Week	Frequency
Week 1	12
Week 2	10
Week 3	19
Week 4	13

5 The heights of a group of boys and girls are shown on this dual bar chart.

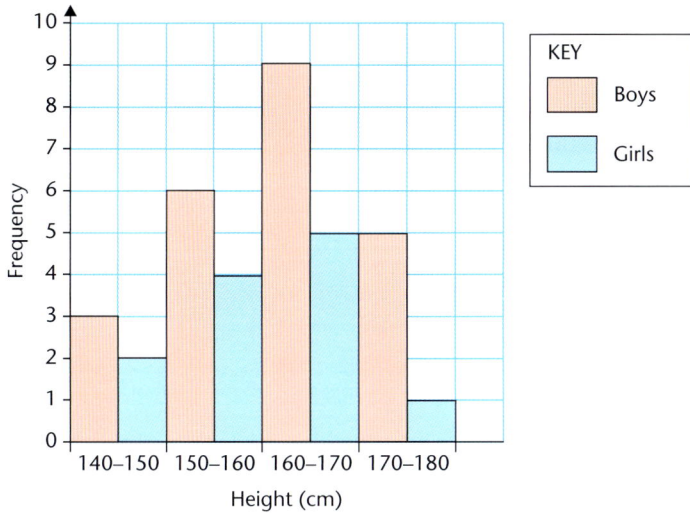

a How many boys and girls are in the group altogether?
b Which is the most common height for the girls?
c How many more boys than girls are there in the 160–170 cm height range?
d In which height range is the difference between the numbers of boys and girls the smallest?

6 Carmen recorded the numbers of cars, vans and trucks that passed her house at 6 a.m., 7 a.m., 8 a.m. and 9 a.m. one morning.
The table shows her results.
Draw a composite (stacked) bar chart to show this information.

	6 a.m.	7 a.m.	8 a.m.	9 a.m.
Cars	1	4	6	5
Vans	2	3	2	4
Trucks	3	1	2	0

Pie charts

A **pie chart** is another way to display data from a frequency table.

The angle of the sector is proportional to the frequency.

The sectors show the proportions not the actual numbers.

A pie chart shows proportions.

You can write these as fractions or percentages.

To find out the number of data items in each group, you need to know the total number of data items.

46 STATISTICAL DIAGRAMS

Example 46.4

Question

The table shows data from a survey about which mathematics topic students preferred.

Draw a pie chart using this data.

Topic	Frequency
Number	8
Algebra	2
Geometry	4
Handling data	6

Solution

To draw a pie chart for this data, you need to work out the angle for each sector.

Topic	Frequency	Calculation	Angle
Number	8	$\frac{8}{20} \times 360$	144°
Algebra	2	$\frac{2}{20} \times 360$	36°
Geometry	4	$\frac{4}{20} \times 360$	72°
Handling data	6	$\frac{6}{20} \times 360$	108°
Total	20		360°

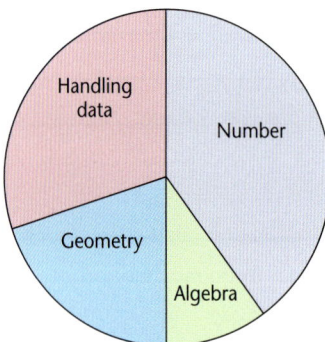

Note

You should check that the angles add up to 360° and measure all sectors accurately when you draw the pie chart.

Example 46.5

Question

This pie chart shows the age groups of the people who entered a shop one morning.

a What was the most common age group?
b What fraction of the people were aged over 70?
c A total of 180 people entered the shop.
  How many people were aged 51–70?

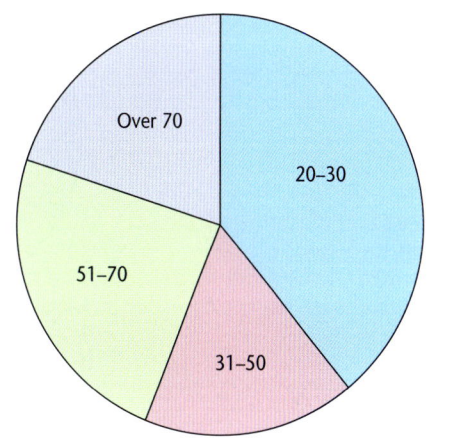

Solution

a The most common age group has the largest sector angle.
  The most common age group was 20–30.
b The sector angle for over 70 is 72°.
  The fraction is $\frac{72}{360} = \frac{1}{5}$.
c The sector angle for 51–70 is 86°.
  $\frac{86}{360} \times 180 = 43$
  There were 43 people aged 51–70.

Exercise 46.2

1 The table shows the lengths of 60 pea pods.

Length (mm)	Frequency
50–59	6
60–69	20
70–79	14
80–89	12
90–99	8
Total	60

Draw a pie chart showing the lengths of the pea pods.

2 This pie chart shows the first letter of the names of 200 students.

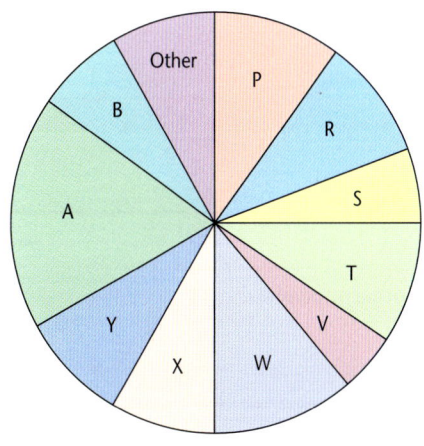

a What was the most common letter?
b How many names started with
  i P
  ii V?

3 Nesrin did a survey to find the number of bedrooms people had in their home.
The pie chart shows her results.

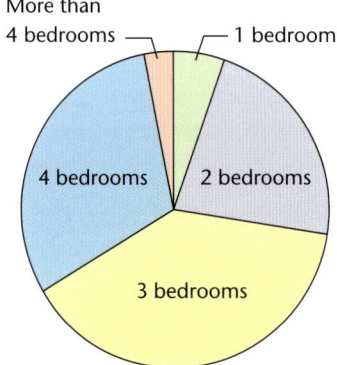

a What fraction of people had three bedrooms?
b 16 people had two bedrooms in their home.
How many people did she ask altogether?

Scatter diagrams

A **scatter diagram** is used to find out whether there is a **correlation**, or relationship, between two sets of data.

The data is presented as pairs of values, each of which is plotted as a coordinate point on a graph.

Here are some examples of what a scatter diagram could look like and how it might be interpreted.

Strong positive correlation

Here, one quantity increases as the other increases.

This is called **positive correlation**.

When the points are closely in line, the correlation is strong.

Weak positive correlation

These points also display positive correlation.

The points are more scattered so the correlation is weak.

Strong negative correlation

Here, one quantity decreases as the other increases.

This is called **negative correlation**.

Again, the points are closely in line so the correlation is strong.

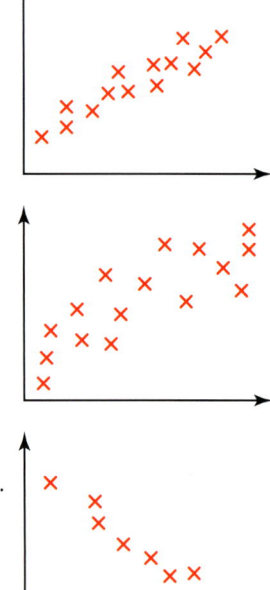

Scatter diagrams

Weak negative correlation

These points also display negative correlation.

The points are more scattered so the correlation is weak.

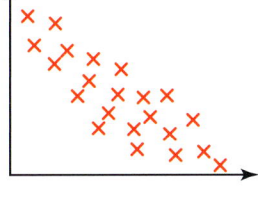

No correlation

When the points are totally scattered and there is no clear pattern, there is no correlation between the two quantities.

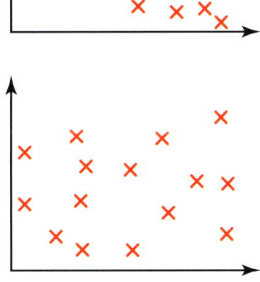

Line of best fit

If a scatter diagram shows correlation, you can draw a **line of best fit** on it.

This line goes in the same general direction as the points and has approximately the same number of points on each side of the line.

The line of best fit should extend across the full data set.

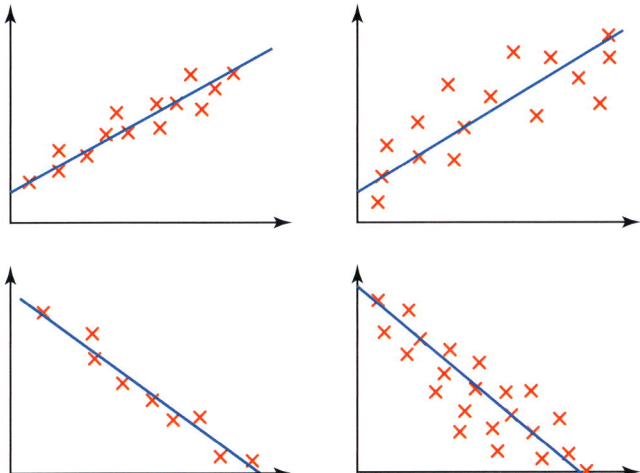

You cannot draw a line of best fit on a scatter diagram with no correlation.

You can use a line of best fit to estimate a value when only one of the pair of quantities is known.

An estimate should only be made if the given value is within the range of the data used in the scatter diagram.

If it is outside this range, the estimate will be unreliable.

46 STATISTICAL DIAGRAMS

Example 46.6

Question

The table shows the heights and weights of 12 newborn giraffes.

Height (cm)	150	152	155	158	158	160	163	165	170	175	178	180
Weight (kg)	56	62	63	57	64	62	65	66	65	70	66	67

a Draw a scatter diagram to show this data.
b Comment on the strength and type of correlation between these heights and weights.
c Draw a line of best fit on your scatter diagram.
d Another newborn giraffe is 162 cm tall. Use your line of best fit to estimate its weight.

Solution

a, c

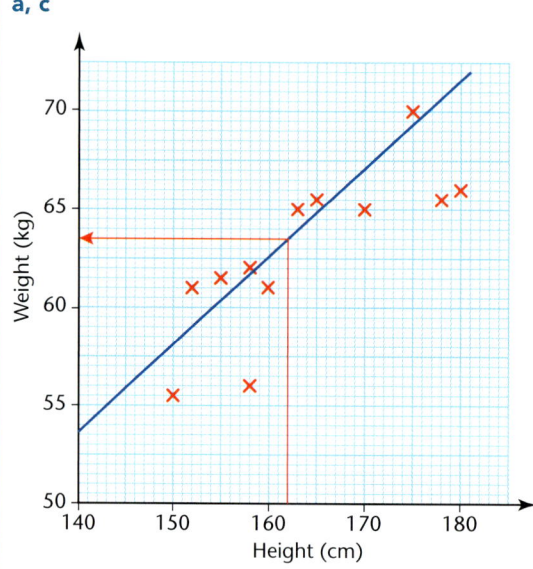

b Weak positive correlation.
d Draw a line up from 162 cm on the height axis to meet your line of best fit.

Now draw a horizontal line and read off the value where it meets the weight axis.

The newborn giraffe's probable weight is about 63 kg.

Scatter diagrams

Exercise 46.3

1 The table shows the maths and science marks of eight students in their last examination.

   a Draw a scatter diagram to show this information, with the maths mark on the horizontal axis.
   b Describe the correlation shown in the scatter diagram.
   c Draw a line of best fit.
   d Use your line of best fit to estimate
      i the science mark of a student who scored 40 in maths
      ii the maths mark of a student who scored 75 in science.

Student	Maths mark	Science mark
A	10	30
B	20	28
C	96	80
D	50	55
E	80	62
F	70	70
G	26	38
H	58	48

2 The table shows the amount of fuel left in the tank of a car after different numbers of kilometres travelled.
   a Draw a scatter diagram to show this information, with the number of kilometres on the horizontal axis.
   b Describe the correlation shown in the scatter diagram.
   c Draw a line of best fit.
   d Use your line of best fit to estimate the number of litres left after 70 kilometres.

Number of kilometres	Number of litres
20	7
40	5.2
60	4.2
80	2.6
100	1.2
120	0.4

3 The table shows the marks of 15 students taking Paper 1 and Paper 2 of a maths examination. Both papers were marked out of 40.

Paper 1	Paper 2
36	39
34	36
23	27
24	20
30	33
40	35
25	27
35	32
20	28
15	20
35	37
34	35
23	25
35	33
27	30

   a Draw a scatter diagram to show this information.
   b Describe the correlation shown in the scatter diagram.
   c Draw a line of best fit on your scatter diagram.
   d Joe scored 32 on Paper 1 but was absent for Paper 2.
      Use your line of best fit to estimate his score on Paper 2.

4 The table shows the engine size and fuel consumption of nine cars.

Engine size (litres)	Fuel consumption (km/l)
1.9	17
1.1	21
4.0	11
3.2	14
5.0	9
1.4	21
3.9	13
1.1	24
2.4	17

a Draw a scatter diagram to show this information.
b Describe the correlation shown in the scatter diagram.
c Draw a line of best fit on your scatter diagram.
d A tenth car has an engine size of 2.8 litres.
  Use your line of best fit to estimate the fuel consumption of this car.

Cumulative frequency diagrams

A **cumulative frequency diagram** is a way to display a large set of grouped continuous data.

It can be used to estimate the median of the data and to give information about the spread of the data.

This table shows the distribution of the heights of 60 plants.

Height h (cm)	Frequency
$0 < h \leqslant 10$	1
$10 < h \leqslant 20$	3
$20 < h \leqslant 30$	9
$30 < h \leqslant 40$	25
$40 < h \leqslant 50$	11
$50 < h \leqslant 60$	6
$60 < h \leqslant 70$	3
$70 < h \leqslant 80$	1
$80 < h \leqslant 90$	1

The data has been combined in the cumulative frequency table.

Cumulative frequency diagrams

The cumulative frequency is the total of all the frequencies up to and including the selected group.

Height h (cm)	Cumulative frequency
$h \leq 0$	0
$h \leq 10$	1
$h \leq 20$	4
$h \leq 30$	13
$h \leq 40$	38
$h \leq 50$	49
$h \leq 60$	55
$h \leq 70$	58
$h \leq 80$	59
$h \leq 90$	60

This row is included in the table to show that there are no plants with a height of 0 cm.

This tells you that there are 38 plants that have a height of 40 cm or less.

It is the sum of all the frequencies of the groups up to and including this one.

$1 + 3 + 9 + 25 = 38$

The cumulative frequency values can be used to plot a cumulative frequency diagram.

The cumulative frequency values are plotted at the upper value of each interval, so the points plotted are (0, 0), (10, 1), (20, 4) and so on.

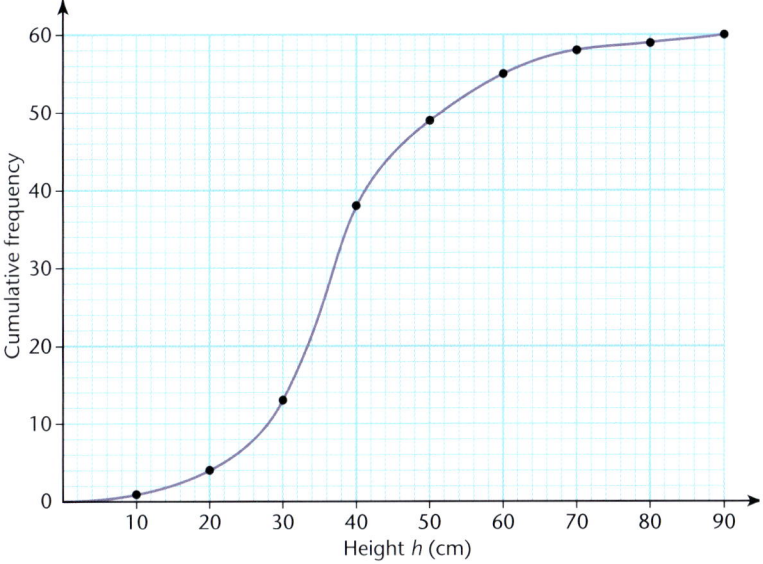

Example 46.7

Question

The table shows the distribution of the masses of 100 apples.

Mass m (g)	50 < m ⩽ 100	100 < m ⩽ 150	150 < m ⩽ 200	200 < m ⩽ 250	250 < m ⩽ 300
Frequency	15	18	34	25	8

a Make a cumulative frequency table for the data.
b Draw a cumulative frequency diagram for the data.
c Estimate the number of apples with a mass of
  i less than 80 g
  ii greater than 190 g.

Solution

a

Mass m (g)	m ⩽ 50	m ⩽ 100	m ⩽ 150	m ⩽ 200	m ⩽ 250	m ⩽ 300
Cumulative frequency	0	15	33	67	92	100

There are no apples with a mass less than or equal to 50 g, so this column is added to the cumulative frequency table.

b

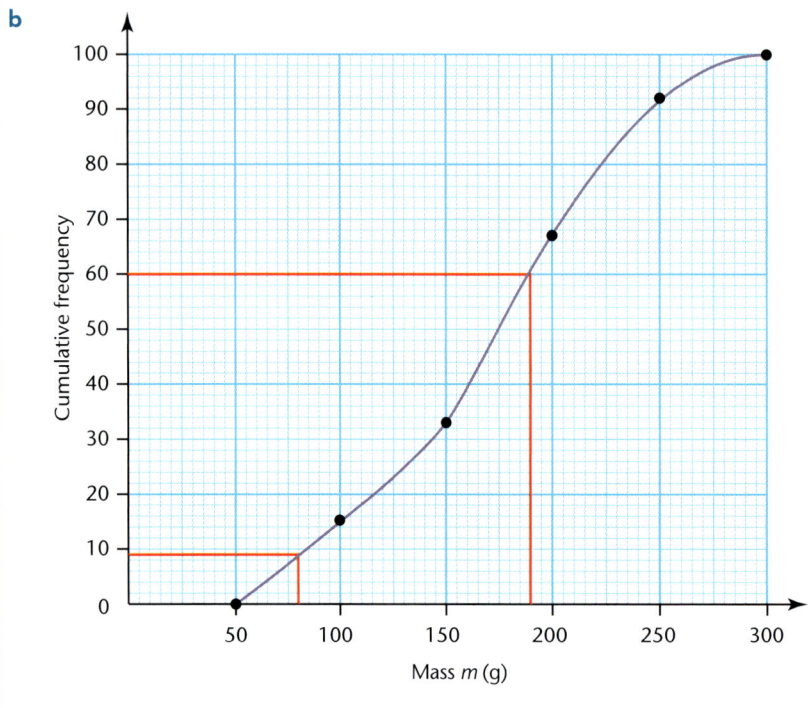

c i To find an estimate of the number of apples with a mass less than 80 g, read from the graph at this point.
There are 9 apples with a mass less than 80 g.
ii To find an estimate of the number of apples with a mass greater than 190 g, read from the graph at this point.
There are 60 apples with a mass *less than* 190 g.
So there are 100 − 60 = 40 apples with a mass greater than 190 g.

Exercise 46.4

1 The table shows the times taken for the 200 runners in a race.

Time in minutes (t)	Number of runners
$0 < t \leqslant 60$	0
$60 < t \leqslant 80$	10
$80 < t \leqslant 100$	37
$100 < t \leqslant 120$	72
$120 < t \leqslant 140$	55
$140 < t \leqslant 160$	18
$160 < t \leqslant 180$	7
$180 < t \leqslant 200$	1

a Make a cumulative frequency table for the data.
b Draw a cumulative frequency diagram for the data.
c Estimate the number of runners who took
   i   less than 90 minutes
   ii more than 150 minutes.

2 The table shows the lengths, in hours, of the lives of 90 lamp bulbs.

Life of lamp bulbs in hours (t)	Number of lamps
$0 < t \leqslant 600$	0
$600 < t \leqslant 625$	3
$625 < t \leqslant 650$	18
$650 < t \leqslant 675$	29
$675 < t \leqslant 700$	25
$700 < t \leqslant 725$	13
$725 < t \leqslant 750$	2

a Make a cumulative frequency table for the data.
b Draw a cumulative frequency diagram for the data.
c Estimate the number of lamp bulbs that lasted for more than 680 hours.

Median, quartiles and percentiles

A cumulative frequency diagram can be used to find the median value of the distribution.

Look back at the cumulative frequency table and diagram on page 495.

There were 60 plants, so the median value is the height of the 30th plant.

The median can also be called the 50th **percentile** because 50% of the data has a value less than or equal to that value.

The median height is approximately 37 cm.

> **Note**
> If you were using a list of values, then for an even number of values the median would be halfway between the middle two values.
>
> However, when using a cumulative frequency diagram for grouped data, the values of the individual items are not known.
>
> Any results are therefore approximate and it is reasonable to use the 50th percentile.

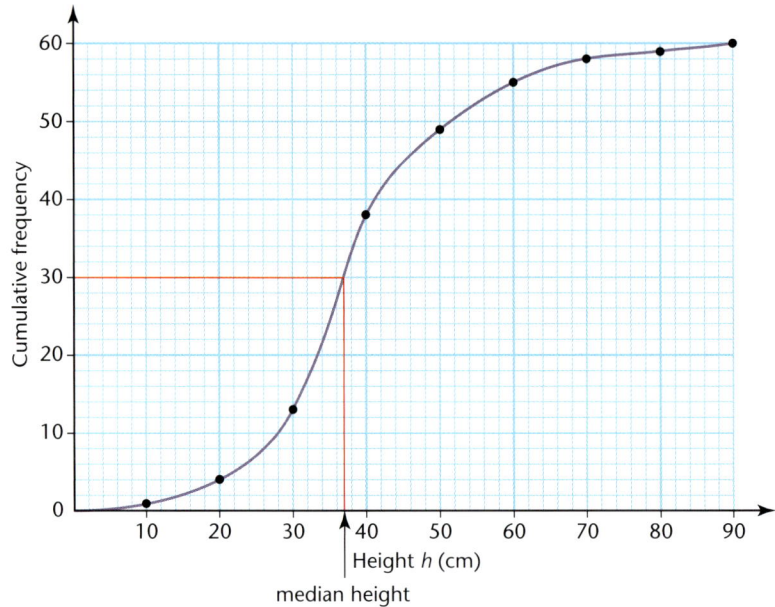

The **lower quartile** (LQ) of the data is the 25th percentile.

25% of the data has a value less than or equal to this value.

The lower quartile is the height of the 15th plant.

The **upper quartile** (UQ) of the data is the 75th percentile.

75% of the data has a value less than or equal to this value.

The upper quartile is the height of the 45th plant.

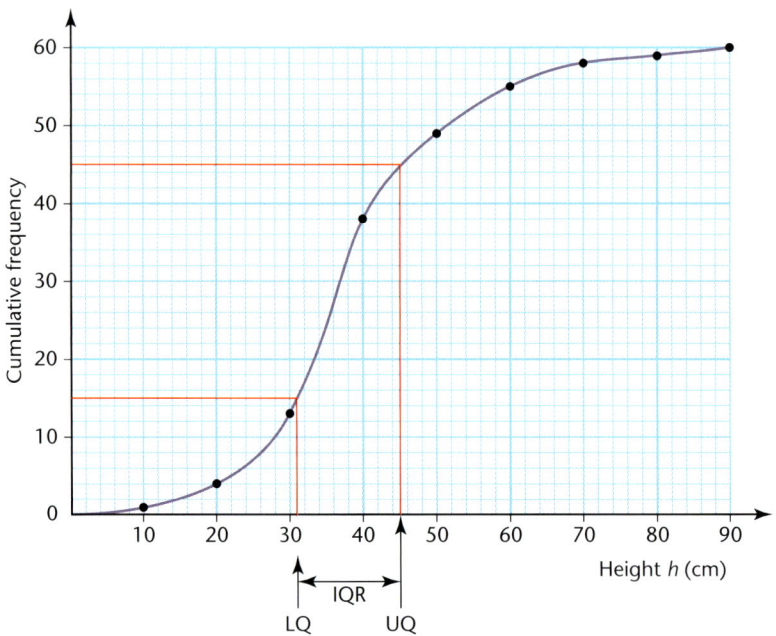

Median, quartiles and percentiles

The **interquartile range** (IQR) is a measure of the spread of the middle 50% of the data.

The interquartile range is the difference between the upper quartile and the lower quartile. (IQR = UQ – LQ).

Lower quartile = 31 cm

Upper quartile = 45 cm

Interquartile range = 45 – 31 = 14 cm

Other percentiles can be found from the data.

For example, 10% of the data has a value less than or equal to the 10th percentile.

For this data, the 10th percentile is the height of the 6th plant and is 24 cm.

Example 46.8

Question

For the data in Example 46.7, estimate

a   the median

b   i   the lower quartile
    ii the upper quartile
    iii the interquartile range

c   the 90th percentile.

Solution

The required values can be read from the cumulative frequency diagram.

a   There are 100 values, so the median is the 50th value.
    Median = 175 g

b   i   The LQ is the 25th value.
        LQ = 132 g
    ii The UQ is the 75th value.
        UQ = 215 g
    iii IQR = 215 – 132 = 83 g

c   90th percentile = 243 g

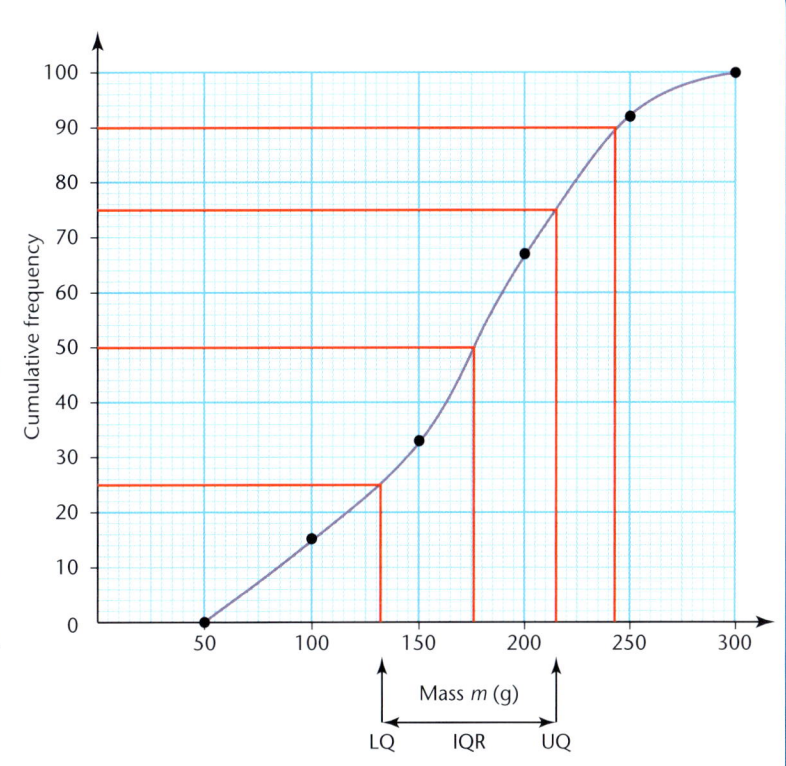

Exercise 46.5

1 The table shows the distribution of the heights of 120 people.

Height h (cm)	Number of people
$0 < h \leq 130$	0
$130 < h \leq 140$	5
$140 < h \leq 150$	12
$150 < h \leq 160$	26
$160 < h \leq 170$	35
$170 < h \leq 180$	23
$180 < h \leq 190$	15
$190 < h \leq 200$	4

a Make a cumulative frequency table for the data and use it to draw a cumulative frequency diagram.
b Use your cumulative frequency diagram to estimate
   i the median height
   ii the lower quartile
   iii the upper quartile
   iv the interquartile range.

2 The cumulative frequency table shows information about the masses of 200 potatoes.

Mass m (grams)	$m \leq 0$	$m \leq 50$	$m \leq 100$	$m \leq 150$	$m \leq 200$	$m \leq 250$	$m \leq 300$	$m \leq 350$
Cumulative frequency	0	16	38	81	143	183	196	200

a Draw a cumulative frequency diagram for this data.
b Use your diagram to estimate the median and interquartile range of these masses.
c Estimate the 80th percentile.

3 The cumulative frequency table shows information about the yield, in litres of milk, from 120 cows.

Yield x (litres)	$x \leq 5$	$x \leq 10$	$x \leq 15$	$x \leq 20$	$x \leq 25$	$x \leq 30$	$x \leq 35$
Cumulative frequency	0	10	33	68	94	111	120

a Draw a cumulative frequency diagram for this data.
b Use your diagram to estimate the median and interquartile range of these amounts.
c Estimate the number of cows that produced more than 28 litres of milk.
d Estimate the 20th percentile.

Median, quartiles and percentiles

4 The cumulative frequency diagram shows the distribution of the speeds of 100 cars on each of two roads.

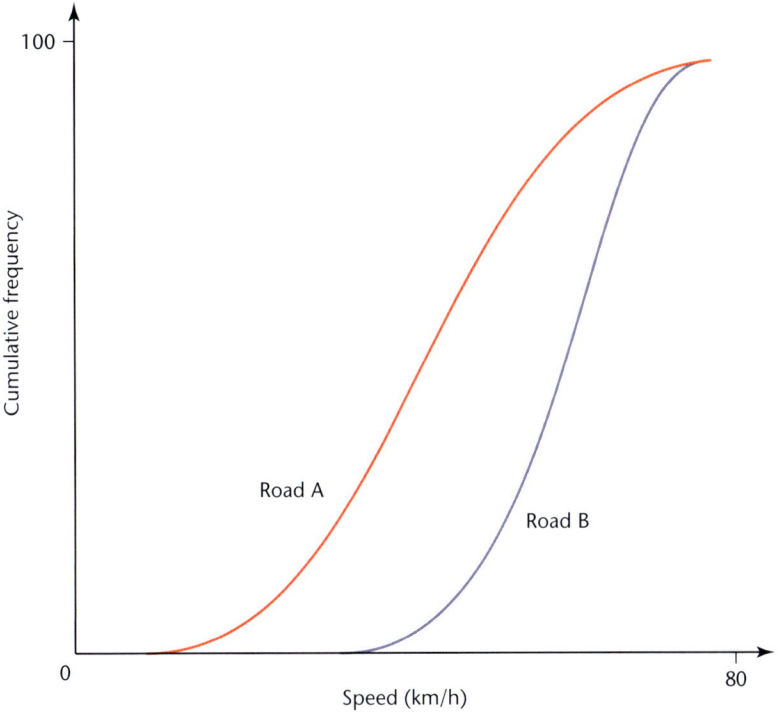

a On which road are the cars travelling faster, on average?
State your reasons.
b On which road are the cars' speeds more consistent?
State your reasons.

5 The cumulative frequency diagram shows the distributions of the masses of 160 panthers and 160 leopards in a big cat sanctuary.

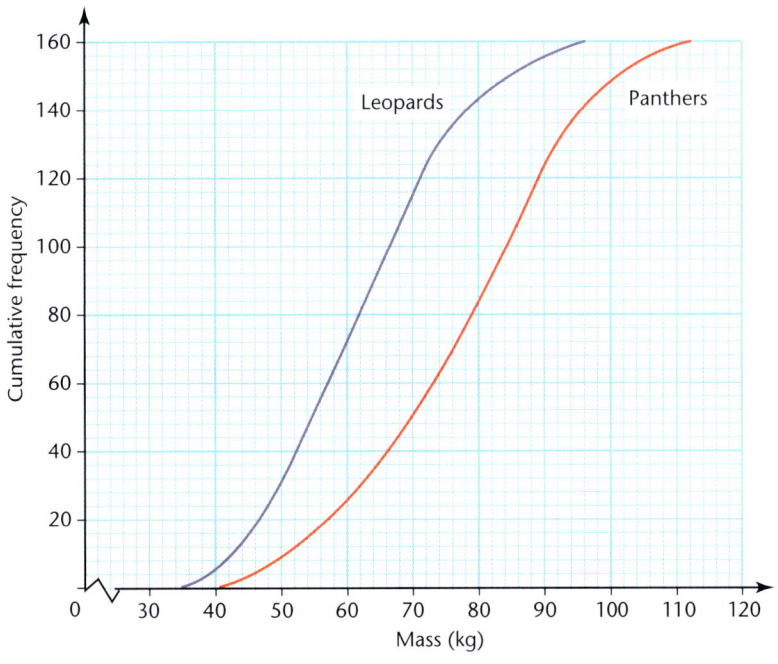

a Estimate the median mass of
   i the leopards
   ii the panthers.
b Estimate the interquartile range for
   i the leopards
   ii the panthers.
c Make two comments comparing the masses of the leopards and the masses of the panthers.

46 STATISTICAL DIAGRAMS

Histograms

A **histogram** is a frequency diagram that is used to represent continuous data.

The area of each bar is used to represent the frequency of the group.

The width of the bar is proportional to the class width.

The vertical scale on a histogram is the **frequency density**.

$$\text{Frequency density} = \frac{\text{frequency}}{\text{class width}}$$

The units of frequency density depend on the units of the data.

The table shows the distribution of the ages of people in a sports club.

Age in years	Frequency
11–15	7
16–18	10
19–24	15
25–34	20
35–49	12
50–64	7

The histogram below shows this distribution.

The upper boundary for the 11–15 age group is the 16th birthday. So the boundaries of this group are 11 and 16.

This applies similarly to the other groups, so the boundaries of the bars in the histogram are at 11, 16, 19, 25, 35, 50 and 65.

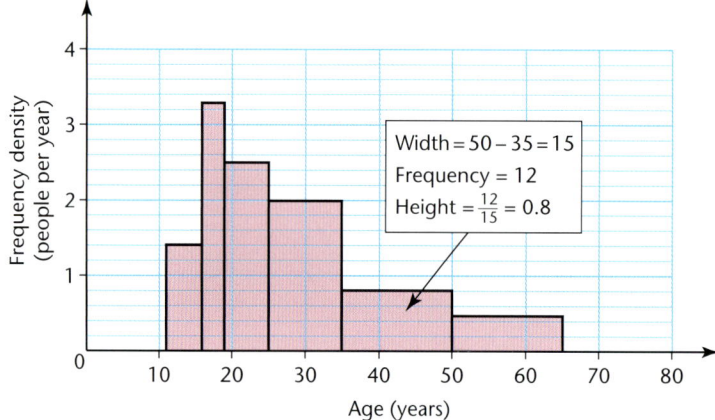

The calculation of the frequency density for the 35–49 group has been added to the histogram.

A similar calculation is done for each group.

For this histogram, the units of frequency density are people per year.

Note that the bars are not equal in width because the class widths are not equal.

Histograms

Example 46.9

Question

An airline investigated the ages of passengers flying between London and Johannesburg.

The table shows the findings.

a Draw a histogram to represent this data.
b Estimate the number of passengers aged between 60 and 80.

Age A (years)	Frequency
$0 < A \leq 20$	28
$20 < A \leq 30$	36
$30 < A \leq 40$	48
$40 < A \leq 50$	20
$50 < A \leq 70$	30
$70 < A \leq 100$	15

Solution

a To draw a histogram you must first calculate the frequency density.

Age A (years)	Frequency	Class width	Frequency density (people per year)
$0 < A \leq 20$	28	20	$28 \div 20 = 1.4$
$20 < A \leq 30$	36	10	$36 \div 10 = 3.6$
$30 < A \leq 40$	48	10	$48 \div 10 = 4.8$
$40 < A \leq 50$	20	10	$20 \div 10 = 2$
$50 < A \leq 70$	30	20	$30 \div 20 = 1.5$
$70 < A \leq 100$	15	30	$15 \div 30 = 0.5$

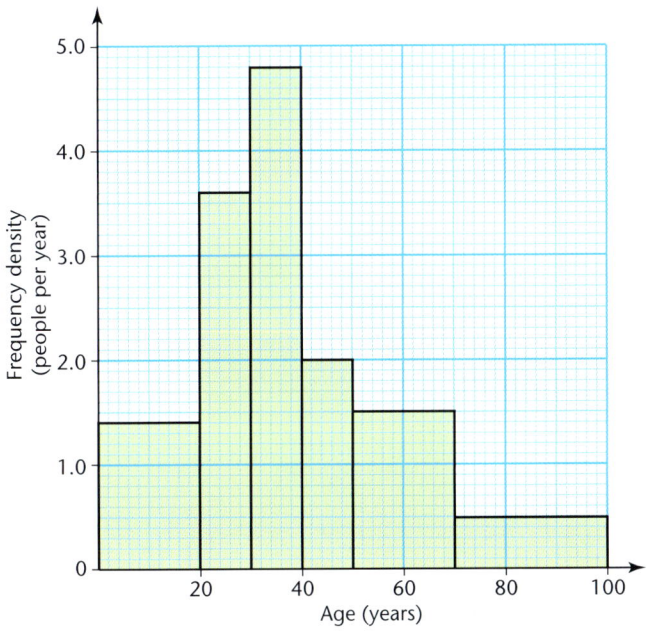

Note

If a range includes parts of a class, then the calculation only gives an estimate, as we do not know how the data is distributed within that group.

b An estimate of the number of people can be found using frequency = frequency density × class width.

The range 60 to 80 contains 10 years from the 50 to 70 group and 10 years from the 70 to 100 group, so add the areas of those two sections of the histogram.

Number of people = 1.5 × 10 + 0.5 × 10 = 20

There are approximately 20 people aged between 60 and 80.

46 STATISTICAL DIAGRAMS

Exercise 46.6

1 200 people were surveyed to find the distances they travelled to work. The table shows the results.

Distance d (km)	Frequency
$0 < d \leqslant 5$	3
$5 < d \leqslant 10$	9
$10 < d \leqslant 15$	34
$15 < d \leqslant 20$	49
$20 < d \leqslant 30$	17
$30 < d \leqslant 40$	41
$40 < d \leqslant 60$	27
$60 < d \leqslant 100$	20

Draw a histogram to show this information.

2 The table shows the results of a survey to find the areas, to the nearest hectare, of 160 farms.

Area A (hectares)	Frequency
$1 < A \leqslant 4$	29
$4 < A \leqslant 8$	18
$8 < A \leqslant 12$	34
$12 < A \leqslant 16$	26
$16 < A \leqslant 24$	28
$24 < A \leqslant 30$	11
$30 < A \leqslant 34$	8
$34 < A \leqslant 40$	6

Draw a histogram to show this information.

3 The age of each person on a bus was recorded. The table shows the results.

Age A (years)	Frequency
$0 < A \leqslant 10$	0
$10 < A \leqslant 20$	2
$20 < A \leqslant 30$	3
$30 < A \leqslant 45$	8
$45 < A \leqslant 50$	5
$50 < A \leqslant 70$	18
$70 < A \leqslant 100$	12

Draw a histogram to show this information.

4 A group of students were asked how long it took them to travel from home to school one day. The results are shown in the histogram.

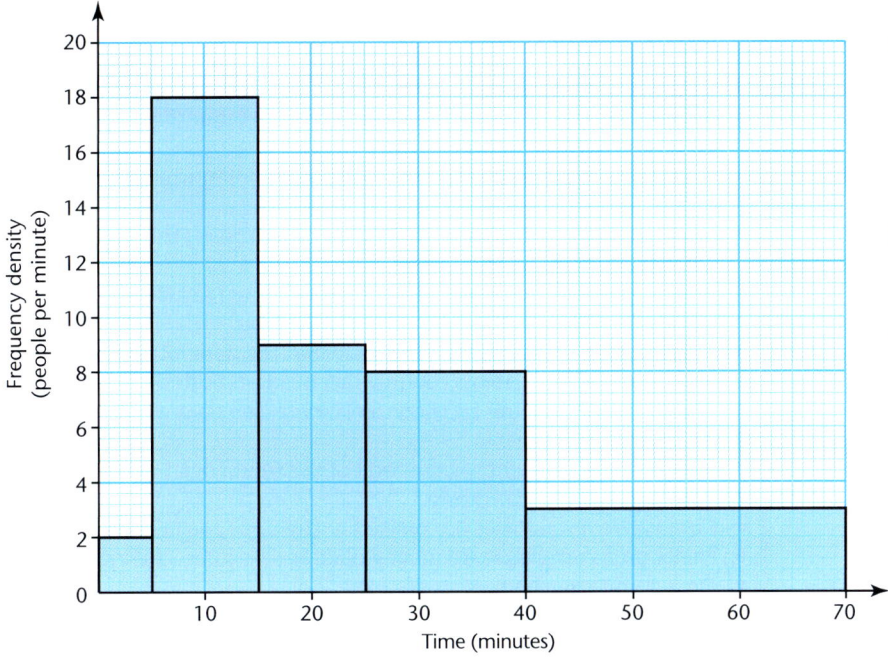

a How many students took between 15 and 25 minutes to travel from home to school?
b Work out the total number of students asked.
c Estimate the number of students who took between 30 and 60 minutes to travel from home to school.

5 The members of a gym club were weighed. The histogram represents their masses.

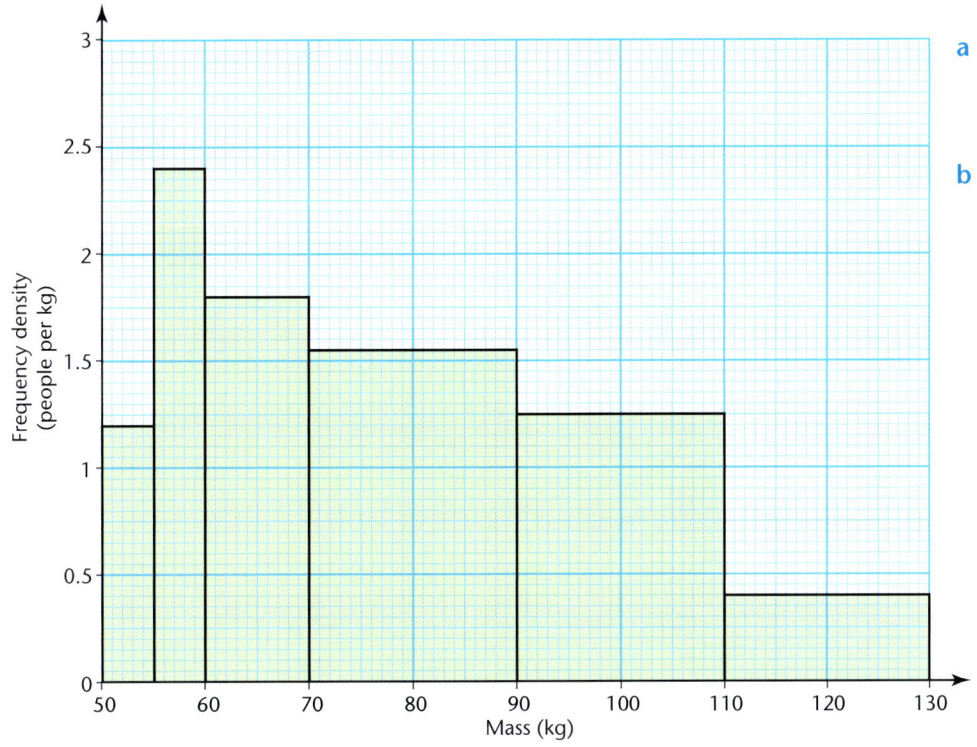

a How many members of the gym were weighed?
b Calculate an estimate of their mean mass.

46 STATISTICAL DIAGRAMS

> **Key points**
> - Pictograms, bar charts, pie charts and frequency diagrams can all be used to display data.
> - You need to know how to extract data from all forms of statistical diagram.
> - A scatter diagram can be used to identify a relationship (correlation) between two sets of data.
> - Positive correlation is shown by data points going from bottom left to top right (//).
> - Negative correlation is shown by data points going from top left to bottom right. (\\).
> - When there is no clear pattern in the points, we say there is no correlation.
> - A line of best fit is a straight line drawn in the general direction of the points, with points roughly balanced on both sides.
> - The line of best fit can be used to estimate one data value when given another.
> - A cumulative frequency is the total of the frequencies up to and including a particular group.
> - Cumulative frequencies are plotted at the upper value of each group.
> - Points on a cumulative frequency diagram are joined by a curve.
> - On a cumulative frequency diagram:
>   - the median of the data is the 50th percentile
>   - the lower quartile of the data is the 25th percentile
>   - the upper quartile of the data is the 75th percentile.
>   - interquartile range = upper quartile − lower quartile.
> - A histogram is a grouped frequency diagram where the area of each bar represents a frequency.
> - The vertical scale of a histogram is the frequency density.
> - Frequency density = $\dfrac{\text{frequency}}{\text{class width}}$.

REVIEW EXERCISE 9

1 200 people were asked their favourite colour.
The table shows the relative frequencies of their answers.

Favourite colour	Blue	Yellow	Purple	Red
Relative frequency	0.12		0.34	0.4

Work out the number of people who said that yellow was their favourite colour. [3]

2 A bag contains 10 counters of which 8 are blue and 2 are white.
Two counters are taken from the bag at random without replacement.
  a Complete the tree diagram to show the possible outcomes and their probabilities. [1]

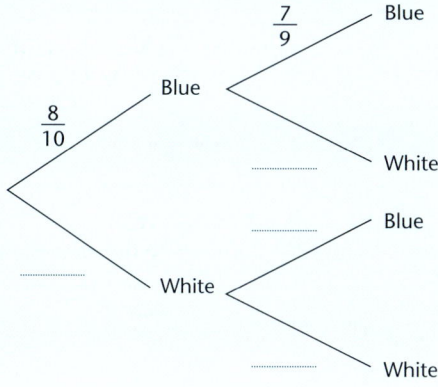

  b Find, as a fraction, the probability that
    i both counters are blue, [1]
    ii one counter is blue and the other is white. [2]

Cambridge O Level Mathematics Syllabus D (4024) Paper 12 Q20, June 2016

3 Jenny asked 60 people how many movies they had each watched in the last month.
The table summarises her results.

Number of movies	0	1	2	3	4	5	6
Frequency	p	14	15	7	q	5	2

The mean number of movies watched is 2.3.
Find the value of p and the value of q. [3]

Cambridge O Level Mathematics Syllabus D (4024) Paper 22 Q2b, June 2018

REVIEW EXERCISE 9

4 The results showing the number of people travelling in 30 cars are shown in the table.

Number of people	1	2	3	4	5
Frequency	6	11	7	4	2

   a Write down the mode. [1]
   b Work out the range. [1]
   c Work out the median. [1]
   d A pie chart is drawn to represent the data in the table.
      Work out the angle in the pie chart for the sector representing 3 people. [2]

5 The table and histogram show some information about the times taken by a group of students to travel to school one day.

Time (t minutes)	$0 < t \leq 10$	$10 < t \leq 20$	$20 < t \leq 30$	$30 < t \leq 60$	$60 < t \leq 120$
Frequency	28	40	52	18	m

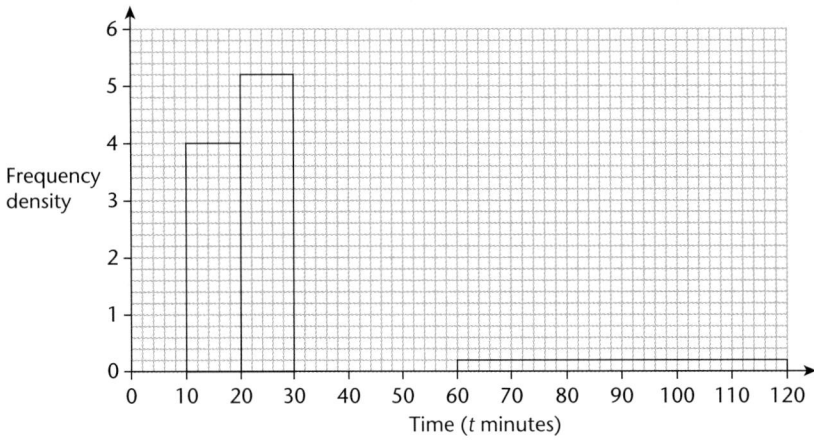

   a Complete the histogram. [2]
   b Find the value of m. [1]
   c Work out the fraction of students who took more than half an hour to travel to school. [2]

Cambridge O Level Mathematics Syllabus D (4024) Paper 12, Q23, June 2016

6 Class A and class B both take the same mathematics test.
The results for Class A have a median of 64 and an interquartile range of 34.
The results for Class B have a median of 53 and an interquartile range of 41.
   a Which class, on average, performed better in the test? [1]
   b Which class had the more consistent test results? [1]

GLOSSARY

acceleration: The rate of change of speed with time is the acceleration of an object.

acute angle: An acute angle lies between 0° and 90°.

adjacent: Adjacent means 'next to'. In a right-angled triangle, the adjacent is the side which is next to the angle.

allied angles: The adjacent angles made between two parallel lines crossed by a transversal.

area: The area of a two-dimensional shape is the amount of flat space inside the shape.

bar chart: A bar chart is a chart that uses rectangular bars to display data. The height of each bar represents the frequency.

base: The number in an index notation that is multiplied by itself.

bearing: The angle a direction makes, measured clockwise from north.

bias: An unfair selection that means certain outcomes have a higher chance of occurrence.

biased: Data is biased if some outcomes occur more or less than would be expected.

capacity: The amount that can be contained.

centre of enlargement: The point from which the enlargement is drawn.

centre of rotation: The centre of rotation is the 'pivot' point about which an object is rotated.

certain: If an event is certain, its probability is 1.

circle: A circle is a shape that is made up of all the points that are a specific distance from the centre.

circumference: The circumference is the perimeter of a circle.

coefficient: The number written in front of a letter.

co-interior angles: The adjacent angles made between two parallel lines crossed by a transversal.

column vector: A way of writing the distance and direction that a shape moves in a translation.

commission: An additional percentage of sales, added to basic wage.

common factor: A common factor of two or more numbers is a number that divides exactly into all the numbers.

common multiple: A common multiple of two numbers is a number that is a multiple of both of them.

complement: A set containing the elements in a universal set, but not in the set under consideration.

compound interest: A form of interest where the interest is added to the original amount before calculation, so that the amount of interest also increases each year.

compound shape: Any shape that is made up of two or more shapes.

cone: A cone is a three-dimensional shape that has two faces: a flat circular base and a single curved face forming a single vertex. It could be described as a circular-based pyramid.

congruent: Congruent shapes are exactly the same shape and size – they are identical.

construct: Draw accurately using only a pair of compasses and a straight edge.

continuous data: Continuous data does not have gaps in the possible values.

correlation: Correlation is when there appears to be a link between how one measure changes when another measure changes.

cosine: The cosine of an angle, cos θ, in a right-angled triangle is the ratio of the side adjacent to the angle and the hypotenuse. Cos θ = adjacent/hypotenuse.

cosine rule: A formula involving the cosine of an angle, used in non-right-angled triangles.

cube: A solid with six faces, all of which are squares.

cuboid: A prism with a rectangular cross-section.

cumulative frequency diagram: A way to display a large set of grouped continuous data.

cyclic quadrilateral: A four-sided shape with all its vertices on a circle.

cylinder: A cylinder is a three-dimensional shape with a constant circular cross-section.

deceleration: Negative acceleration.

decimal place: The number of decimal places of a decimal is the number of digits after the decimal point.

denominator: The denominator is the bottom number in a fraction.

dependent events: Two events are dependent when the outcome of the second event is affected by the outcome of the first.

GLOSSARY

diagonal: A line joining two corners of a polygon.

diameter: A diameter is a line which passes through the centre of a circle and joins two points on the circumference.

direct proportion: Two quantities are in direct proportion if, when one increases, the other increases by the same multiple.

direction: The motion aspect of vector notation.

discount: Reduction in the price of something.

discrete data: Discrete data does not include any in-between values; it has gaps.

disjoint: Two sets with no elements in common.

domain: Set of starting values for a function.

edge: In a three-dimensional shape, an edge forms the boundary between two faces.

elements: The individual members of a set.

empty set: A set with no elements.

enlargement: An enlargement changes the size of an object. When a shape is enlarged, the image is mathematically similar to the object but a different size.

equilateral triangle: An equilateral triangle has three equal angles (all 60°) and three sides of equal length.

equivalent fractions: Fractions that are equal, despite having different numerators/denominators.

event: An event is any of the possible outcomes from an experiment.

exterior angle: An angle made from outside a polygon.

face: In a three-dimensional shape, a face is a surface that forms a part of the boundary of the shape. For example, a cube has six faces.

factor: A factor of a number divides into that number exactly. For example, the factors of 18 are 1, 2, 3, 6, 9 and 18.

fair: All possible outcomes have an equal chance of occurrence.

finite set: A set with a fixed number of elements.

frequency density: The vertical scale on a histogram.

frustum: The part of a solid between the base and a plane parallel to the base.

function: An expression, rule or law that defines a relationship between one variable and another.

gradient: The gradient is a measure of how steep a line is. The gradient of the line joining two points can be calculated as the change in the y-values divided by the change in the x-values.

gross pay: What one earns.

hemisphere: Half of a sphere.

highest common factor (HCF): The highest common factor (HCF) of two numbers is the greatest integer that divides exactly into both numbers.

histogram: A frequency diagram that is used to represent continuous data.

hypotenuse: The hypotenuse is the longest side of a right-angled triangle.

image: When a shape undergoes a transformation, the resulting shape is the image.

impossible: An event with probability 0 is impossible.

improper fraction: In an improper fraction, the numerator is greater than its denominator.

independent events: Two events are independent when the outcome of the second event is not affected by the outcome of the first.

index: The power to which a number is raised.

index notation: A way of noting numbers that have been multiplied by themselves multiple times.

indirect proportion: Two quantities where when one quantity increases, the other decreases.

inequality: An expression that compares the size of two values, showing if one is less than, or greater than, the other.

interior angle: An angle inside a polygon.

interquartile range: A measure of the spread of the middle 50% of the data.

intersection: The intersection of two sets is all the elements that are in both sets.

invariant: Anything that stays the same when a transformation is performed.

inverse: The opposite of another operation.

irregular polygon: If a polygon does not meet the criteria of equal angles and lengths to be regular, it is irregular.

isosceles triangle: A triangle with two sides and two angles equal.

like terms: Like terms use only the same letter or the same combination of letters.

line: A line is a straight one-dimensional figure with no thickness and extending infinitely in both directions. It is often called a straight line.

line of best fit: A line drawn on a scatter diagram that can be used to make predictions about two variables that have a linear relationship.

Glossary

line of symmetry: Divides a shape into two identical parts.

line segment: A line segment is part of a line. It has a beginning and an end.

linear sequence: A sequence with a constant increase.

lower bound: The lower bound is the least possible value that the true measurement could be.

lower quartile: The lower quartile of the data is the 25th percentile.

lowest common multiple (LCM): The lowest common multiple of two numbers is the least integer that is a multiple of both numbers.

magnitude: The length aspect of vector notation.

maps: A way of describing how a transformation puts an object onto the image.

mean: The mean is found by adding together all of the data values and then dividing this total by the number of data values.

median: The median is the middle value when the data set is organised in order of size.

mirror line: The line in which a shape is reflected.

mixed number: A mixed number is made up of a whole number and a fraction.

modal class: The class with the highest frequency.

mode: The mode is the value occurring most often in a data set.

multiple: The multiple of a number is the result when you multiply that number by a positive integer.

multiplier: The multiplier is the number you are multiplying by.

mutually exclusive events: Two events are mutually exclusive if they both cannot happen at the same time.

negative correlation: Negative correlation occurs when points are roughly in a line from top left to bottom right of a scatter diagram.

negative number: A value which is less than zero.

net: A flat shape that will fold to make a three-dimensional shape.

numerator: The numerator is the top number in a fraction.

object: A shape that is transformed to give the image.

obtuse angle: An obtuse angle lies between 90° and 180°.

opposite: In a right-angled triangle, the opposite is the side which is opposite the angle.

order of rotational symmetry: The number of times that a shape fits onto itself in one complete turn.

outcomes: Possibilities resulting from an event.

parabola: The shape of a quadratic graph.

parallel: A pair of parallel lines can be continued to infinity in either direction without meeting.

per cent: One part of every 100.

percentile: Percentile is the value below which a percentage of data falls.

perfect square: A number generated by multiplying two equal integers by each other. For example, 16 is a perfect square because it can be expressed as 4 × 4.

perimeter: The perimeter of a shape is the distance all the way around the edge of the shape.

perpendicular: Two lines are perpendicular if they meet at right angles.

perpendicular bisector: A line which passes through the midpoint of a line segment at right angles.

pictogram: A pictogram is a chart that uses pictures or symbols to display data.

pie chart: A way to display data from a frequency table.

pilot survey: A test of a designed questionnaire/survey given to a few people.

plane: A two-dimensional figure with length and width but no height.

plane of symmetry: Divides a solid into two identical parts.

point: An exact location without dimensions.

polygon: A polygon is a closed two-dimensional shape made up of straight lines.

position vector: Starts at the origin and is the vector equivalent of coordinates.

positive correlation: Positive correlation is when points are roughly in a line from bottom left to top right of a scatter diagram. As one quantity increases, the other one does too.

positive number: A value which is greater than zero.

prime number: A prime number is a number with exactly two factors: one and itself.

principal: The amount initially invested.

prism: A three-dimensional shape that has a constant cross-section.

GLOSSARY

proper fraction: In a proper fraction, the denominator is greater than its numerator.

pyramid: A three-dimensional solid shape, with a base and triangular faces.

quadrilateral: A quadrilateral is a four-sided polygon (a four-sided plane shape).

questionnaire: A way of collecting data that poses questions to a sample.

random sample: A sample where there is an equal chance of selection, due to elimination of bias.

range: Range is a measure of the spread of a data set. The range is the difference between the highest and lowest data values.

range (function): A set of finishing values for a function.

reciprocal: The reciprocal of a number is 1 divided by that number.

recurring decimal: A decimal with digits that continue forever in a pattern.

reflection: A reflection is a 'flip' movement about a mirror line. The mirror line is the line of symmetry between the object and its image.

reflex angle: A reflex angle is an angle between 180° and 360°.

regular polygon: A shape where all angles are of equal size and all sides are an equal length.

relative frequency: The average number of times an outcome occurs across a number of repeated trials.

right angle: A 90° angle.

right-angled triangle: A triangle where one of the internal angles is 90°.

rotation: A rotation is a 'turning' movement about a specific point known as the centre of rotation.

salary: A fixed amount of money earned each month or year from a job, paid monthly or more regularly.

sample: A smaller selection of the population selected to gather statistical information.

sample space diagram: A way of calculating all possible outcomes.

scalar: A quantity with magnitude but no direction.

scale drawing: In a scale drawing, the scale is the ratio or multiplier that tells you how much bigger (or smaller) the original is than the drawing.

scale factor: The number used to multiply lengths for the enlargement.

scalene triangle: A scalene triangle has three sides of different lengths and three different angles.

scatter diagram: A scatter diagram is drawn using pairs of data as coordinates.

sector: A sector is part of a circle that looks like a piece of pie; it is enclosed by two radii and part of the circumference.

seller: Individual or organisation selling an item.

sequence: A sequence is a collection of terms arranged in a specific order.

significant figure: The first significant figure of a number is the first non-zero digit in the number. The second significant figure is the next digit in the number and so on.

similar: Two shapes are similar if corresponding sides are in proportion and corresponding angles are equal; the shapes are enlargements of each other.

simple interest: Calculated as a percentage of an amount.

simplest form: A fraction is in its simplest form when it can't be simplified or cancelled any further.

simplify: To simplify an expression, you write it in an equivalent way but using smaller numbers or fewer terms.

simultaneous equations: Two equations with multiple unknown variables of the same value.

sine: The sine of an angle, sin θ, in a right-angled triangle is the ratio of the side opposite the angle and the hypotenuse.

sine rule: A formula involving the sine of an angle, used in non-right angled triangles.

solid: A three-dimensional shape.

sphere: A shape with only a curved surface.

subset: A set is a subset when all of its elements are also elements of a larger set.

subtended: Formed at a point by lines from two other points.

supplementary: Angles which add to 180°.

surface: A two-dimensional collection of points.

take-home pay: What one actually earns, after taxation.

tangent: In this context, a tangent is a straight line touching a curve.

tangent (trig): The ratio opposite/adjacent is called the tangent of the angle.

terminating decimal: A terminating decimal has digits that do not continue forever.

term-to-term rule: A term-to-term rule describes how to use one term in a sequence to find the next term.

Glossary

the difference of two squares: The difference of two squares is a squared number subtracted from another squared number; you get only two terms because the middle terms cancel each other out.

translation: A translation is a sliding movement.

triangle: A triangle is a three-sided polygon.

union: The union of two sets is all the elements in each set or both sets.

universal set: A set from which all other sets will be taken.

upper bound: The upper bound is the greatest possible value that the true measurement could be.

upper quartile: The upper quartile of the data is the 75th percentile.

vertex: A vertex of a shape is a point where two sides meet.

wage: Amount earned, typically according to hours or days worked, and paid monthly or more regularly.

y-intercept: The point at which a straight line meets the y-axis.

INDEX

A
acceleration   226–8
accuracy
   appropriate   68–9, 70, 111–12
   checks   109–11
   limits   70–5
addition
   of algebraic fractions   160–1
   of column vectors   435–6
   of fractions   43–4
   of measurements   71–3
   of negative numbers   34–5, 36–7
   of numbers in standard form   59–60
   order of operations   37, 106
algebra
   algebraic manipulation   141–58
   completing the square   157–8
   expansion of brackets   144, 147–50
   factorisation of expressions   144–7, 150–7
   and indices   162–5
   introduction to   137–40
   and letters for unknowns   137–8
   quadratics   154–6
   rules of   137–8
   simplification   141–3, 162–3
   substituting numbers into   138–40
   see also equations
algebraic fractions   159–61
   addition   160–1
   simplification   159–61
   solution of equations involving   189–90
   subtraction   160–1
alternate segment theorem   332–4
angles   272–4, 312–22

acute   173, 289
alternate   314
alternate segment theorem   332–4
between a line and a plane   404–7
of a circle   325–34
co-interior (allied)   315
corresponding   314
of depression   385–7
of elevation   385–7, 401
exterior   277, 320–1
finding in three dimensions   398–404
formed by straight lines   312–14
formed within parallel lines   314–16
interior   277, 320–1
measurement   287–90
obtuse   173, 287–8, 387–9
opposite   328–31
of a pie chart   487–9
of a polygon   320–2
of a quadrilateral   318–19
reflex   173, 288–9
right angles   273, 325–7, 331–2
and similar shapes   301–2
supplementary   315, 328–31
of a triangle   316–17
and trigonometry   378, 383–404
arc   325–6
arc length   353–6
area   88
   of a circle   352–3, 356–7, 359
   of compound shapes   367–71
   measures   336–7
   of a parallelogram   343–5
   and Pythagoras's theorem   374–5
   of a rectangle   340–1

   of a sector   353–6
   of similar shapes   304–6
   of a trapezium   345–7
   of a triangle   341–3, 348–50, 397–8
   under a speed-time graph   229–32
asymptotes   252–4
averages   468–78, 471–82
   mean   469–82, 505
   median   468–9, 471–9, 481–2, 494, 497–501
   mode   468–9, 471–5, 477–8

B
bar charts   483–7
   composite (stacked)   484–5
   dual (side-by-side)   484, 487
bases   52–3
bearings   274–5, 298–300
bias   449–51, 463, 465
bounds of measurement, lower/upper   70–4
brackets   106
   expansion   144, 147–50, 171–2
   and factorisation   146, 152–6, 179–80
   and order of operations   37
   and sets   6
   solving equations with   171–2, 179–80

C
calculators   106–15
   and accuracy checks   109–11
   display interpretation   111–12
   and exponential growth   126
   and fractions   106, 114–15
   and order of operations   106–7
   and pi (π)   350
   and standard form   107–9
   and time calculations   112–14

capacity, units of   335–6
categorical data   462–82
centre
    of a circle   285–6
    of enlargement   415
    of rotation   411
certain events   446
chords   285–6, 323–5, 331–3
    equal   323
chunking (division)   41
circle theorems   323–34
    angles in a circle   325–32
    symmetry properties   323–5
circles   285–6
    alternate segment theorem   332–4
    angles   325–34
    arc length   353–6
    area   352–3, 356–7, 359
    centre   285–6
    circumference   285–6, 325–8, 350–2
    diameter   285–6, 350–3
    radius   285–6, 331–3, 350–4, 356–7, 359–60, 362–5
    sector area   353–6
circumference   285–6, 325–8, 350–2
clocks, 24-hour   116–17
coefficients   151–3, 155–7
column vectors   428, 430–1, 433, 435–9, 440
    addition/subtraction   435–6
    magnitude   438–9
    notation   430–1
combined events, probability of   453–5, 457–9
command words   xiv
commission   100
common denominators   174, 189
common factors   4–5, 151, 155–7, 174
    and factorisation   144–6
    highest common factor (HCF)   4–5
common multiples   4–5

compasses   290–1
complement, of sets   9–11
completing the square   157–8, 181–3
composite functions   261–2
compound measures   87–9
compound shapes, area/volume   367–71
cones   279, 280
    surface area   364–7
    volume   361–4, 367–8
congruence   282–5
    in transformations   408, 411, 413
    of triangles   307
constructions *see* geometrical constructions
conversion graphs   217–19
coordinate geometry   263–9
    general form of the equation of a straight line   265–7
    gradient of a straight-line graph   263–7
    line segments   264–5
    parallel lines   267–9
    perpendicular lines   267–9
correlations   490–4
cosine (cos)   378, 381, 383
    for obtuse angles   387–9
cosine rule   389, 394–6
cross-sections
    decreasing   361
    of a prism   279, 356–7
cube roots   17–19
    index   56–7
cubes (numbers)   17–19
cubes (shapes)   278, 291
    nets   280–1, 291, 293–4
    volumes in similar shapes   304–5
cubic graphs   237–9, 251–2
    'double bends'   238–9, 251–2
cubic sequences   205–7
cuboids   398–400, 402
    diagonals   189
    nets   291–4

planes of symmetry   310
    properties   278, 279
    volume   356
cumulative frequency diagrams   494–501
currency conversion   86, 124–5, 217–18
cylinders   279, 356–61, 369

D
data
    averages   468–82
    categorical   462–82, 483–7
    collection   462–8
    continuous   477–82, 494–7, 502–5
    discrete   477–82
    grouped   462–82, 477–82, 494–7
    grouped continuous   494–7
    numerical   462–82
    and statistical diagrams   483–506
data collection sheets   463–4
data sets, large   474–7
decagons   277
decay   244–6
decimal places, rounding to   62–3
decimals   23–6
    division   41–2
    and dot notation   24
    and fractions   23–6, 29
    hours written as   119–20
    multiplication   40
    ordering   28, 30
    and percentages   25–6, 91, 92–3
    recurring   23–5
    terminating   23–5
denominators   20–1
    common   174, 189
    rationalisation of   132–4
density   75, 88–9
dependent events, probability of   459–61
diagonals   189, 277

INDEX

diameter 285–6, 350–3
difference
   of measurement 71–3
   of two squares 148–51, 180
dimensions 272
direction (vectors) 430–2
discounts 100–1
distance 87, 229–32
distance–time graphs, rate of change 223–6
division
   and chunking 41
   of decimals 41–2
   and estimation 67–8
   of fractions 47–8
   of integers 41
   long division 41
   of measurements 73–5
   of negative numbers 35–7
   and number facts 110
   of numbers in index form 54
   of numbers in standard form 59–60
   and order of operations 37, 106
   and prime numbers 3
   of a quantity in a given ratio 80–1
domains 256–60

E

edges 278
elements, of sets 6–12
enlargement 282–3, 415–20, 421, 423–6
   centre of 415
   with a fractional scale factor 416–18
   with a negative scale factor 418–20
equation of a straight line, general form 265–7
equations 168–90
   and changing the subject of formulas 185–9
   and cubic graphs 237–9
   and exponential graphs 244–6

and graphs 233–46
involving algebraic fractions 189–90
involving a bracket 171–2
involving fractions 173–5
involving the unknown on both sides 172
linear 170–1
linear simultaneous 175–8
quadratic 179–84
and quadratic graphs 233–7
and reciprocal graphs 239–40
simultaneous 175–8
writing 169–70
writing formulas 168–9
estimation 61–9
   and accuracy 68–9
   of angles 287–8
   of answers to problems 65–8
   from a population 450
   of the gradient to a curve 246–7
   of length 61
   and rounding 62–6
events
   certain 446
   combined 453–5, 457–9
   dependent 459–61
   impossible 446
   independent 456–7
   mutually exclusive 454
exchange rates 86, 124–5, 217–18
exponential decay 127–8
exponential graphs 244–6, 254–5
exponential growth 126, 127–8
exponential sequences 205–7

F

faces 278
factor tree method 3
factorisation
   of algebraic expressions 144–5
   of expressions of the form $a^2 + 2ab + b^2$ 153–4
   of expressions of the form $a^2x^2 - b^2y^2$ 150–1

   of expressions of the form $ax^3 + bx^2 + cx$ 157
   of expressions of the form $ax + bx + kay + kby$ 146–7
   of expressions of the form $x^2 + bx + c$ 151–3
   of quadratic expressions 152, 154–6, 179–81
factors 2–5
   common 4–5
   prime 2–3, 4
fairness 449–50, 452–6
flow rate 86–7
FOIL mnemonic 148
force 88
formulae
   changing the subject of 185–9
   and letters for unknowns 137–8
   proportion as 209–13
   and sequences 203–4
   writing 168–9
four operations 31–49
   addition of fractions 43–4
   addition of negative numbers 34–5, 36–7
   combination 36–7
   division of decimals 41–2
   division of fractions 47–8
   division of integers 41
   division of negative numbers 35–7
   multiplication of decimals 40
   multiplication of fractions 45–6
   multiplication of integers 39
   multiplication of negative numbers 35–7
   and numbers below zero 31–5
   order of 37–9
   subtraction of negative numbers 34–5, 36–7
fractional indices 56–7
fractional scale factors 416–18
fractions 20–3, 25–6
   addition 43–4
   on calculators 106, 114–15

and decimals 23–6, 29
division 47–8
equations involving 173–5
equivalent 22–3, 29, 43
improper 20, 43–4
and mixed numbers 20, 43–4
multiplication 45–6
ordering 29–30
and percentages 25–6, 91–2
proper 20
of a quantity 21
and rationalisation of denominators 132–4
simplest form 22–3
see also algebraic fractions
frequency density 502–3, 505
frequency tables 463–4, 474–82, 486–9, 494, 502–4
frustums 280, 367–70
function notation 256
functions 256–62
composite 261–2
domain 256–60
graphs of 233–47
inverse 258–60
range 256–60

G

geometrical constructions 287–94
and compasses 290–1
measurement of angles 287–90
nets of 3-D shapes 291–4
geometrical terms 272–86
angles 272–3
bearings 274–5
circles 285–6
congruence 282–5
dimensions 272
lines 272, 273–4
nets of 3-D shapes 280–1
polygons 277–8
quadrilaterals 276–7
similarity 282–5
solids 278–80

triangles 275
geometry
coordinate 263–9
vector 439–43
see also mensuration
gradients
of distance–time graphs 223–4
estimation 246–7
of parallel lines 267–9
of perpendicular lines 267–9
of speed–time graphs 226–7
of straight–line graphs 263–7
graphs
area under a speed-time graph 229–32
conversion 217–19
cubic 237–9, 251–2
exponential 244, 254–5
families of 248
of functions 233–47
and inequalities 193–8
linear 248
in practical situations 217–32
quadratic 233–5, 249–50
rate of change on distance–time 223–6
rate of change on speed–time 226–9
reciprocal 239–40, 252–4
sketching curves 248–55
solving equations with 235–7
straight-line 263–4
travel 220–2
greater than 27–8, 191–9
grouped data 462–82, 494–7
growth 244–6
exponential 126, 127–8

H

hemispheres 279
hexagons 277, 320–1
highest common factor (HCF) 4–5
histograms 502–5
hours 114

hour/minute conversions 119–20
written as decimals 119–20
hypotenuse 374–5, 378–9, 404

I

image and object, congruence 408, 411, 413
impossible events 446
independent events 456–7
index notation 52
indices 52–7, 162–5
and algebraic expressions 164–5
division of numbers in index form 54
fractional 56–7
laws of 55, 56, 59–60, 162, 164–5
multiplication of numbers in index form 53–4
negative 55
simplification of algebraic expressions using 162–3
simplification of numbers with 52
see also standard form
inequalities 27–8, 191–9
representing regions satisfying more than one equality 195–8
showing regions on graphs 193–8
solution 191–3
integers 1, 45
division 41
multiplication 39
ordering 27
intercept 265–9
interest
compound 102–5, 126
simple 102–5
interquartile range (IQR) 499
intersection, of sets 8–9
invariance 408
inverse functions 258–60

INDEX

inverse operations, as checks 110–11
inverse proportion 82–3, 208, 210–13

K
kites, properties 276

L
length
    estimation 61
    finding in three dimensions 398–404
    and Pythagoras's theorem 375–6
    and trigonometry 381–2
    units of 335–6
less than 27–8, 191–9
like terms 141
line segments 264–5
    length 264–5
    midpoint 264
line symmetry 249–50, 252, 307–8, 310, 383
linear graphs 248
linear sequences 202
lines 272, 273–4
    angle between a line and a plane 404–7
    angles formed by straight lines 312–14
    angles formed within parallel lines 314–16
    general form of the equation of a straight line 265–7
    parallel 267–9, 273, 314–16
    perpendicular 267–9, 273, 323, 333
    see also specific lines
lines of best fit 491–4
longitude 120–1
loss 95–6, 100–1
lowest common multiple (LCM) 4–5

M
magnitude (length) (vectors) 430–2, 438–9
mapping 408

mapping diagrams 256–9
maps 295–7
    scales 77–8, 297
mark schemes xii
mass 88–9
    units of 335–6
mean 469–82, 505
measurement
    addition and subtraction 71–3
    bounds of 70–4
    division 73–5
    multiplication 73–5
    sums and differences 71–3
measures, compound 87–9
median 468–9, 471–9, 481–2, 494, 497–501
mensuration 338–71
    arc length 353–6
    area of a circle 352–3
    area of compound shapes 367–71
    area of a parallelogram 343–5
    area of a rectangle 340–1
    area of shapes made from rectangles and triangles 348–50
    area of a trapezium 345–7
    area of a triangle 341–3
    circumference of a circle 350–2
    perimeter of a 2-D shape 338–40
    sector area 353–6
    surface area of a cone 364–7
    surface area of a prism 359–61
    surface area of a pyramid 364–7
    surface area of a sphere 364–7
    volume of compound shapes 367–71
    volume of a cone 361–4
    volume of a prism 356–9
    volume of a pyramid 361–4
    volume of a sphere 361–4
midpoints, of line segments 264
minutes 114

hour/minute conversions 119–20
mirror lines 408
mixed numbers 20, 43–4
modal class, of grouped data 478–9, 481–2
mode 468–9, 471–5, 477–8
money 122–5
    currency conversion 124–5
    value for money 122–3
multiples 4–5
    common 4–5
    lowest common multiple (LCM) 4
multiplication
    of decimals 40
    and expansion of brackets 144, 147–9
    of fractions 45–6
    grid method 39
    of integers 39
    long multiplication 39
    of measurements 73–5
    of negative numbers 35–7
    of number facts 110
    of numbers in index form 53–4
    of numbers in standard form 59–60
    and order of operations 37, 106
    and simplification of algebra 141–2
    of a vector by a scalar 435
multipliers 81–2
    percentage increase/decrease using 97–8
mutually exclusive events 454

N
negative indices 55
negative numbers 31–6
    addition 34–5
    division 35–6
    multiplication 35–6
    subtraction 34–5

negative numerical terms, factorisation of quadratic expressions   153, 156
negative scale factors   418–20
nets, of three-dimensional objects   280–1, 291–4
newtons (N)   88
notation
   dot notation for decimals   24
   function   256
   index (standard form)   52
   for sets   6, 7
   for vectors   430–5
nth term, the   203–7
number facts, as accuracy checks   110
number lines   34
number patterns   200–2
numbers   1–5
   common factors   4–5
   common multiples   4–5
   irrational   1, 129–34, 350
   mixed   20, 43–4
   natural   1
   negative (below zero)   31–6
   prime   2–3
   prime factors   2–3
   rational   1, 129–34
   reciprocals   2
   simplification of numbers with indices   52, 53
   types of   1–2
   whole *see* integers
numerators   20–1
numerical data   462–82
numerical expressions, simplification   164–5

O

object and image, congruence   408, 411, 413
octagons   277
order of operations (calculators)   106–7
order of rotational symmetry   308–9
ordering   27–30
   of decimals   28, 30
   of fractions   29–30
   of inequalities   27–8
   of integers   27
   of percentages   30
outcomes   446
   see also events

P

parabolas   233
parallel lines   267–9, 273
   angle properties   314–16
parallelograms   276
   area of   343–5
   similar   303
   transformations   409
pay
   gross   100
   rates of   86–7
   take-home   100
pentagons   277, 320
per cent   25, 91
percentages   25–6, 91–105
   and decimals   25–6, 91, 92–3
   and discounts   100–1
   expressing one quantity as a percentage of another   93–5
   and finding the original quantity   98–9
   and fractions   25–6, 91–2
   and interest   102–5
   and loss   95–6, 100–1
   ordering   30
   percentage change   95–8, 102
   and profit   95–6, 100–1
   of a quantity   91–3
   and salaries/wages   100
percentiles   497–501
perfect squares   153–4
perimeters   338–40
perpendicular bisectors   269, 273, 331
perpendicular lines   267–9, 273, 323, 333
pi (π)   350–4, 356–7, 359–60, 362, 364–5
pictograms   483–7
pie charts   487–90
pilot surveys   465
planes   272, 278
   angle between a line and a plane   404–7
   of symmetry   310
points   272
polygons   277–8
   angles in   320–2
   convex   320
   irregular   277
   regular   277
population density   75, 89
populations, estimation from   450
position vectors   437–8
powers   16–19
   cubes   17–19
   indices   56–7
   order of operations   37, 106–7
   squares   16–17, 106–7, 148–9, 205–7
   see also index notation; standard form
pressure   88
prime factors   2–3, 4
prime numbers   2–3
principal   103
prisms   279
   cross-section   279, 356–7
   rectangular   279
   surface area   359–61
   triangular   279
   volume   356–9
probability   446–61
   of an event not occurring   448–9
   of combined events   453–5, 457–9
   of dependent events   459–61
   estimation from a population   450
   and independent events   456–7
   and relative frequency   450–3

INDEX

of a single event 446–7
and tree diagrams 457–9
profit 95–6, 100–1
proportion 81–3, 208–14
   direct 81, 208, 210, 213–14
   as a formula 209–13
   indirect/inverse 82–3, 208, 210–13
protractors 287–9
pyramids 279, 291, 402–3, 406
   nets of 280–1, 291, 293
   surface area 364–7
   volume 361–4
Pythagoras's theorem 374–7, 394
   and finding lengths and angles in three dimensions 398–404
   and line segments 264–5
   and problem-solving 377

Q

quadratic equations, solution
   by completing the square 181–3
   of the form $ax^2 + bx + c = 0$ by factorisation 180–1
   of the form $x^2 + bx + c = 0$ by factorisation 179–80
   using the quadratic formula 183–4
quadratic expressions 150
   factorisation 152–6
quadratic formula 183–4
quadratic graphs 233–7, 249–50
quadratic sequences 205–7
quadrilaterals 276–7
   angles 318–19, 328–31
   cyclic 328–31
   and transformations 410
   see also parallelograms; rectangles; squares; trapezium
quantities
   division in a given ratio 80–1
   expression as a percentage of another 93–5
   finding the original 98–9
   fraction of a 21
   percentage of a 91–3
   and proportion 81
   using ratio to find an unknown 78–9
quartiles 497–501
questionnaires 465
questions, leading 465

R

radius 285–6, 331–3, 350–4, 356–7, 359–60, 362–5
random sampling 463
range 256–60, 468–75, 478–9, 481–2, 503
rate of change
   on distance–time graphs 223–6
   on speed–time graphs 226–9
rates 86–90
   common measures of 86–7
   and density 88–9
   of exchange 86
   of flow 86–7
   of pay 86–7
   and population density 89
   and pressure 88
   and speed 87
ratio 76–81
   division of a quantity in a given ratio 80–1
   simplification 76–7
   use to find an unknown quantity 78–9
   writing in the form $1:n$ 77–8
reciprocal graphs 239–40, 252–4
reciprocals 2, 47–8
rectangles 276
   area 340–1, 348–50, 359–60
   perimeter 339–40
   similar 301–3
   symmetry 307
   and transformations 410, 411, 417, 419
reflection 408–10, 421, 425
relative frequency, and probability 450–3
rhombus, properties 276
roots 16–19
   cube roots 17–19, 56–7
   square roots 16–17, 56, 67–8, 106–7, 129–32
rotation 411–13, 421–2, 425
   angle of 411, 421
   centre of 411
   direction of 411
rounding 62–6
   and accuracy checks 109
   to a given number of decimal places 62–3
   to a given number of significant figures 63–6

S

salaries 100
sample space diagrams 453
samples 463
   random 463
scalars, multiplying a vector by 435, 440
scale drawings 295–300
   bearings 298–300
   and maps 295–7
scale factors 415–20
   fractional 416–18
   negative 418–20
   and similar shapes 302–6
scatter diagrams 490–4
seconds 114
sectors 285–6
   area 353–6
   and pie charts 487–9
segments 285–6, 328
   alternate segment theorem 332–4
semi-circles 325–7
sequences 200–7
   cubic 205–7
   exponential 205–7
   and formula 203–4
   linear 202
   and the nth term 203–7
   and number patterns 200–2

Index

quadratic 205–7
and the term-to-term rule 200–2
sets 6–15
  complement 9–11
  definition 6, 12–13
  disjoint 7, 9
  elements 6–12
  empty 9, 11
  finite 6, 12
  infinite 12
  intersection 8–9
  notation 6, 7, 11
  relationship between 8–9
  subsets 7–8, 11
  three-set problems 13–15
  union 9
  universal 6–9
  Venn diagrams 7–8, 12
shapes 272, 275–85
  area 340–50, 352–6, 367–71
  circumference 350–2
  compound 348–50, 367–71
  congruence 282–5
  construction with compasses 290–1
  nets of 280–1, 291–4
  perimeter 338–40
  similar 282–5, 301–6
  surface area 359–61, 364–7
  symmetry 307–11
  three dimensional 291–4, 398–404
  transformations 408–27
  volume 356–9, 361–4, 367–71
  see also specific shapes
significant figures, rounding to a given number of 63–6
similarity 282–5, 301–6, 415
simplest form 22–3
simplification
  of algebraic expressions 141–3, 162–5
  of algebraic fractions 159–61
  using indices 52–3, 162–5

  of ratios 76–7
  of surds 129–30
simultaneous equations 175–8
sine (sin) 378–9, 383
  for obtuse angles 387–9
sine rule 389–93
'SOHCAHTOA' mnemonic 378
solids 278–80
  symmetry 310–11
  see also three-dimensional objects
speed 87, 223–9
  average 87
speed–time graphs
  area under 229–32
  gradients 226–7
  rate of change on 226–9
spheres
  properties 279
  surface area 364–7
  volume 361–4
spread 469, 471, 494
square roots 16–17
  and calculators 106–7
  and the estimation of answers 67–8
  index 56
  and surds 129–32
squares (numbers) 16–17
  and calculators 106–7
  and expansion of brackets 148–9
  sequences 205–7
squares (quadrilaterals) 374–5
  properties 276
standard form 58–60
  calculations with numbers in 59–60
  and calculators 107–9
  large numbers 58, 59
  small numbers 58–9
statistical diagrams 483–506
  bar charts 483–7
  cumulative frequency diagrams 494–501

  histograms 502–5
  and the median 497–501
  and percentiles 497–501
  pictograms 483–7
  pie charts 487–90
  and quartiles 497–501
  scatter diagrams 490–4
straight lines
  angles formed by 312–14
  general form of the equation of 265–7
  gradients 263–7
straight–line graphs 263–4
subsets 7–8, 11
substitution 138–40
  and harder numbers 139–40
  and simultaneous equations 176–8
subtraction
  of algebraic fractions 160–1
  of column vectors 435–6
  of measurements 71–3
  of negative numbers 34–5, 36–7
  of numbers in standard form 59–60
  order of operations 37, 106–7
sums, of measurement 71–3
surds 129–34
  and manipulation of expressions of the form $a + b\sqrt{c}$ 130–2
  and rationalisation of denominators 132–4
  and simplification 129–30
surface area 359–61, 364–7
  of a cone 364–7
  of a prism 359–61
  of a pyramid 364–7
  of a sphere 364–7
surfaces 272, 278
surveys 463–4
  pilot 465
syllabus mapping xii
symmetry 307–11

INDEX

circles 323–5
line 249–50, 252, 307–8, 310, 383
plane of 310
and properties of shapes and solids 310–11
rotational 308–9, 310

T

tally charts 463
tangent (tan) 378–9
tangents 246–7, 331–3
  distance–time graphs 223–4
  from an external point 323
  speed–time graphs 227
taxation 100
term-to-term rule 200–2
  three-dimensional objects
  lengths and angles 398–404
  nets 291–4
  properties 272
  see also solids
three-set problems 13–15
time 87, 116–21
  24-hour clock 116–17
  calculating with 117–19
  on calculators 112–14
  hour/minute conversions 119–20
  hours written as decimals 119–20
  time zones 120–1
  timetables 118–19
transformations 408–27
  combining 425–7
  enlargement 282–3, 415–20, 421, 423–6
  language of 408
  recognition and description 420–5
  reflection 408–10, 421, 425
  rotation 411–13, 421–2, 425
  translation 413–15, 421–2, 426, 428–30
translations 413–15, 421–2, 426

and vectors 428–30
trapezium 276
  area 345–7, 357
  isosceles 276
  transformations 408, 410
travel graphs 220–2
tree diagrams
  for combined events 457–9
  for dependent events 460–1
triangles 275
  adjacent sides 378–9
  angle properties 316–17
  area 341–3, 348–50, 397–8
  congruence 307
  construction given three sides 290–1
  equilateral 275, 317
  general formula for the area of 397–8
  height 380
  isosceles 275, 310, 317, 333, 383
  non-right-angled 383, 389–98
  opposite sides 378–9
  right-angled 275, 317, 374–84, 386–9, 397–404
  scalene 275, 317
  similar 301–4
  symmetry 307
  transformations 408–10, 412–14, 416–26
  see also Pythagoras' Theorem; trigonometry
trigonometry 374, 378–407
turning points 249–50
two-way tables 465–8

U

union, of sets 9
units of measure 335–7
  units of capacity 335–6
  units of length 335–6
  units of mass 335–6
unknowns

on both sides of the equation 172
representation by letters 137–8
using ratios to find 78–9

V

value for money 122–3
vectors 428–43
  addition of column vectors 435–6
  column 428, 430–1, 433, 435–9, 440
  general 431
  geometry 439–43
  magnitude 430–2, 438–9
  multiplication by a scalar 435, 440
  notation 430–5
  position 437–8
  subtraction of column vectors 435–6
  and translations 428–30
Venn, John 7
Venn diagrams 7–8, 12, 455
vertex 278
volume 88–9
  of compound shapes 367–71
  of a cone 361–4
  measures 336–7
  of a prism 356–9
  of a pyramid 361–4
  of similar shapes 304–6
  of a sphere 361–4
  see also capacity

W

wages 100

Y

y-intercept 265–9

Z

zero, rounding to a given number of significant figures 63